Ordering in Strongly Fluctuating Condensed Matter Systems

NATO ADVANCED STUDY INSTITUTES SERIES

A series of edited volumes comprising multifaceted studies of contemporary scientific issues by some of the best scientific minds in the world, assembled in cooperation with NATO Scientific Affairs Division.

Series B: Physics

RECENT VOLUMES IN THIS SERIES

Volume 41 – Fiber and Integrated Optics
edited by D. B. Ostrowsky

Volume 42 – Electrons in Disordered Metals and at Metallic Surfaces
edited by P. Phariseau, B. L. Györffy, and L. Scheire

Volume 43 – Recent Advances in Group Theory and Their Application to Spectroscopy
edited by John C. Donini

Volume 44 – Recent Developments in Gravitation – *Cargèse* 1978
edited by Maurice Lévy and S. Deser

Volume 45 – Common Problems in Low- and Medium-Energy Nuclear Physics
edited by B. Castel, B. Goulard, and F. C. Khanna

Volume 46 – Nondestructive Evaluation of Semiconductor Materials and Devices
edited by Jay N. Zemel

Volume 47 – Site Characterization and Aggregation of Implanted Atoms in Materials
edited by A. Perez and R. Coussement

Volume 48 – Electron and Magnetization Densities in Molecules and Crystals
edited by P. Becker

Volume 49 – New Phenomena in Lepton-Hadron Physics
edited by Dietrich E. C. Fries and Julius Wess

Volume 50 – Ordering in Strongly Fluctuating Condensed Matter Systems
edited by Tormod Riste

Volume 51 – Phase Transitions in Surface Films
edited by J. G. Dash and J. Ruvalds

This series is published by an international board of publishers in conjunction with NATO Scientific Affairs Division

A Life Sciences B Physics	Plenum Publishing Corporation London and New York
C Mathematical and Physical Sciences	D. Reidel Publishing Company Dordrecht and Boston
D Behavioral and Social Sciences	Sijthoff International Publishing Company Leiden
E Applied Sciences	Noordhoff International Publishing Leiden

Ordering in Strongly Fluctuating Condensed Matter Systems

Edited by

Tormod Riste
Institutt for Atomenergi
Kjeller, Norway

PLENUM PRESS • NEW YORK AND LONDON
Published in cooperation with NATO Scientific Affairs Division

Library of Congress Cataloging in Publication Data

Nato Advanced Study Institute on Strongly Fluctuating Condensed Matter Systems, Geilo, Norway, 1979.
Ordering in strongly fluctuating condensed matter systems.

(NATO advanced study institutes series: Series B, Physics; v. 50)
Includes index.
1. Phase transformations (Statistical physics)–Congresses. 2. Fluctuations (Physics)–Congresses. 3. Order-disorder models–Congresses. I. Riste, Tormod, 1925- II. Title. III. Series.
QC176.8.P45N39 1979 530.4'1 79-28173
ISBN 0-306-40341-2

Proceedings of the NATO Advanced Study Institute on Strongly Fluctuating Condensed Matter Systems, held in Geilo, Norway, April 16–27, 1979.

A Division of Plenum Publishing Corporation
227 West 17th Street, New York, N.Y. 10011

Printed in the United States of America

PREFACE

This NATO Advanced Study Institute held at Geilo, Norway, April 16th-27th 1979, was the fifth in a series devoted to the subject of phase transitions and instabilities. The application to NATO for the funding of this ASI contained the following paragraphs:

"Traditionally one has made a clear distinction between solids and liquids in terms of positional order, one being long-ranged and the other at most short-ranged. In recent years experiments have revealed a much more faceted picture and a less sharp distinction between solids and liquids. As an example one now has 3-dimensional (3-D) liquids with 1-D density waves and 3-D solids with 1-D-liquid molecular chains. The subsystems have the common feature of low-dimensional systems: a strong tendency for fluctuations to appear. Although the connection between fluctuations and dimensionality, and the suppression of long-range order by fluctuations, was pointed out as early as 1935 by Peierls and by Landau, it is in the last five years or so that theoretical work has gained momentum. This development of understanding started ten years ago, however, much inspired by the experimental work on 2-D spin systems.

Introductory lectures will give a review and catalogue of theories for strongly fluctuating systems and phase transitions in terms of models, interactions, types of ordering, dynamics and physical systems for which they apply. These lectures will be followed by presentation of results (obtained by diffraction experiments) that bring out the different manifestations and behaviour of fluctuations and order in extreme cases: a dipolar magnet with mean-field behaviour and very weak fluctuations, as contrasted to a smectic A liquid crystal with strong fluctuations and no long-range positional order.

Low-dimensional magnetic systems will be treated, starting with the 2-D Ising model and its experimental confirmation. This will be followed by discussions of different 1-D and 2-D model systems that lead to phase transitions with or without long-range order, including predictions on their critical behaviour and the spectrum of excitations. Tests of these predictions by computer simulation and by experiments on real physical systems will be reviewed, including the recent exciting results on solitons in a 1-D ferromagnet.

For low-dimensional magnets also dilute systems will be discussed, with emphasis on the percolation problem: the breaking up of long-range order into finite clusters, their freezing as $T \to 0$ and their dynamics. Related to this problem is that of spin-glasses, for which the evidences for possible time ordering will be reviewed and discussed with reference to the latest theoretical and experimental results.

The discussion of non-magnetic systems susceptible to strong fluctuations will start with a review of the Kosterlitz method and its application to melting and surface roughening. A colour movie will display the results of extensive computer simulation of the latter problem. The topological or bond orientational order used here will be carried over to smectics whose different classes will be reviewed. Recent light-scattering experiments will be reported, they will demonstrate the potentials of these substances in the investigations of 2-D melting and similar problems in physically realizable systems. Other physical systems to be discussed are physisorbed gas layers and mercury chain compounds. The attention will there be directed especially towards orientational instabilities, dislocations, chain ordering transitions and solitons.

The purpose of this ASI is to provide the educational forum outlined above. By bringing together theorists and experimentalists employing different techniques, with ample time for discussions after lectures, in the leisure time or during common sport activities, one intends at creating an atmosphere that will foster the development of further work in the field of phase-transitions in condensed-matter systems."

To a very large extent the above remarks could as well describe the school which was held rather than the plans for it. In retrospect, however, the school served purposes beyond those envisaged. It became apparent, for example, that ideas developed in one field of the school were often immediately applicable to another. A fine example of this was the realization, which dawned during the meeting, that resuls developed to explain properties of an ice-rule ferroelectric were directly applicable to models of the roughening transition and crystal growth.

A major achievement of the lecturers was to draw together the various concepts of the funding application into a coherent, pedagogical course. A very important classification of the subject matter is provided by the concepts of upper- and lower-critical dimensionality (UCD and LCD). For a system above UCD fluctuations are unimportant and mean-field treatments of phase-transitions can be used. Between UCD and LCD the full apparatus of the renormalisation group theory is applicable. At the below LCD, fluctuations are so important that conventional long-range order does not occur. To a large extent the school concerned itself with phenomena falling in the latter category.

Participants were introduced to the blossoming number of nonlinear excitations which are expected to occur at and below LCD. Most of the discussion of these excitations, which include solitons, breathers, walls, dislocations, disclinations and vortices, was of a theoretical nature. However, experimental contributions concerned with the observation of solitons in one-dimensional magnets, with the behaviour of one-dimensional chain compounds and with disclinations in smectic liquid crystals, helped to relate these ideas to the real world, as did the results of several computer simulations. Even so, it was clear that many of the more recent theoretical predictions are awaiting experimental confirmation. It is to be hoped that the clear theoretical exposition which this school achieved will add momentum to the efforts of experimetalists and will contribute to a more uniform progress of theory and experiment.

Another very important aspect for the subject of the school was the distinction between quenched and annealed systems. Systems containing quenched disorder, which include such different materials as ordinary glass, rubber and polymers, random conductors (Fermi glasses), dilute magnets and spin glasses, exhibit stronger fluctuations than annealed systems. Of particular interest are the fluctuations associated with percolation, which are predominantly governed by cluster statistics. The experimental discovery that the scaling field for two-dimensional percolation is the one-dimensional coherence length, provides a striking demonstration for the essentially one-dimensional nature of ramifield two-dimensional clusters.

The list of contents distinguishes between invited lectures and contributed seminars. Invited lectures are intended to constitute the tutorial framework of the subject. Seminars generally serve the purpose of demonstrating points raised in the lectures.

The Netherlands-Norwegian Reactor School was responsible for the organization of the institute. The programme committee joins the other participants in expressing their sincere thanks to Eigil Andersen and Gerd Jarrett for their careful planning and creative assistance, upon which largely rested the success of this study institute.

J. Feder
P.A. Fleury
R. Pynn
T. Riste
H. Thomas

May, 1979

CONTENTS

ORDERING IN STRONGLY FLUCTUATING SYSTEMS

INTRODUCTORY COMMENTS

S.F. Edwards

Cavendish Laboratory
Madingley Road
Cambridge CB3 OHE

INTRODUCTION

This lecture is intended to pose problems rather than present solutions, so I shall avoid low dimensional systems in which the strong fluctuations are amenable to calculation, as later papers will show in detail. Rather I will concentrate on three dimensional problems which are fluctuating in a different way from the familiar fluctuations of equilibrium in which there is a specified Hamiltonian and all states are accessible. This is not only because they are covered later, but also because it is worth emphasizing that they are rather a rare phenomenon in nature, and the intense effort they attract is because they are the simplest of systems although not of course simple.

This leads one to ask: what systems follow those of simple equilibrium in a hierarchy of complexity. The simplest transport problems are those which rely again on equilibrium properties, but require time dependent equilibrium correlation functions, so I class them with equilibrium functions. The next simplest systems are these:

1.1 Systems in which some degrees of freedom are specified, and are not permitted to take up all values within overall energy conservation. In some of these the specification of the frozen degrees of freedom is explicit, and therefore exact formulae, the analogue of $e^{-H/\kappa T}$ of Gibbs, can be derived. Such systems are usually difficult to obtain in nature, but can be thought of as 'Theorists' Ideal Glass', or TIG for short. Although exact formal solutions can be given to these problems, the derivation of their thermodynamic functions is one stage more complex than for normal systems, and several subsequent papers will be discussing various examples.

1.2 The other case is the analogue of the TIG in transport theory and the simplest case must be that of steady states, far from equilibrium. The typical situation here is that of some energy input into the system, coupled with an energy output. If and when a balance is obtained, a distribution function will exist and correlation functions invariant under time displacement. The formal solution of such problems is not clear, though practical approximations are available in some cases.

In both of these cases highly ordered states can occur, with transitions of unknown complexity, whilst being in conditions remote from conventional statistical equilibrium. They represent the next stage of development of statistical physics.

2. A theorist's ideal glass

I call the system a theorist's, because in practice it is very difficult to control the great urge all physical systems have to move towards equilibrium. For example it is quite impossible to cool a vitreous system so quickly that its structure at a final low temperature is the same as that at the initial high temperature. The spin glass is nearer to the ideal, a rubber very close to the ideal. Let us consider a system which permits the degrees of freedom to be separated into a slow group X and a fast group x, and suppose it has a Hamiltonian H(x,X), and is in equilibrium at T_i.

Then the probability of finding X is

$$P(X)=\int e^{-H(x,X)/kT_i}dx/\int e^{-H(y,Y)/kT_i}dydY \tag{2.1}$$

Now suppose we quench it (we theorists) instantly to T_f. Now X is frozen at the values prevailing when $T = T_i$ and the free energy is a function of these X i.e.

$$e^{-F(X)/kT_f} = \int e^{-H(Z,X)/kT_f}dZ \tag{2.2}$$

The experimental value of the Free energy is

$$F_{expt} = \int F(X)P(X)dX \tag{2.3}$$

$$= -kT_f\int \frac{(\log \int e^{-H(z,X)/kT_f}dz)\int e^{-H(x,X)/kT_i}dx}{\int e^{-H(y,Y)/kT_i}dydY}dX \tag{2.4}$$

This problem can therefore be given a formal solution, more complex than the formula of Gibbs

$$F = -kT \log \int e^{-H/kT}$$

but nevertheless explicit, and as Brout noted several years ago, these frozen degrees of freedom open a rich new seam of important physical problems. A device to bring (2.4) closer to the Gibbs form is to use replication i.e. to note that

$$A^n = 1 + n \log A + O(n \log A)^2 \tag{2.5}$$

so that if one defines $F(n, T_i T_f)$ by

$$e^{-\{F(n)/kT_f + F(T_i)/kT_i\}} = \int e^{-\sum_{\alpha=0}^{n} H(Z_\alpha, X)/kT_\alpha} \, dX \prod_{\alpha=0}^{n} dZ_\alpha \tag{2.6}$$

where $T_o = T_i$, $T_\alpha = T_f$ $\qquad \alpha = 1 \ldots\ldots\ldots n$

then F_{expt} is the coefficient of n in F(n). One can picture the problem of equation (2.6) as n + 1 replicas

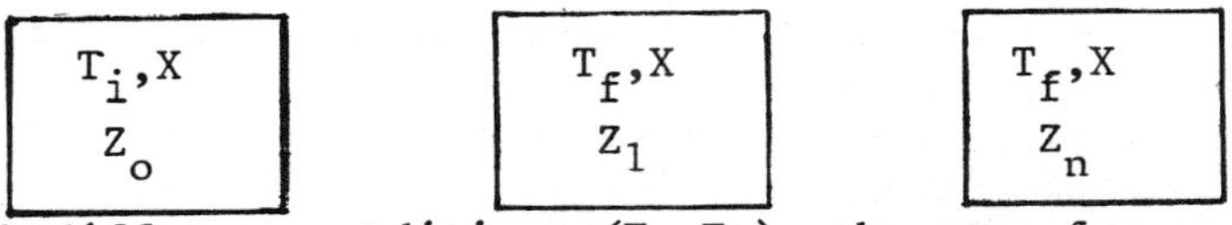

with different conditions (T_i, T_f), the same frozen variable (X), and other free degrees of freedom (Z).

A very clear version of this arises in rubbers. If one has a melt of very long polymer molecules and irradiates it (or promotes some chemical reaction - but irradiation is instantaneous whereas a reaction normally takes time and builds in a history which spoils the TIG concept), then cross links are formed which are permanent. They can be expressed as saying chain $R_i(S_i)$ (S label of the mathematically an arc length) crosslinks with chain $R_j(S_j)$ at points S_i^j, S_j^i respectively, and these are the specification of the rubber, the analogue of X.

The temperature could be changed, but in this example it is more interesting to shear the system and get the picture:

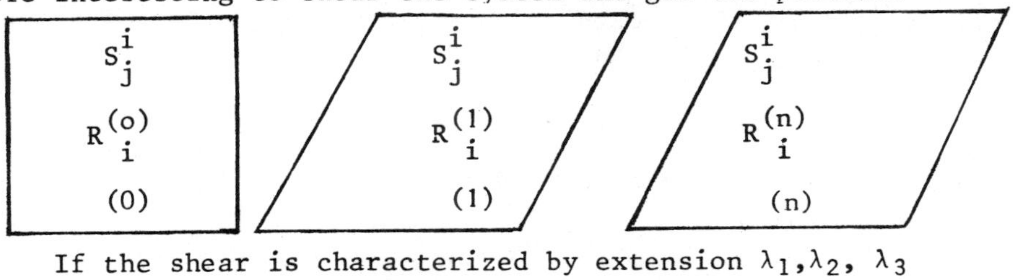

If the shear is characterized by extension $\lambda_1, \lambda_2, \lambda_3$

Then F(n) is F(n, $\lambda_1\ \lambda_2\ \lambda_3$) and $F_{expt} = F_{expt}\ (\lambda_1\ \lambda_2\ \lambda_3)$.

In this way a rigorous theory of rubber elasticity can be derived, the energetics being as in (2.6) but the cross-linking constraint appearing as

$$\prod_{ij} \prod_{\alpha=0}^{n} \partial\left(R_i^\alpha\,(S_i^j) - R_j^\alpha\,(S_j^i)\right)$$

the λ appearing in the specification of the box through which one integrates.

A simpler case is the quenched dilute magnetic alloy, the 'spin glass', which, in its TIG formalisation has $T_i=\infty$ putting the magnetic ions completely at random sites R_{ij} and

$$e^{-F(R_i)/kT} = \int e^{-\sum_{ij} J_{ij}\sigma_i\sigma_j/kT}$$

where J_{ij} is $J(R_i-R_j)$. However this problem suffers from two difficulties. One is that it is not clear whether an "ordered" state is achieved in three dimensions via a transition i.e. it might be analogous to a one dimensional or to a three dimensional ferro or anti ferro magnet. Secondly if the replica method is used, it is not obvious that (F(n) will take the form F_0+nF_1, so that one has also to take the limit $n \to 0$. Whereas for n an integer the formulation is thermodynamic and the various rules of thermodynamics such as the positive definiteness of entropy and minimization of the free energy apply, as $n \to 0$ these results will not obtain, and indeed one should expect the exact opposite: a maximum can be expected for the free energy since the entropy is minimized but the imposition of the fixed distribution rather than permitting the system to take up the maximum space in phase space. The $n \to 0$ limit seems to work in practice both here and elsewhere e.g. in Regge theory, and in de Gennes' derivation of an ε expansion for the excluded walk problem. It is interesting to note that for the problem of rubber elasticity the problem does not seem to arise since when F is calculated, it has precisely the form $F_0 + nF_i$, but in general calculation must be completed with n an integer and then the limit taken. It is worth noting that the replication idea can be useful and quite straightforward in other areas of classical statistical mechanics, and also in quantum problems. In ferromagnetism for instance the simplest breaking of symmetry involves the introduction of a mean field in a definite direction, whose magnitude is calculated self consistently. An alternative is to note that

$$e^{-nF/kT} = \int e^{-\sum_{\alpha=1}^{n} \frac{H^{(\alpha)}}{kT}} \prod d x^\alpha$$

where n is any positive integer. One has then

$$e^{-nF/kT} = \int e^{-\sum_{\alpha=1}^{n} J_{ij}\sigma^\alpha_i\sigma^\alpha_j/kT}$$

For $T<T_c$ one can take the ordered state as one in which dipoles in the different replicas are correlated, so write

$$e^{-\sum_\alpha\sum_{ij} J_{ij}\sigma^\alpha_i\sigma^\alpha_j/kT} = e^{-\sum_{\alpha,\beta}\sum_i q\sigma^\alpha_i\sigma^\beta_i - [\sum_{ij}\sum_\alpha J_{ij}\frac{\sigma^\alpha_i}{kT}\sigma^\alpha_j - \sum_i\sum_{\alpha,\beta} q\sigma^\alpha_i\sigma^\beta_i]}$$

This can be evaluated by expanding the part in the square bracket and recontracting, giving bounds on the free energy, so that q is determined variationally. The answer, not surprisingly, is the mean field theory, but whereas a complete calculation must give F independent of n, it is very easy to get n dependent bits to any particular order of accuracy (In this particular problem the ambiguity is removed by letting $n \to \infty$).

So far the discussion has been classical, for classical statistical mechanics is simpler than quantum statistical mechanics in that a knowledge of the Hamiltonian reduces the Classical problem to a quadrature (albeit in many dimensions). The quantum problem requires a knowledge of eigenvalues of the Hamiltonian, an intermediate problem which has to be solved before reaching the same mathematical problem as the classical case. It is therefore not surprising that physically cruder problems than those above can give rise to difficult problems in random systems. In particular the Anderson problem provides a remarkable bridge between different mathematical formalisms, and still has unresolved features. The problem, an aspect of what Anderson calls the 'Fermi Glass', is to find the levels of the Schrodinger Equation in a random potential. Suppose a single electron experiences a fixed, statistically defined potential V(r) giving rise to a hierarchy of density matrices, the simplest of which is

$$\rho(E,r,r') = \langle \sum_n \psi_n(r)\psi^+_n(r')\delta(E-E_n)\rangle$$

$\mathrm{Tr}\rho = N(E)$, the density of states. The average <......> is taken over the distribution of V. Anderson showed in 1958 that the eigenfunctions ψ_n changed their character at some critical energy E_c from being extended to localized states. The precise specification of the change at E_c is not agreed, except that N(E) is continuous.

The conductivity is zero for $E<E_c$, but it is not agreed as to whether it falls to zero abruptly in a first order transition, or continuously as a second order transition. The dimensional dependence is also strange, being the opposite of ferromagnetism: in one dimension all states are localised, in four dimensions none are, presumably the states become mixed at two dimensions, and cease to be mixed at four. The Anderson problem can be expressed as a 'first quantisation' problem in terms of path integrals, or a 'second quantisation' problem using an $n \to o$ method. If one uses the path

integral

Then $\langle G\rangle = \int e^{-\frac{i}{\hbar}\int_o^t \frac{m\dot{r}^2}{2}dt - \frac{i}{\hbar}\int_o^t V(r(\tau)d\tau} \delta r \qquad \rho = \mathrm{Im}\, G$

which formally solves the problem. In particular if V has a gaussian distribution with

$$\langle V \rangle = 0$$

and $\langle V(r)\, V(r')\rangle = W(r-r')$, then

$$\langle G\rangle = \int \delta r e^{-\frac{i}{\hbar}\int_o^t \frac{m\dot{r}^2}{2} d\tau - \frac{1}{\hbar^2}\int_o^t\int_o^t W(r/\tau_1)-V(\tau_2)d\tau_1, d\tau_2}$$

and various approximation techniques can be commenced. However one can also formulate the problem as a quantum field theory since it is a one electron problem one need only consider one E and omit time. Then the Lagrangian is

$$\psi(E - \frac{\hbar^2}{2m}k^2)\psi^+ + \Psi V \Psi^+$$

but it cannot be easily averaged, for

$$G = \langle \int \psi\psi^+ e^{\frac{i}{\hbar}\int L} / \int e^{\frac{i}{\hbar}\int L} \rangle$$

contains V in the denominator. If one replicates the system one gets

$$\sum \psi^\alpha (E- \frac{\hbar^2}{2m} k^2)\psi^{+\alpha} + \sum \psi^\alpha \psi^{+\alpha} V$$

and if one now uses $\int \psi^{(1)}\psi^{(1)+} e^{\frac{i}{\hbar}\int L}$

the denominator can be omitted as $n \to o$ since it is unity. Hence one can average to get

$$\langle G\rangle = \int \psi^{(1)}\psi^{(1)+} e^{[\frac{i}{\hbar}\int\sum_\alpha \psi^\alpha (E- \frac{\hbar^2}{2m}k^2)\psi^{+\alpha} - \frac{1}{\hbar^2}\sum_{\alpha\beta}\int\int \psi^\alpha\psi^{\alpha+}W\psi^{\beta+}\psi^\beta]}$$

Both of these methods have been used in Anderson problem, and the excluded volume problem of random walks (where $i/\hbar$ is replaced by $3/2\ell$, ℓ the step length, and $\frac{1}{\hbar^2}W(r-r')$ becomes the selfrepulsion). Although they are of course formally identical they suggest different methods of approximation, and at the present time one can get a second order transition out of the first formalism, and a first order transition out of the second, so clearly more thought is required here. Critical indices are also different by the two methods.

A related more difficult problem is the generalization of the Anderson problem to electron-electron interactions. For example there are oxides say Fe_2O_3 which in the uniform state do not have an integral number of conduction electrons per atom. If some oxygen atoms are replaced by fluorine atoms the number of electrons can be varied continuously and the Anderson problem becomes entwinned with several other solid state theory problems.

To summarise this section, I have described several systems in which strong disorder is present, but which are capable in principle of experimental realisation to TIG specification. These systems have important experimental properties, for example can have phase changes which are quite different from proper equilibrium systems, and can have important equations of state.

3. Systems far from equilibrium

A transport situation can be too complex to be studied theoretically, but it is possible to conceive of systems which have well defined properties and are remote from equilibrium.

Take any Hamiltonian system and add to it a source and a sink of energy, associated with different space scales. Then energy will cascade through the Hamiltonian modes of the system from the source to sink, and if the source is independent of the fluctuations of the Hamiltonian modes, but the sink depends on their magnitude, the agitation of the system will rise until the input of energy equals the output, and thereafter the system is steady. The system will now have correlation functions for density and velocity:

$$\langle \rho_{\underline{k}\omega}\rho_{\underline{k}'\omega'}\rangle = M(\underline{k}\omega)\delta(\underline{k}+\underline{k}')\delta(\omega+\omega') \tag{3.1}$$

and higher correlations. In addition at any given point there will be a probability of finding density, velocity and energy fluctuations e.g.

$$\langle u_{\underline{k}\omega}\, u_{\underline{k}'\omega'}\rangle = Q\ (\underline{k},\omega)\delta(\underline{k}+\underline{k}')\delta(\omega+\omega') \tag{3.2}$$

say, choosing in all these examples systems which are homogeneous for simplicity.

To give a simple example consider the Ehrenfest problem of particles scattering isotropically from fixed dilute centres with a cross section/unit volume of $\alpha(\mathbf{v})$. Suppose they are accelerated by gravity, g, and dissipate energy by friction $\mu\mathbf{v}$. Then if the probability of finding a particle with $\mathbf{v}$ in δ direction θ, ϕ is P.

$$\left(\frac{\partial}{\partial t} + \mu\,\frac{\partial}{\partial v}\cdot v + \alpha(v)L(\theta,\phi) + g\,\frac{\partial}{\partial v}\right) P = 0 \tag{3.3}$$

where ℓ is Legendres operator. The steady state P_o has $\partial P_o/\partial t=0$ and the problem is reduced to a differential equation, a version of the Fokker Planck equation, but one which has a unique P_o which has no relationship with the Maxwell -Boltzmann distribution. More realistic versions of this problem appear in plasma physics such as give rise to the Druvesteyn distribution. This class of problems are summarized by

$P_o(v)$ and $< \underline{v}(t)\underline{v}(o) >$

and there is no spatial distribution. More interesting are problems where the particles interact, and a particularly interesting case is that of turbulence. In this problem a homogeneous cascade problem can exist if an incompressible fluid is tirred randomly on a very long length scale, whilst viscosity removes energy which disappears from the scene. The Navier-Stokes equations

$$\frac{\partial u}{\partial t} + u\cdot\nabla u - \frac{1}{\rho}\nabla p - \nu\nabla^2 u = f \tag{3.4}$$

are conveniently put into Fourier form

$$\frac{\partial u_k}{\partial t} + \nu k^2 u_k + \sum M_{kj\ell}\, u_j u_\ell = f_k(t) \tag{3.5}$$

where $f_k(t)$ is a random stirrer

$$< f_k(t) f_{k'}(0) > \sim h_k \delta(t-t')\delta(\underline{k}+\underline{k}') \tag{3.6}$$

$$h_k \sim h\,\delta(\underline{k}) \tag{3.7}$$

There will be a steady state $P([\,u\,])$ and correlation functions, in particular

$$< u_k(t) u_{k'}(0) > = \delta(k+k')\, Q_k(t) \tag{3.8}$$

and at a point, if $u(\underline{r}_1) = v_1$

$$P(v_1) = \int \delta(v_1 - u(\underline{r}_1))\, P\,([u]) \Pi du_k \tag{3.9}$$

is the probability of finding a velocity of magnitude v, at $\underline{r}_1$ which must be independent of $\underline{r}_1$. Similarly the pressure (determined from $\nabla u = 0$) and its fluctuations have distributions and correlations.

This problem is peculiar since it is deficient in dimension, and provided the solution exists the static correlation can be written down from dimensional analysis

$$Q_k(0) = q_k \propto h^{2/3} k^{-11/3} \tag{3.10}$$

and time scales must come out in terms of the dimensionless quantity $h^{1/3} k^{2/3} t$.

In (d) dimensions

$$q_k = (k^{-(d)})\left(\frac{h^{2/3}}{k^{2/3}}\right) \tag{3.11}$$

or

$$q_k d^{(d)} k = \frac{h^{2/3}}{k^{5/3}}\, dk \tag{3.12}$$

Although there are papers in the literature trying to use ε

expansions on the problem of turbulence, the independence of the right hand side of (3.12) on d suggests that such attempts will not be successful. Simple theories of turbulence always run into divergence difficulties which (after much algebra) are effectively dominated by the small k region of the integral

$$\int q_k d^{(d)}k \tag{3.13}$$

so unlike say ferromagnetism where d>4 is a soluble theory, there is no known dimensionality in which any method of solving turbulence can be said to be demonstrated, perhaps even demonstrable.

The dimensionality argument says nothing about the time dependence, or single point functions

$$Q_k(t) = q_k \phi_k(t) \ , \ \phi_k(0) = 1$$

$$u(r_1) = v_1 \quad : \quad P(u_1) \tag{3.14}$$

I have studies this problem and guess that

$$\phi_k(t) = e^{-c|k|t^{3/2}} \tag{3.15}$$

$$P(v_1) \propto e^{-\theta|v_1|^3} \tag{3.16}$$

but my present point is that there are new indices here awaiting discovery.

For the static problem P([u]) will satisfy a Liouville-Fokker-Planck type equation

$$\frac{\partial P}{\partial t} + \sum_k h_k \frac{\partial^2}{\partial u_k \partial \underline{u}_k} P + \sum_k \frac{\partial}{\partial u_k}(\nu h^2 u_k + M u_j u_i) P = 0 \tag{3.17}$$

i.e. $$D([u])P = 0 \tag{3.18}$$

where D is a differential operator in the space of all the u_k. This function if derived should give q_k and $P(U_1)$. That q_k is known provides a stringent test of turbulence theories, and although several different methods are now available, there are drawbacks in all of them, as explained in the book of D.C. Leslie. In equilibrium one has for an equation like the Boltzmann equation, conservation laws for number, momentum and energy, and a bound on the entropy.

Equally equation (3.18) has the equivalent in that energy input must equal energy output, but the entropy, conventionally defined and still representing the negative of the information content is steadily produced in the steady state, but unless one is close to equilibirum is not controlled by an H theorem i.e. all one can say is that

$$(S)_{input} - (S)_{output} = (S)_{transport} \tag{3.19}$$

and (S) transport > 0. It is also the case that the fully developed turbulence has no kind of coupling constant and so is the analogue of a critical phenomenon. Presumably one could drive energy through a normal statistical mechanical system at critical and again produce a crop of new indices. The absence of any coupling constant provides a severe difficulty in the chain of development.

Liouville (many degree of **freedom**) equation

$$\longrightarrow \text{Boltzmann (few degree freedom) equation} \longrightarrow \text{Solution} \tag{3.20}$$

because whereas in our first example (3.3) the first transition was easy to make and (3.3) resulted, an equation of manageable complexity for turbulence the first stage is difficult to construct. The problem is that any attempt to expand in terms of one or two degree of freedom distributions cannot be based on a coupling constant. Hence it is not easy to see how it can be terminated. Suppose one made some expansion in terms of a distribution q_k and a lifetime W_k. Then P will be given as a functional expansion in these variables. Suppose this expansion is terminated and one wishes to choose a 'best' q_k and W_k. The present author has tried to maximize

$$\int P \log P \ \Pi \, du$$

where P_1 is a solution of (3.18), but a functional of q_k and W_k which are to be determined. The point I am trying to emphasize is that there are too many solutions of (3.18) not too few, once some expansion method is adopted and one needs some criterion for the 'best'. Other methods such as minimizing $\langle (\text{Eq } 3.5)^2 \rangle$

do not seem so fruitful. This brings us to the ultimate problem in this field which is to understand mean flow under turbulent conditions.

If $\langle u(\underline{r},t) \rangle = U(\underline{r},t)$

then (3.18) say is replaced by equations for U and $u - U$, and by eliminating $u - U$ one can in principle produce a kind of turbulent hydrodynamics of an even more non-linear variety than the usual kind. Although the theoretical physicist may recoil in horror from this problem, it is important to realise that it is the most commonplace of all theoretical problems: how does water move when it moves quickly.

It is also the concluding problem in the sequence of complexity.

Liouville $\longrightarrow$ Hydrodynamics $\longrightarrow$ Turbulent hydrodynamics

there is no further stage of "turbulent turbulent" hydrodynamics.

PHASE TRANSITIONS IN LOW-DIMENSIONAL SYSTEMS AND RENORMALIZATION GROUP THEORY

A.P. Young
Dept. of Mathematics
Imperial College
London SW7

1. PHASE TRANSITIONS AND SOME SIMPLE SPIN MODELS

These two lectures are intended as an introduction to some of the model systems and theoretical techniques which will be discussed by other speakers at this school.

One of the simplest models which exhibits interesting properties has a classical spin vector $\vec{S}_i$ at each lattice site with an isotropic interaction between neighbours so the Hamiltonian is

$$\mathcal{H} = - \sum_{<ij>} J_{ij} \vec{S}_i \cdot \vec{S}_j \qquad (1)$$

Theorists like to allow the vector $\vec{S}_i$ to have an arbitrary number of components, n, and we will adopt the convention that $S_i^2 = 1$. For n = 3 the model is the classical Heisenberg model while for n = 2 it is called the classical planar spin model (or sometimes XY model) which will be extensively discussed at this school. If n = 1 the spin variable can only take one of two discrete values ±1 and eq. (1) then corresponds to the spin-½ Ising model. It is of crucial importance to distinguish between spin dimensionality n and the dimensionality of the lattice upon which the spins are situated, which will be denoted by d.

The traditional picture[1,2] of a second order phase transition is that below a certain critical temperature T_c the system develops an 'order parameter' ψ and so spontaneously lowers its symmetry. In the spin models introduced ψ would be the magnetization $\langle S \rangle$ while for a fluid close to its critical point ψ would be the difference in density between its gas and liquid phases. The order parameter vanishes continuously at T_c and, for $|t| << 1$ where $t = (T-T_c)/T_c$, is found to obey a power law of the form

$$\psi \sim |t|^{\beta} \tag{2}$$

For $T > T_c$, $\psi = 0$ of course. The simplest type of theories, known as mean field theories (see §2) predict that $\beta = \frac{1}{2}$. Experimentally the values of β are lower, between 0.3 and 0.4, in agreement with more sophisticated theories[2,3].

As T approaches T_c the response of the system to an external field which couples to the order parameter is very large and the corresponding susceptibility, χ, diverges at T_c. We define an exponent γ by

$$\chi \sim t^{-\gamma} \tag{3}$$

the mean field prediction being $\gamma = 1$. Related to a diverging susceptibility is a correlation length ξ, which also diverges at T_c. In fact one can consider this to be the most fundamental feature of a critical point. Normally correlations decay exponentially for $T > T_c$ so

$$\lim_{r\to\infty} \langle\psi(r)\psi(0)\rangle \sim e^{-r/\xi} \tag{4}$$

defines ξ and for $t << 1$, ξ varies as

$$\xi \sim t^{-\nu} \tag{5}$$

which defines an exponent ν. Precisely at T_c correlations decay with a power law and an exponent η, usually very small, is defined by

$$\langle\psi(r)\psi(0)\rangle_{T=T_c} \sim r^{-(d-2+\eta)} \tag{6}$$

Mean field values of ν and η are $\frac{1}{2}$ and 0 respectively.

2. FLUCTUATIONS AND THE LOWER CRITICAL DIMENSION

The approximation in mean field theory (MFT) is that it neglects all 'short range order' in the system. For example, in our model spin system the Hamiltonian (1) is replaced by

$$\mathcal{H}^{MF} = - \sum_{\langle ij\rangle} J_{ij}[\langle\vec{S}_i\rangle \cdot \vec{S}_j + \langle\vec{S}_j\rangle \cdot \vec{S}_i - \langle\vec{S}_i\rangle \cdot \langle\vec{S}_j\rangle] \tag{7}$$

where the last term avoids double counting of the energy. Hence every spin sees only the *average* effect of its neighbours, fluctuations about this average value being neglected. One approach which attempts to do better was initiated by Bethe and Peierls[4] for the Ising model. They argued that whereas mean field theory considers a single spin in an effective field one could improve on this by treating a finite cluster exactly and representing the rest by a suitably chosen effective medium. The next level of approximation is therefore to consider a pair as the unit cluster, which gives the Bethe-Peierls (BP) approximation (for further discussion see ref. 5). BP predicts a lower T_c than MFT and if each spin interacts equally with z neighbours, the fractional change in T_c is of order $1/z$, but critical exponents have their MF values. The inverse susceptibility is sketched in fig. 1 in BP and MF approximations. It is a straight line in MFT but acquires some curvature in BP. Also sketched in fig. 1 is an exact curve for χ^{-1} (which could be estimated from high temperature series expansions or renormalization group methods). Notice that T_c (exact) is somewhat lower than T_c(BP) and the curve is horizontal at T_c corresponding to $\gamma > 1$. The temperature range over which the curve starts to flatten as T_c is approached is called the 'critical' region and from an estimate due to Ginzburg[6] one finds the size of the critical region is of order $1/z^2$ in three dimensions ($z^{-2/(4-d)}$ in d-dimensions). For large z the size of the critical region is therefore small.

Considerable insight into the difference between actual values of critical exponents and their mean field values has been gained by the renormalization group ideas of K.G. Wilson[7]. According to this approach critical exponents would have their mean field values in space dimensionalities greater than four, although, of course, T_c is reduced from its mean field estimate in all dimensions. Four dimensions is called the 'upper critical dimensionality' (UCD) and will be discussed in detail in Dr. Als-Nielsen's lectures.

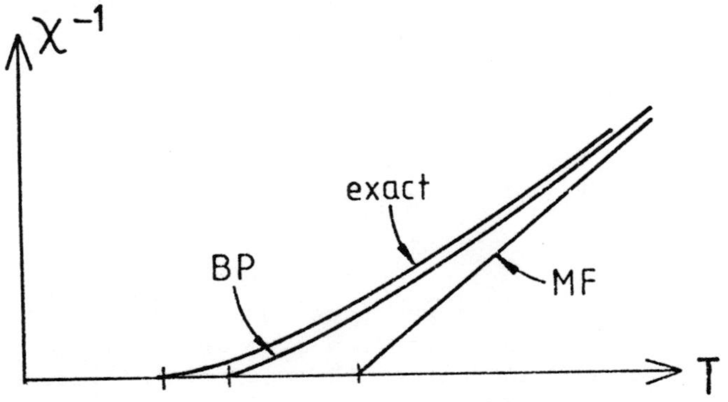

Fig. 1. A sketch of inverse susceptibility in mean field (MF) and Bethe-Peierls (BP) approximations compared with an exact curve.

Fluctuations become increasingly important as the dimensionality is reduced, as is illustrated by non-mean field exponents below the UCD. However fluctuations also reduce T_c and if the dimensionality is low enough they are so important that they wipe out the ordering completely. This dimensionality is known as the 'lower critical dimension' (LCD). In this situation the error in mean field theory is much more important than between the LCD and UCD where it is only badly wrong in a small critical region. At or below the LCD mean field theory predicts a transition which does not occur at all. In this situation one has to resort to more sophisticated techniques, such as the renormalization group, which will be introduced in §5. The lectures at this meeting on strongly fluctuating systems will mainly discuss systems at or below their LCD.

3. VALUES OF LOWER CRITICAL DIMENSIONALITY FOR SOME SPIN MODELS

Having introduced the concept of the LCD we shall now investigate its value for the spin models introduced earlier. The simplest case is the Ising model ($n = 1$) so we will treat it first. In two space dimensions many of its properties have been evaluated exactly[8]. A transition is found at temperature T_c where, for a square lattice $T_c/T_c^{MF} \sim 0.57$. Below T_c there is long range order which implies that the LCD is less than 2. The one-dimensional model with nearest neighbour interactions is sufficiently simple that some of its properties will be worked out exactly here.

First of all it is useful to note the identity

$$\exp(KS_iS_j) = \cosh K\,[1 + wS_iS_j] \tag{8}$$

where $K = J/k_BT$, J is the interaction between a neighbouring pair, k_B is Boltzmann's constant and $w = \tanh K$. Equation (8) holds because S_iS_j can only take the two values ± 1. The partition function of the Ising chain can therefore be written as

$$\begin{aligned} Z &= \sum_{\{S_i=\pm 1\}} \exp[\sum_i KS_iS_{i+1}] \\ &= (\cosh K)^{N_b} \sum_{\{S_i=\pm 1\}} \prod_i [1 + wS_iS_{i+1}] , \end{aligned} \tag{9}$$

where N_b is the number of interactions, which is equal to N for a cyclic chain and N-1 for a chain with free ends. If the number of spins, N, is large one can neglect the possible difference between N_b and N. Expanding the bracket in (6) one obtains a sum of 2^N terms each of which is a product of factors like (wS_iS_{i+1}). When the trace over the spins is carried out only terms where each spin variable occurs an even number of times will contribute.

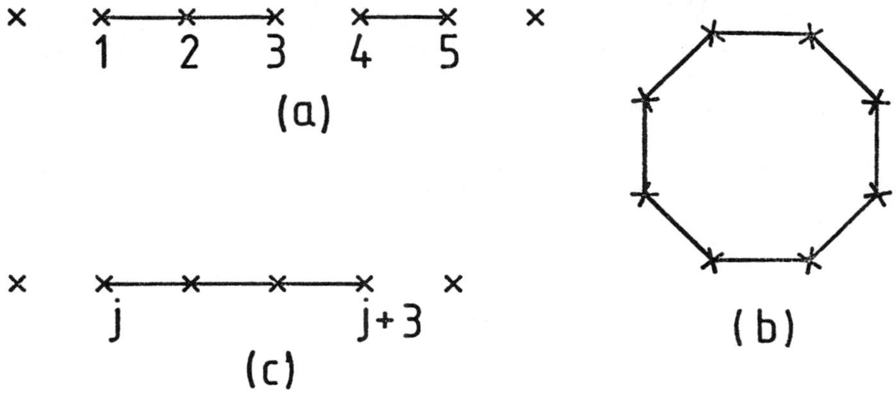

Fig. 2. (a) A diagram representing $S_1S_2S_2S_3S_4S_5$ in an expansion of the partition function Z of the Ising chain, (b) a contribution to Z for a cyclic chain with 8 spins, (c) the only diagram contributing to the correlation function for spins three lattice spacings apart.

It is useful to represent these terms by diagrams. For each factor S_iS_{i+1} we draw a line between points i and i+1 and associate a value of w with it. There can at most be one line between a neighbour pair. Fig. 2a shows a diagram representing $S_1S_2S_2S_3S_4S_5$ which has a factor w^3 associated with it but which vanishes when the trace is taken. In general only diagrams with an even number of lines entering each lattice point contribute to Z.

For a chain with free ends the only possibility is no lines at all so

$$Z = (\cosh K)^N \qquad \text{(free ends)} \tag{10a}$$

whereas for a cyclic chain there is another contribution from the diagram with a line between every pair of sites. This is shown in fig. 2b for N = 8 and gives a contribution w^N (apart from the overall factor of $(\cosh K)^N$) so

$$Z = (\cosh K)^N[1 + w^N] \qquad \text{(cyclic chain)} \tag{10b}$$

Since $w \leq 1$ we can write the free energy per site, f, in both cases as

$$f = -k_BT \ln[\cosh K] \tag{11}$$

for large N. This is a perfectly smooth function of temperature with no evidence of a phase transition. To confirm the absence of a phase transition we shall now show that the correlation length is finite at all temperatures except T = 0.

The correlation function C(n) between two spins a distance n lattice spacings apart is defined to be

$$C(n) = \langle S_j S_{j+n} \rangle$$
$$= \frac{1}{Z} \sum_{\{S_i = \pm 1\}} S_j S_{j+n} \exp[K \sum_{\ell} S_\ell S_{\ell+1}] \tag{12}$$

which is independent of j in a translationally invariant system. The numerator is evaluated by using (8) and expanding the resulting product. Since there is an extra factor $S_j S_{j+n}$ then, in the diagrams, sites j and j+n must have an odd number of lines emerging from them whereas other sites must still have an even number. There is consequently just one contributing diagram, which has a continuous line from j to j+n and is shown in fig. 2c for n = 3. This gives a contribution w^n so we find

$$C(n) = w^n \tag{13}$$

which is of the form (4) at all n with a correlation length given by

$$\xi^{-1} = -\ln w \tag{14}$$

which simplifies as $T \to 0$ to

$$\xi = \exp(2J/k_B T) \tag{15}$$

The correlation is therefore always finite at positive T but diverges exponentially as $T \to 0$. The susceptibility χ is given by the fluctuation-dissipation theorem as

$$\chi = \frac{1}{k_B T} \sum_{n=-\infty}^{\infty} C(n)$$
$$= \frac{1}{k_B T} \exp(2K) \tag{16}$$

which is plotted in fig. 3. In view of the exponential divergence of ξ and χ at T = 0 one sometimes talks of a pseudo-transition at T = 0.[9] We conclude, therefore, that the LCD of the Ising model is d = 1.

The situation for n > 1 has been somewhat more controversial. The one-dimensional model with classical spins has been solved exactly by Fisher[10] for n = 3 and Stanley[11] for arbitrary n. The

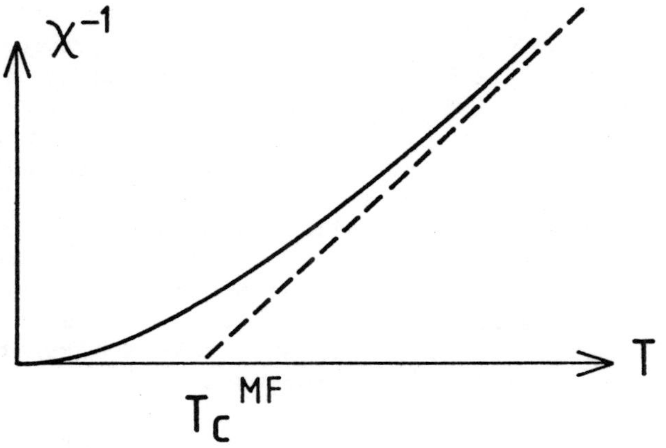

Fig. 3. Sketch of the inverse susceptibility for the one-dimensional Ising model. The dashed curve is the mean field theory prediction.

correlation length is found to vary as

$$\xi = \frac{2J}{(n-1)k_BT} \tag{17}$$

at low temperatures, which diverges as $T \to 0$ but only as T^{-1} not exponentially. Later we shall indicate that exponential increase is characteristic of systems at their LCD.

What makes models with $n > 1$ different from the Ising model is that the former have a continuous rotational symmetry. If we consider ferromagnetic state with magnetization in a certain direction then the state with magnetization rotated infinitesimally away from this direction has the same energy. If now a small periodic variation in the transverse spin component is applied, i.e. $S^\perp(\vec{r}) \sim S^\perp(\vec{k})\cos(\vec{k}.\vec{r})$ then for $k \to 0$ this reduces to a uniform rotation so the energy $E_\perp(k)$ must tend to zero as $k \to 0$. A term linear in k is forbidden by inversion symmetry so we expect $E_\perp(k) \sim k^2[S^\perp(k)]^2$. Since the energy is very small, long wavelength fluctuations are classical so the mean energy per mode is $\sim k_BT$. Hence the mean square transverse fluctuation is given by

$$\langle (S^\perp(k))^2 \rangle \sim \frac{k_BT}{k^2} \tag{18}$$

and the fluctuation in $S^\perp$ on a single site i is

$$\langle (S_i^\perp)^2 \rangle \sim k_BT \int_0^\Lambda \frac{d^dk}{k^2} \tag{19}$$

where Λ is of order an inverse lattice spacing. The integral on the right of eq. (19) converges for $d > 2$ but diverges for $d \leq 2$. Since fluctuations on a site must be finite there is an inconsistency in the latter case which implies that our assumption of long range order must be incorrect for $d \leq 2$. In other words the LCD is $d = 2$. This argument has been made rigorous by Mermin and Wagner[12] who showed that $<(S^{\perp}(k))^2>$ diverges at least as fast as k^{-2}. Since small transverse fluctuations are called spin waves we can say that at any finite temperature there are always enough spin waves to destroy long range order for $d \leq 2$. This argument depends on the existence of a continuous rotational symmetry. If there is some anisotropy, so the order parameter can only take a discrete set of values, the LCD is one, as for the Ising model.

The absence of long range order does not necessarily imply there is no phase transition and a transition does occur at finite T_c for $n = 2$ in the sense that $\chi = \infty$ for all $T < T_c$, although $<\underset{\sim}{S}> = 0$ (see for example my other lecture at this School). However for $n > 2$ the susceptibility is finite at all non-zero T as will be discussed in §5. From a practical point of view it is worth noting that even a very small amount of anisotropy can lead to a sizeable T_c for, let us say, the two-dimensional Heisenberg model ($n = 3$) and in fact[13]

$$\frac{T_c}{J} \sim [\ln(J/D)]^{-1}, \tag{20}$$

for $D/J \to 0$, where D is the anisotropy energy. This makes analysis of experiments on quasi two-dimensional magnetic systems difficult. Furthermore if there is a weak three-dimensional coupling J_3 between planes then T_c is still given by (20) but with D replaced by J_3.

4. FLUCTUATIONS

Fluctuations which are responsible for the behaviour of systems in low dimensionality can generally be divided into two types:-

(i) Small deviations from an aligned state, which become important at the LCD because there are many of them.

(ii) Fluctuations which represent a large deviation from an aligned state and can therefore be important even when they are small in number.

We have already encountered an example of type (i) fluctuations in our discussion in §3 of the two-dimensional Heisenberg model where they are, of course, spin waves. Several lecturers at this school (Axe, McTague and Villain) will be discussing low dimensional solids where the corresponding oscillations are long wavelength phonons. The LCD in this case is also two, as was first noted in the 1930s by Landau and Peierls.

We shall now show that it is fluctuations of type (ii) that destroy long range order for the one-dimensional Ising model. It is useful to rewrite the model in continuum form, defining a variable m(x) which represents the local magnetization averaged over a distance of several lattice distances but small compared with the scale over which fluctuations are assumed to occur. The effective Hamiltonian for long wavelength fluctuations then assumes the Landau-Ginzburg form

$$\mathcal{H} = \frac{1}{2} \int dx [rm^2 + (\frac{dm}{dx})^2 + \frac{u}{2} m^4] \quad (21)$$

where $r \propto T - T_c^{MF}$, $u > 0$ and we have rescaled m so that, for convenience, the coefficient of the gradient term is unity. The partition function is obtained by integrating $e^{-\beta\mathcal{H}}$ over all configurations m(x), i.e.

$$Z = \int \mathcal{D}m \, e^{-\beta\mathcal{H}} \quad (22)$$

where the notation $\mathcal{D}$ indicates a functional integral. Mean field theory assumes that the integral in (22) is dominated by the single configuration m(x) which gives the largest contribution. For $T > T_c^{MF}$ this is $m(x) = 0$ while for $T < T_c^{MF}$ it is $m(x) = \pm m_0$, independent of x, where

$$m_0^2 = |r|/u. \quad (23)$$

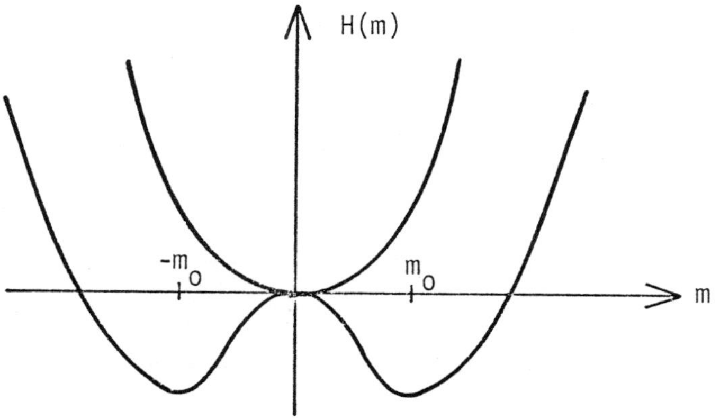

Fig. 4. Form of the Landau free energy, $\mathcal{H}(m)$. Curves (i) and (ii) are for $T > T_c^{MF}$ and $T < T_c^{MF}$ respectively, where T_c^{MF} is the transition temperature in mean field theory.

The form of $\mathcal{H}(m)$ is illustrated in fig. 4 for configurations with m independent of x. For $T << T_c^{MF}$ $\mathcal{H}(m)$ has the form of a deep double well potential with minima at $\pm m_0$. Whereas MFT assumes the system is everywhere in a configuration corresponding to the bottom of a single one of these wells the functional integral (22) also has sizeable contributions from configurations where $m(x) = m_0$ for some ranges of x and $m(x) = -m_0$ for other values of x with crossover regions in between. Such a configuration is sketched in fig. 5 and the region where m(x) is varying between $-m_0$ and $+m_0$ is known as a kink. The form of a single kink can be obtained analytically by solving the Euler-Lagrange equation

$$\frac{d^2m}{dx^2} = rm + um^3, \tag{24}$$

which is the condition for an extremal configuration, subject to boundary conditions that $m \to \pm m_0$ as $x \to \pm\infty$. The solution is

$$m(x) = m_0 \tanh [(x - x_0)/\Delta x] \tag{25}$$

where $(\Delta x)^{-2} = |r|/2$. The energy ΔE of a kink is finite and is found to be

$$\Delta E = \frac{2}{3}\frac{r^2}{u} \tag{26}$$

Notice that $\Delta E \sim 1/u$ so the kink solution can never be obtained by perturbation theory in u. Type (ii) fluctuations in general represent non-perturbative effects.

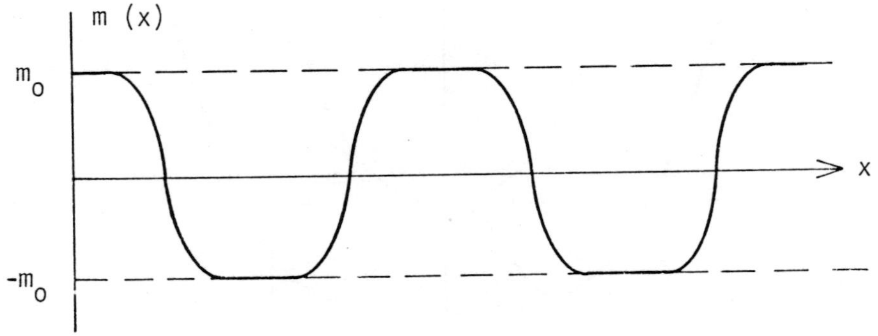

Fig. 5. Sketch of a configuration m(x) which gives an important contribution to the partition function of an Ising chain illustrating regions where $m(x) = \pm m_0$ separated by randomly positioned 'kinks' which interpolate between these two values.

If there are several kinks but with separation much larger than Δx the total energy should just equal the sum of the energies of the individual kinks given by (26). (In general this is not true because eq. (24) is non-linear.) Therefore at low temperatures the density of kinks is given by the Boltzmann factor

$$n_{kink} \sim \exp(-\Delta E/k_B T) \ . \qquad (27)$$

Since the correlation length is of order the mean separation between kinks it diverges exponentially at low temperature in agreement with our exact calculation in §3. The advantage of the present approximate discussion is that it shows up the nature of the fluctuations which destroy long range order. A much more detailed investigation of the role of kink solutions is given in ref. 14. A somewhat related one-dimensional Hamiltonian, known as the Sine-Gordon model, will be discussed in Dr. Steiner's lectures. The configurations which interpolate between different minima of the Landau free energy are, in this case, known as solitons.

One can think of the kink solution as corresponding to a domain wall. In one dimension the energy of creating a domain of linear dimension L is just $2\Delta E$. In d-dimensions the corresponding energy is proportional to the area of the domain wall, $\sim L^{d-1}$, so large domains do not occur at low temperatures for $d > 1$, (this is why the LCD is equal to one). However for Ising models with interactions of random sign, known as spin glasses, (see Dr. Binder's lectures) the domain wall energy is considerably less[15]. In general it should be (roughly) independent of L in the LCD and some calculations suggest[15,16] that the LCD might be as high as four.

To conclude, the LCD is that dimensionality where the system is unstable with respect to creation of a sufficiently large number of low energy excitations (e.g. spin-waves or domain walls) that an ordered ground state breaks up at finite temperature.

5. INTRODUCTION TO THE RENORMALIZATION GROUP[7]

A second order phase transition is characterized by a correlation length which diverges as T approaches T_c. This large correlation range is the heart of the difficulty in developing a theory of critical phenomena but it also gives rise to one simplification which is exploited in the renormalization group approach and which will now be explained.

Consider, for concreteness a square lattice, fig. 6, and decompose it into elementary squares such that each site lies at the corner of one and only one square. These are shown by the solid lines in fig. 6. We now associate with each square a single spin variable, related in some as yet unspecified way to the elementary

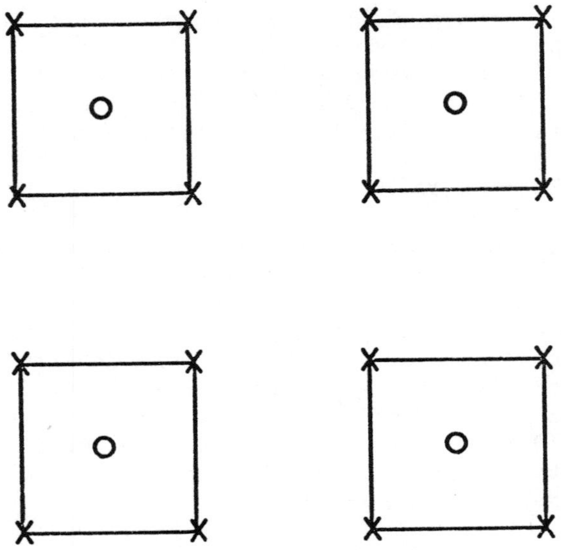

Fig. 6. The crosses represent spins on a square lattice. The circles represent block spins for each of the elementary squares marked by solid lines.

spins. One then tries to evaluate effective interactions between these 'block' spins such that (i) the partition function remains unchanged and (ii) the correlation function at large distances for the block spins is the same as for the original spins. Thus, for long wavelength phenomena, the new problem with block spins is equivalent to the original one. The block spins also form a square lattice but with lattice spacing increased (rescaled) by a scale factor b, which in the present example is two. This change from original spins to block spins is known as a renormalization group transformation (the nomenclature is actually not very appropriate but has arisen from earlier somewhat related work in quantum field theory). Later we shall construct such a transformation in a simple case but for the time being let us assume it can be done, within a reasonable approximation, and investigate the consequences.

The correlation length ξ remains the same, by construction, so denoting quantities in the rescaled lattice by a prime we have

$$\xi' = \xi \tag{28}$$

The lattice spacing, a, has been increased by b so

$$a' = ba. \tag{29}$$

It is generally necessary to consider several types of interaction, for example nearest and next nearest neighbour, four spin terms etc., which we denote by $\{K\}$. These determine, in a complicated way, a dimensionless correlation length $\xi_r(= \xi/a)$, the number of lattice spacings over which spins are correlated. From (28) and (29) we have

$$\xi'_r = \frac{1}{b}\xi_r \tag{30}$$

so, relative to the new lattice spacing, the correlation length has decreased by b.

A very important circumstance is when the set of rescaled interaction parameters is the same as the original set. This is called a *fixed point* of the transformation and we denote the corresponding interaction values by $\{K^*\}$. At a fixed point the dimensionless correlation length must remain unchanged, i.e.

$$\xi'_r = \xi_r \qquad \text{(at a fixed point)} \tag{31}$$

which is inconsistent with (30) unless ξ_r is infinity or zero. The second possibility arises for example, at infinite temperature but the first is more interesting because it is just the condition for a critical point.

Further information can be gleaned by considering what happens when the interactions differ only slightly from their fixed point values. Denoting this difference by δK_α (α indicates a particular interaction) we can linearize the transformation (which in its full form is highly non-linear) about the fixed point, obtaining formally,

$$\delta K'_\alpha = R_{\alpha\beta}\delta K_\beta \tag{32}$$

The matrix $R_{\alpha\beta}$ is diagonalized so its eigenvalues λ_μ and eigenvectors e_μ are known. Hence, expending the interactions as $\{\delta K\} = a_\mu e_\mu$, the coefficients a_μ transform as

$$a'_\mu = \lambda_\mu a_\mu. \tag{33}$$

The dependence of λ_μ on the scale factor b can be obtained by noting that a scale change of b followed by one of b' is equivalent to a single one of bb'. Therefore

$$\lambda_\mu(b)\lambda_\mu(b') = \lambda_\mu(bb') \tag{34}$$

which is satisfied if

$$\lambda_\mu = b^{y_\mu} \tag{35}$$

where the exponents of y_μ are independent of b.

After n iterations

$$a_\mu^{(n)} = b^{ny_\mu} a_\mu \tag{36}$$

which become very large if $y_\mu > 0$ and e_μ is termed a *relevant* eigenvector. On the other hand if $y_\mu < 0$, $a_\mu^{(n)} \to 0$ as $n \to \infty$ so e_μ is called *irrelevant*. To be at a critical point it is not necessary that the parameters have their fixed point values but merely that the coefficients of all the relevant eigenvectors are zero, the coefficients of irrelevant operators are driven to zero anyway by the transformation. The locus of all critical points is known as the critical surface. We show in fig. 7 a possible situation with two parameters, and, at fixed point P, one relevant and one irrelevant eigenvalue. The critical surface is the thick line which coincides with the direction of the irrelevant eigenvector close to P, where the linearized transformation is valid, but deviates from this direction further from P. It is useful to draw a curve, known as a trajectory, through the discrete set of points specifying the parameters after successive iterations. Some possible trajectories are shown in fig. 7.

At a second order transition we know that the temperature must equal its critical value T_c so we expect the coefficient of one of the relevant eigenvectors to be proportional to $T - T_c$. In a magnetic system we also have to set the external field h equal to zero so the coefficient of another relevant eigenvector must be proportional to h. An ordinary critical point is completely determined by

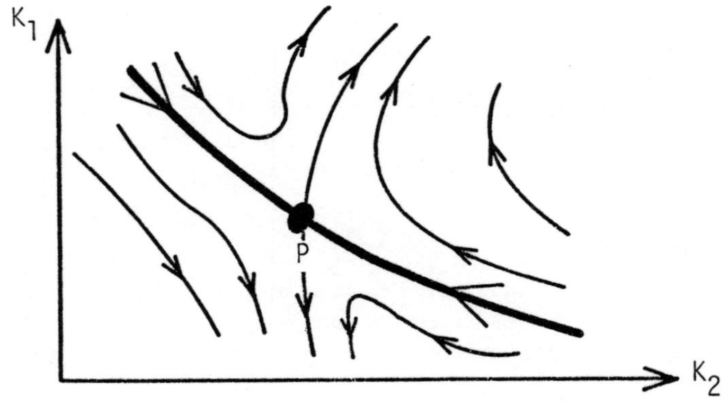

Fig. 7. Hypothetical renormalization group trajectories with two parameters K_1 and K_2 and a fixed point P. At P there is one relevant and one irrelevant eigenvector and the critical surface is marked by the thick line.

specifying h and T so we conclude that it is described by a fixed point with two relevant directions. A fixed point with three relevant directions would correspond to a tricritical point, which does not concern us here. A fixed point with no relevant eigenvectors would be a sink to which all trajectories in a given phase would flow, and is associated with zero rather than infinite correlation length.

The exponent y_T corresponding to the eigenvector proportional to $t = (T - T_c)/T_c$ is directly related to the correlation length exponent ν, eq. (5). To see this note that eq. (30), together with $t' = b^{y_T} t$ imply

$$(b^{y_T} t)^{-\nu} = t^{-\nu}/b$$

or

$$y_T = 1/\nu \tag{37}$$

One can also show that the magnetic field exponent y_H is given by

$$y_H = \frac{1}{2}(d + 2 - \eta) \tag{38}$$

where η is defined by eq. (6). These exponents are characteristic of systems close to any part of the critical surface and depend only on the fixed point. Thus systems with many different Hamiltonians can have the same critical behaviour, which is known as universality.

To illustrate these ideas we apply them to the Ising chain which was studied in §3. The number of degrees of freedom is reduced by carrying out the trace over all spins except for every b-th spin which is left unchanged. These form the rescaled lattice as shown in fig. 8 for b = 3. This procedure, which is known as decimation[9,17], is somewhat different from the block spin idea introduced earlier and has problems for d > 1. In one dimension, however, the transformation is fine and we use it here because of its simplicity.

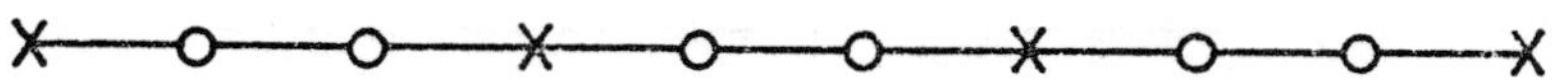

Fig. 8. Crosses denote spins that remain after a decimation transformation with b = 3. The circles indicate spins that are traced over.

Let us denote by i one of the spins that is to be traced over and consider first of all b = 2. This spin enters in the partition function as

$$\exp[KS_i(S_{i-1} + S_{i+1})].$$

We evaluate the trace over S_i and write the result as an effective interaction between S_{i-1} and S_{i+1}, i.e.

$$\sum_{S_i=\pm 1} \exp[KS_i(S_{i-1}+S_{i+1})] = \exp[K'_0+K'S_{i-1}S_{i+1}] \tag{39}$$

K'_0 is a spin independent term which gives an additive contribution to the free energy but is otherwise unimportant in what follows. From equation (8) it is easy to show that

$$w' = w^2 \tag{40a}$$

and, for general b,

$$w' = w^b \tag{40b}$$

where w' = tanh K'. Equation (40b) is our desired recursion formula. There are fixed points at w = 0, corresponding to T = 0 where ξ is zero, and at w = 1, or T = 0, where we showed earlier that ξ = ∞. Trajectory flows are shown in fig. (9a).

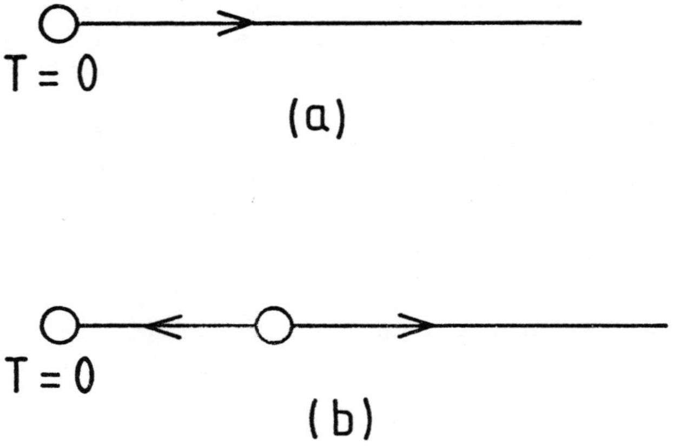

Fig. 9. (a) Trajectory flows for the one-dimensional Ising model, (b) Flows for the n component spin model where n > 2 and also d > 2. For n > 2 but d ≤ 2 the flows are as in (a).

It is useful to write $b = e^{\delta\ell}$ and consider the limit $\delta\ell \to 0$, corresponding to an infinitesimal phase change so eq. (40b) becomes the differential equation

$$\frac{dw}{d\ell} = w \ln w \tag{41}$$

Defining a reduced temperature $\overline{T}$ by $\overline{T} = K^{-1}$ one finds that for $\overline{T} << 1$ equation (41) can be written as

$$\frac{d\overline{T}}{d\ell} = \frac{\overline{T}^2}{2}\,. \tag{42}$$

There is no term linear in $\overline{T}$ on the RHS of (42), the significance of which will appear shortly.

The exponential divergence of the correlation length at low T can be demonstrated just from (42). One integrates up to a value ℓ^* where $\overline{T}(\ell^*)$ is some number x, sufficiently small that corrections to (42) are unimportant but much larger than the actual temperature $\overline{T}$ ($\equiv \overline{T}(\ell = 0)$). One obtains

$$\ell^* = 2/\overline{T} \tag{43}$$

independent of x. Denoting the dimensionless correlation length at $\overline{T} = x$ by $\xi_r(x)$ from eq. (30) we have

$$\begin{aligned} \xi_r &= \xi_r(x)e^{\ell^*} \\ &= \exp(2/\overline{T})\,\xi_r(x) \end{aligned} \tag{44}$$

which correctly predicts an exponential increase of the correlation length.

It is interesting to contrast the result with the situation when a term linear in $\overline{T}$ appears on the RHS of (42), i.e.

$$\frac{d\overline{T}}{d\ell} = \overline{T} + O(\overline{T}^2) \tag{45}$$

Repeating the above steps one finds

$$\ell^* = \ln(x/\overline{T}) \tag{46}$$

so

$$\xi_r = \frac{x}{\overline{T}}\,\xi_r(x) \tag{47}$$

and the correlation length diverges but only as T^{-1}.

The Ising chain can of course be solved more easily by the direct method used in §3. The power of the renormalization group lies in its being able to construct recursion relations, analogous to eq. (42), valid in certain ranges of temperature or dimensionality, where no exact solution is available.

An important illustration of this is provided by the n-component spin model with $n > 1$ for which the only exact solution is in one dimension. It is however possible[18] to construct a low temperature renormalization group equation analogous to (42). The calculations are beyond the scope of these lectures so we will simply quote the result which is

$$\frac{d\bar{T}}{d\ell} = (2 - d)\bar{T} + (n - 2)\bar{T}^2 + \ldots . \tag{48}$$

There is a fixed point when the RHS vanishes. This always occurs at $T = 0$ and, for $d > 2$, there is another fixed point at

$$\bar{T}^* = \frac{d - 2}{n - 2} . \tag{49}$$

Temperature is a relevant variable at this latter fixed point so trajectory flows for $d > 2$ are as in fig. (9b). Linearizing about T^* and using (37) one obtains

$$\nu = \frac{1}{d - 2} .$$

For $d < 2$ the trajectories have the same form as for the one-dimensional Ising model shown in fig. (9a).

Equation (49) shows that the method only works for $n > 2$ and the special case of $n = 2$ will be discussed in my other lecture at this school. For $n > 2$, we note that $T_c \to 0$ continuously as $d \to 2$. Furthermore in $d = 2$ the correlation length diverges exponentially at low T,

$$\xi \sim \exp\left[\frac{1}{(n-2)\bar{T}}\right] \tag{50}$$

because the recursion relation has no term linear in T. In general the LCD is characterized by a vanishing of this linear term in the low temperature renormalization group, and hence by an exponentially increasing correlation length. In $d = 1$ the recursion relation has the form (45) and leads to a correlation length diverging as T^{-1} in agreement with exact solutions.

For $n = d = 2$ the recursion relation is

$$\frac{d\bar{T}}{d\ell} = 0 \tag{51}$$

which can be shown to be valid to all orders in $\bar{T}$. This implies that every temperature is a fixed point, i.e. critical point, the significance of which will be described in my other lecture.

REFERENCES

1. L.D. Landau and E.M. Lifshitz, "Statistical Physics", 2nd ed. Pergamon (1969).
2. H.E. Stanley, "Introduction to Phase Transitions and Critical Phenomena", O.U.P. (1971).
3. G.A. Baker, B.G. Nickel and D.I. Meiron, Phys. Rev. B17, 1365 (1978).
4. H. Bethe, Proc. Roy. Soc. A150, 552 (1935).
5. R. Kikuchi, Phys. Rev. 81, 988 (1951).
6. V.L. Ginzburg, Sov. Phys. - Solid St. 2, 1824 (1960).
7. K.G. Wilson and J. Kogut, Phys. Rep. 12C, 75 (1974);
   S.K. Ma, Rev. Mod. Phys. 45, 589 (1973); S.K. Ma 'Modern Theory of Critical Phenomena' (Benjamin) (1976); G. Toulouse and P. Pfeuty 'Introduction au Groupe de Renormalisation' (Presses Universitaires de Grenoble) (1975); M.E. Fisher, Rev. Mod. Phys. 46, 587 (1974).
8. L. Onsager, Phys. Rev. 65, 117 (1944).
9. D.R. Nelson and M.E. Fisher, Ann. Phys. (NY) 91, 226 (1975).
10. M.E. Fisher, J. Math. Phys. 5, 944 (1964).
11. H.E. Stanley, Phys. Rev. 179, 570 (1969).
12. N.D. Mermin and H. Wagner, Phys. Rev. Lett. 17, 1133 (1966).
13. J.M. Kosterlitz and D.J. Thouless, Journal of Low Temperature Physics (to be published).
14. J.A. Krumhansl and J.R. Schrieffer, Phys. Rev. B11, 3535 (1975).
15. P. Reed, M.A. Moore and A.J. Bray, J. Phys. C 11, L139 (1978).
16. A.J. Bray and M.A. Moore (to be published).
17. M.N. Barber, J. Phys. C. 8, L203 (1975); L.P. Kadanoff and A. Houghton, Phys. Rev. B11, 377 (1975).
18. A.M. Polyakov, Phys. Lett. 59B, 79 (1975); E. Brezin, and J. Zinn-Justin, Phys. Rev. B14, 3110 (1976); A.A. Migdal, Sov. Phys. JETP 42, 413, 743 (1976).

REAL-SPACE RENORMALIZATION-GROUP METHOD FOR QUANTUM SYSTEMS

R.Jullien, K.A.Penson+, P.Pfeuty and K.Uzelac*

Laboratoire de Physique des Solides

Bât.510, Université Paris-Sud

91405 ORSAY, France

In contrast with classical systems, quantum systems exhibit fluctuations already at T = 0 due to the nature of quantum mechanics. A wide range of quantum systems show interesting transitions at T = 0 by varying a given parameter. This should be compared with the transitions in temperature of classical systems and, in many cases, a rigorous mapping has been established. The equivalent of a T = 0 quantum system in D dimension is generally a D+1 classical system. This comes from the fact that a quantum hamiltonian contains its own dynamics and thus the time plays the role of an extra dimentionality So, an interesting first step to understand the physics of quantum systems is to study their ground state properties. This is either an interesting problem in itself or an indirect way to study the classical equivalent.

We present here a real-space renormalization-group method able to construct approximately the low lying states of a quantum system and to derive any mean value of interest in the ground state such as the order parameter or the correlation functions. This method has been first developped by Drell et al. [1] who studied quantum field theories on lattices after the pioneering work of Wilson [2] on solving the Kondo problem. We have tested the method on the quantum Ising model with a transverse field in 1D [3] for which an exact solution exists [4]. This model, described by the hamiltonian:

$$H = -\sum(JS_i^xS_{i+1}^x + hS_i^z) \tag{1}$$

where S_i^x and S_i^z are Pauli matrices on each sites of an infinite chain :

$$S^x = \begin{pmatrix} 0 & 1 \\ 1 & 0 \end{pmatrix} \qquad S^z = \begin{pmatrix} 1 & 0 \\ 0 & -1 \end{pmatrix} \tag{2}$$

exhibits at $T = 0$ a transition from a low field regime $h/J < 1$ with a degenerate ground state and a finite x magnetization $\langle S^x \rangle \neq 0$ to a large field regime $h/J > 1$ with a singlet ground state and a zero x-magnetization $\langle S^x \rangle = 0$. The critical behavior close to $(h/J)_c = 1$ is strictly equivalent [5] to the critical behavior of the classical Ising model in 2D, the transverse field playing the role of the temperature in the classical analog.

Let us choose this simple example here to describe the method and then we will show how the method can be extended to the same model in 2D for which no exact solution is known. Our method is an iterative procedure :

(i) We first cut the chain into independent blocks of n_s sites. The block hamiltonian H_{block} is solved exactly after observing that it acts inside two independent subspaces ε^+ and ε^- corresponding to spin combinations with respectively an even or an odd number of down spins (the hamiltonian flips always two spins together). Then we retain the ground state of each subspace ε^+ and ε^- we call respectively $|+\rangle_{block}$ and $|-\rangle_{block}$ with energies E_+ and E_-. We observe that they are the two lowest eigenstates of H_{block}.

(ii) Then we introduce a new spin $\underset{\sim}{S}'$ per block, the eigenstate of S'^z being respectively $|+\rangle$ and $|-\rangle$. Neglecting the other excited states, H_{block} can be written simply as $-h'S'^z$ by introducing a new field:

$$h' = \frac{1}{2}(E_- - E_+) \tag{3}$$

Doing so, we drop a constant term $1/2(E_-+E_+)$ which can be kept at each step to get the ground state energy per site.

(iii) Taking the matrix elements of original spin operators between the new block states $|+\rangle$ and $|-\rangle$ we get spin recursion relation:

$$S_p^x = \xi_p \, S'^x \tag{4}$$

for the pth spin in the block. This relation is used to rewrite the original interblock interaction. This gives a new constant J'

$$J' = J \, \xi_1^2 \tag{5}$$

Thus, after the blocking procedure, the new hamiltonian has the same form as (1) but with new parameters h' and J', while the inter-site distance is multiplied by n_s. This procedure is repeated until we reach a "fixed point". Recursion relations (3) and (5) alow us to study the behavior of the unique dimensionless parameter h/J during the iterative process. We observe that :

(i) if the original value of h/J is smaller than a critical value $(h/J)_c$ $h \to 0$ and $J \to J^\infty \neq 0$. We reach the stable fixed point $(h/J)^* = 0$ in which the hamiltonian reduces to an Ising chain without field with a doublet ground state.

(ii) if the original value of h/J is greater than $(h/J)_c$, $h \to h^\infty \neq 0$, $J \to 0$ we reach the stable fixed point $(h/J)^* = \infty$ in which the hamiltonian reduces to a set of free spins in a z-field yielding to a singlet ground state with a gap $G = 2h^\infty$. We can extract an exponent s for the gap :

$$G \sim \left\{\frac{h}{J} - \left(\frac{h}{J}\right)_c\right\}^s \tag{6}$$

The critical value $(h/J)_c$ corresponds to an unstable fixed point starting with $h/J = (h/J)_c$ both h and J rescale with the same factor. At $(h/J)_c$, we can define a z-exponent giving the way the energy rescales :

$$J^{(n+1)}/J^{(n)} = n_s^{-z} \tag{7}$$

Not only the method allow us to study the flow diagram in the parameter space, it permits also an evaluation of the magnetization components and spin correlation functions by integrating the spin recursion relations. For example the x-component of the magnetization is obtained after integrating formula (4) up to the fixed point.

$$\langle S_i^x \rangle = \xi_{p_o}^{(o)}\, \xi_{p_1}^{(1)} \ldots \xi_{p_{n-1}}^{(n-1)} \langle S_j^{x(n)} \rangle \tag{8}$$

In the fixed point $\langle S_j^{x(n)} \rangle = 1$ if $h/J < (h/J)_c$ and $\langle S_j^{x(n)} \rangle = 0$ if $h/J > (h/J)_c$. Thus $\langle S_i^x \rangle$ is zero above the transition while $\langle S_i^x \rangle$ is given by an infinite product, which can be numerically evaluated, below the transition. The values p_0, $p_1 \ldots$ depend on the original position i of the spin. This site-dependent result is an artifact of our blocking procedure. The best result is obtained by taking $p_0 = p_1 = p_2 \ldots = n_s/2$ if n_s even, $(n_s+1)/2$ is n_s odd ; this avoids edge effects. Close to $(h/J)_c$ we can extract an exponent β :

$$\langle S^x \rangle \sim \left\{\left(\frac{h}{J}\right)_c - \frac{h}{J}\right\}^\beta \tag{9}$$

At the unstable fixed point, ξ_o obtained for $p = n_s/2$ or $p = (n_s+1)/2$, tends to a constant value which allow us to extract an exponent d_x related with the spin dilatation :

$$\xi_o = n_s^{-d_x} \tag{10}$$

Then the exponent η_x giving the power law decay of the spin-spin correlation function $\langle S_i^x S_j^x \rangle \sim R^{-\eta_x}$ at the transition is given by $\eta_x = 2d_x$. More on the correlation functions is obtained in reference 3. The exponent ν giving the divergence of the coherence length near the transition can be also extracted. Results for the exponents are given in table I. The scaling law $s = \nu z$ is verified while the scaling law $\eta_x = 2\beta/\nu$ is only approximatively verified. On these results we see that increasing the size of the block gives a good improvement for the "magnetic exponent" β and η_x while the "thermal exponents" s, ν, z converge only very slowly to the exact values. Improvements on s, ν and z can be obtained by retaining much more levels at each step. By taking four levels we get already the exact result s = 1 [6].

This method has been also applied to the some model in 2D [7] equivalent to the classical Ising model in 3D for which no exact solution is known. In the ferromagnetic case the results for the location of the transition and the exponents in the case of hexagonal and triangular lattices are summarized in table II. There is a conjecture that the transition must be at $(h/J)_c = z'-1$ where z' is the number of nearest neighbours in the structure. The exponents are compared with estimates given by Le Guillou and Zinn-Justin [8] for the classical 3D Ising model. The best results are obtained in the case of the less compact hexagonal lattice. In the case of the triangular lattice an improvement is obtained by increasing the size of the block but even for a block of seven sites the results are not so good. This is due to important edge effects in such a compact structure.

An interesting case is the antiferromagnetic Ising model with a transverse field on a triangular lattice. Here the behavior is expected to be completely different than in the ferromagnetic case due to the complete frustration of the system in the classical zero field limit h = 0 for which the ground state has a finite entropy [9] . It is important to choose a block having all the features of the complete systems i.e. frustration and right symetries. A convenient choice is the hexagonal block of seven sites. By applying the method [10] we find again a three fixed point picture with $(h/J)_c = 1.41$. So we expect that for $h/J < (h/J)_c$ the ground state remains highly degenerate without order. For $h/J > (h/J)_c$ there is an opening of a gap with s = 0.8 with exponential decay for the correlation function. The results for $(h/J)_c$ and s are strongly different than in the ferromagnetic case and are much more like 1D results. For $h/J < (h/J)_c$ the x-x-correlation function has a power law decay and we find $\eta = 0.32$ (exact result for h = 0 gives $\eta = 0.5$ [11]). For $h/J = (h/J)_c$, η jumps to a different value $\eta = 0.68$ This model corresponds to a 3d classical system made of antiferromagnetic triangular layers connected together by nearest neighbour vertical ferromagnetic bounds [12]. In such a system, where the ground state has a degeneracy proportional to $N^{2/3}$, we predict a transition at a finite T_c to a low temperature phase without order but with a power law behavior for the correlation functions as for the classical XY model in 2D.

n_s	2	3	4	5	6	7	exact
$(h/J)_c$	1.2768	1.1547	1.1057	1.0797	1.0638	1.0530	1
s	0.805	0.82	0.835	0.845	0.855	0.86	1
ν	1.47	1.31	1.24	1.20	1.18	1.16	1
z	0.55	0.63	0.675	0.705	0.725	0.74	1
β	0.40	0.18	0.185	0.15	0.155	0.145	0.125
η_x	0.55	0.29	0.32	0.29	0.30	0.275	0.25

Table 1 : Location of the transition and exponents obtained for the transverse Ising model in 1D

Lattice	Triangular		Hexagonal	
Blocks	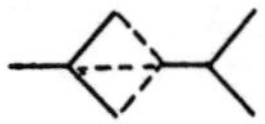	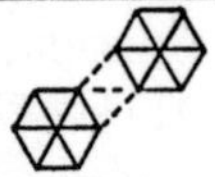		3D Ising model [8]
$z'-1$	5	5	2	
$(h/J)_c$	4.1177	4.7578	1.9685	
s	0.50	0.53	0.64	$0.625<s<0.64$
z	0.32	0.54	0.73	1
β	0.78	0.50	0.29	$0.31<\beta<0.33$
η_x	0.809	0.845	0.617	$1.016<\eta_x<1.030$

Table 2 : Location of the transition and exponents obtained for the ferromagnetic transverse Ising model on triangular and hexagonal lattice in 2D

A new interesting feature of the method when applied to the Ising model on a triangular lattice is that for h = 0 the method can be slightly modified to give the main features of the exact solution. In this case we recover the right ground state energy per site -J and we can construct a ground state with a finite entropy (lower than the exact value) which contains a subset of components of the exact solution having the scaling properties imposed by the renormalization group transformation.

So this simple minded method appears to be powerfull. We are presently applying it to many other cases. We have studied the isotropic X-Y model in a transverse field in 1D [13] and we are now extending this study to higher dimensionalities [14]. By including an imaginary longitudinal field in the transverse Ising model in 1D we can study the Yang-Lee edge singularity of the classical 2D Ising model [15]. We are studying also fermion systems [16] and we expect to study disordered quantum systems.

REFERENCES

+On leave from the Institute of Theoretical Physics, Freie Universität, Berlin, West Germany ; partly supported by the Deutsche Forschungsgemeinschaft and C.I.E.S.

*On leave from the Institute of Physics of the University, Zagreb Yougoslavia.

1. S.D.Drell, M.Weinstein and S.Yankielowicz, Phys.Rev. D 14, 487 (1976)
2. K.G.Wilson, Rev.Mod.Phys. 47, 773 (1975)
3. R.Jullien, P.Pfeuty, J.N.Fields and S.Doniach, Phys.Rev. B 18, 3568 (1978)
4. P.Feuty, Ann.Phys. 57, 79 (1970)
5. M.Suzuki, Prog.Theor.Phys. 46, 1337 (1971)
6. R.Jullien, J.N.Fields and S.Doniach, Phys.Rev. B 16, 4889 (1977)
7. K.A.Penson, R.Jullien and P.Pfeuty, to appear in Phys.Rev. B (1 May 1979)
8. J.C.Le Guillou and J.Zinn-Justin, Phys.Rev.Lett. 39, 95 (1977)
9. G.H.Wannier, Phys.Rev. 79, 357 (1950)
10. K.A.Penson, R.Jullien and P.Pfeuty, to appear in J. of Phys. C
11. J.Stephenson, J. of Math.Physics 11, 420 (1970)
12. M.Suzuki, Prog.Theor.Phys. 56, 1454 (1976)
13. R.Jullien and P.Pfeuty, to appear in Phys.Rev. B (1 May 1979)
14. K.A.Penson, R.Jullien and P.Pfeuty, to be published
15. K.Uzelač, P.Pfeuty and R.Jullien, to be published
16. R.Jullien, P.Pfeuty, J.N.Fields and S.Doniach, proceeding of the conference on rare earths, St Pierre de Chartreuse (1978), to be published in Le Journal de Physique (France) April 1979.

UPPER MARGINAL DIMENSIONALITY. CONCEPT AND EXPERIMENT

Jens Als-Nielsen

Risø National Laboratory
DK-4000 Roskilde, Denmark

Ib Laursen

Lab. Electrophysics
Technical University
DK-2800 Lyngby, Denmark

1. Phenomenological Description[1)]

A broad class of second-order phase transitions have the same qualitative features. In order to describe these it is convenient to use the nomenclature of one particular type of system, and we shall chose the language of magnetic systems. It is emphasized that simply by changing a few words the following description may be adapted to describe any second-order phase transition; for example, in the case of the gas-liquid transition around the critical point insert "density" for "magnetization", "pressure" for "magnetic field", and "compressibility" for "susceptibility".

The origin of magnetization is the atomic spin S. To be more specific, the magnetization M_r around r is the thermal average of S_r times $g\mu_B$, that is, $M_r \equiv g\mu_B \langle S_r \rangle$. For simplicity we set $g\mu_B = 1$ in the following. At high temperatures disorder prevails and the spatial correlation between the spins is only of short range. As the temperature is lowered the size of correlated regions of the spins grows and grows, and eventually at the critical temperature spontaneous ordering sets in - that is, $M \neq 0$.

Let H be the field conjugate to the order parameter M. For a ferromagnet this is just an ordinary uniform field, for an antiferromagnet it is the staggered field corresponding to M being the staggered magnetization. The response to the field is linear at small fields.

$$M = M^{H=0} + \chi H. \quad (1)$$

There is an important relation between the susceptibility χ and the spin-spin correlation function $\langle S_0 S_r \rangle$. The latter is decomposed into two parts describing short-range correlations and long-range order; that is, by definition

$$\langle S_0 S_r \rangle \equiv g(\vec{r}) + \langle S \rangle^2 = g(\vec{r}) + M^2. \quad (2)$$

One can show that χ and $g(\vec{r})$ are related by

$$\chi = \sum_{\vec{r}} g(\vec{r}). \quad (3)$$

This relation expresses the fact that one obtains a large response to a small field if the system is "cooperative", that is, when $g(\vec{r})$ is of long range. It is useful to generalize Eqs. (1) and (3) in the following manner[3]. Suppose that the field varies sinusoidally in space with a wavevector $\vec{q}$, $H_r = H_q \exp(i\vec{q}\cdot\vec{r})$. The response M_r will then also vary sinusoidally with the same wavevector, $M_r = M_q \exp(i\vec{q}\cdot\vec{r})$ with M_q and H_q being related by the wavevector-dependent susceptibility χ_q:

$$M_q = M_q^{H=0} + \chi_q H_q. \quad (1')$$

It should be noted that by definition the largest response occurs for q = 0 and $M_{q\neq 0}^{H=0} = 0$. The generalization of Eq. (3) is then

$$\chi_q = \sum_{\vec{r}} g(\vec{r}) \exp(i\vec{q}\cdot\vec{r}). \quad (3')$$

The peak of χ_q around q = 0 has a half-width, ξ^{-1}, and from Eq. (3') we identify ξ as the correlation range of the short-range order. Often it is easiest to derive χ_q theoretically and then to find $g(\vec{r})$ by Fourier inversion, but more importantly χ_q is the quantity which is directly measured in a scattering experiment when the radiation couples to the order parameter. Why? Because the phase difference between the waves scattered from an element around 0 and and element around $\vec{r}$ is $\exp(i\vec{q}\cdot\vec{r})$ with $\vec{q} = \vec{k}_i - \vec{k}_f$, the difference between incident and scattered wavevectors. The wave amplitude from element 0 is proportional to M_0, whereas the wave amplitude from element $\vec{r}$ is proportional to M_r, so the scattering cross section is simply proportional to χ_q when $q \neq 0$. At $\vec{q} = 0$ the scattering is a superposition of the $\chi_{q=0}$ and Bragg scattering originating from the $M^{H=0}$ term in Eq. (1); the latter is proportional to M^2. We have summarized this discussion in Fig. 1, showing the wavevector- and temperature-dependent susceptibility diverging as $T \to T_c$, $q \to 0$ and the onset of spontaneous long-range order. The behaviour near T_c is often given by power laws as indicated in Fig. 1 with t being the reduced temperature $t = |T-T_c|/T_c$.

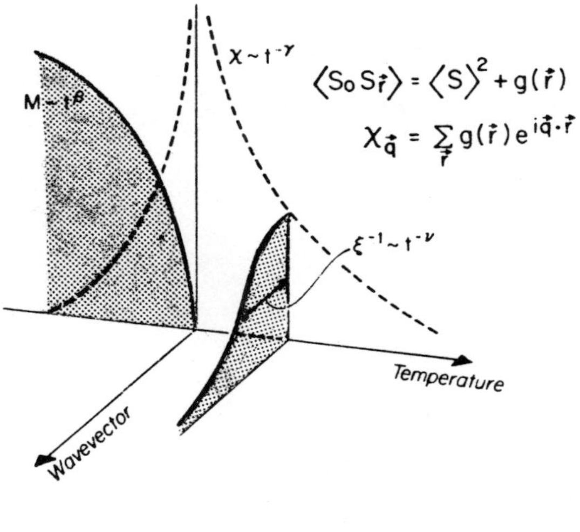

Fig. 1. Order parameter M as well as the wavevector-dependent susceptibility χ_q are directly determined in a scattering experiment. The width of χ_q is the inverse correlation range.

2. Mean Field Theory

We shall now present by far the simplest approximate calculation of χ_q, that is, the mean field theory[4]. Let us first assume that the spins of the system do not interact at all. In that case there are no correlations and the response function, denoted χ^0, does not depend on q. In an applied conjugate field H_i^{app} the order parameter at position $\vec{r}_i$ would be $M_i = \chi^0 H_i^{app}$. For simplicity we now consider temperatures above T_c.

We assume, and this is the essential approximation, that the interaction between S_i and its neighbors is equivalent to a molecular field, $M_i = \chi^0\{H_i^{app}+H_i^{mf}\}$, and that the molecular field H_i^{mf} is of the form $H_i^{mf} = \sum_j f(\vec{r}_{ij})\langle S_j\rangle$. That is, only the average interaction of the j neighbors is taken into account; fluctuations are neglected. Let H_i^{app} be sinusoidal in space, that is, $H_i^{app} = H_q \exp(i\vec{q}\cdot\vec{r}_i)$, and correspondingly $M_i = M_q \exp(i\vec{q}\cdot\vec{r}_i)$. Recalling that $\langle S_j\rangle \equiv M_j$, we find above T_c

$$M_q \exp(i\vec{q}\cdot\vec{r}_i) = \chi^0[H_q \exp(i\vec{q}\cdot\vec{r}_i) + \sum_j f(\vec{r}_{ij}) M_q \exp(i\vec{q}\cdot\vec{r}_j)] \qquad (4)$$

and thereby

$$\chi_q = \chi^0[1 - \alpha_q \chi^0]^{-1} \qquad (5)$$

with

$$\alpha_q = \sum_{\vec{r}} f(\vec{r}) \exp(i\vec{q}\cdot\vec{r}). \qquad (6)$$

For the case of a simple ferromagnet α_q has its maximum for q = 0 and expansion around q = 0 yields

$$\alpha_q = \alpha_0[1 - a^2(\vec{q})], \quad \text{with } a^2(0) = 0. \qquad (7)$$

From Eq. (5) we find the critical point by noting $\chi_{q=0} \to \infty$ when $\alpha_0\chi^0 \to 1$. For a degenerate ground state, χ_0 varies as 1/T and thence one obtains $\alpha_0\chi^0 = T_c/T$; thus Eq. (5) can be rewritten in the form

$$\chi_q \sim [t + a^2(\vec{q})]^{-1}, \quad t = (T-T_c)/T_c. \qquad (8)$$

If the interaction function $f(\vec{r})$ is isotropic, the long-wavelength form of $a^2(\vec{q})$ becomes particularly simple, $a^2(\vec{q}) = \xi_0^2 q^2$, and in that case χ_q has a Lorentzian line shape:

$$\chi_q \sim [\xi^{-2} + q^2]^{-1}, \quad \xi = \xi_0 t^{-1/2}. \tag{9}$$

From Eqs. (8) and (9) we find the mean field values of the critical exponents: $\gamma = 1$ and $\nu = \frac{1}{2}$. A similar analysis below T_c yields the primed exponents[5] $\gamma' = \gamma = 1$, $\nu' = \nu = \frac{1}{2}$.

3. Ginzburg Criterion

It is clear that the essential approximation above is the neglect of fluctuations in the molecular field acting on a given site. On the other hand, the fluctuations become more and more pronounced as the temperature approaches T_c. In order to obtain a self-consistent picture, Ginzburg stated that below T_c[6] the fluctuations of M averaged over a suitable region Ω (to be specified below) must be small compared to the value of M itself, that is,

$$(\delta M)_\Omega^2 \ll M_\Omega^2 .$$

Let us assume that we have divided Ω into N identical lattice cells. One then has for the mean square fluctuation amplitude

$$\begin{aligned}(\delta M)_\Omega^2 &= \langle [\sum_\Omega (S_i - \langle S\rangle)]^2 \rangle \\ &= \langle \sum_{\substack{i\in\Omega \\ j\in\Omega}} (S_i - \langle S\rangle)(S_j - \langle S\rangle)\rangle \\ &= N \sum_\Omega [\langle S_0 S_i\rangle - \langle S\rangle^2].\end{aligned}$$

If the sum had been over the entire crystal, then by Eq. (3) it would simply be equal to $\chi_{q=0}$.

In order to assess the importance of fluctuations in the mean field theory, the region Ω must be chosen appropriately. On physical grounds it is clear that fluctuations are important over linear dimensions of the order of ξ, the correlation range. Indeed, it is an essential feature of our present understanding of phase transitions that ξ is the only length in the problem. Therefore, we take $\Omega = \Omega_\xi$, the region of correlated spins[7]. As $T \to T_c$, $\Omega_\xi \to \infty$, but the sum becomes a constant fraction, F, of χ, that is

$$(\delta M^2)_{\Omega_\xi} = FN(\Omega_\xi)\chi_{q=0}(t), \tag{11}$$

It is easily shown that F is independent of temperature, see e.g. Ref. 1.

For the mean square order parameter one has

$$M^2_{\Omega_\xi} = <\sum_{\Omega_\xi} S_i>^2 = N^2<S>^2 = N^2M^2.$$

Thus the Ginzburg criterion may be rewritten as

$$F\chi(t) << N(\Omega_\xi)M^2(t). \tag{12}$$

The number of spins $N(\Omega_\xi)$ within a correlated region Ω_ξ is, of course, proportional to the volume of Ω_ξ.

If the interaction $f(\vec{r})$ is of short range, for example, between nearest neighbors only, the Fourier transform of $f(\vec{r})$ will, in the long-wavelength limit, be of the form $\alpha_q \simeq \alpha_0(1-\xi_0^2 q^{*2})$, where q^{*2} is of the form $q^{*2} \equiv \sum_{i=1}^d p_i q_i{}^2$. The region of critical fluctuations in q space therefore scales as ξ^{-1} in all d directions, and by Fourier inversion it follows that $\Omega_\xi = \xi^d$.

We have already found that the mean field critical exponents in this case are $\beta = \frac{1}{2}$, $\gamma = 1$, $\nu' = \frac{1}{2}$, and Eq. (12) gives $d^* = 4$. Consequently, the mean field theory is not self-consistent for, say, the three-dimensional Ising model, and indeed one finds experimentally (β-brass[8]) as well as theoretically, using series expansion techniques[9], that $\gamma = 1.25$, $\nu = 0.64$, and $\beta = 0.30$. When d is even further away from d^*, for example, $d = 2$, we would expect even larger deviations from the mean field exponents. Indeed, experiments on the 2d antiferromagnet K_2CoF_4[10], which is a model system of the 2d Ising model, as well as Onsager's exact solution of the 2d Ising model[2], give exponents very far from mean field behavior: $\gamma=7/4$, $\nu=1$, $\beta=1/8$.

Let us now consider an Ising ferromagnet (ferroelectric) where the magnetic (electric) dipole moments are only coupled by the dipolar interaction, that is, the interaction between two spins pointing in the z direction and situated at the origin and at $\vec{r} = (x,y,z)$ is $f(\vec{r}) = (3z^2-r^2)/r^5$. In the long-wavelength limit the Fourier transform of $f(\vec{r})$ is of the form

$$\alpha_q = \alpha_0[1 - a_1q^2 - a_2(q_z/q)^2 + a_3q_z{}^2]. \tag{13}$$

The coefficients a_1, a_2 and a_3 are readily calculated when the lattice and the size of the magnetic moments are specified. As an example we consider in the following section the material $LiTbF_4$. This form of α_q is very peculiar since the limiting value of α_q when $q \to 0$ depends on the direction of $\vec{q}$: if $q \to 0$ along the x axis $\alpha_q \to \alpha_0$, but if $\vec{q} \to 0$ along the z axis the limiting value is $\alpha_0(1 - a_2)$. Any limiting value between these two extremes is obtained by choosing the appropriate direction of $\vec{q}$.

As χ_q is of the form

$$\chi_q{}^{-1} \sim 1 + [(\xi q)^2 + g(q_z\xi^2/q\xi)^2 - (b_3/\xi)^2(q_z\xi^2)^2] \tag{14}$$

it is evident that q_z scales as ξ^{-2}, whereas modulus q scales as ξ^{-1}. This is most clearly demonstrated pictorially[11] in Fig. 2. Here the half-contour of χ_q defined by the surface in q space where $\chi_q = \frac{1}{2}\chi_{q_x\to 0, q_y=q_z=0}$, is sketched in the q_x-q_z plane. The entire surface is generated by rotation of this contour around the z axis. The contour intersects the q_x axis at ξ^{-1}, and if we omit the term $b_3{}^2/\xi^2$, which becomes negligible near T_c, the contour starts out from the origin with a slope of $g^{-\frac{1}{2}}\xi^{-1}$. The maximum dimension along the q_z axis is $g^{-\frac{1}{2}}\xi^{-2}$, so the contour is therefore not only shrinking as $T \to T_c$, due to the fact that $\xi \to \infty$, but it is also changing shape in becoming more and more confined to the x-y plane. We infer immediately that the correlation range along the z axis, ξ, is superdiverging, that is $\xi_{\parallel} \sim \xi^2$. The correlated regions are long rods along the Ising axis of length $\xi_{\parallel}$ and with a diameter of ξ. We thence have $\Omega_\xi = \xi\cdot\xi\cdot\xi_{\parallel} \sim \xi^4$, and the marginal dimensionality is $d^* \doteq 3$. The three-dimensional ferro-

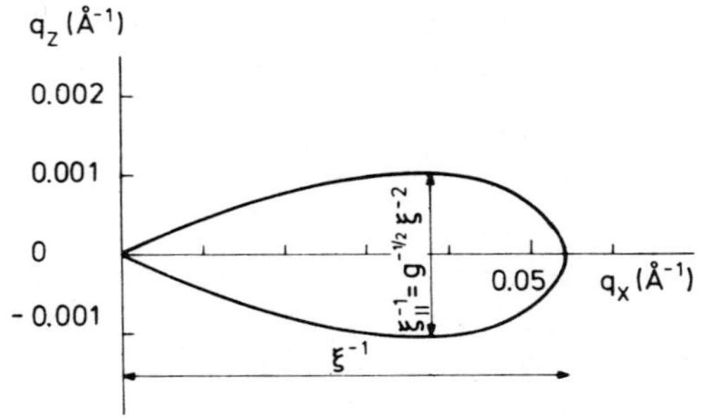

Fig. 2. Half-contour of $\chi(\vec{q},T)$ for the dipolar-coupled, uniaxial ferromagnet. The contour gives the points in the q_x-q_z plane for which $\chi(q_x,0,q_z,T) = (\frac{1}{2})\lim_{q_x\to 0}\chi(q_x,0,0,T)$. Note the scale difference on the q_z and q_x axes.

magnet or ferroelectric exhibits critical behavior of marginal dimensionality, and we expect "almost" mean field behavior. The first theoretical treatment of the critical behavior for this system being at marginal dimensionality was given by Larkin and Khmel'nitskii in 1969 using perturbation expansion techniques[12]. It is also possible to calculate the exact critical behavior in this case using renormalization group theory[13], essentially because RG uses mean field theory as the "basis", and for marginal dimensionality this "basis" is already quite close to the correct solution. The results are summarized below using the notation $t = (T-T_c)/T_c$, subscripts + and - referring to above and below T_c and v_o as volume per spin:

Table 1

Critical behavior of the dipolar coupled Ising ferromagnet.

Quantity	Units of	Critical behavior
Susceptibility	$\chi^{o}(T_c)=(g\mu_B)^2S^2/kT_c$	$\chi_{\vec{q}=0}(t,0) = \Gamma_{\pm}t^{-1}\lvert \ell n t\rvert^{1/3}$
Specific heat	k_B/v_o	$C(t,0)=A_{\pm}\lvert \ell n t\rvert^{1/3}$
Zero-field magnetization	$g\mu_B S/v_o$	$m(-t,0) = B(-t)^{\frac{1}{2}}\lvert \ell n(-t)\rvert^{1/3}$
Critical isotherm magnetization	same	$m(0,h) = Dh^{1/3}\lvert \ell n h\rvert^{1/3}$
Field vs. magn. (equation of state)	$kT_c/g\mu_B S$	$h=(t/r)\tilde{m}+(\Gamma B^2)^{-1}\tilde{m}^3$ with $\tilde{m}=m(h,t)/[\ell n\chi(h,t)/\chi^{o}]^{1/3}$

4. Experiments on $LiTbF_4$

The magnetic Tb ions ($\mu = 9\mu_B$) in the compound $LiTbF_4$ form a lattice of tetragonal structure as shown in Fig. 3. Measurements of the magnetic susceptibility[14] for fields along and perpendicular to the tetragonal axis show that crystalline field is very strong and confines the spins to be along the tetragonal axis. The crystal field splitting of the J = 6 level can also be deduced from analysis of the longitudinal susceptibility and has further been confirmed by inelastic neutron scattering[15].

It turns out that the level diagram consists of an almost doubly degenerate ground state with the first excited state about 200 K above the ground state. The crystal becomes a ferromagnet at a temperature T_c = 2.87 K. The Curie temperature θ calculated by summing all dipolar interactions, cf. the relation $\chi/\chi^{o} = \theta/T$ from the previous section, becomes θ = 3.97 K. The most detailed experimental proof that $LiTbF_4$ represents a

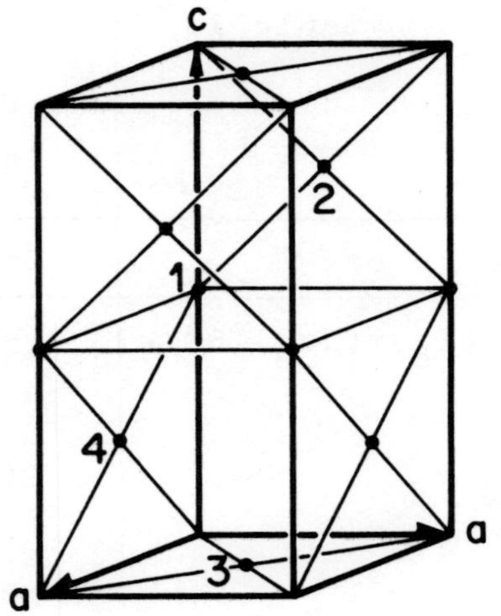

Fig. 3. The magnetic structure of $LiTbF_4$. Only the positions of the magnetic Tb ions are shown. Lattice parameters are a = 5.181 Å, c = 10.873 Å and the magnetic moment is 8.9 μ_B per Tb ion.

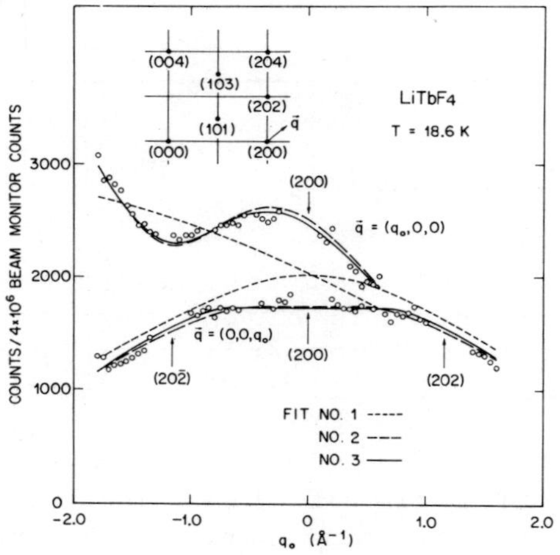

Fig. 4. Longitudinal (upper) and transverse (lower) scans through the lattice point (2,0,0) at T = 6.5 T_c. Nuclear Bragg peaks have been omitted. The dashed line gives the wavevector dependence of the squared formfactor and the polarization factor $1-(Q_z/Q)^2$ in the magnetic neutron scattering cross section. The dot-dashed curves would result if the dipolar force was the only coupling in $LiTbF_4$. Including exchange interactions gives only a marginal improvement of the fit (full lines).

model example of the dipolar-coupled Ising ferromagnet is obtained by measuring the wavevector-dependent susceptibility over several Brillouin zones at a temperature well above the Curie temperature and compare the results with those derived from mean field theory[16]. This comparison is shown in Fig. 4.

In Fig. 5 we show the specific heat of β-brass[17], the model example of the [3d] Ising system with short range interactions ($d^* = 4$) compared to that of $LiTbF_4$[18] and to the mean field theory. Clearly the specific heat resembles the mean field behavior much more than β-brass. The finite specific heat above T_c and the cusp below T_c turns out to be very accurately described by the R.G. relations as given in table 1 - the leading singularity is the logarithmic correction to mean field behavior. Recently the equation of state has been measured very accurately[19] and here again the results are in accordance with R.G. theory and the perturbation expansion theory of Larkin and Khmel'nitzkii. The results are summarized in Fig. 6.

The basic reason for the Ising, dipolar-coupled ferromagnet being a system at upper marginal dimensionality is the peculiar wavevector-dependent susceptibility, and this can be measured directly by critical magnetic neutron scattering. Fig. 7 shows the intensity versus wavevector for scans along the Ising axis at $T = T_c$. From eq. (14) we see that for $\xi \to \infty$ the width of q_z scans at fixed q_x is $g^{-\frac{1}{2}} q_x^{\,2}$ and the data of Fig. 7 confirm indeed this simple behavior. A best fit value of g is g = 1.3±0.1 Å^{-2}, remarkably close to the unrenormalized value of a_2/a_1 = 1.56 Å^{-2} from eq. (13). A typical mean field result is the ratio of 2 between the amplitudes, Γ_+/Γ_-, of the susceptibility above and below T_c. This result is aintained <u>at</u> marginal dimensionality. Fig. 8 shows the temperature dependence of χ_q for fixed values of q as we pass through the critical temperature T_c. The full lines through the data points above T_c are imaged with a scale factor of two to the full lines below T_c and the agreement with the data below is striking. In contrast similar data for β-brass give an asymmetry factor around 5 (lower part of Fig. 8).

The temperature dependence of the correlation range ξ is obtained by q_x-scans at q_z = 0. Results at different temperatures are shown in Fig. 9. The full lines

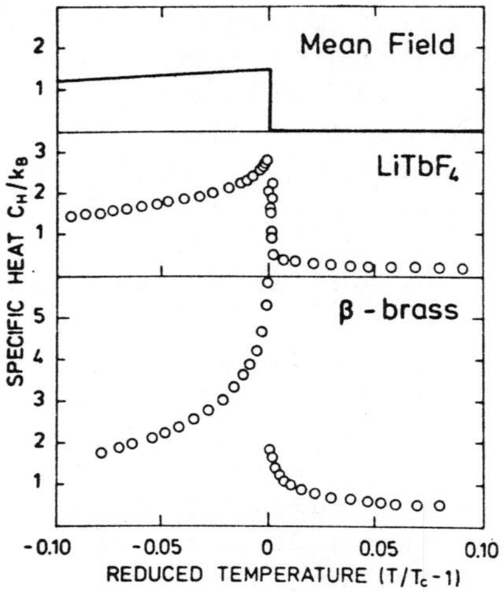

Fig. 5. The specific heat of β-brass the model example of the 3d Ising system, diverges at T_c due to critical fluctuations, whereas the mean field theory predicts a discontinuity. In the dipolar coupled Ising ferromagnet $LiTbF_4$ the specific heat divergence is very weak indicating that the fluctuations are marginal rather than critical.

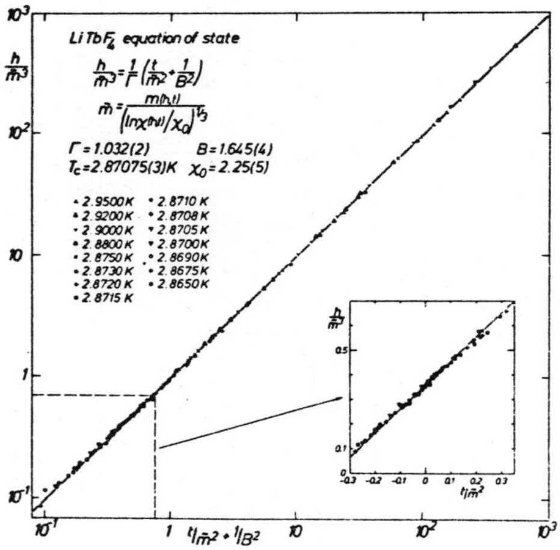

Fig. 6. Scaled representation of magnetization versus field and temperature agree with the logarithmic corrections predicted by Larkin and Khmel'nitskii and R.G. theory. (From ref. 19).

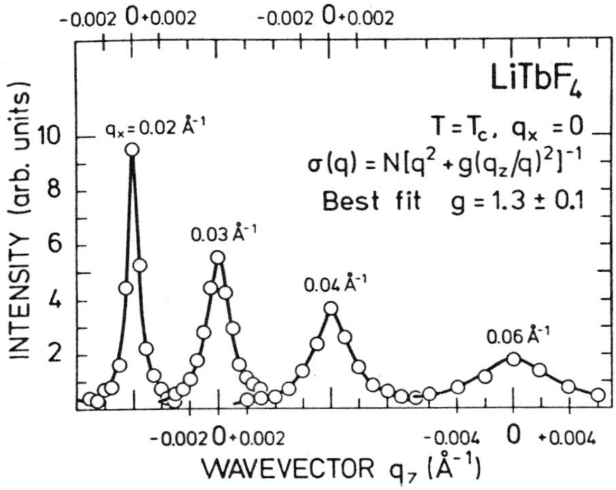

Fig. 7. Scans along the Ising axis at different values of q_x at $T \simeq T_c$ show the peculiar wavevector dependence of $\chi_{\vec{q}}$ for the dipolar coupled Ising ferromagnet. Full lines represent a best fit of the cross section folded with the experimental resolution. Two parameters, g and the over-all scale factor, describe this family of curves.

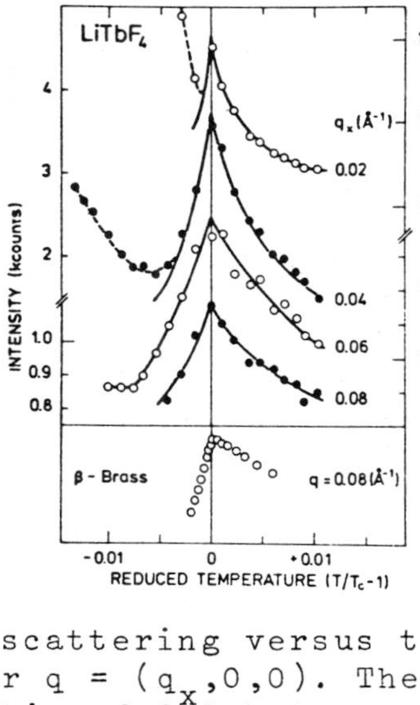

Fig. 8. Critical scattering versus temperature at fixed wavevector $q = (q_x, 0, 0)$. The asymmetric shape with a ratio of 2:1 between equicritical scattering above and below T_c is in accordance with mean-field and RG theory. The rise of intensity at low temperatures (dashed curve) is due to domain scattering. Left scales refer to filled signatures; right scales to open signatures.

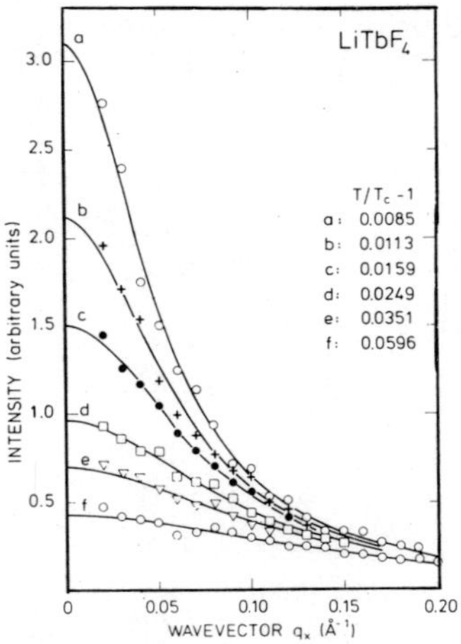

Fig. 9. q_x-scans at $q_z = 0$ give a Lorentzian cross section of width ξ^{-1}. Full lines represent the cross section folded with the experimental resolution.

are best fit values of the cross section, essentially given by eq. (14), folded with the experimental resolution function. In the fit only ξ is fitted since we know the value of g from Fig. 7 and the term $(b/\xi)^2$ in eq. (14) is negligible. The resulting values of the squared correlation range vs. temperature[11] are given in Fig. 10.

The striking feature of this figure is the line marked RG threading through the data points in the critical region. This line represents the prediction of renormalization group theory <u>without any adjustable</u> parameters[20]. Aharony and Halperin[20] have shown that the RG equations imply the following relation between the correlation range ξ and the specific heat C:

$$\xi^2 \xi_{\parallel} t^2 C = \frac{3}{32\pi} \ln(t/t_o) \tag{15}$$

Here $\xi_{\parallel}$ is the superdiverging correlation length $g^{\frac{1}{2}}\xi^2$, cf. Fig. 2. The specific heat data[18] including the correction term t_o inserted in eq. (15) give the line marked RG in Fig. 10. The equation of state data of Fig. 6 can also be used to predict $\xi^2(T)$ and the result[19] agrees accurately with the experimental data.

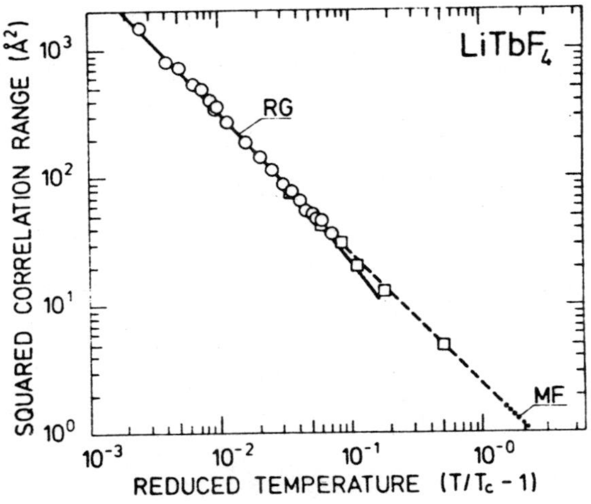

Fig. 10. The square of the correlation range ξ^2 versus the reduced temperature $t = (T/T_c - 1)$. The full line RG represents the prediction of renormalization-group theory.

5. Conclusion

We conclude that the critical behavior of the dipolar-coupled Ising ferromagnet is now understood in minute detail. On the theoretical side the perturbation theory of Larkin and Khmelit'zkii agree with the renormalization group approach, which in this case does not require any further approximations as the system is at upper marginal dimensionality. On the experimental side it has been possible by neutron scattering to confirm the expectations from susceptibility measurements that $LiTbF_4$ is a model example of the dipolar-coupled Ising ferromagnet. Detailed and accurate studies of the specific heat, spontaneous magnetization and equation of state as well as the correlation range by neutron scattering in the critical region have revealed a striking agreement with the theoretical results.

References

1. The following 3 sections follow the exposition given in J. Als-Nielsen and R.J. Birgeneau, Am. J. Phys. 45, 554 (1977).

2. H.E. Stanley, Introduction to Phase Transition and Critical Phenomena (Oxford U.P., London, 1971).

3. W. Marshall and R.D. Lowde, Rep. Prog. Phys. 31, 705 (1968).

4. J.S. Smart, Effective Field Theories of Magnetism (Saunders, Philadelphia, 1966).

5. Below the critical temperature the critical indices for susceptibility and correlation range are conventionally denoted γ' and ν'.

6. V.L. Ginzburg, Sov. Phys. Solid State 2, 1824 (1960).

7. R. Bausch, Z. Phys. 254, 81 (1972).

8. J. Als-Nielsen and O.W. Dietrich, Phys. Rev. 153, 706 (1967); 153, 711 (1967); 153, 717 (1967); J. Als-Nielsen, ibid. 185, 664 (1969); J.C. Norwell and J. Als-Nielsen, Phys. Rev. B2, 277 (1970); O. Rathmann and J. Als-Nielsen, ibid. 9, 3924 (1974).

9. M.E. Fisher and R.J. Burford, Phys. Rev. 156, 583 (1967).

10. E.J. Samuelsen, Phys. Rev. Lett. 31, 936 (1973), H. Ikeda, I. Hatta, A. Ikushima, and K. Hirakawa, J. Phys. Soc. Jpn. 39, 827 (1975).

11. J. Als-Nielsen, Phys. Rev. Lett. 37, 1161 (1976).

12. A.I. Larkin and D.E. Khemel'nitzkii, Zh. Eksp. Teor. Fiz. 56, 2087 (1969) [Sov. Phys. JETP 29, 1123 (1969)].

13. A. Aharony, Phys. Rev. B8, 3363 (1973), 9, 3946(E) (1974).

14. L.M. Holmes, T. Johansson, and H.J. Guggenheim, Solid State Commun. 12, 993 (1973).

15. L.M. Holmes, H.J. Guggenheim, and J. Als-Nielsen, in Proceedings of the Ninth International Conference on Magnetism, Moscow, U.S.S.R., 1973 (Nauka, Moscow, U.S.S.R., 1974), Vol. VI, p. 256.

16. J. Als-Nielsen, L.M. Holmes, and H.J. Guggenheim, Phys. Rev. Lett. 32, 610 (1974); L.M. Holmes, J. Als-Nielsen, and H.J. Guggenheim, Phys. Rev. B12, 180 (1975).

17. The specific heat data of Ashman and Handler, Phys. Rev. Lett. 23, 642 (1969) were analyzed by Baker and Essam, J. Chem. Phys. 55, 861 (1971) and the data in Fig. 3 is taken from the latter reference.

18. G. Ahlers, A. Kornblitt, and H.J. Guggenheim, Phys. Rev. Lett. 34, 1227 (1975).

19. R. Frowein, J. Kötzler and W. Assmus, Phys. Rev. Lett. 42, 739 (1979).

20. A. Aharony and B.I. Halperin, Phys. Rev. Lett. 35, 1308 (1975).

LOWER MARGINAL DIMENSIONALITY.
X-RAY SCATTERING FROM THE SMECTIC-A PHASE OF LIQUID CRYSTALS.

Jens Als-Nielsen
Risø National Laboratory
DK-4000 Roskilde, Denmark

J.D. Litster, R.J. Birgeneau, M. Kaplan and C.R. Safinya
Dept. of Physics
Massachusetts Institute of Technology
Cambridge, Mass. 02139, U.S.A.

1. Introduction

In the previous chapter we have seen that the critical fluctuations associating a second order phase transition, expressed in terms of a response function $\chi(\vec{q})$ directly measurable by scattering spectroscopy, may be treated in a self-consistent and correct way by simple mean field theory if the dimensionality exceeds a certain marginal dimensionality d*. In particular we stressed that d* depends on the volume in reciprocal space available to the critical fluctuations, and systems with $d \leq d^*$ can indeed be studied experimentally. In this chapter we shall discuss the opposite limit: That the fluctuations are so strong that they prevent the onset of true long range order at all. Let us look upon an example. We postulate an ordered crystalline state in d dimensions and we want to calculate the mean squared fluctuations $\langle u^2 \rangle$. If $\langle u^2 \rangle$ diverges in an infinite sample the postulated long range order cannot take place. First, consider a sinusoidal fluctuation with wavevector $\vec{q}$ and amplitude $u_{\vec{q}}$. For simplicity let us assume that the medium is elastically isotropic. The free energy density is then of the form $f_q = \frac{1}{2}Bq^2u_q^2$. Note that the dimension of the stiffness constant B is energy per $(\text{length})^d$. Let the system be enclosed in a box of linear dimension L. The total free energy is $F = L^d \Sigma_q f_q$, and by equipartion we find for the thermal average value of u_q^2:

$$\langle u_q^2 \rangle = kT/(L^d B q^2) \tag{1}$$

and

$$\langle u^2 \rangle = \sum_q \langle u_q^2 \rangle = \left(\frac{L}{2\pi}\right)^d \frac{kT}{L^d B} \int_{q_{min}}^{q_{max}} q^{-2} d^d q \tag{2}$$

The minimum wavevector $q_{min} = 2\pi/L$ whereas the maximum wavevector $q_{max} = 2\pi/a$ where a is the interatomic distance. The integral depends on the dimensionality d:

d:	1	2	3
Integral:	$L/2\pi$	$2\pi \ln(L/a)$	$8\pi^2/a$

$\langle u^2 \rangle$ diverges with the sample size for $d < 2$. At $d = 2$ it is a weak logarithmic divergence and we have in this case a lower marginal dimensionality $d^+ = 2$.

In the following sections we shall see that a phase, the smectic A phase, occurring in some liquid crystal materials has $d^+ = 3$, and we shall describe a high resolution X-ray diffraction experiment designed to study the long range, but not infinitely long range, correlations in this phase.

2. The Nematic and Smectic A Phases of Liquid Crystals[1)]

The materials called "liquid crystals" consist of long rod-like molecules, typically with two benzene rings in the centre and a hydro-carbon tail sticking out in each end. We shall consider the molecules as being rotational symmetric around the long axis and also as being symmetric about a plane perpendicular to the axis through the centre of the molecule. In a continuum description of the liquid crystal the average orientation of the molecules is given by the vector function $\hat{n}(\vec{r})$. The symmetry implies that $\hat{n}(\vec{r})$ and $-\hat{n}(\vec{r})$ describe the same physical state. In the isotropic liquid phase there is no preferred molecular orientation, but as the temperature is lowered a phase appears with a preferred molecular orientation - the nematic phase. In applying a magnetic or electric field the molecular response depends on whether the field is along or perpendicular to the molecular axis. The anisotropic susceptibility can be used to create a nematic single-domain by applying a moderate magnetic field. The anisotropy of the dielectric constant implies that the thermal fluctuations of the director field $\hat{n}(\vec{r})$ scatter visible light very strongly so the nematic phase appears turbid. In further lowering the temperature some liquid crystals exhibit a new liquid phase, the smectic A phase. Here the molecules are well aligned as in the nematic phase, but in addition the molecular centres start to correlate along the molecular axis. This is seen very clearly by X-ray diffraction: By shining monochromatic X-rays on the smectic A phase a strong diffraction spot appears at a wavevector transfer $\vec{k}_{inc} - \vec{k}_{diff} = q_0\hat{n}$. However, no diffraction spots appear when the direction of $(\vec{k}_{inc} - \vec{k}_{diff})$ differs from the molecular axis, and one con-

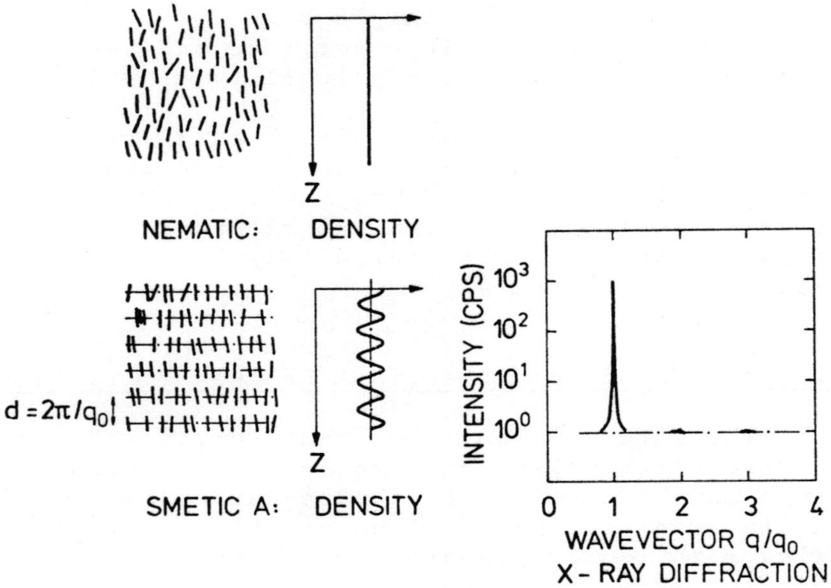

Fig. 1. In the nematic phase (upper left) the molecules are aligned but the density is structureless. In the smectic A phase the density forms a wave with wavevector along the molecular axis. The sinusoidal shape of the density wave is reflected in the diffraction pattern by the absence of higher order reflections.

cludes readily that the structure is liquid-like in directions perpendicular to $\hat{n}$.

Another striking fact is the lack of higher order reflections for scattering vectors along $\hat{n}$. The second order reflection, $\vec{k}_{inc} - \vec{k}_{diff} = 2q_0 \hat{n}$, is typically 10^{-4} times less intense than the first order peak. The smectic A phase must therefore be considered as a one-dimensional sinusoidal density wave, and it is the long range character of this density wave which is of central interest in the present context. In particular we shall see that the thermal fluctuations are strong enough to just prohibit true long range order in an infinite system - the smectic A phase is a system at lower marginal dimensionality.

First we shall calculate the mean squared fluctuation of the spacing between the SmA planes and see that

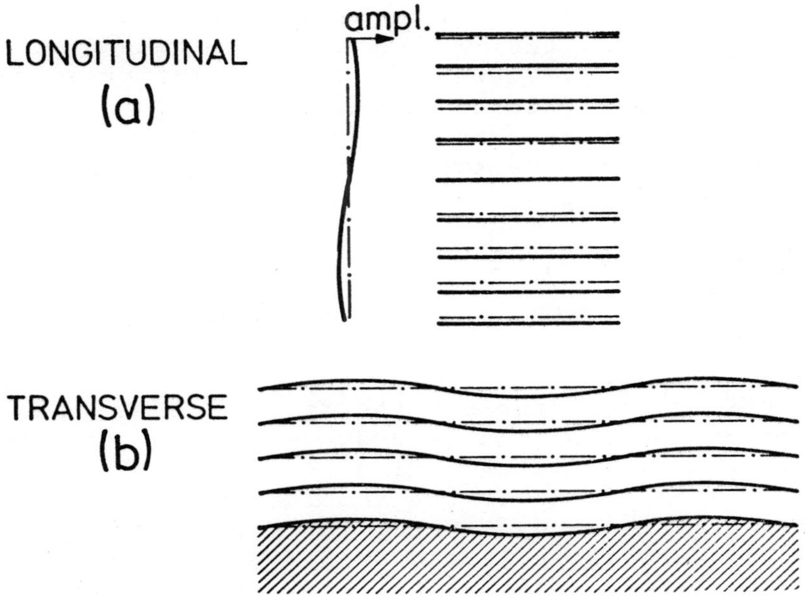

Fig. 2. The two fundamental modes of the smectic A phase. The energy density is proportional to $q_{\parallel}^2$ for the longitudinal mode but to $q_{\perp}^4$ for the transverse mode.

it diverges logarithmically with sample size. To this end we need to set up an expression for the free energy of the distorted SmA phase. In Fig. 2 we consider two fundamental modes with wavevectors $q_{\parallel}$ and $q_{\perp}$, respectively. The energy density associated with the dilation and compression in the longitudinal mode is of the usual elastic form $\frac{1}{2}Bq_{\parallel}^2 u_q^2$. In the transverse mode, the so-called undulating mode, the planar spacing is everywhere at its optimal value and to second order in $q_{\perp}$ this mode costs no elastic energy to set up.

However, in this mode the director field $\hat{n}(\vec{r})$ varies in space with a corresponding cost of energy. For simplicity we assume that the molecules are rigidly perpendicular to the planes. The director fluctuation energy density must in general depend on the derivatives of the vector field, i.e. on $(\nabla\cdot\hat{n})$ and $(\nabla\times\hat{n})$. It is readily seen that the line integral for the undulator mode along a closed path c, $\int_c \hat{n}\cdot d\vec{s}$, is zero, so $\nabla\times\hat{n}$ vanishes. Since $\hat{n}$ and $-\hat{n}$ describe the same state the first non-vanishing term in $\nabla\cdot\hat{n}$ must be $(\nabla\cdot\hat{n})^2$. Consequently the energy density of director fluctuations in the SmA phase is of the form $\frac{1}{2}K(\nabla\cdot\hat{n})^2$. Noting that the amplitude $u(\vec{r})$ and $\hat{n}$ are related by $\hat{n} = -\nabla u$ we find the undulating single-q mode energy density to be of the form $\frac{1}{2}Kq_\perp^4 u_q^2$. The constant K is called the Frank elastic constant for splay mode. By combining the two modes the energy density at fluctuation $\vec{q}$ becomes

$$f_{\vec{q}} = \tfrac{1}{2}(Bq_\parallel^2 + Kq_\perp^4)u_q^2 \qquad (3)$$

We note that the ratio $\lambda = (K/B)^{\frac{1}{2}}$ has the dimension of length, and it has a direct physical significance. Suppose that the undulating lines in Fig. 2 denote the director field $\hat{n}(\vec{r})$ and not the SmA planes and let the cross hatched bottom indicate an undulating boundary condition. As the distance from this boundary is increased the amplitude of the undulation decreases with a characteristic decay length. It is shown in deGennes' book[1] that the decay length or penetration depth is λ. The Frank elastic constant K is the same in the nematic and SmA phases, but the stiffness constant B vanishes as the nematic phase is approached[8].

In the smectic A material which we shall consider later in detail, 8OCB, the N-A transition is of second order, or very nearly so, and from light scattering experiments one finds $\lambda q_o = 1.52\, t^{-0.133}$, t being the reduced temperature $(T_c^o/T - 1)$[9].

In analogy with the derivation of eq. (2) we readily determine $\langle u^2\rangle$:

$$\langle u^2\rangle = \frac{kT}{8\pi^3 B}\, 2\pi \int \frac{dq_\parallel\, q_\perp dq_\perp}{q_\parallel^2 + \lambda^2 q_\perp^4} = \frac{kT}{8\pi^3 B\lambda}(2\pi)^2 \ln\frac{L}{a} \qquad (4)$$

that is $\langle u^2\rangle$ diverges logarithmically with the sample size L and true long range order is not possible. This problem was first discussed by Peierls in 1934[2] and by Landau in 1937[3]. We shall call the peculiar order of the SmA phase for the Landau-Peierls state.

3. The Correlation Function in the Harmonic Approximation

Since true long range order does not exist the diffraction line in Fig. 1 must, at least in principle, differ from the Bragg line profile as it occurs from a 3-dimensional crystal. In this section we shall evaluate the line profile as outlined for the SmA phase by Caillé[4] and for the analogous 2-d crystal problem by Mikeska and Schmidt and by Imry and Gunther[5].

We find it convenient first to give a list of notations to be used in the evaluation:

$\hat{z}$ unit vector along average molecular axis

d average spacing between neighbouring planes

n enumerates different planes

$\vec{R}_i, \vec{\rho}, u_n(\vec{\rho})$ The ith molecular position in plane number n is $\vec{R}_i = \vec{\rho}+\{nd+u_n(\vec{\rho})\}z$. Here $\vec{\rho}\cdot\hat{z} = 0$ and $u_n(\vec{\rho})$ denotes the instantaneous deviation at $\vec{\rho}$ of the n'th plane.

$\vec{q}$ Wavevector transfer in the scattering process. $\vec{q}$ is decomposed into $q_{\parallel}$ along $\hat{z}$ and $q_{\perp}$ perpendicular to $\hat{z}$.

$\langle A\rangle$ Thermal average of the quantity A.

In a scattering process with wavevector transfer $\vec{q}$ the cross section $S(\vec{q})$ is

$$S(\vec{q}) = \langle\sum_i e^{i\vec{q}\cdot(\vec{R}_i-\vec{R}_o)}\rangle$$

$$= \sum_n \int d\vec{\rho}\; e^{iq_{\parallel} nd}\; e^{i\vec{q}_{\perp}\cdot\vec{\rho}} \langle e^{iq_{\parallel}(u_n(\vec{\rho})-u_o(o))}\rangle \qquad (5)$$

Since in general the structure factor can be considered as the Fourier transform of a pair correlation function $G(\vec{r})$, we define $G(\vec{r})$ from eq. (5) by

$$G(\vec{r}) = \langle e^{iq_{\parallel}(u(\vec{r})-u(o))}\rangle \qquad (6)$$

where we now use a continuum description of the fluctuations $u(\vec{r})$ rather than the discrete set $u_n(\vec{\rho})$. The structure factor becomes

$$S(\vec{q}) = \int G(\vec{r})\; e^{i\vec{q}\cdot\vec{r}}\; d\vec{r} \qquad (7)$$

In the <u>harmonic</u> approximation $G(\vec{r})$ becomes

$$G(\vec{r}) \simeq e^{-\frac{1}{2}q^2<[u(\vec{r})-u(o)]^2>} \tag{8}$$

In the appendix we evaluate the average value in detail. Here we insert the results and find

$$G(z,\vec{\rho}=0) \sim \left(\frac{d^2}{\lambda z}\right)^{\eta} e^{-\eta C} \tag{9a}$$

$$G(z=0,\vec{\rho}) \sim \left(\frac{4d^2}{\rho^2}\right)^{\eta} e^{-2\eta C} \tag{9b}$$

$$G(z,\vec{\rho}) \sim e^{-2\eta C} \left(\frac{4d^2}{\rho^2}\right)^{\eta} e^{-\eta E_1\left(\frac{\rho^2}{4\lambda z}\right)} \tag{9c}$$

with

$$\eta \equiv q^2 kT/(8\pi B\lambda) \tag{10}$$

$$C \equiv 0.5772\ldots \text{ (Eulers constant)}$$

$$E_1(-x) \equiv -\int_x^\infty e^{-t}/t \, dt \quad \text{(the exponential integral)}$$

Equation (9a) and (9b) follows from (9c) using the asymptotic expressions of the $E_1(x)$ function.

The exponential factor $e^{-\eta C}$ is the analogue of the conventional Debye-Waller factor since the exponent is proportional to the square of the wavevector transfer and the remaining part of η resembles $<u^2>$ as calculated in section 1 for crystal structures.

The essential result is that the correlation function along the molecular axis decays slowly as $(1/z)^{\eta}$ the exponent η being of the order 0.1-0.2. It means that the diffraction line profile along $\hat{z}$ is of the form $(q_{\parallel}-q_o)^{-2+\eta}$. Similarly, the correlation within the plane decays as $(1/\rho)^{2\eta}$ and the diffraction line profile is of the form $q_{\perp}^{-4+2\eta}$ for $q_{\parallel} = q_o$.

In summarizing, the lack of true long range order is expected to manifest itself in the detailed line profile of the diffraction peak in Fig. 1. The diffraction peak should display power law tails with a characteristic doubling of the exponent as one goes from the longitudinal profile to the transverse profile. Measured exponents can be compared with a calculation in the harmonic approximation giving η in terms of the penetration depth λ, measurable by light scattering, and the stiffness constant B. In particular, η is expected to increase as the temperature approaches the transition to the nematic phase.

4. Experiment and Analysis

In the experiment we investigated the smectic A phase of octyloxycyanobiphenyl, hereafter called 8OCB. A single-domain sample was grown in situ by slow cooling (0.1°C/h) from the nematic phase in a homogeneous magnetic field (4.8 kGauss) and the diffraction of X-rays from the smectic A planes was examined. The resolution must be very good in order to determine whether the scattering cross section reflects true long order (Bragg scattering) or the expected Landau-Peierls behaviour. High angular resolution can be obtained by using perfect crystals together with the essentially monochromatic X-rays from a characteristic line as only X-rays with incidence angle θ fulfilling the Bragg condition $\lambda = 2d\sin\theta$ will pass the "crystal-collimator". In this way angular resolution of the order of 0.1 mrad can be obtained with a decent beam width of approximately 1 mm whereas collimation using a slit or a multislit system (Soller collimator) in practise would be coarser by an order of magnitude. However, in addition to the requirement that the resolution must be narrow it must also be essentially "tail-less" in order to ensure that the tails observed of the SmA line are not due to a resolution effect. In this respect the crystal-collimator using a single Bragg reflection is not very effective because of the thermal diffuse scattering superimposed on the Bragg scattering. The relative weight of these two components depends on the geometrical collimation of the beam incident on the crystal-collimator. In addition, the line profile of Bragg scattering from an ideal perfect crystal has got $1/q^2$-like tails beyond the Darwin width (see e.g. appendix A in ref. 10). As pointed out by Bonse and Hart[6] some years ago a solution to the tail-problem is to use several Bragg reflections in succession. If the tails of the scattering profile from a single reflection varies as q^{-2} then after m reflections the tails vary as q^{-2m} whereas the loss in the Bragg peak intensity is quite small.

Our set-up to study the SmA line profile is depicted in Fig. 3. The $K_{\alpha 1}$ X-rays from a 12 kW rotating

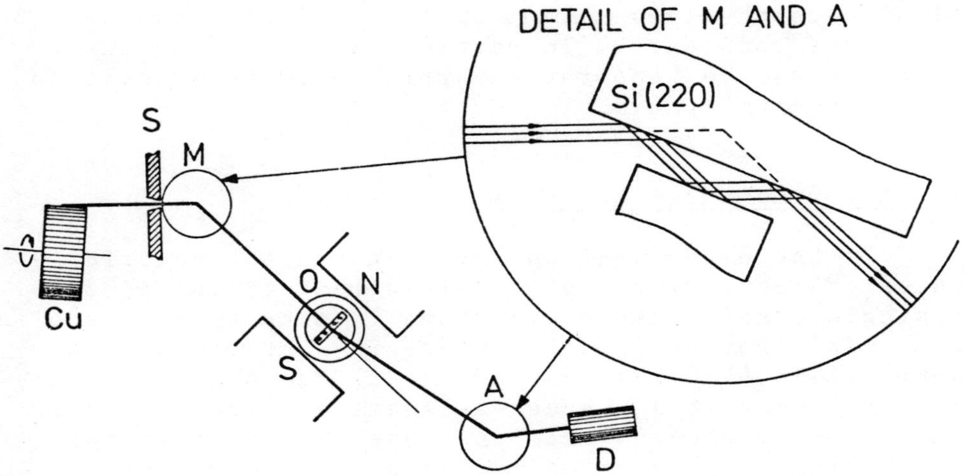

Fig. 3. General lay-out of the experimental set-up. $CuK_{\alpha 1}$ X-rays, collimated by the channel-cut crystal M, are diffracted from the planes of a magnetic field oriented SmA phase of 8OCB. The direction of the diffracted ray is determined by the channel-cut crystal A.

Cu anode generator were Bragg reflected three times in a channel-cut monolithic perfect Si crystal using the (220) reflection. The direction of the scattered X-rays was determined by a similar analyzer crystal.

There are two contributions to the resulting wave-vector resolution. One is the spectral width of the $CuK_{\alpha 1}$ line, the other is the Darwin width of the monochromator and analyzer crystals. The Darwin width from a perfect crystal reflection originates from the depletion of the incident beam as it penetrates into the crystal implying that only a finite number of planes are effectively contributing to the Bragg scattering and the coherence is therefore not perfectly sharp.

The two effects can be separated by recording the direct beam profiles for right-left and right-right scattering in the monochromator and analyzer, respectively. Using this information one can then calculate the line profile for Bragg scattering for a single crystal sample.

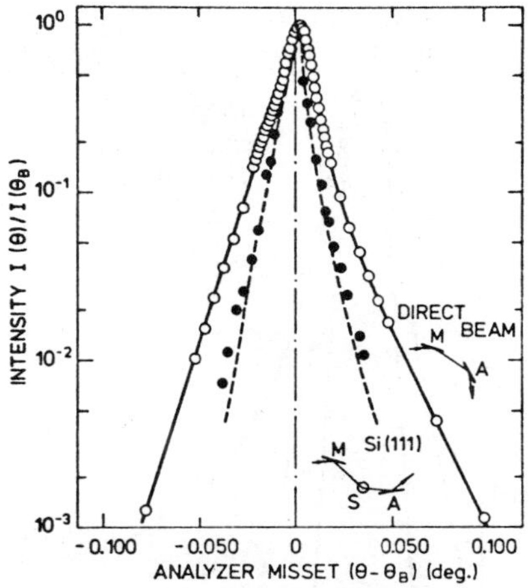

Fig. 4. The profile of the direct beam in the dispersive orientation of the analyzer crystal determines the line width of $CuK_{\alpha 1}$ radiation. This in turn can be used to calculate the Bragg profile of the Si(111) reflection in the indicated orientation (dashed curve).

In Figs. 4 and 5 are shown the two direct-beam profiles. Fig. 4 also shows the observed Bragg profile from a Si(111) single crystal replacing the liquid crystal sample, and the agreement with the calculated profile is acceptable. We can therefore with a considerable degree of confidence simulate the line profile if SmA had this true long range order. This is shown as the dashed line in Fig. 5. The observed SmA profile also given in Fig. 5 is clearly very different from the simulated Bragg profile - for a considerable wavevector interval the difference is more than an order of magnitude. This is then the evidence for the Landau-Peierls state of the SmA phase and we shall now proceed to analyze the data in more detail.

So far we have only considered the wavevector resolution in one direction, that is along the wavevector $\vec{q}_0$ describing the density wave. However, the perpendicular wavevector resolution cannot be neglected. It factorizes into a component perpendicular to the scat-

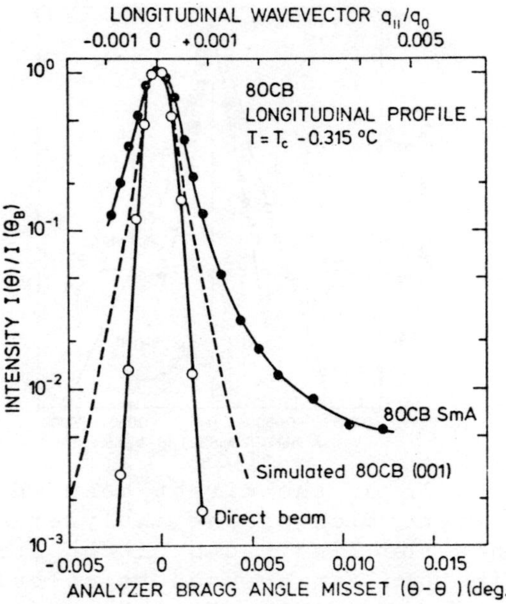

Fig. 5. The profile of the direct beam is the non-dispersive orientation of the analyzer crystal (open circles), the simulated Bragg profile from the (001) reflection in SmA with true long range order (dashed curve), and the actual line profile from SmA 8OCB. Note that the intensity scale is logarithmic.

tering plane and a component in the scattering plane. The first component is determined by the heights of the X-ray spot and the different slits along the beam path. It is made comparatively broad (0.046 $Å^{-1}$ full width at half maximum) in order to obtain a decent intensity from the SmA phase. The second component is negligible because the scattering angle is small, $k/q_o \sim 20$. However, the SmA single domain is unfortunately not perfect. The normal to the smectic planes within the scattering volume has an angular spread δ of the order of 1° causing effectively a perpendicular wavevector resolution width of $q_o\delta$.

In summarizing the discussion of wavevector resolution we find that the 3-dimensional wavevector resolution function $R(\vec{q}-\vec{q}_o)$ can be factorized into

$R_1(q_x)R_2(q_y)R_3(q_z-q_0)$. The longitudinal part $R_3(q_z-q_0)$ can be accurately calculated from the measured direct beam profiles obtained in the two possible analyzer orientations. The transverse, vertical part $R_2(q_y)$ is given by slit heights along the beam path. The transverse horizontal part $R_1(q_x)$ is dominated by the mosaic spread of the SmA sample.

The relation between the measured line profile, the scattering cross section and the experimental resolution is well-known. When the spectrometer is set at wave-vector transfer $\vec{q}$ the intensity $I(\vec{q})$ is the folding of the cross section $S(\vec{q})$ and the experimental resolution $R(\vec{q})$:

$$I(\vec{q}) = \int S(\vec{q}\,')R(\vec{q}-\vec{q}\,')d\vec{q}\,' \tag{11}$$

Usually the cross section $S(\vec{q}\,')$ is known explicitly, and one calculates numerically the folding integral varying free parameters in $S(\vec{q}\,')$ until a best fit to the observed intensity $I(\vec{q})$ is obtained. In the present case the cross section diverges at a certain point and that in itself complicates the numerical evaluation, and furthermore we only know the cross section by its Fourier-transform. An elegant solution to these problems is to utilize the folding theorem of Fourier transforms. Let the Fourier transform of $I(\vec{q})$ and of the resolution function $R(\vec{q})$ be $\mathcal{J}(\vec{r})$ and $\mathcal{R}(\vec{r})$, respectively. The folding theorem then states that $\mathcal{J}(\vec{r}) = G(\vec{r})\mathcal{R}(\vec{r})$, and the folded intensity is calculated as

$$I(\vec{q}) = \int G(\vec{r})\mathcal{R}(\vec{r})\, e^{iq\cdot\vec{r}}\, d\vec{r} \tag{12}$$

The resolution functions $R_1(q_x)$ and $R_2(q_y)$ are approximated by Gaussians. The width of $R_1(q_x)$ is the mosaic spread of the SmA crystal and it turns out to depend on the temperature. The Fourier transforms are, of course, also Gaussians. The longitudinal resolution function $R_3(q_z)$ was approximated by the folding of a Gaussian of width σ_z and an exponential because that gives a convenient form of the Fourier transform as the product of a Gaussian and a Lorentzian. The explicit forms and parameters of the resolution function is given in the table below. The pair correlation function $G(\vec{r})$ is given in eq. (9c) and since λ is determined by light scattering experiments ($\lambda q_0 = 1.52\ t^{-0.133}$) the only free parameter in fitting the line shape is the exponent η.

Table 1

The experimental resolution function.

q-space	r-space
$R_1(q_x) = \exp(-q_x^2/2\sigma_x^2)$	$\exp(-\sigma_x^2 x^2/2)$
reduced temp. 9×10^{-4} 5.9×10^{-4} 4×10^{-6}	
$\sigma_x/q_o\ 10^3$ 6.45 4.2 4.9	
$R_2(q_y) = \exp(-q_y^2/2\sigma_y^2)$	$\exp(-\sigma_y^2 y^2/2)$
$\sigma_y/q_o = 0.22$	
$R_3(q_z) = \int \exp(-q_{\parallel}^2/2\sigma_z^2)$ $\times\exp(\sqrt{2}\,\lvert q_z-q_{\parallel}\rvert/\sigma\sigma_z)\,dq_{\parallel}$	$\dfrac{\exp(-z^2\sigma_z^2/2)}{(1+\sigma^2\sigma_z^2 z^2/2)}$
$\sigma_z/q_o = 2.2\times10^{-4}$, $\sigma = 1.54$	

5. Results and Conclusions

Fig. 6 shows a comparison of the experimental data and the calculated longitudinal line profiles at a temperature 0.31 ^{o}C below the transition temperature T_c and at a temperature very close to T_c. The Caillé correlation function, eq. 9c, derived in the harmonic approximation accounts very well for the data, even near T_c where the harmonic approximation is expected to break down. We also have got data for an intermediate temperature and the results are given in Table 2. From eq. (10) we note that the ratio of $\lambda q_o/\eta$ is expected to be independent of temperature and as shown in Table 2 this is indeed the case. From $\lambda q_o/\eta \sim 20$ we find K_1 $8\cdot10^{-7}$ dyne, which is consistent with common values in liquid crystals[1)].

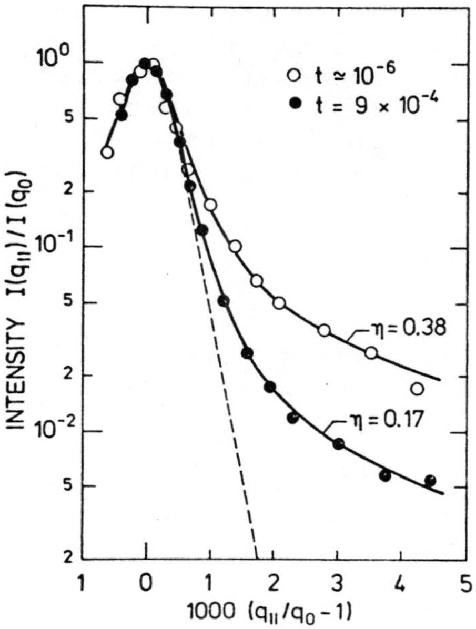

Fig. 6. Comparison of experimental data at two reduced temperatures t with the harmonic theory of Caillé using η as the only adjustable parameter.

Table 2

Exponent η and penetration depth λ.

t	9×10^{-4}	5.9×10^{-4}	$\sim4\times10^{-6}$
η	0.17 ± 0.02	0.23 ± 0.02	0.38 ± 0.06
$\lambda q_0/\eta$	22.7 ± 2.7	19.1 ± 1.7	21.0 ± 6.0

The data of Fig. 6 shows unambiguously that the smectic A state is not an infinite stacking of planes with a well defined lattice spacing. This is consistent with the theoretical model of the SmA state as described by the free energy density of eq. (3). Furthermore, the X-ray line profiles can be evaluated in this model in the harmonic approximation, and we find good agreement between the calculated line profile and the data both with respect to the temperature dependence and the magnitude of the parameter η describing the wings. We must therefore conclude that there is considerable experimental evidence for the statement that the smectic A phase is a state of matter at lower marginal dimensionality.

Acknowledgements

The channel-cut perfect crystals of Si were made by A. Lindegaard-Andersen and S. Mathiesen from the Technical University, Lyngby.

We are grateful to Michael Stephen for helpful conversations on the scattering cross section from the Landau-Peierls state.

The X-ray source and the spectrometer were granted by the Danish National Science Foundation.

Appendix: Calculation of $<[u(\vec{r})-u(0)]^2>$ in the SmA Phase

In order to calculate the thermal average value of $(u(\vec{r})-u(0))^2$ we introduce the Fourier transform amplitudes u_q:

$$u(r) = \sum_{\vec{q}} u_q \, e^{i\vec{q}\cdot\vec{r}}$$

It then follows that $u(0)-u(\vec{r}) = \sum_{\vec{q}} u_{\vec{q}}(1-e^{i\vec{q}\cdot\vec{r}})$ and the squared difference becomes

$$(u(\vec{r})-u(0))^2 = 2\sum_q u_q^2 \, (1-\cos(\vec{q}\cdot\vec{r}))$$

The thermal average values of u_q^2 follows readily from the free energy expansion in eq. (3) and equipartion

$$<u_q^2> = \frac{1}{V} \frac{kT}{(Bq_{\parallel}^2 + Kq_{\perp}^4)}$$

We then find for the exponent in equation (8) with $q \simeq q_o$:

$$\tfrac{1}{2}q_o^2<[u(\vec{r})-u(0)]^2> = \frac{kTq_o^2}{8\pi^3 B} \int \frac{1-\cos(\vec{q}\cdot\vec{r})}{q_{\parallel}^2 + \lambda^2 q_{\perp}^4} \, d^3q$$

Evaluation of the integral I.

First integrate over $q_{\parallel}$ using Gradshteyn and Ryzhik formula 3.7237[7])

$$\int_{-\infty}^{\infty} \frac{1-\cos(a(b-x))}{x^2+c^2} \, dx = \frac{\pi}{c} \, (1-e^{-ac}\cos(ab))$$

to get

$$\int_0^{\infty} \frac{1-\cos(q_{\parallel} z+\vec{q}_{\perp}\cdot\vec{\rho})}{q_{\parallel}^2 + \lambda^2 q_{\perp}^4} \, dq_{\parallel} = \tfrac{1}{2}\frac{\pi}{\lambda q_{\perp}^2} \, [1-e^{-\lambda q_{\perp}^2 z}\cos(\vec{q}_{\perp}\cdot\vec{\rho})]$$

Then with θ being the angle between $\vec{q}_{\perp}$ and $\vec{\rho}$

$$\int_0^{2\pi} [1-e^{-\lambda q_{\perp}^2 z}\cos(\vec{q}_{\perp}\cdot\vec{\rho})]d\theta = 2\pi - e^{-\lambda q_{\perp}^2 z} \int_0^{2\pi} \cos(q_{\perp}\rho\cos\theta)d\theta$$

Here we use the integral representation of the Bessel function:

$$J_0(x) = \frac{1}{\pi}\int_0^\pi \cos(x\cos\theta)d\theta$$

and find

$$I(\rho,z) = \tfrac{1}{2}\pi\cdot 2\pi \int_{q_{min}}^{q_{max}} \frac{1-e^{-\lambda q_\perp^2 z} J_0(q_\perp \rho)}{\lambda q_\perp^2} q_\perp dq_\perp$$

Let us first for simplicity consider $\rho = 0$ and utilize $J_0(0) = 1$:

$$I(0,z) = \pi^2 \int_{q_\perp = q_{min}}^{q_{max}} \frac{1-e^{-\lambda q_\perp^2 z}}{\lambda q_\perp^2} q_\perp dq_\perp$$

Integrating $\int_\varepsilon^K (1-e^{-x})\frac{dx}{x}$ by parts we find

$$\int_\varepsilon^K (1-e^{-x})\frac{dx}{x} = (1-e^{-x})\ln x\Big]_\varepsilon^K - \int_\varepsilon^K e^{-x}\ln x\, dx$$

$$\to \ln K + C \text{ for } K \to \infty,\ \varepsilon \to 0$$

where C is the Euler constant 0.5772...

$$I(0,z) = \frac{\pi^2}{\lambda}\{\ln(\lambda q_{max}^2 z) + C\} \text{ and finally}$$

$$\tfrac{1}{2}q_0^2 <[u(0,z)-u(0,0)]^2> = \frac{kTq_0^2}{8\pi B\lambda}\{\ln(\lambda q_{max}^2 z) + C\}$$

Denoting the dimensionless prefactor η:

$\eta \equiv kTq_0^2/(8\pi B\lambda)$ we find

$$\tfrac{1}{2}q_0^2 <[u(0,z)-u(0,0)]^2> = \ln(\lambda q_{max}^2 z)^\eta + \eta C$$

By utilizing the series expansion form of $J_0(x)$ and of $E_1(x)$

$$J_0(x) = \sum_{n=0} \frac{x^{2n}(-1)^n}{2^{2n} n!n!} \quad \text{and}$$

$$E_1(x) = -C - \ln x - \sum_{n=1}^{\infty} \frac{(-1)^n z^n}{n(n!)}$$ one arrives finally at the general expression given in eq. (9c).

References

1. P.G. de Gennes, "The Physics of Liquid Crystals" (Clarendon, Oxford, 1974)

2. R.E. Peierls, Helv. Phys. Acta 7, Suppl. 11, 81 (1934)

3. L.D. Landau, in "Collected Papers of L.D. Landau" edited by D. ter Haar (Gordon and Breach, New York, 1965), p. 209

4. A. Caillé, C.R. Acad. Sc. Paris 274, B891 (1972)

5. H.-J. Mikeska and H. Schmidt, J. Low Temp. Phys. 2, 371 (1970)
   Y. Imry and L. Gunther, Phys. Rev. B3, 3939 (1971)
   Y. Imry, Proceedings from Third Int. Inst. in Surface Science (1977)

6. U. Bonse and M. Hart, Appl. Phys. Lett. 7, 238 (1965)

7. I.S. Gradshteyn and I.M. Ryzhik, "Table of Integrals, Series and Products" (Academic Press, New York, 1965)

8. H. Birecki, R. Schaetzig, F. Rondelez and J.D. Litster, Phys. Rev. Lett. 36, 1376 (1976). See also Fig. 10 in Litster's proceedings of this conference.

9. J.D. Litster, J. Als-Nielsen, R.J. Birgeneau, S.S. Dana, D. Davidov, F. Garcia-Golding, M. Kaplan, C.R. Safinya and R. Schaetzing, to be published in Colleque Bordeaux issue of J. de Physique 1979.

10. B. Warren, "X-ray Diffraction", Addison-Wesley 1973.

CRITICAL FLUCTUATIONS UNDER SHEAR FLOW

D. BEYSENS

Service de Physique du Solide et de Résonance Magnétique

CEN Saclay, BP n°2 - 91190 Gif-sur-Yvette, France

The behaviour of fluctuations in non equilibrium systems is of considerable interest, especially in critical systems where the fluctuations are very sensitive to small disturbances. We consider here the influence of a shear flow (dissipative process) on a critical binary fluid of Cyclohexane-Aniline.

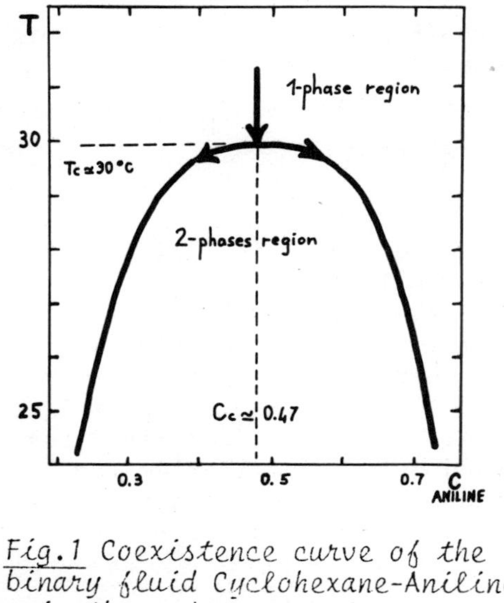

Fig.1 Coexistence curve of the binary fluid Cyclohexane-Aniline. c is the order parameter. Experiments were carried out at constant critical concentration.

This system belongs to the class of fluids (dimensionality d=3, upper critical dimensionality UCD=4). The order parameter (scalar, n=1) is the concentration c of one of the components (Fig.1). Theoretical values of the critical exponents can be obtained either by high temperature series expansions, or ε-expansions around UCD=4 using renormalization group (RG) calculations. The susceptibility $\chi_{\vec{q}}$ for a given wavevector $\vec{q}$ diverges versus the reduced temperature $t=T/T_c-1$ as $\chi_{\vec{q}} \sim t^{-\gamma}$, with $\gamma=1.24$. The correlation length varies as $\xi = \xi_o t^{-\nu}$ with $\nu=0.63$. In the system we studied, $\xi_o \simeq 2.5$ Å, so at $t \simeq 3.10^{-5}$ ($T-T_c \simeq 10$mK), $\xi \sim 1000$ Å. The typical relaxation time τ of the order parameter, at the scale $q \simeq \xi^{-1}$, is $\tau = \frac{5\pi\eta\xi^3}{kT} \simeq 30$ msec, at the same temperature (η is the shear viscosity, k is the Boltzmann constant).

The shear flow was produced in a rectangular pipe (Fig.2).

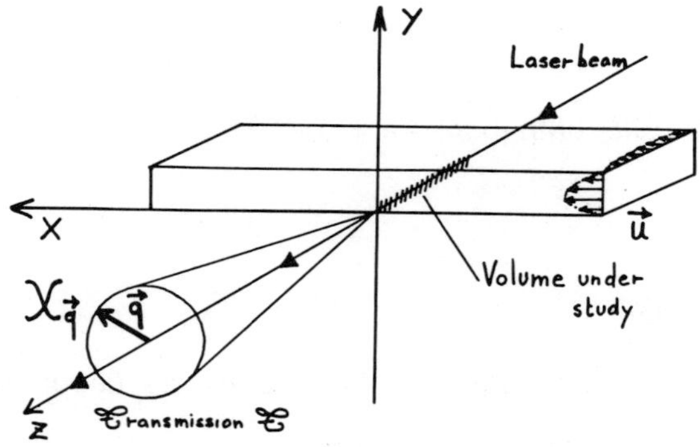

Fig.2 Experimental

Close to a side wall the Poiseuille velocity ($\vec{u}$) distribution leads to a nearly constant shear rate $s = \partial u_x/\partial y$, the effect of $\partial u_x/\partial z$ beeing small. s could vary between 0-1000 sec^{-1}. The extra-heating due to the shear flow was shown to be lower than 0.1 mK and was thus negligibly small (thermal regulation : ±0.2mK over more than one hour).

We used light scattering techniques (Fig.2) in order to measure (i) the turbidity $\bar{\tau}$ related to the mean susceptibility $\bar{\tau} \propto \int_{\vec{q}} \chi_{\vec{q}} \cdot d^3q$ and to the intensity $\mathcal{T}$ of the transmitted light $\bar{\tau} \propto -\ell_N \mathcal{T}$, and (ii) the scattered intensity at given $\vec{q}$: $I_{\vec{q}} \propto \chi_{\vec{q}}$.

1. TURBIDITY : T_c CHANGE

Figs.3 and 4 show that (i) the effect of shear is visible only close to T_c, i.e. for a temperature range $T < T_s$ where T_s is the cross-over temperature below which the fluctuations "feel" the shear. T_s can be defined as the temperature where the lifetime of the fluctuations τ is equal to the characteristic time of the shear s^{-1} : $T_s = T(s\tau = 1)$.(ii) The shear lowers $\bar{\tau}$ (Fig.3) as if T_c were also lowered, i.e. as if $T_c = T_c(s)$. We checked from Fig.3 that all transmission curves could be fitted by the equilibrium (s=0) curve if the scale $T - T_c(o)$ was changed by $T - T_c(s)$, so that $T_c(s) = T_c(o) - 1.8\times10^{-4} s^{0.53}$. Experimental data for various shears are reported in Fig.5.

Moreover, the actual transition temperature $T_c(s)$ can be accurately determined by the divergence of $\chi_{\vec{q}}$, or the zero value of $\mathcal{T}$. Fig.4d shows unambiguously that T_c is *actually* shifted by the shear.

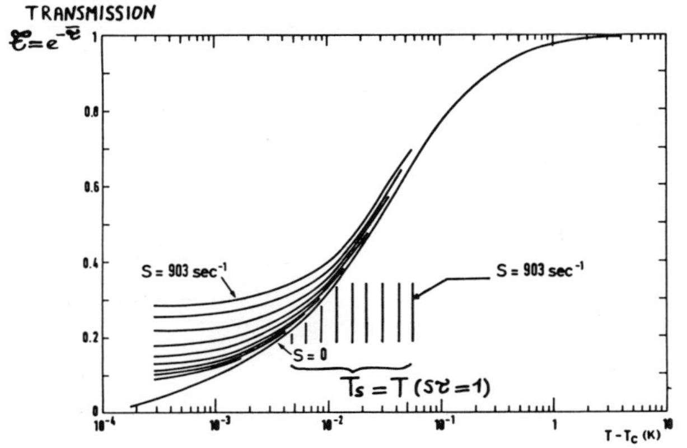

Fig.3 Different transmission curves versus T-T_c, and cross-over temperatures T_s, for the shears : 903,510,288,163,92,52,29,16,9, and 0 sec^{-1}.

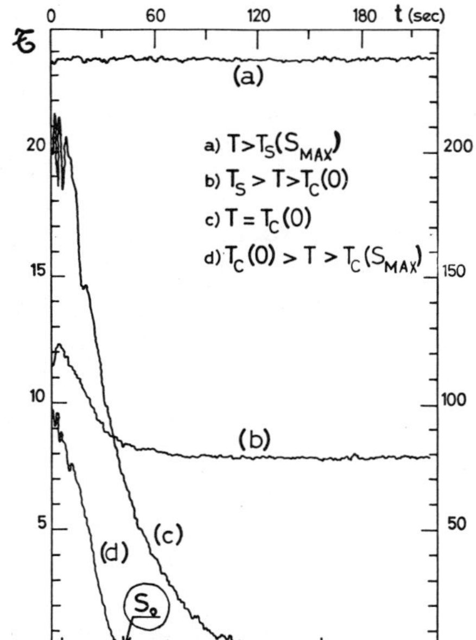

Fig.4 Variation of τ with shear at different temperatures (d) corresponds to temperatures lower than the critical temperature at equilibrium (S=0).

The value s_o at which $\mathcal{E}$ goes to zero measures the critical temperature $T=T_c(s_o)$, and the corresponding data agree well with those deduced from Fig.3, as is shown in Fig.5.

Thus the first conclusion is that the shear *actually* lowers T_c. If T_c is defined by the temperature at which $\xi^{-1}=0$, these results show that at constant temperature T, the shear reduces the fluctuation extent :

$$\xi(T,s) \ < \ \xi(T,o) \ .$$

2. THE SCATTERED LIGHT : ANISOTROPY ; MEAN-FIELD ; LOWERING OF UCD ?

The susceptibility in the $\vec{q}$ space becomes increasingly anisotropic as the shear increases. Moreover, in accordance with the turbidity data, $\langle\chi_{\vec{q}}\rangle$ decreases with increasing shear. Since $\chi_{\vec{q}}$ is the Fourier transform of the correlation function in the

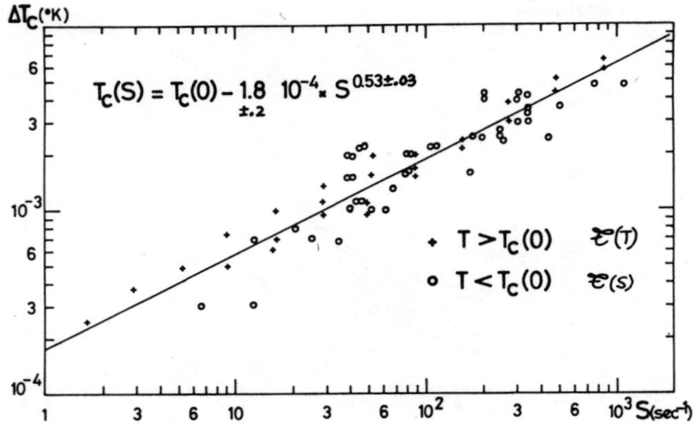

Fig.5 T_c *shift versus shear obtained from the methods of Figs 3,4.*

real space, these data show that fluctuations are anisotropic ($\xi_x > \xi_y$) and are suppressed below their equilibrium values : $\langle\xi(T,s)\rangle < \langle\xi(T,o)\rangle$.

For the same temperature T, could this anisotropy lower the UCD ? In other words, is the lowering of the fluctuations along the shear direction sufficiently high to induce a mean-field behavior of the susceptibility ? For small values of q, and assuming the anisotropy to be expanded in $q_x^p\, s^m$, the mean-field behavior will be relevant if the contribution from the fluctuations is negligible, even at $T_c(s)$, i.e. if $\int_o^{1/\xi_o} d^3q/q_x^{2p}$ remains finite. This leads to $p < 3/2$. The condition is experimentally verified when considering Fig.6 and plotting the anisotropy on log-log scales, leading to the conclusion that it is merely the reduction of the order parameter fluctuations which involves the mean-field behavior.

Fig.7 shows the results of an experiment performed at very small angle ($\theta=2°$, $q=5200\ cm^{-1}$). A cross-over occurs at T_s , between the regular variation ($s\tau < 1$, $T > T_s$, $\gamma = 1.24$) and a mean-field behavior ($s\tau > 1$, $T < T_s$, $\gamma = 1$). For the highest shears, a rounding-off is visible (see below).

3. DISCUSSION

Through a theoretical approach, Onuki and Kawasaki reached the same general conclusions at the same time. They performed both static and dynamic R-G calculation up to order ε. Let us compare both conclusions : (i) T_c change

Experimental : $\Delta T_c(s) = -(1.8\pm0.2)\times10^{-4}\ s^{0.53\pm0.03}$

Theoretical : $\Delta T_c(s) = -1.3\times10^{-4}\ s^{1/3\nu}$ with $1/3\nu = 0.529$.

The agreement is reasonably good.

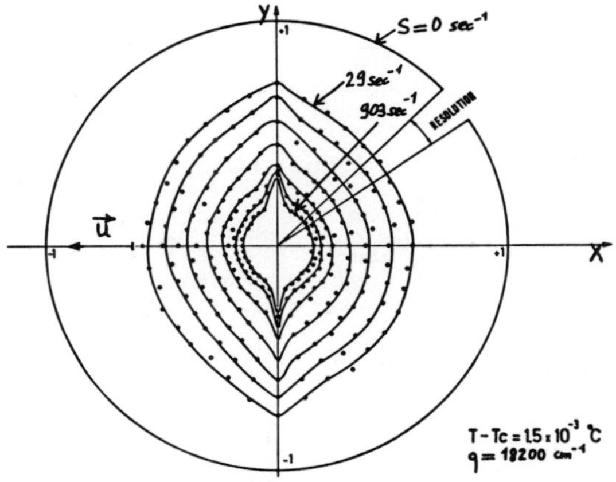

Fig. 6 *Reduced scattered intensity (corrected for the transmissions)* $\frac{\chi_{\vec{q}}(T,s)}{\chi_{\vec{q}}(T,o)}$ *for the same shears than in Fig. 3.*

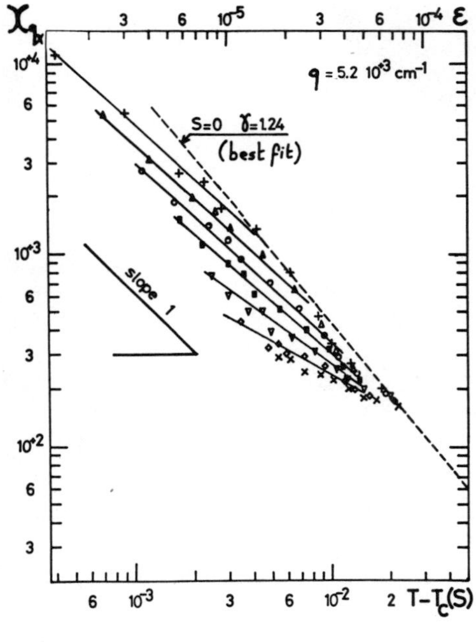

(ii) Susceptibility χ_{q_x}

Exp : $\chi_{q_x}^{-1} \propto [T-T_c(s)]^{+1} + \Theta_{exp}[T-T_c(s),s]$ (Fig.7)

Θ_{exp} represents the rounding-off for the highest shears.

Th. :

$\chi_{q_x}^{-1} \propto A_{th} S^{0.14}[T-T_c(s)]^{+1} + B_{th}\, S^{8/15}|q_x|^{2/5} (+q^2, \text{negligible})$

Fig. 7 *Susceptibility in the flow direction versus* $T-T_c(s)$ *for shears* : ×903; ◇ 510; ▽ 288; ■ 163; ● 92; △ 52; + 29 sec^{-1}. *The dotted line is the best fit to experimental results with S=0. It corresponds to the behavior at equilibrium* $\simeq (T-T_c)^{-1.24}$.

The problem is that $\Theta_{exp} << B_{th}\ S^{8/15}\ |q_x|^{2/5}$. This point needs further investigations and is now underway (q_x and s dependence).

(iii) Susceptibility χ_{q_y} : "Moustache effect".

In the shear direction the scattered intensity is much stronger than expected (Fig.8) and varies strongly with $T-T_c(s)$. This point is also under further study.

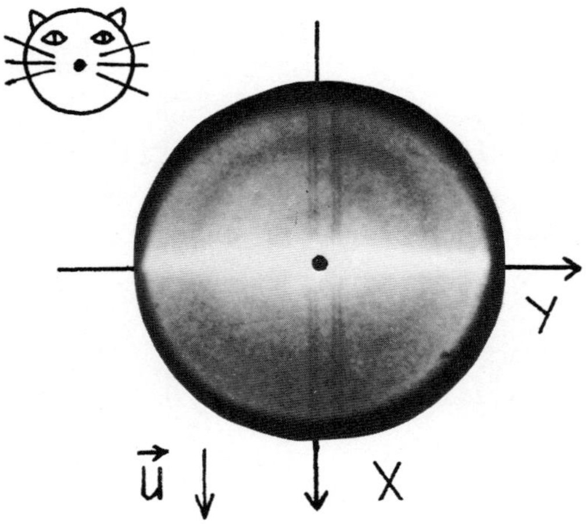

Fig.8 "Moustache Effect"

(iv) Correlation function $<\chi_{\vec{q}}(t)\ \chi_{\vec{q}}(t+t')>$.

Due to the Doppler broadening, it is only in the y direction that the dynamics of the fluctuations can be detected. We have already detected a strong effect, but its interpretation would require a better understanding of point (iii).

Many other points should be studied in order to fully understand these unexpected phenomena. So ...

4. "MORALITE"

... It seems that the field of critical fluctuations out of equilibrium - here under shear - which involves critical statics and critical dynamics is a particularly rich field for both experimentalists and theoricists.

ACKNOWLEDGMENTS

I would like to thank P. Berge who suggested this kind of experiments started in 1976, L. Boyer for help at the beginning of this work, and M. Gbadamassi who contributed to all the presented results.

BIBLIOGRAPHY

Binary fluids at equilibrium :

- P. Calmettes, Thesis (Paris VI, 1978)-Saclay ref.DPhT/PSRM/1545

Binary fluids out of equilibrium :

- D. Beysens, M. Gbadamassi and L. Boyer, to appear in Phys. Rev. Lett. Saclay ref.DPhT/PSRM/1568
- D. Beysens and M. Gbadamassi, submitted,Saclay ref.DPhT/PSRM/1591
- A. Onuki and K. Kawasaki, to appear in Ann. Phys. (N.Y.), Progr. Theor. Phys. Supplement, and Phys. Lett.A.

LIFSHITZ POINTS IN ISING SYSTEMS WITH COMPETING INTERACTIONS

R. Liebmann
Institut für Theoretische Physik
Universität Frankfurt
6000 Frankfurt 1, Robert-Mayer-Str. 8
Fed. Rep. Germany

I. Introduction

A few years ago, Hornreich, Luban and Shtrikman[1] introduced a new multicritical point, the Lifshitz point (L.P.) denoting the triple point between the disordered, uniformly ordered and modulated ordered phases. Lifshitz points have been suggested for a variety of differedt systems including magnetic alloys[1,2], liquid crystals[3], TTF-TCNQ[4], and perovskites[5]. Recently, some very interesting experiments on liquid crystals[6] and on a structural phase transition[7] in Rb Ca F_3 have been published.

The model systems we have chosen for the study of Lifshitz points are simple hypercubic Ising systems with competing nearest and next nearest neighbour exchange interactions. As a first step we discuss the spin correlation function for the one-dimensional case. Then we calculate phase diagrams for two-dimensional systems using (a) the method of Müller-Hartmann and Zittartz[3] (MZ) to determine the transition temperature via the vanishing of the interface free energy (b) a Migdal-Kadanoff[9,10] bond moving scheme and (c) Monte Carlo simulations. It is shown that in two-dimensional uniaxial Ising systems a Lifshitz point exists at non-zero temperatures, whereas in systems with identical competing interactions along each of its cartesian axes the lower critical dimension d_ℓ for a Lifshitz point is $d_\ell \geq 2$.

II. One-dimensional Ising systems with competing interactions

We consider the one-dimensional Ising model

$$H = -J_1 \sum_{i=1}^{N} s_i\, s_{i+1} - J_2 \sum_{i=1}^{N} s_i\, s_{i+2} \,, \qquad (1)$$

where the interaction between nearest neighbours is taken to be ferromagnetic ($J_1 > 0$) and between next nearest neighbours antiferromagnetic ($J_2 < 0$). The spins $s_i = \pm 1$ are sited on N lattice points and the lattice constant is normalized to unity. Using a transfer-matrix method the spin correlation function $G(r) = \langle s_i\, s_{i+r} \rangle$ can be calculated exactly. The interesting result is that $G(r)$ decreases exponentially with increasing r only for

$$\cosh\text{-}K_1 > e^{-2K_2} \qquad (2)$$

where $K_i = J_i / k_B T$. For stronger antiferromagnetic coupling an oscillation described by a continuously varying wave vector q is superimposed to the exponential decrease of $G(r)$. In Fig. 1, where the dimensionless temperature $1/K_1$ is plotted versus the ratio $p = -J_2/J_1$ of the coupling constants, the full curve is the border line (eq. 2) of the region with $q \equiv 0$ and the dashed curves are lines of constant non-zero values of q. The existence of two regimes with monotonous e.g. oscillating decrease of the spin correlation for d=1 indicates already possible Lifshitz points for $d > 1$.

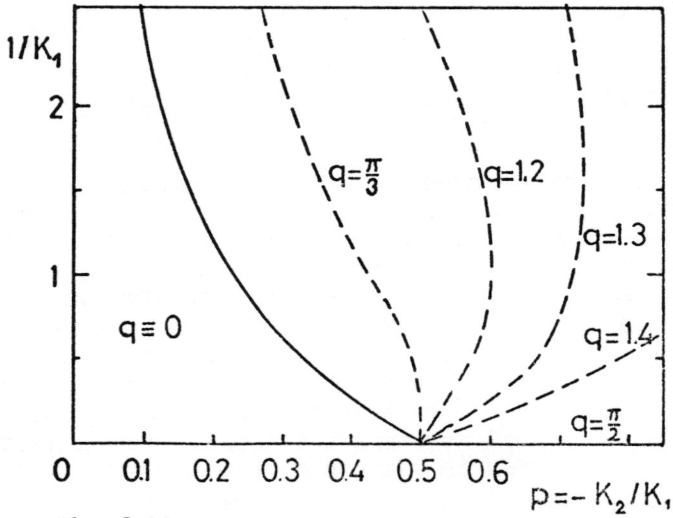

Fig. 1 : The full curve is the border between the regions of monotonous and oscillating decrease of the spin correlation function for d = 1 . The dashed curves are lines of constant wavevector in the oscillating regime.

III. Two-dimensional Ising systems

In this section we calculate the boundary of the ferromagnetic region in the phase diagram of a two-dimensional square lattice using (a) the method of Müller-Hartmann and Zittartz[8] to determine T_c of a ferromagnet via the vanishing of the interface free energy and (b) the Migdal-Kadanoff bond-moving scheme[9,10] and compare it with results of (c) Monte Carlo simulations.

In the uniaxial case (where we have competing interactions in the uniaxial direction and only ferromagnetic nearest neighbour interactions $K_\perp$ along the perpendicular direction) the interface free energy per spin σ is given by[8]:

$$\sigma = 2T(K_1 + 2K_2) + T \ln \operatorname{tgh} K_\perp \quad . \qquad (3)$$

The first term on the right hand side of eq. (3) is the energy $\Delta E_1 = 2(K_1 + 2K_2)$ needed to create a straight "domain wall" perpendicular to the uniaxial direction at $T = 0$. The second term arises from a certain subset[8] of more complicated interfaces. The boundary of the ferromagnetic region in the phase diagram is determined by $\sigma = 0$ i.e.:

$$\sinh\left(2\left[K_1 + 2K_2\right]\right) \sinh(2K\ ) = 1 \qquad (4)$$

For different orientation of the domain wall the MZ method cannot be carried out as easily but eq. (4) can be proven to give a lower limit for T_C. The resulting phase diagram is the dotted curve in Fig. 2 and comparison with the Monte Carlo calculations suggests that eq. (4) is indeed a very good approximation. In the isotropic case analogous to the uniaxial case we replace K_1 by $K_1 + 2K_2$ in both directions and obtain for the boundary of the ferromagnetic phase

$$\sinh^2\left(2\left[K_1 + 2K_2\right]\right) = 1 \quad . \qquad (5)$$

This result, the dotted line in Fig. 3, is exact for $p = 0$ and $p = 0.5$ and we find again reasonable agreement with the Monte Carlo results.

In this isotropic case the phase boundary was also calculated by the Migdal-Kadanoff bond-moving approximation. This method requires the one-dimensional recursion $K_i' = R_\lambda\{K_i\}$, where λ is the length rescaling factor. We have calculated this recursion with a transfer matrix method decimating always pairs of spins. The recursion creates three additional couplings and is closely related to that of the braced ladder model[11]. The Migdal-Kadanoff recursion for $d = 2$ is then given by

$$K'_{i,x} = \lambda \cdot R_\lambda\{K_{i,x}\} \ , \quad K'_{i,y} = R_\lambda\{\lambda K_{i,y}\} \quad . \qquad (6)$$

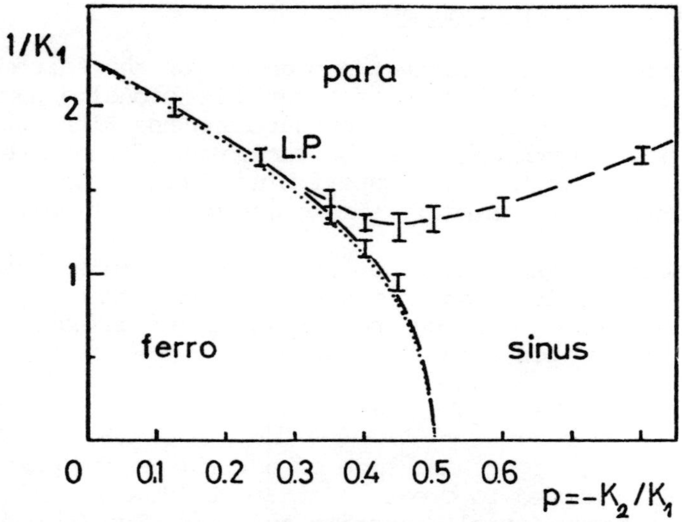

Fig. 2 : Phase diagram in the uniaxial case (d = 2). The dotted line is obtained from the MZ-method. The error bars indicate the Monte Carlo results; the dashed lines are guides to the eyes.

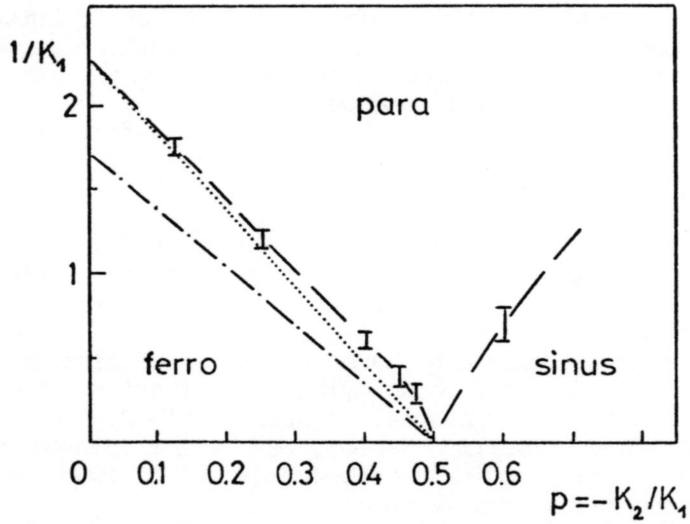

Fig. 3 : Phase diagram as in Fig. 2, but for the isotropic case. The dot-dashed line is obtained from the Migdal-Kadanoff method.

In the limit $\lambda \to 1$, where these equations describe the isotropic situation, we obtained the dot-dashed curve in Fig. 3. It is again nearly linear, only the slope of the curve is a little too small.

As a third method a Monte Carlo simulation for a 40×40 two-dimensional Ising system with competing interactions was performed using 500-2000 "Monte Carlo steps per spin". The phase transition temperatures have been determined in the standard way by the anomalous behaviour of the energy, specific heat and the order parameter (magnetization or amplitude of the magnetization in the sinusoidal phase). The Monte Carlo results for the uniaxial and isotropic case are shown in Fig. 2 and 3. The error bars indicate th range of temperatures, in which the anomalous behaviour has been found. The dashed lines are guides to the eyes. For the uniaxial case Fig. 2 demonstrates the existence of a L.P. at a non-zero temperature. The transition line from the ferromagnetic phase to the sinusoidal phase may represent only an upper bound, because this transition is possibly of first order, and we approached it by using the ferromagnetic ground state as the initial configuration of our runs. In the isotropic case, see Fig. 3, the triple point between the three phases appears to move to zero temperature with $p=0.5$. Because in the Monte Carlo method the relaxation times to equilibrium become exceedingly long for very low temperatures, the data for $p=0.45$ and 0.475 are to be taken as upper limits for the three transition temperatures. At the point $T=0$, $p=0.5$ the groundstate is infinitely degenerate and it should be considered further wether this triple point is a L.P. in the usual sense.

IV. Conclusions

Our main conclusion is that in two-dimensional Ising systems, there exists a Lifshitz point at a non-zero temperature in the uniaxial case while no L.P. at $T \neq 0$ is expected to occur in the isotropic situation. This result differs from the lower critical dimension d^{ℓ} expected for a L.P. in systems where the order parameter has a continuous symmetry

$$d^{\ell}_{\text{isotr.}} = 4 \quad \text{and} \quad d^{\ell}_{\text{uni.}} = 2.5 \quad .$$

We believe that this result should encourage an experimental search for a L.P. in two-dimensional Ising systems.

Acknowledgements

This report is the result of joint work by the author, R.M. Hornreich, H.G. Schuster, and W. Selke. A detailed paper will be published in Z. Physik 1979. We thank K. Binder for helpful discussions.

References

1 R.M. Hornreich, M. Luban and S. Shtrikman, Phys.Rev.Lett. 35, 1678 (1975); Phys.Lett. A55, 269 (1975); Physica 86A, 465 (1977); J.Mag.Magnetic Mat. 7, 121 (1978).

2 A. Yoshimori, J.Phys.Soc.Jap. 14, 805 (1959); J. Villain, J. Chem.Phys.Solids 11, 303 (1959); T.A. Kaplan, Phys.Rev. 116, 888 (1959); Phys.Rev. 124, 329 (1961).

3 J. Chen, T.C. Lubensky, Phys.Rev. A14, 1202 (1977); A. Michelson, Phys.Rev.Lett. 39, 464 (1977); R. Blinc, F.C. de Sa Baretto, Phys.Stat.Sol.

4 E. Abrahams, I.E. Dzyaloshinskii, Solid State Commun. 23 883 (1977).

5 A. Aharony, A.D. Bruce, Phys.Rev.Lett. 42, 462 (1979).

6 G. Sigaud, F. Hardouin, M.F. Achard, Solid State Commun. 23, 35 (1977).

7 J.Y. Buzare, J.C. Fayet, W. Berlinger, K.A. Müller, Phys.Rev. Lett. 42, 465 (1979).

8 E. Müller-Hartmann and Zittartz, Z.Physik B27, 261 (1977).

9 A.A. Migdal, Zh.Eks.Teor.Fiz. 69, 1457 (1975).

10 L.P. Kadanoff, Ann.Phys. 100, 359 (1976).

11 D.R. Nelson and M.E. Fisher, Ann.Phys. 91, 226 (1975).

ELEMENTARY EXCITATIONS IN MAGNETIC CHAINS

Jacques Villain

Department de Recherche Fondamentale
Laboratoire de Diffraction Neutronique
Centre d'Etudes Nucléaires, 85 X, 38041 Grenoble Cedex

ABSTRACT

One-dimensional magnets exhibit elementary excitations which can be observed at low temperature. The theory of these excitations is presented here in a tutorial way in certain simple cases, namely i) Magnons of XY like chains ii) solitary excitations of Ising-like chains iii) solitons of XY like chains in a field or anisotropy. Experimental examples like TMMC, $CsNiF_3$ and $CsCoCl_3$ are considered.

I. INTRODUCTION

One-dimensional models are often used by theorists (and sometimes by experimentalists) in order to test rather general concepts on a more or less solvable model. For instance, one-dimensional spin glasses have recently been observed (Wiedemann, Burlet 1978) and 1-D media are more or less the only systems where spin glass ordering can easily be analysed theoretically (e.g. Derrida et al. 1978).

But one-dimensional systems also have special features in particular an extremely intense "short range" order. In the case of magnetic chains with short range interactions (the case of interest in this leture) it can be shown that the spin pair correlations have an exponential decay when the distance p is long:

$$\langle \vec{S}_m \cdot \vec{S}_{m+p} \rangle \sim \exp - k|p| \qquad \text{(I-1)}$$

but the correlation length 1/p can be very large at low temperature T. It is about 500 Å for $(CH_3)_4NMnCl_3$ (alias TMMC) at T = 1K.

This strong "short range" order has certain interesting consequences. For instance, Villain (1977) has pointed out that quasi 1-D helimagnets can have an ordering of the chirality of the helix (i.e. a positive or negative pitch) without magnetic long range order. This lecture will be focussed on another problem, namely elementary excitations.

The standard method of treating elementary excitations in magnets uses the fact the the system is ordered (see e.g. Herpin 1968, Chapter XI). Magnons have been experimentally observed, in the absence of long range magnetic order, in the antiferromagnetic chain TMMC (Hutchings et al. 1977) and in the ferromagnetic chain $CsNiF_3$ (Steiner et al. 1975). *

The theory of magnons in the XY chain will be discussed in Section II. Magnons are delocalized excitations, the amplitude of which can change continuously in classical mechanics. The standard theory assumes a harmonic system.

In addition to magnons, "solitary excitations" can be observed in certain magnetic chains. They are localized and their amplitude is quantised even in classical mechanics. They can only exist in anharmonic systems. This distinction between harmonic, delocalized excitations and solitary excitations may be encountered in everyday life. For instance, if a violin string is excited with a bow, the result is a delocalized, harmonic, continuously variable wave. If it is excited with a pair of scissors the result is an integer number of localized cuts, which testify to the anharmonic nature of the string.

According to usual definition (Scott et al. 1972) a solitary excitation is called a soliton if it propagates with a constant velocity in the absence of collisions and if its velocity and shape are not affected by collisions with other solitons except during the collision itself.

Solitary excitations probably do not exist in XY chains like $CsNiF_3$ in zero field. They do exist in Ising chains at low temperature (Fig. 1) and have a well known effect (Landau, Lifshitz): they destroy long range order at any non-vanishing temperature in 1-D systems with short range interactions. Propagation of such excitations will be studied in Section IV. A slightly more complicated case is the XY chain with a weak uniaxial anisotropy (Fig. 1c) or the XY antiferromagnetic chain in a magnetic field (Fig. 1d). It

* The word "(anti) ferromagnetic" is used for "(anti) ferromagnetically coupled" and does not imply long range order.

will be considered in Sections III and V.

The Heisenberg model (isotropic, 3-D spins) will generally be ignored in this lecture. Solitons have been predicted in the Heisenberg chain (Tjon, Wright 1977) although their existence and stability has no topological foundation and can only be derived from complicated anharmonic equations. This situation is in contrast with XY systems where solitary excitations are associated with 2π rotation (Fig. 1e) or π rotation (Fig. 1c), for example. Magnons are also a difficult subject in the Heisenberg chain because of the non-Abelian character of the 3-D rotation group. However, in additon to numerical calculations (e.g. Blume et al. 1975, Loveluck, Windsor 1978, Müller, Beck 1978) there exist analytical calculations (Mikeska, Patzak 1977, Reiter, Sjölander 1977).

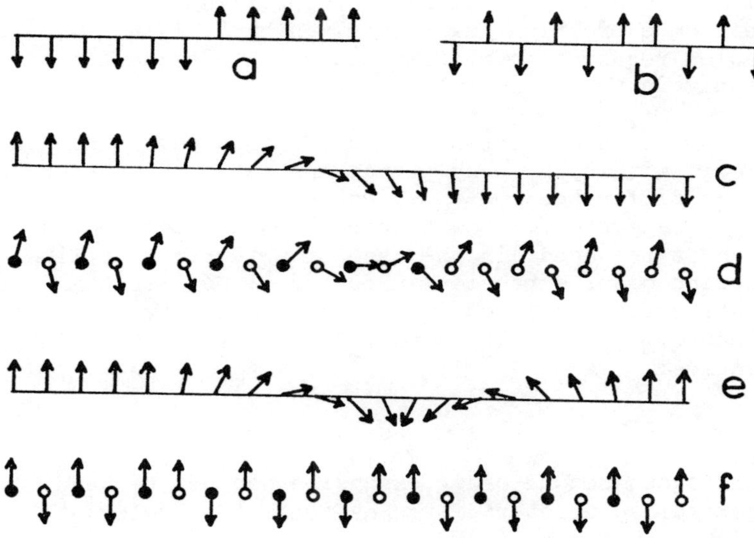

Fig. 1 Walls (alias solitons) a) in the ferromagnetic Ising chain b) in the antiferromagnetic Ising chain ($CsCoCl_3$) c) in the ferromagnetic XY chain with a uniaxial anisotropy. d) in the antiferromagnetic XY chain in a magnetic field (TMMC) e) in the ferromagnetic XY chain in a field ($CsNiF_3$) f) two walls (see § IV-2).

II. MAGNONS IN XY-LIKE MAGNETIC CHAINS

This section is devoted to chains of spins with an easy magnetisation plane and no prefered orientation in the plane. $CsNiF_3$ is an excellent example (Steiner et al. 1975). The antiferromagnetic chain TMMC can also be considered as an XY system at low temperature (Walker et al. 1972, Hone, Pires 1977), let us say below 5K (Boucher et al. 1979). Electron resonance experiments of Magarino et al. (1978) show evidence of a very weak in-plane anisotropy, which can be neglected in most cases.

We consider a chain of N identical spins:

$$\vec{S}_m = \left(\sqrt{s^2 - S_m^{z\,2}}\,\cos\phi_m,\ \sqrt{s^2 - S_m^{z\,2}}\,\sin\phi_m,\ S_m^z\right) \tag{II-1}$$

This expression is a classical approximation, correct for large s. A quantum expression has been given by Villain (1974) but its exploitation requires approximations which are quantitatively correct only in the classical limit.

Nearest neighbour interactions are assumed. Since they should be isotropic within the easy plane, the interaction energy depends only on the difference $(\phi_m - \phi_{m-1})$ rather than on the two angles ϕ_m and ϕ_{m-1} separately. Low temperatures are assumed, so that the energy can be replaced, in the case of a ferromagnet like $CsNiF_3$, by its second order expansion in powers of S_m^z, S_{m-1}^z and $(\phi_m - \phi_{m-1})$. For an antiferromagnetic chain the small quantity is $(\phi_m - \phi_{m-1} - \pi)$ rather than $(\phi_m - \phi_{m-1})$. It is therefore convenient to introduce the variables:

$$\psi_m = \phi_m + m\pi \tag{II-2}$$

and an analogous second order expansion can now be used. In view of the periodic character of the system, it is relevant to use the Fourier transforms:

$$S_k^z = N^{-\frac{1}{2}} \sum_{m=o}^{N-1} S_m^z \exp ikm, \text{ etc}$$

The approximate Hamiltonian for $CsNiF_3$ is:

$$\mathcal{H} = 2J \sum_k \left[s^2(1-\cos k)\phi_k^*\phi_k + (\eta - \cos k) S_{-k}^z S_k^z \right] \tag{II-3}$$

where the single-ion anisotropy $(\eta-1)$ has its origin in the spin-orbit interaction. For the antiferromagnetic chain TMMC the anisotropy originates from dipole interactions. In one dimension it is not unreasonable to neglect dipole interactions between next nearest neighbours and further; the appropriate Hamiltonian for TMMC is:

$$= 2|J| \sum_k \; s^2(1-\cos k)\psi_k\psi_k+(1+\alpha\cos k)S^z_{-k}S^z_k \qquad \text{(II-4)}$$

where the anisotropy parameter $(1-\alpha)$ is small. S^z_m and ϕ_m (or ψ_m) are conjugate variables, as shown for instance by Villain (1974) or Halperin and Saslow (1977):

$$[S^z_m,\phi_p] = [S^z_m,\ \psi_p] = i\hbar\delta_{mp} \;;\; [S^z_m,S^z_p] = [\phi_m,\phi_p] = 0 \qquad \text{(II-5)}$$

The one-dimensional systems described by (II-3) and (II-4) are seen to be analogous to a harmonic phonon or magnon system, despite the lack of long range order. The statistical mechanics is therefore straightforward, and was treated in detail by Villain (1974).

There is an important difference between the above treatment and the standard spin wave theory in ordered systems. Instead of the spin operators, the quantities that appear in (II-3) and (II-4) are the z component of the spin operators and the polar angles ϕ_m. The quantities which are observed by inelastic neutron scattering are the pair correlations of the spin operators (Marshall, Lovesey 1971). This difficulty is frequently encountered in 1-D systems: the operators appropriate for the description of experiments are not the operators appropriate for the description of quasi-particles.

A similar problem occurs for the elementary excitations of the Ising chain (Villain 1975) and of the XY chain of spins $\frac{1}{2}$ (Capel, Perk 1977 and refs therein).

In the case of the classical XY chain, eqs (II-3) and (II-4), it is not too difficult to evaluate the neutron cross section $S(k,\omega)$ (Villain 1974). The following qualitative results are rather intuitive. $S(k,\omega)$ contains two components.

- An "out of plane" component (O P C) resulting from fluctuations S^z_k along the hard axis. The O P C is a narrow line in the spectrum because S^z_k is a linear combination of a creation and a destruction operator of a magnon, as is easily seem from (II-3) or (II-4).

- An "in-plane" component (I P C) resulting from fluctuations

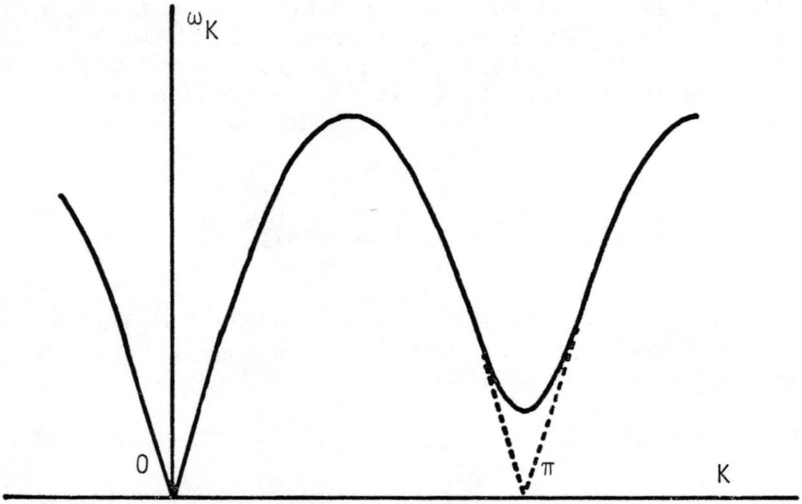

Fig. 2 Magnon spectrum for TMMC. Near the zone boundary $k=\pi$, a sharp, but weak OPC is expected at finite frequency (full curve). The IPC is overdamped but more intense (dotted curve).

S_k^x and S_k^y in the easy plane. This component results from many-magnon processes because magnon creation and destruction operators involve ϕ_k rather than S_k^x and S_k^y. Therefore, the I P C is much broader than the O P C. This effect has been observed in $CsNiF_3$ (Steiner et al. 1975).

In TMMC, the anisotropy is very small. The spin wave spectrum is shown by Fig. 2. If anharmonic terms are added to (II-4), they produce a very high damping in the linear region of the curve because energy momentum conservation conditions are always fulfilled. It is therefore not too surprising that no sharp OPC was detected by neutron scattering (Hutchings et al. 1977, Villain et al. 1979). A sharp OPC (but much less intense than the IPC) should be observable at the zone boundary and in its vicinity (Fig. 3). Very recent neutron experiments (Heilmann et al. 1979) are apparently in agreement with this prediction.

Another theoretical prediction (Villain 1974) which is beyond current possibilities is that the IPC should also become narrower in the flat regions of the dispersion curve of an XY chain. Similar predictions have been made by Reiter and Sjölander (1978) for Heisenberg chains. Neutron spin echo techniques might be appropriate tests of these predictions.

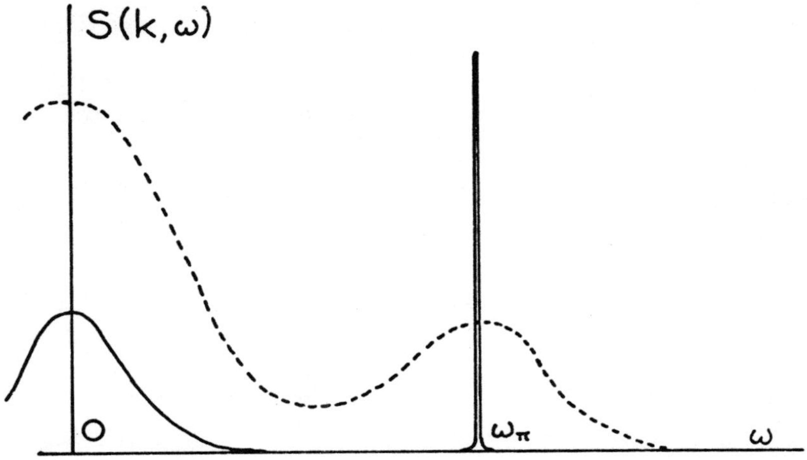

Fig. 3 Expected neutron cross section at the zone boundary for TMMC at T = 5K (full curve) and T = 10K (dotted curve) according to Villain et al. (1979).

Other theoretical methods have been developed by Nelson and Fisher (1977).

III. THE ANISOTROPIC, CLASSICAL XY CHAIN

III.1 The Model

A more complicated case is an XY chain subject to an anisotropy or a magnetic field. A large amount of experimental work has been recently devoted to this subject (e.g. Steiner's lecture, or Boucher et al. 1979).

Time-dependent properties will be outlined in Section V. This Section is devoted to a time-independent model subjected to a Hamiltonian:

$$\mathcal{H} = -2Js^2 \sum_m \cos(\phi_m - \phi_{m-1}) - A \sum_m \cos q\phi_m \qquad \text{(III-1)}$$

where A is a magnetic field (q=1), a uniaxial anisotropy (q=2) or a q-fold anisotropy (q=4 or 6).

At low temperature $K_BT \ll J s^2$ a good approximation for a ferromagnet (J > 0) is:

$$\mathcal{H} = Js^2 \sum_m (\phi_m - \phi_{m-1})^2 - A \sum_m \cos q\phi_m \qquad \text{(III-2)}$$

In this case the value q = 2 can always be assumed. If $q \neq 2$, one can use the variables $\theta_m = q\phi_m/2$.

If A is small, expression (III-2) can be replaced by its continuum approximation:

$$\mathcal{H} = \int_0^N dm \, [Js^2 (\frac{d\phi}{dm})^2 + 2A\sin^2\phi] \qquad \text{(III-3)}$$

where q has been replaced by 2 and an additive constant has been neglected.

III.2 Intuitive Analysis at Low Temperature for q = 2 and J = 0.

The correct mathematical method for classical chains with short range interactions uses the transfer operator (e.g. Blume et al. 1975). The highest 2 or 3 eigenvalues are usually necessary and can be calculated numerically (e.g. Hijmans et al. 1978), by approximate methods like the WKB approximation (Krumhansl, Schrieffer 1975, Nakamura et al. 1978) by perturbation theory (Borsa et al. 1978, Nakamura et al. 1978) and sometimes exactly.

The transfer operator technique is not very transparent and will not be used here. Our survey will rather be based upon intuitive considerations.

The chain has 2 ground states: i) $\phi_m = 0$, ii) $\phi_m = \pi$. These ground states are separated by a potential barrier, whose height w_1 will be determined in the next paragraph. Intuitively, the chain is expected to look like:

an Ising chain if $K_B T \ll w_1$

an XY chain if $K_B T \gg w_1$

For $K_B T \ll w_1$ there are two types of thermal fluctuations i) Fluctuations of ϕ_m around the value 0 or π, without crossing the potential barrier. The decay of the correlation function $\langle \sin\phi_m \sin\phi_{m+p} \rangle$ is produced by these fluctuations. For large $|p|$, this decay can be shown to be exponential

$$\langle \sin\phi_m \sin\phi_{m+p} \rangle \sim \exp - \kappa_2 |p| \qquad \text{(III-4)}$$

ii) Fluctuations across the potential barrier. For instance $\phi_m \approx 0$ in the left hand part of the chain and $\phi_m \approx \pi$ in the right hand part (Fig. 1c). This defines a wall, similar to Ising walls (Fig. 1a) but broader. The minimum energy of the wall is the energy w_1 of the potential barrier. The decay of the correlation function $\langle \cos\phi_m \ \cos\phi_{m+p} \rangle$ is produced by walls:

$$\langle \cos\phi_m \ \cos\phi_{m+p} \rangle \sim \exp - \kappa_1 \ |p| \qquad \text{(III-5)}$$

Expression (III-4) decays more rapidly with p than (III-5) since its decay is due to "easy" fluctuations which do not demand the high energy w_1 of a wall. When K_BT becomes larger than w_1, both correlation lengths $1/\kappa_1$ and $1/\kappa_2$ are expected to merge into a single one $1/\kappa$.

These simple arguments will be put into a more quantitative form in the following paragraphs.

III.3 Energy of a Wall

The structure of a wall results from a minimisation of (III-2) or (III-3) with respect to ϕ_m:

$$Js^2 \ \frac{d^2\phi}{dm^2} = A \sin 2\phi \qquad \text{(III-6)}$$

with the conditions: i) $\phi(-\infty) = 0$; ii) $\phi(+\infty)=\pi$; iii) ϕ is an increasing function of m. It is convenient to consider on open chain rather than a cyclic chain.

Far from a wall, the "sine Gordon" equation (III-6) can be linearized. Its solution is an exponential and it can be shown that the argument of the exponential is the same in the correlation function (III-4). Hence:

$$\kappa_2 = \frac{1}{\lambda_0} = \sqrt{2A/Js^2} \qquad \text{(III-7)}$$

The solution of (III-6) which satisfies the required conditions is (Scott et al. 1973, Herpin 1968):

$$\phi_m = 2 \tan^{-1} (\exp - \kappa_2 m) \qquad \text{(III-8)}$$

The energy can be deduced from (III-3) and (III-8) (Herpin 1968)

$$w_1 = 8Js^2\kappa_2^2 \int_{-\infty}^{\infty} dm \frac{\exp-2\kappa_2 m}{(1+\exp-2\kappa_2 m)^2}$$

$$w_1 = 4\sqrt{2AJs^2} \qquad \text{(III-9)}$$

III.4 Number of Walls

The number of walls per atom can be identified with the inverse correlation length κ_1 which appears in (III-5). It is expected to be dominated by a factor $\exp - \beta w_1$ typical of uniaxial systems. The correct expression which can be deduced from the transfer matrix method is (Mikeska 1978 and refs therein):

$$\kappa_1 = 8\,\kappa_2^{3/2}\sqrt{\beta Js^2/2\pi}\,\exp - \beta w_1 \qquad \text{(III-10)}$$

This expression is correct if 3 conditions are satisfied: i) few walls, $\kappa_1 << \kappa_2$. ii) classical mechanics. iii) broad walls $\kappa_2 << 1$.

The derivation of (III-10) is tedious and will not be given, but an intuitive justification of the prefactor $\sqrt{\beta}$ will be outlined. The average number of walls over a length of ℓ sites ($\kappa_1\ell << 1$) is $\ell\kappa = Z_1/Z_0$ where Z_0 and Z_1 are respectively the partial partition functions with 0 walls and with 1 wall.

The low temperature behaviour of Z_0 is dominated by a factor $\beta^{-\ell/2}$ arising from the ℓ degrees of freedom. Z_1 only contains a factor $\beta^{-(\ell-1)/2}$ because one degree of freedom is soft, namely the position of the wall centre. The walls can indeed be moved continuously if K_BT is much larger than the wall pinning energy which is of order $w_1 \exp - \kappa_2$ (Friedel 1978).

III.5 Antiferromagnets in a Magnetic Field

The most spectacular application of the above considerations is probably the effect of a magnetic field on a quasi-1-D antiferromagnet: The Néel temperature increases, as discovered by Dupas and Renard (1976). Hijmans et al.(1978) observed the effect in several materials and obtained a quantitative fit with numerical calculations. In TMMC, T_N increases by a factor 3 under a field of 80 000 Ø, but there is a quantitative disagreement with theory. According to Loveluck (private communication) direct measurements of the correlation lengths (Boucher et al. 1979) do agree with theory.

The theory (Villain, Loveluck 1977) can be outlined as follows:

a) the Néel temperature (Scalapino et al. 1975) is given by:

$$2\ |J'|\ z\ \chi(T_N) = 1 \qquad \text{(III-11)}$$

where $\chi(T)$ is the staggered susceptibility of an isolated chain and J' is the interchain interaction, here treated in M.F.A. z is the number of chains interacting with a given chain.

b) The spins of an antiferromagnet orient perpendicular to the field. Two orientations are possible, therefore the system experiences an effective anisotropy. The effective anisotropy constant A is an even, analytical function of H which vanishes for H = 0, therefore it is proportional to $H^2/|J|$, where the denominator $|J|$ is justified by dimensionality arguments:

$$A \sim H^2|J| \qquad \text{(III-12)}$$

c) The staggered susceptibility for n-dimensional spins is proportional to:

$$\chi(T) \sim 1/nK_B T\kappa_1(T) \qquad \text{(III-13)}$$

For $H \simeq K_B T$, a dramatic increase of T_N results from eqs. (III-11, 13, 10, 9, 7 and 12).

For a Heisenberg chain the relative increase of T_N cannot exceed $\sqrt{3}$. The value $\sqrt{2}$ given by Villain and Loveluck results from a MFA equation which differs from (III-11).

IV. A SIMPLE DYNAMICAL MODEL: THE ALMOST - ISING ANTIFERROMAGNETIC CHAIN

IV.1 The Model

We consider the following Hamiltonian (Villain 1975)

$$\mathcal{H} = -2J \sum_{m-o}^{N-1} S_m^z\, S_{mH}^z - 2\varepsilon\ J \sum_{m=o}^{N-1} (S_m^x S_{m+1}^x + S_m^y S_{m+1}^y) \qquad \text{(IV-1)}$$

with $J < 0$ and $|\varepsilon| << 1$. This model is appropriate for $CsCoCl_3$ (Hirakawa et al. 1978). As a first approximation ε can be neglected. The lowest excited states (Fig. 1a) are states with one wall centred at (p+1/2). They can be designated by the symbol $|p+1/2)\rangle$. They have all the same energy $4\ |J|\ s^2$.

When ε is taken into account, the states $|p+1/2$ are no longer eigenstates. The eigenstates, of course, are:

$$|k\rangle^{\pm} = N^{-\frac{1}{2}} \sum_{p=1}^{N/2} |2p \pm \tfrac{1}{2}\rangle \exp ikp \tag{IV-2}$$

where k is a multiple of $4\pi/N$ inside the first Brillouin zone $(-\pi < k < \pi)$. In our 1975 paper, different notations were used in order to account for periodic boundary conditions. Here, boundary condtions will be ignored for simplicity.

It can be shown (Villain 1975) that these states $|k\rangle$ correspond to a wall propagating with a constant velocity

$$\dot{x}(p) = \frac{4}{\hbar} \varepsilon J \sin k \tag{IV-3}$$

The energy now depends on k:

$$E_k = 4 \ |J| \ s^2(1+2\varepsilon \cos k) \tag{IV-4}$$

The analogy between these propagating walls and particles is obvious. In particular, for $k \approx 0$ or π, one can define an effective mass $M^* = \hbar^2/2 \ |\varepsilon J s^2|$.

IV.2 Collisions Between Two Solitons

Consider states with two solitons: $|\ 2m+\frac{1}{2},\ 2p-\frac{1}{2}\ \rangle$ (Fig. 1f). If collisions are neglected, the states:

$$|k,k'\rangle = N^{-1} \sum_{m,p} |2m+\tfrac{1}{2},\ 2p-\tfrac{1}{2}\rangle \ \exp ik(p+m)$$

are approximate eigenstates.

Collisions are produced by off-diagonal elements $\langle k'',k'''|\mathcal{H}|\ k,k'\rangle$. The non-vanishing matrix elements satisfy either the "normal" condition:

$$k + k' = k'' + k''' \tag{IV-5}$$

or the "Umklapp" condition:

$$k + k' = k'' + k''' \pm 2\pi \tag{IV-6}$$

In addition, energy must be conserved except during the collision:

$$E_k + E_{k'} = E_{k''} + E_{k'''} \tag{IV-7}$$

According to the definition of Scott et al. (1972) propagating walls may be called solitons if the system of equations (IV-7), (IV-4 and (IV-5 or 6) has only the trivial solution k" = k, k"' = k'. This system implies:

$$[\cos \tfrac{1}{2} (k-k') \pm \cos \tfrac{1}{2} (k''-k''')] \cos \tfrac{1}{2}(k+k') = 0 \qquad \text{(IV-8)}$$

where the minus sign applies to normal processes and the plus sign to Umklapp processes. If $(k+k') \neq \pm\pi$, equations (IV-8) has no solution satisfying (IV-6), and no solution satisfying (IV-5) except the trivial solution. If $k+k' = \pm\pi$, there are an infinite number of solutions. We conclude that propagating walls can reasonably be called solitons because the probability that they meet another soliton satisfying the condition $k+k' = \pi$ is very weak. For large soliton densities, however, this conclusion is modified by 3-soliton collisions.

Propagating walls in the antiferromagnetic Ising chain should give rise to a characteristic spectrum in neutron inelastic scattering. However, experiments in $CsCoCl_3$ have been negative (Hirakawa et al. 1978, Regnault, Delapalme, Rossat-Mignod, Henry unpublished) probably because narrow solitons are strongly pinned by lattice defects.

V. PROPAGATION OF BROAD WALLS

In Section III the statics of broad walls was studied in Section IV the dynamics of narrow walls was considered. The dynamics of broad walls is a slightly more complicated problem which will be briefly analysed in this Section. Physically, it corresponds to an XY ferromagnet or antiferromagnet in a field or in an anisotropy. The algebra is analogous in all cases, and for definiteness the case of an XY ferromagnet with a uniaxial anisotropy will be considered as in Section III. The Hamiltonian is the sum of (II-3) and an anisotropy (last term of Hamiltonian III-1 with q = 2).

$$\mathcal{H} = Js^2 \sum_m (\phi_m - \phi_{m-1})^2 - A \sum_m \cos 2\phi_m - 2J \sum_m (S^z_m S^z_{m-1} - \eta S^{z^2}_m) \qquad \text{(V-1)}$$

The equation of motion for ϕ_m is easily written with the help of (II-5):

$$\ddot{\phi}_m = - 4J^2s^2[\,(4\eta +2)\phi_m - 2(1+\eta)(\phi_{m+1}+\phi_{m-1})+\phi_{m+2}+\phi_{m-2}\,]$$

$$- 4JA[\,2\eta\sin 2\phi_m - \sin 2\phi_{m+1} - \sin 2\phi_{m-1}\,]$$

In the continuum limit, this equation reduces to the time-dependent sine Gordon equation:

$$\frac{\partial^2\phi}{\partial_m{}^2} - \frac{1}{c^2}\frac{\partial^2\phi}{\partial_t{}^2} = \tfrac{1}{2}\,\kappa_2^2 \sin 2\phi \qquad \text{(V-2)}$$

where κ_2 is given by (III-7) and c is the magnon velocity in the absence of in-plane anisotropy (A = 0):

$$c^2 = 8J^2s^2(\eta-1) \qquad \text{(V-3)}$$

Eq. (V-2) can be transformed into the time-independent sine Gordon equation (III-6) if one uses the variable:

$$p(m,t) = \frac{m - vt}{\sqrt{1-v^2/c^2}} \qquad \text{(V-4)}$$

and $\phi_m(t)$ is given by (III-8), where m is to be replaced by p. Notice the relavistic character of the transformation (V-4). These results mean that an isolated broad wall propagates with a constant velocity v which cannot exceed the magnon velocity.

Many wall solutions of (V-2) can be exhibited (Hirota 1972, Scott et al. 1973 and refs. therein) and it can be shown that sine Gordon walls are solitons according to the above definition, Sections I and IV.

In a ferro- or antiferromagnet of the XY type with a weak anisotropy, the relaxation time of the spatial Fourier transform of $\langle\cos\phi_m \cos\phi_{m+p}(t)\rangle$ for $k\approx 0$ and $k\approx\pi$ respectively is directly related to the number and velocity of solitons. It might be measured by neutron scattering. Low temperatures are of course necessary. An antiferromagnetic chain like TMMC in a magnetic field is also a possible candidate. The only apparently successful experiment was made by Steiner and Kjems (1978) following a proposal of Mikeska (1978) for the ferromagnetic chain $CsNiF_3$ in a field. In this case the weakly inelastic soliton contribution has to be separated from a completely elastic Bragg line and an inelastic magnon contribution.

ACKNOWLEDGEMENT

The author is indebted to J. Loveluck for critical reading of the manuscript, and to his colleagues of the Département de Recherche Fondamentale (C.E.N. Grenoble) (especially L.P. Regnault) for useful discussions.

REFERENCES

This list is not exhaustive. Certain important References have been omitted when they are available in a more recent work listed below, for instance in the review by Steiner et al. (1976).

Blume, M., Heller, P., Lurie, N.A. 1975, Phys.Rev. B 11, 4483
Borsa, F., Boucher, J.P., Villain, J., 1978, J. Appl.Phys. 49,1327
Boucher, J.P., Regnault, L.P., Rossat-Mignod, J., Villain, J., 1979, to be published
Capel, H.W., Perk, J.H.H., 1976, Physica 87a, 211
De Jonge, W.J.M., Hijmans, J.P.A.M., Boersma, F., Shouten, J.G., Kopinga, 1978, Phys.Rev. B 17, 2922
Derrida, B., Vanimenus, J., Pomeau, Y., 1978, J.Phys. C 11, 4749
Dupas, C., Renard, J.P., 1976, Solid State Com. 20, 581
Friedel, J., 1978"Extended defects in materials". Trieste lectures
Halperin, B.I., Saslow, W.M., 1977, Phys.Rev. B 16, 2154
Heilmann, Birgeneau, R.J., Endoh, Y., Reiter, G., Shirane, G., Holt, S.L., 1979, Brookhaven preprint No. 25828
A. Herpin, 1968, "Téorie du magnétisem". Presses Universitaires du France, Paris.
Hijmans, J.P.A.M., Kopinga, K., Boersma, F., De Jonge, W.J.M, 1978, Phys.Rev.Lett. 40, 1108
Hirakawa, K., Yoshizawa, H., 1978, Technical report of I.S.S.P., A 907. University of Tokyo
Hirota, R., 1972, J.Phys.Soc.Japan 33, 1459
Hone, D., Pires A., 1977, Phys.Rev. B 15, 323
Hutchings, M.T., Windsor, C.G., 1977, J.Phys. C 10, 313
Krumhansl, J.A., Schrieffer, J.R., 1975, Phys.Rev. B 11, 3535
Landau, L.D., Lifshitz, E.M., 1959, "Statistical Physics", Pergamon Press, London
Loveluck, J.M., Windsor, C.G., 1978, J.Phys. C 11, 2999
Margarino, J., Tuchendler, J., Renard, J.P., 1978. Preprint. See also Phys.Rev. B 14, 865 (1976)
Marshall, W., Lovesey, S.W. 1971, "Theory of thermal neutron scattering", Academic Press, Oxford
Mazenko, G.F., Shani, P.S., 1978, Phys.Rev. B 18, 6139
Mikeska, H.J., Patzak, E., 1977, Z.Phys. B 26, 253
Mikeska, H.J., 1978, J.Phys. C 11, L 29
Müller, G., Beck, H., 1978, J.Phys.C 11, 483
Nakamura, K., Sasada, T., 1978, J.Phys. C 11, 331
Nelson, D.R., Fisher, D.S., 1977, Phys.Rev. B 16, 4945

Patkos, A, Rujan, P., 1978,ITP-Budapest report No. 386, to be published in Z. Phys.
Reiter, G., Sjölander, A., 1977, Phys.Rev.Lett. 39, 1047
Scalapino, D.J., Imry, Y., Pincus, P., 1975, Phys.Rev.B 11, 2042
Scott, A.C., Chu, F.Y.F., Mc Laughlin, D.W., 1973, Proc. IEEE 61, 1443
Steiner, M., Dorner, B., Villain, J., 1975, J.Phys. C 8, 165
Steiner, M., Kjems, J., 1978, J. Physique Lettres 39, L 493
Steiner, M, Villain, J., Windsor, C.G., 1976, Adv.Phys. 25, 87
Tjon, J., Wright, J., 1977, Phys.Rev. B 15, 3470
Villain, J., 1974, J. Physique 35, 27, 1975, Physica B 79, 1, 1977, Communication at the 13th Conference of Statistical Mechanics, Haifa
Villain, J., Loveluck, J.M., 1977, J. Physique 38, L 77
Villain, J., Boucher, J.P., Regnault, L.P., Rossat-Mignod, J.M., 1978, Bulletin du Département de Recherche Fondamentale No. 13, (note CEA N-2074), Centre d'Etudes Nucléaires de Grenoble
Walker, L.R., Dietz, R., Andres, K., Darack, 1972, Solid State Com. 11, 593
Wiedenmann, A., Burlet, P., 1978, J. Physique 39, C6-720

EXPERIMENTAL STUDIES OF LINEAR AND NONLINEAR MODES IN 1-D-MAGNETS

M. Steiner

Hahn-Meitner-Institut, Glienickerstr. 100
1000 Berlin 39
Germany

ABSTRACT

Collective excitations are an essential part of the dynamics of magnets even with only short range order. Recently it has become clear that not only the well known spin waves have to be considered, but in certain cases nonlinear modes like solitons too. In this review experimental results obtained by inelastic neutron scattering for the linear as well as for the nonlinear modes in 1-D-Magnets are discussed and compared with theory. It is shown that most of the properties of the linear excitations are understood whereas in the case where nonlinear modes contribute the dynamics are not fully understood yet. Results for the following systems are discussed: TMMC (x-y-antiferromagnet), CPC (S = 1/2-Heisenberg-antiferromagnet), $CsNiF_3$ (x-y-ferromagnet). In all these systems the linear excitations have been studied in great detail whereas nonlinear modes have been observed in $CsNiF_3$ only.

INTRODUCTION

One dimensional (1-D-) magnets have as all 1-D-systems the characteristic property that they do not show a phase transition for $T > 0$. Their properties are governed by short range order (SRO), which varies with time and temperature. The SRO is described by a characteristic length, the correlation length ξ, which describes the exponential decay of the correlation along the 1-D-chain. Therefore the region over which the system can be considered to be "ordered" is given by $\xi(T)$. The fact that no long range order (LRO) exists implies that $\langle M \rangle \equiv 0$. The SRO however can be described by the spin pair correlation function $G(r,t)$:

$$G(r,t) = \langle S(0,0)\cdot S(r,t)\rangle$$

which varies with temperature. G(r,t) contains the quasistatic or equal time (t = 0) properties as well as the dynamic (t ≠ 0) properties. In this paper only that part of G(r,t) will be considered which contains the collective excitations. A detailed discussion of all aspects of G(r,t) can be found in the review by Steiner et al.[1] As collective excitations we consider such which involve an in phase movement of a certain number of spins: Spin waves or linear excitations are extended, whereas the nonlinear excitations are localized: they contain an infinite or a finite number of spins respectively.

Since the basic theory is given by Villain,[2] the theoretical results will be taken from there. The Hamiltonian which describes the 1-D-magnets is the following:

$$\mathcal{H} = -2J \sum_{i=1}^{N} \vec{S}_i \vec{S}_{i+1} + A\sum_{i=1}^{N} (S_i^z)^2 \qquad (1)$$

The systems which will be considered here correspond to:

TMMC :	$J < 0$;	$A > 0$;	$S = 5/2$
CPC :	$J < 0$;	$A = 0$;	$S = 1/2$
$CsNiF_3$:	$J > 0$;	$A > 0$;	$S = 1$

As will be seen the possibility to add a Zeeman term in (1) via an external field is a great advantage of the magnetic systems since it modifies dramatically the properties of 1-D magnets: it is only due to this term that the nonlinear excitations become important. Generally collective excitations are expected to be stable as long as the SRO is strong enough: it is the 1-D-character of the systems that makes SRO very strong over a large temperature range so that the excitations can be studied in detail. And due to the one-dimensionality nonlinear modes can be studied theoretically and can be described easily in terms of quasiparticles.

The aim of the experiments is to explore under which conditions collective excitations exist and to test the theoretical predictions. All together this leads to detailed understanding of the collective excitations in 1-D-magnets.

In Section I it will be shown how one obtains 1-D-magnets in real 3-D-crystals and what the principles of the experimental methods used are. In Section II the linear excitations in zero and finite external field are discussed. Section III deals with the nonlinear excitations briefly summarizing the theoretical background before discussing the experimental results for $CsNiF_3$.

I. REAL SYSTEMS; EXPERIMENTAL METHODS

1-D-Magnets in 3-D-Crystals

The magnetic interaction in insulating magnetic crystals, which are considered here only, is usually well described by the Heisenberg superexchange (SE). The main property of this mechanism is its short range and its strong dependence on the geometry of the superexchange path, since it relies on the overlap of electronic orbitals: typically the interaction is reduced by an order of magnitude by inserting one more nonmagnetic ion into the bond path. Strength and sign of the interaction often vary strongly with the bond angle. It is therefore understandable that one can produce systems in which the magnetic interaction J is strong along one lattice direction and weak, J', in the two other directions. Thus, such systems are good candidates for 1-D-magnets as long as $J \gg J'$. In Fig. 1 two typical crystal structures of 1-D-magnets are shown. Both clearly demonstrate the ideas outlined about SE above. Figure 1(a) shows the structure of TMMC and $CsNiF_3$ (hexagonal $P6_3/mmc$) whereas CPC has a different structure (for details, see Ref. 1). Another effect in the crystal is mostly responsible for the anisotropy term A in (1). Due to the local symmetry of the magnetic ion preferred directions for the magnetic moments might exist which are described by the anisotropy constant A. Having these ideas in mind it is possible to find very good realizations of 1-D-magnets which often very closely approach the simple Hamiltonian (1). A selection of such systems is shown in Table I (from Steiner et al.[1]).

Since these systems are 3-D-crystals a phase transition to a three dimensional magnetic ordered state occurs at sufficiently low temperatures. But as long as one stays above this phase transition the system behaves as a 1-D-magnet. The 1-D-character is the

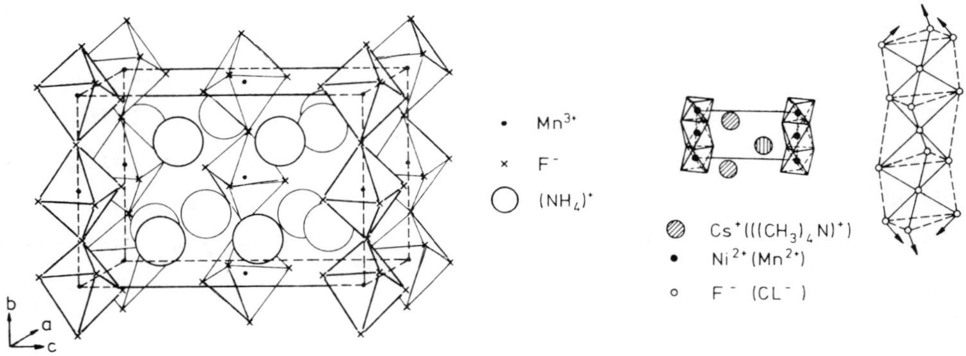

Fig. 1: The crystal structure of $(NH_4)_2MnF_5$ (left) and $CsNiF_3$ or TMMC (right).

Table I. Some 1-D-magnetic systems and their properties (from Ref. 1)

Substance	S	J/k[K]	T_c, T_N[K]	T_c/θ_p	Model
$CuSO_4.5H_2O$	1/2	-1.45	0.03	0.02	Heisenberg
$CuSeO_4.5H_2O$	1/2	-0.8	0.045	0.056	Heisenberg
$Cu(NH_3)_4SO_4.H_2O$	1/2	-3.15	0.37	0.12	Heisenberg
$Cu(NH_3)_4SeO_4.H_2O$	1/2	-2.36	1.2	?	Heisenberg
$Cu(NH_3)_4(NO_3)_2$	1/2	-3.70	1.2	?	Heisenberg
$CuCl_2.2NC_5H_5$	1/2	-13	1.7	0.13	Heisenberg
$KCuF_3$	1/2	-190	38	0.2	Heisenberg
$CsNiCl_3$	1	-13	4.65	0.13	Heisenberg
$RbNiCl_3$	1	-17	11.0	0.24	Heisenberg
$CsMnCl_32H_2O$	5/2	-3.57	4.89	0.12	Heisenberg
TMMC	5/2	-6.5	0.84	0.011	Heisenberg
$CsCuCl_3$	1/2	+ ?	10.4	?	Heisenberg
$(CH_3)_4NNiCl_3$	1	+2	1.2 ?	0.5 ?	Heisenberg
$K_3Fe(CN_6)$	1/2	-0.23	0.129	0.56	Ising
$CsCoCl_3$	1/2	-100 ?	8 ?	0.08 ?	Ising
$(NH_4)_2MnF_5$	1/2	-12	7.5	0.08	Ising
$CsCoCl_3.2H_2O$	1/2		3.8	?	Ising
$CoCl_2.2NC_5H_5$	1/2	+9.5	3.5	0.37	Ising
$CsCoBr_3$			16	?	Ising
$RbCoBr_3$			36	?	Ising
$RbFeCl_3$	2		2.55	?	plan. Heisenberg
$CsNiF_3$	1	11.8	2.61	0.08	plan. Heisenberg

better, the larger θ_p/T_N is: θ_p is the temperature when 1-D-coupling starts to become significant whereas the 3-D-coupling only appears around T_N.

Thus, very good realization of 1-D-magnets are available for very different values of J, A and S in (1).

Experimental Technique

In order to study collective excitations in detail one needs to explore the whole Brillouin zone. Thus, the experimental technique must be able to transfer enough moment ($\vec{Q} \sim 10$ Å^{-1}) and energy ($\Delta E \sim 1TH_z$). This can most conveniently be done by inelastic neutron scattering. Thermal neutrons interacting via their magnetic moments with the magnetic moments in the crystal have the right wavelength and energy to be an ideal microscopic probe for that purpose.

The scattering process is illustrated in Fig. 2 and has to conserve energy as well as momentum:

Momentum conservation	Energy conservation
$\hbar\vec{k}_i + \hbar\vec{Q} = \hbar\vec{k}_f$	$\hbar^2\lvert k_i\rvert^2 = \hbar^2\lvert k_f\rvert^2 + \hbar\omega$

$\vec{k}_i$, $\vec{k}_f$, $\vec{Q}$, $\hbar\omega$ being moment of the incoming and outgoing neutron, the momentum transfer to the crystal and the energy of the excitation respectively. $\hbar\omega$ can be greater or less than zero depending on whether the neutron gained or lost energy. The general cross section for unpolarized neutrons is then given by (Marshall and Lovesey[3])

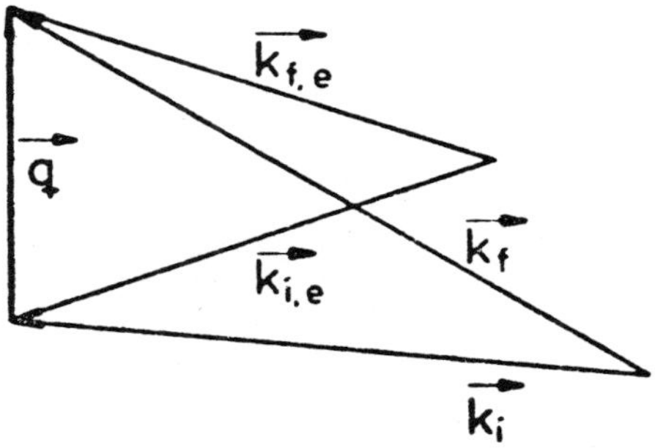

Fig. 2: The scattering triangle defined by $\vec{q}$, $\vec{k}_i$ and $\vec{k}_f$ the momentum transfer, the moment of the incoming and of the outgoing neutron, resp. The neutron momenta for elastic process ($\vec{k}_{i,e}$, $\vec{k}_{f,e}$) are shown as well.

$$S(\vec{q},\omega) \sim \frac{d^2\sigma}{d\omega d\Omega} = \frac{|k_f|}{|k_i|} \left(1.91 \frac{e^2}{mc^2}\right)^2 \sum_{ij} F_j^*(q) F_i(q) \times \tag{2}$$
$$1/2\pi \int_{-\infty}^{\infty} dt e^{i\omega t} \langle S_{i\perp}(0)\, S_{j\perp}(t)\rangle\, e^{i\vec{q}(\vec{R}_i - \vec{R}_j)}$$

$F_j(q)$ being the magnetic form factor. Since $G(r,t) \equiv \langle\vec{S}_i(0).\vec{S}_j(t)\rangle$ inelastic neutron scattering directly measures the space- and time-Fourier-transform of the pair correlation function $G(r,t)$, which is the fundamental quantity for systems with SRO. In addition, due to the dipolar coupling between neutron moment and $\vec{S}_i(t)$ only those spin components contribute which are perpendicular to the momentum transfer $\vec{Q}$ which is indicated by $S_{i\perp}$. There is however one limitation: since the neutron spin is 1/2, the maximum angular momentum change which can take place during the scattering process is $\Delta S = 1$. Thus n-scattering is the ideal tool to study excitations throughout the Brillouin zone. The instrument generally used for such experiments is shown in Fig. 3. By means of analyzing the scattered neutrons with regard to the scattering angle φ and the energy $\hbar\omega$ $S(\vec{q},\omega)$ can be determined.

Other methods like infrared absorption or ferromagnetic resonance can only probe the $q = 0$ region, but yield important

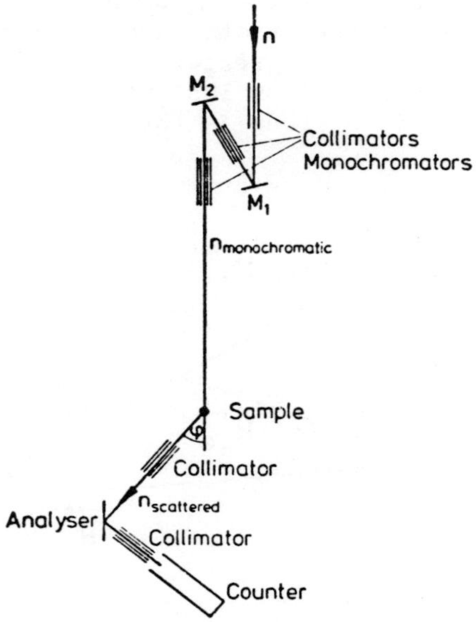

Fig. 3: Neutron flight path in a triple axis neutron spectrometer (IN2 at ILL Grenoble).

additional information however.

II. LINEAR EXCITATIONS

The linear excitations or spin waves are obtained through linearization of the equations of motion being the lowest lying excitations which change the total magnetization by 1. The well known classical spin wave is shown in Fig. 4. As discussed by Villain[2] such spin waves cannot be obtained in 1-D-magnets through the usual theoretical methods used in 3 dimensions due to the lack of LRO.

Dispersion at Low Temperatures

The typical feature of such a spin wave is that its energy $\hbar\omega$ depends on the wavelength. Thus there exists a dispersion relation between $\hbar\omega$ and q_c.

The dispersion relations can easily be calculated (Villain[2]):

Heisenberg AFM

$$\hbar\omega = 4|J|S \sin\pi q_c \tag{3}$$

x-y Ferromagnet

$$\hbar\omega = 2S\{(2J - 2J\cos\pi q_c)(2J - 2J\cos\pi q_c + A)\}^{1/2} \tag{4}$$

Quantum Heisenberg AFM (S = 1/2)

$$\hbar\omega = \pi|J| \sin\pi q_c \tag{5}$$

q_c denotes the distance from the actual Brillouin zone center along the chain direction, which is assumed to be the c-axis. If J, the intrachain interaction, is much larger than the interchain interaction J' then there shouldn't be any measurable dispersion along $\vec{q}_a$ or $\vec{q}_b$. Thus the measurement of the dispersion relation does

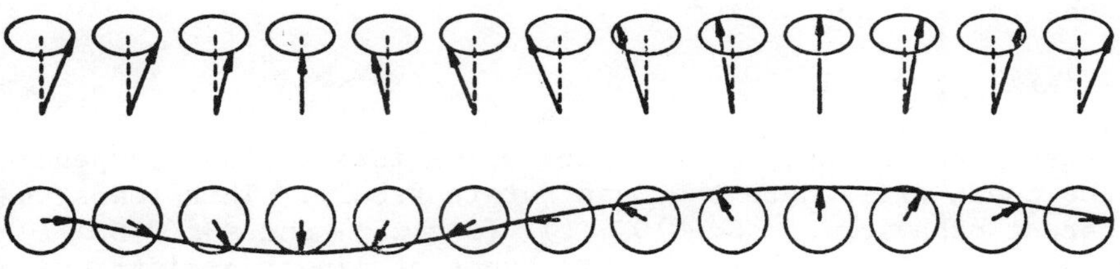

Fig. 4: Classical picture of spin waves in a 1-D-ferromagnet.

not only give the microscopic parameters which describe the systems, J and A, but might give an estimate for the interchain interaction as well being therefore a test for the one dimensionality of the systems.

In Fig. 5(a), (b) and (c) the spin wave peaks in the neutron cross section measured at low temperatures are shown for TMMC,[4] $CsNiF_3$,[5] CPC ($CuCl_2.2NC_5H_5$).[6] All the peaks are well defined and the energy for different wave vectors can be determined very precisely. The resulting dispersion for the three substances is shown in Fig. 6. For TMMC and $CsNiF_3$ the dispersion perpendicular to q_c is shown as well. The fact that there is no measurable dispersion shows that both systems are very good realizations of 1-D-magnets.

TMMC ($S = 5/2$ of the Mn^{2+} ion) has been considered a Heisenberg (isotropic) antiferromagnet up to now as far as the dynamics were concerned, despite the x-y-like anisotropy appearing at $T < 20$ K in the magnetization.[7] Therefore the found spin wave dispersion has been compared with (3). An excellent least squares fit can be obtained with $J/k = -6.5$ K as shown in Fig. 6(a). As Villain has shown, Fig. 2 in Ref. 2, the dispersion relation should split around $q_c = \pi$ due to the x-y-anisotropy. Hutchings et al.[4] did not see this, but recently Heilman et al.[8] have shown that it exists and behaves as predicted. Thus the discrepancy between spin waves and magnetization measurement could be resolved.

The effect of the single site anisotropy which enters the Hamiltonian through the anisotropy A was first studied in the 1-D ferromagnetic chain system $CsNiF_3$[5] where magnetization measurements have shown a very strong x-y-like anisotropy.[9] In this case the dispersion relation (4) is modified by A in such a way that the q^2 dependence of $\hbar\omega$ at small q, typical for an isotropic ferromagnetic chain, is replaced by a linear dispersion relation at small q ($\hbar\omega \sim q$). An excellent fit of (4) to the experimental data, Fig. 6(b), yields

$$J/k = 11.8 \text{ K} \; ; \qquad A/k = 4.5 \text{ K}$$

If we use corrections for the spin $S = 1$ (classical calculations assume $S \to \infty$) of the Ni^{2+} ions only the value for the anisotropy A/k has to be renormalized to:

$$(A/k)_{S=1} = 9 \text{ K} \quad .$$

The strong anisotropy confines the spins into a plane perpendicular to the chain axis where they can rotate freely. This leads to different behavior of $\langle S_r^\alpha(0) S_{r'}^\alpha(t)\rangle$ ($\alpha = x,y$) and $\langle S_r^z(0) S_{r'}^z(t)\rangle$[10] similar to TMMC and therefore two types of linear excitations are expected: the IPC (in-plane correlation) spin wave due to $\langle S^\alpha S^\alpha\rangle$ ($\alpha = x,y$) and the OPC (out of plane) spin wave due to $\langle S^z S^z\rangle$.

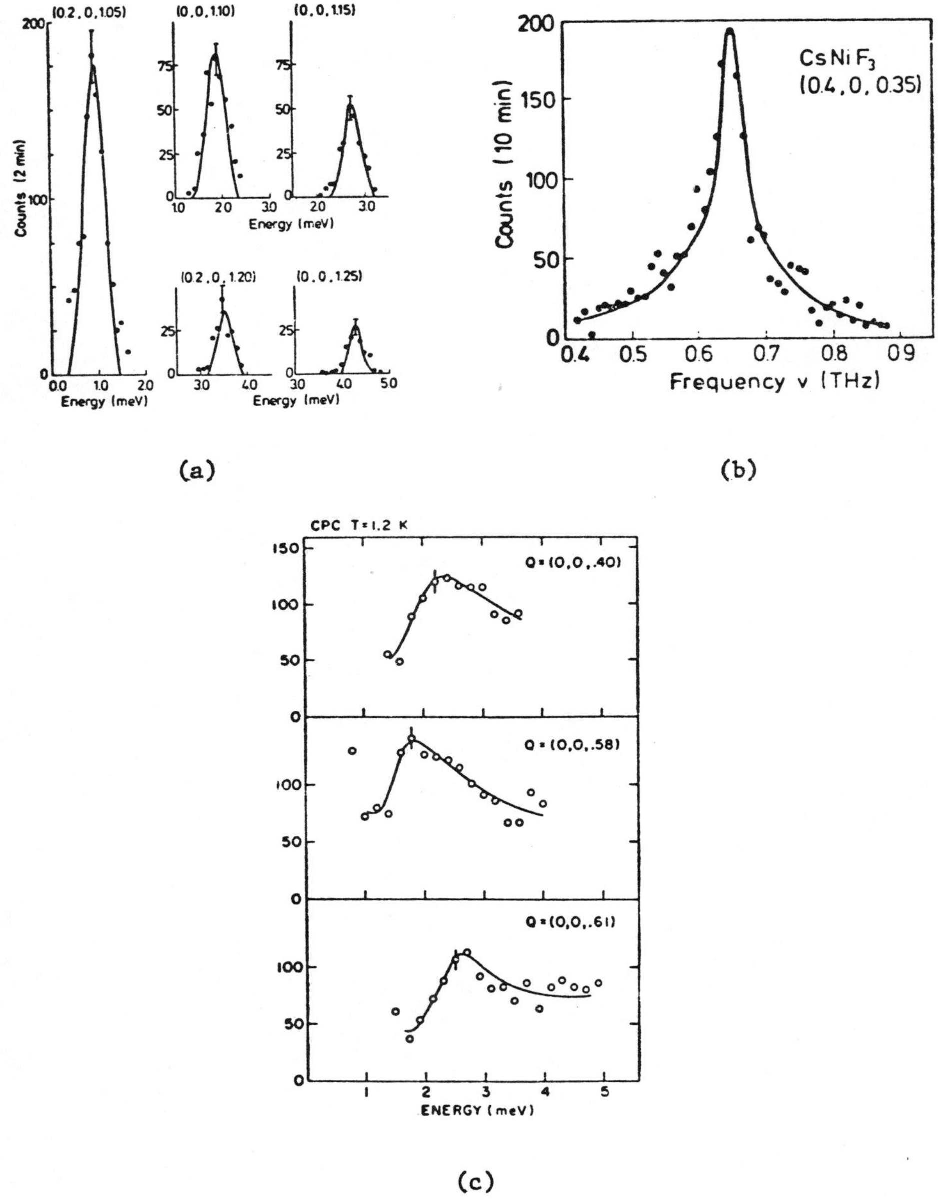

Fig. 5: Spin wave-peaks as measured at low temperatures in (a) TMMC; (b) $CsNiF_3$; (c) CPC.

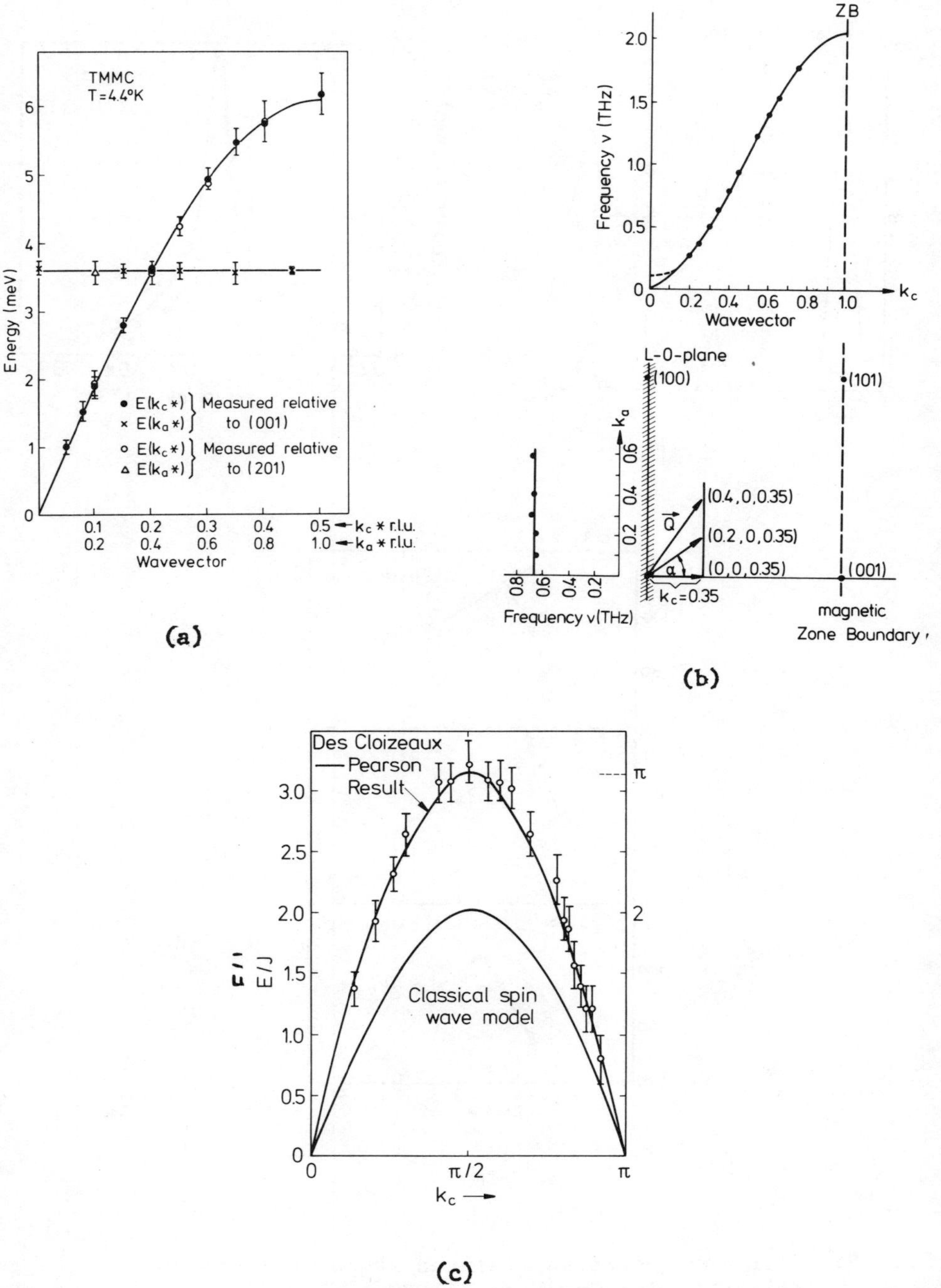

Fig. 6: Spin wave dispersion as measured in (a) TMMC; (b) $CsNiF_3$; (c) CPC. The solid lines are the theoretical results.

Contrary to TMMC in the x-y-ferromagnet the OPC peak has practically the same energy, but is much sharper than the IPC peak. Thus there is only one dispersion curve in agreement with experiment.

CPC is a 1-D-antiferromagnet with no measurable anisotropy and a spin value of the Cu^{2+} ion of $S = 1/2$. The main difference between the classical and the quantum mechanical calculation[11] is the prefactor in the dispersion relation. In Fig. 6(c) it can be seen that this prefactor is essential to describe the measured dispersion using the known value for the exchange interaction $J/k = -13.4$ K. No fit has been made. The classical result is shown for comparison.

Thus, at low temperatures the linear spin wave theory describes the dispersion of the excitations in these model systems very well.

Lineshape of the Linear Excitations

Let us now turn to the fundamental question: under which conditions of q_c and T are spin waves stable? Let us first ask why spin waves were observed at low temperatures as shown in Fig. 5. The simple physical picture to explain this is the following: at low temperatures $T \ll 2S(S + 1)J$ the correlation length ξ is large (TMMC at 1.9 K: $\xi = 42 \cdot c/2$, $c/2$ being the Mn-Mn-distance) whereas the wavelengths of the spin waves studied were smaller than the correlation length. Thus we can say that the spin wave rides on longer waves characterized by ξ as shown in Fig. 7. If this picture were right then spin waves with wavelength $\lambda > \xi$ would be unstable. The correct theoretical conditions says[10]

$$\lambda < 2\pi\xi \qquad \text{or} \qquad q_c > \kappa \quad ; \quad \kappa = \xi^{-1}$$

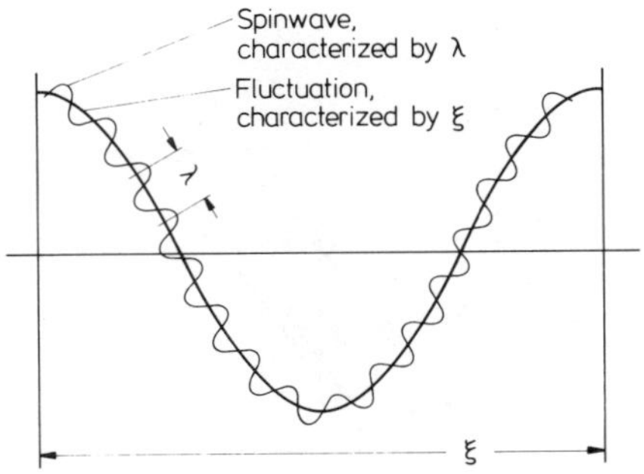

Fig. 7: The two characteristic lengths when a spin wave with wavelength λ is travelling in a system with SRO characterized by ξ.

At low temperatures this condition can easily be fulfilled. If that picture would be correct, we would expect the spin waves to be stable if $q_c > q_{critical}$ being proportional to κ. Since $\kappa \sim T$ at low temperatures (besides for Ising-like systems, which we don't consider here) the stability limit for the spin waves, the critical wave vector $q_{critical}$, would move into the Brillouin zone with increasing temperatures, allowing only short wavelength spin waves to be stable.

Can we give a similar picture for the line shapes of the spin waves? Usually, e.g., in 3-D-systems the line width of a spin wave peak, if measurable, is connected with the lifetime τ of the spin wave. In 1-D-magnets however there is another mechanism which contributes to the line width: the cross section (2) shows that the scattering process and therefore the results of inelastic neutron scattering have to be considered in momentum (q-) space. In order to get the Brillouin zone we have to Fourier transform the real space lattice. In 3-D-systems with long range order, this leads to the well known Bragg spots (δ-functions) in reciprocal space. In a 1-D magnet however with a characteristic length of ξ we have to Fourier transform finite size systems. Therefore the distances $\vec{Q}$ in the reciprocal space can only be determined with an uncertainty of the order of $\pm\,\kappa$. We therefore cannot measure a single spin wave at exactly q_c, but we always measure all spin waves within $q_c \pm \kappa$. This is illustrated in Fig. 8 when $1/\tau$ is small compared to the above described effect. This mechanism yields a line width Δ[10]

$$\Delta = \kappa \frac{\partial\omega}{\partial q_c} \sim T \qquad (6)$$

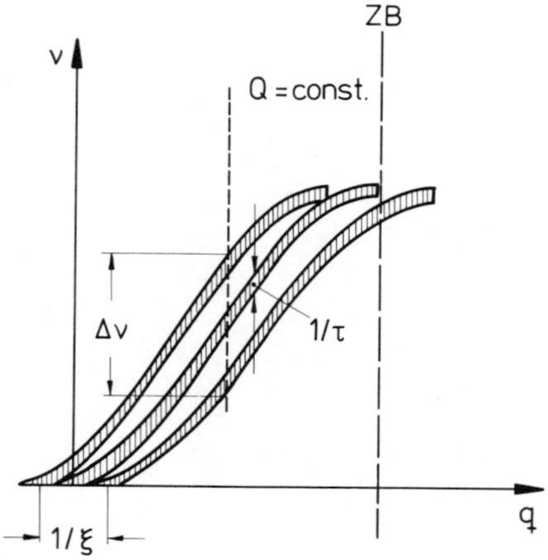

Fig. 8: Influence of the finite correlation length ξ on the measurements of spin waves in 1-D-magnets.

Thus, one would expect a linear increase of Δ with T and a zone boundary narrowing, since $\partial\omega/\partial q_c$ approaches zero at the zone boundary. This consideration applies for the IPC peak at low temperatures since only the S^x and S^y spins are affected by the critical fluctuations yielding ξ: these fluctuations, which have approximately zero energy, cannot overcome the anisotropy energy at low temperatures and thus do not affect the S^z-component of the spins. Therefore the OPC-peak corresponds to a single spin wave and its width should reflect the lifetime: one expects the OPC-peak to be very narrow at low temperatures.

The line width as a function of temperature has been studied in great detail for TMMC. Hutchings and Windsor[12] have measured it between 4 K and 20 K. The experimental results are shown in Fig. 9. Their best fits of a power law in T to the experimental data yield (see Fig. 9)

$$\Delta = 0.093\ T^{0.88} \qquad \text{and} \qquad \Delta = 0.084\ T^{0.97}$$

for $q_c = 0.2$ and $q_c = 0.25$ respectively. These results are in reasonable agreement with the theoretical predictions given above. Thus the correlation length ξ does indeed determine the width of the IPC-peak.

ξ should as well determine the critical wave vector $q_{critical}$. Its effect on the measured spin wave peaks should be the following:

$q_c > q_{critical}$: well defined spin wave peaks

$q_c < q_{critical}$: overdamped spin wave peaks.

This has been studied by Birgeneau and Shirane[13] on TMMC for very small $q_c < 0.1$ r.l.u. and $1.45\ K \leq T \leq 9\ K$. Typical results are

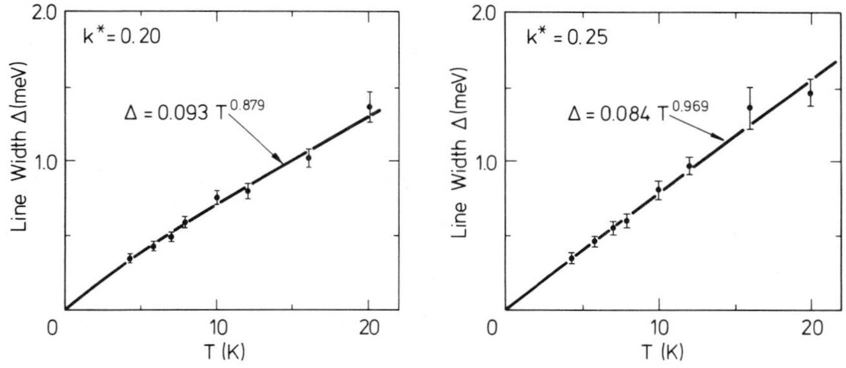

Fig. 9: Temperature dependence of the spin wave line width Δ in TMMC for two wave vectors $q_c = 0.2$ r.l.u. and $q_c = 0.25$ r.l.u.

shown in Fig. 10 for q_c = 0.015 and 0.03 r.l.u., $\kappa = 1/\xi$ is also given in the figure. At 1.45 K the spin wave peaks are quite sharp in both cases, at 4.3 K however the spin wave peak for q_c = 0.015 r.l.u. shows typical overdamped behavior already, whereas for q_c = 0.03 r.l.u. the peak is still reasonably sharp and becomes overdamped at higher temperatures only. Thus $q_{critical}$ defined by that q_c, at which the lineshape changes at a given temperature, increases by a factor of two when going from 4.3 K to 9 K, as κ does in the same temperature range. A detailed analysis shows that

$$q_{critical} \sim 1.5\ \kappa$$

which is again in good agreement with the above picutre.

The first measurements done on the OPC peak in TMMC[8] do indeed show the predicted sharp peak at low temperatures.

The fact that in the ferromagnetic chain system $CsNiF_3$ both the IPC and the OPC have the same energy leads to a peculiar lineshape at arbitrary $\vec{Q}$, since both $\langle S^xS^x + S^yS^y \rangle$ and $\langle S^zS^z \rangle$ are measured at the same time with relative weights given by $\vec{Q}$. In order to study the behavior of the different contributions it is necessary to separate the IPC from OPC peak. This can be done by using the fact that only the spin component perpendicular to $\vec{Q}$ contributes to the scattering. Thus having $\vec{Q}||c^*$ only the IPC peak appears in the spectrum. By adding a larger and larger q_a-component to $\vec{Q}$ the sharp OPC peak should gradually appear. This is demonstrated (T = 4.9 K) in Fig. 11.[5] The used positions in reciprocal space are those indicated in Fig. 5(b). The observed increase of the OPC and the decrease of the IPC contribution are given by the following relations,[5] α being the angle between $\vec{Q}$ and $\vec{c}^*$ ($\alpha = \tan^{-1}(q_a/q_c)$) (see Fig. 5(b))

$$\text{Intensity (IPC)} \sim 1 + \cos^2\alpha$$

$$\text{Intensity (OPC)} \sim \sin^2\alpha$$

By using this formula the three peaks in Fig. 11 could be fitted simultaneously yielding the intensity ratio R for the two contributions. The width of the OPC peak is given by the resolution width alone whereas the IPC has a significant intrinsic width which will be discussed later. The found intensity ratio R agrees very well with the theoretically calculated one.[5] R is determined by the anisotropy due to the following mechanism: in the isotropic case a spin wave corresponds to an in phase precession of the spins on a circular cone. Due to the anisotropy S^z feels a stronger restoring force than S^x and S^y, thus the circular cone is deformed to an elliptical one with smaller amplitude along c than along x and y. The ratio of these amplitudes determines R and is q_c dependent.

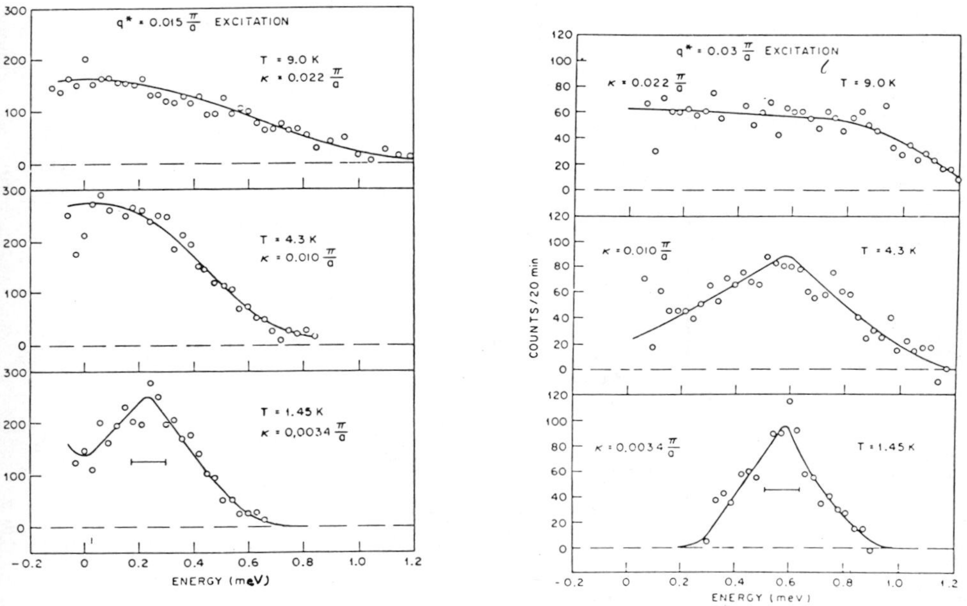

Fig. 10: Crossover from propagating to overdamped behavior of the spin waves in TMMC at q_c = 0.015 r.l.u. and q_c = 0.03 r.l.u.

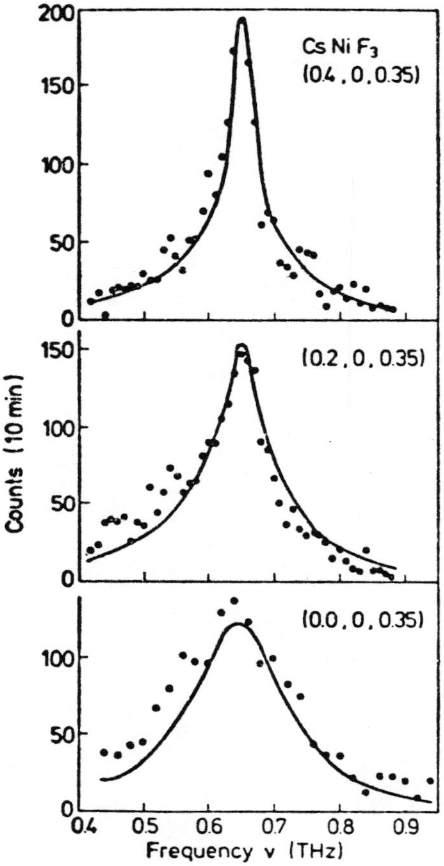

Fig. 11: The appearance of the OPC-peak with increasing q_a in $CsNiF_3$ at 4.9 K.

The smaller q_c is the lower the energy gets for the S^x and S^y components whereas the anisotropy demands a relatively high energy for creating a significant S^z. Thus the amplitude in S^z gets smaller with decreasing q_c whereas the amplitude in S^x and S^y gets larger with decreasing q_c. Finally at $q_c = 0$ the precession degenerates to a rotation in the easy plane without S^z component.

When increasing the temperature the OPC should disappear very rapidly since the lifetime τ of the OPC spin wave decreases rapidly due to increasing importance of thermal fluctuation in S^z. The OPC peak should however show only very small renormalization of the energy.[14] These predictions could be verified experimentally. In Fig. 12 the experimental results for the OPC peak in $CsNiF_3$ measured at $\vec{Q} = (0.4, 0, 0.35)$ are shown. Both the energy's and the intensity's dependence agrees well with the theoretical results.

As discussed before, the IPC peak should be strongly connected with the correlation length ξ. The evolution of the IPC peak with temperature measured at $\vec{Q} = (0, 0, 0.35)$ in $CsNiF_3$ is shown in Fig. 13. The solid lines are results of a least squares fit of a convolution of the resolution function of the instrument with the expected Lorentzian line for the spin wave peak. The resulting temperature dependence of the line width is shown in Fig. 14. It is obvious that the line width increases proportionally to T as expected from (6).

In the quantum system CPC a detailed lineshape study has not yet been done, but there is even at low temperatures a significant difference in the lineshapes between classical and quantum systems. Whereas the lineshape in classical systems is quite symmetric at low temperatures and wave vectors larger than $q_{critical}$, the quantum system shows significant assymmetry in the lineshape; this can be seen in Fig. 15.[8] This assymmetry, especially at low q_c, is due to higher ($S = 1$) excited states which have significant weight to contribute. Several theoretical approaches do predict this contribution.[15,16]

In concluding this part on linear excitations in zero field, four points seem to be important:

a) For classical ($S > 1/2$) systems linear spin wave theory is able to describe all aspects of the spin wave excitations. In quantum systems ($S = 1/2$) different theoretical approaches are needed to describe the behavior of the linear excitations.

b) Planar anisotropy leads to two components in the cross section: the one is very sharp (OPC), the other is wide (IPC).

c) It appears that the correlation length ξ is the characteristic length for the spin waves as well since the width and the

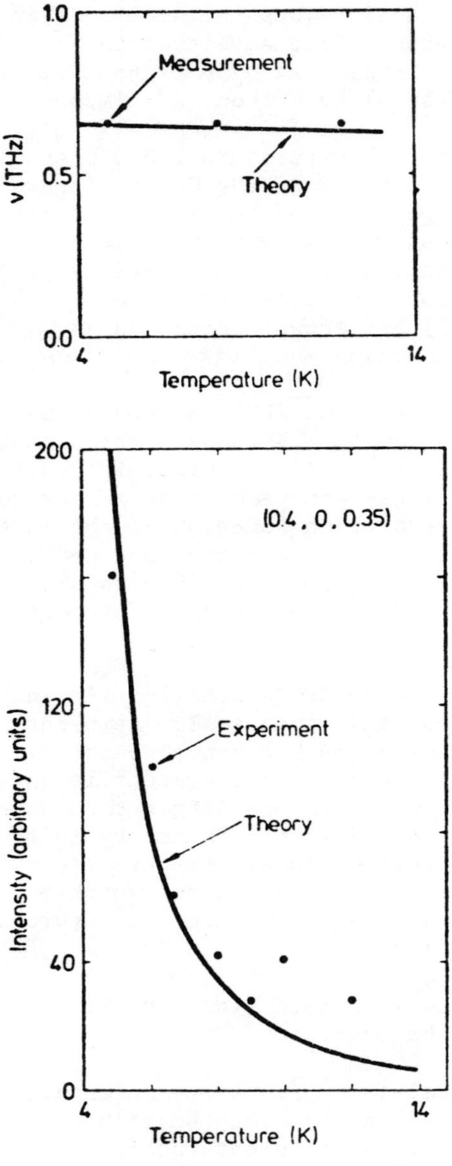

Fig. 12: Temperature dependence of the energy and the intensity of the OPC in $CsNiF_3$. Solid lines are theoretical results.

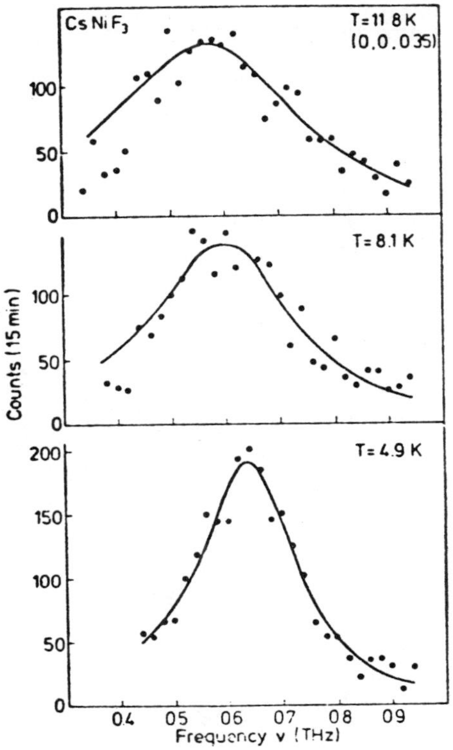

Fig. 13: Temperature dependence of the IPC-peak. Solid lines are results of least squares fits as described in the text.

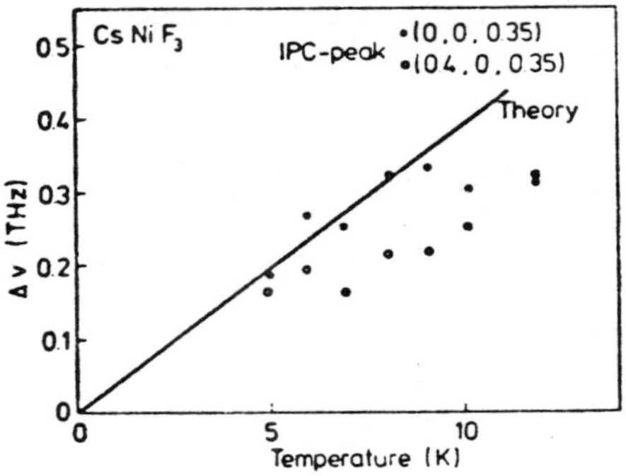

Fig. 14: Temperature dependence of the width of the IPC-peak. Solid line: theoretical prediction.

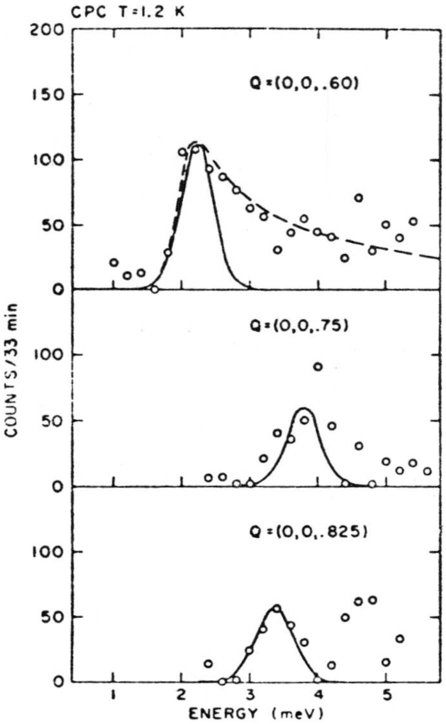

Fig. 15: Comparison of the spin wave peaks with results of linear spin wave theory (solid line).

critical wave vector of the IPC peak are proportional to $\kappa = 1/\xi$.

d) The OPC peak should reflect the true lifetime τ of the spin waves, a measurable $1/\tau$ however makes the intensity of the peak too small to be measured.

Linear Excitations in a Magnetic Field

For both TMMC and CPC the recently published field dependencies are not fully understood and will therefore not be discussed here. We will therefore concentrate on the result of studies on $CsNiF_3$ in a strong magnetic field.[17] The influence of a magnetic field on a ferromagnetic chain system is quite simple in linear theory. The Hamiltonian (1) becomes

$$\mathcal{H} = -2J\sum_i \vec{S}_i\vec{S}_{i+1} + A\sum_i (S_i^{\,2})^2 - g\mu_B H^x \sum_i S_i^{\,x} \qquad (7)$$

and the field being in the easy plane has two effects:

1. there appears a gap at $q_c = 0$;
2. the fluctuations in the systems are strongly suppressed, yielding an effectively ordered 3-D-ferromagnet with planar anisotropy in a field.

The approach 2. can be used for $H > 10$ kG and $T < 6$ K. In this field and temperature region the magnetization is saturated and the spin wave linewidth is resolution limited. Under these conditions the only interesting quantities are the spin wave energy and intensity for different scattering geometries.

In Fig. 16 the spin wave dispersion is shown for H = 0, 4.9 K and H = 41 kG, 4.9 K. The solid lines are fits of the results of linear spin wave theory with correction for finite S[18] yielding

$$J/k = 11.8 \text{ K} ; \quad A/k = 9 \text{ K}$$

with very good agreement between theory and experiment.

This theory predicts for the field dependence of the energy gap E_o at $q_c = 0$ at 4.9 K.

$$E_o^2 = A(1 + \frac{A}{8J}) g\mu_B H + \{1 - (\frac{A}{8J})^2\} (g\mu_B H)^2 \tag{8}$$

In Fig. 17(a) the experimental data together with the theoretical results (without fit) are shown. The agreement between theory and experiment is obviously very good. It is worth noticing here that this $q_c = 0$ spin wave corresponds to $\lambda \to \infty$. Such an excitation would not be stable in the presence of strong fluctuations as at H = 0. In fact, this peak broadens very strongly with decreasing field until it finally disappears.

Linear spin wave theory[18] predicts the intensity ratio for $\vec{Q} || \vec{c}^*$ and $\vec{Q} \perp \vec{c}^*$ as well:

$$J(\vec{Q} || \vec{c}^*) \sim \frac{1}{I(\vec{Q} \perp \vec{c}^*)} \sim \frac{A + g\mu_B H(1 - A/8J)}{(1 + A/8J)\, g\mu_B H} \tag{9}$$

Comparison between theory and experiment is in this case not very easy since several corrections have to be applied to account for absorption in the sample in different scattering geometries and for transmission of the instrument, etc. But despite these complications experiment and theory agree reasonably well as seen in Fig. 17(b). Thus, as in zero field linear spin wave theory accounts for all aspects of the behavior of the spin waves in high field and at low temperatures in $CsNiF_3$, where it behaves as a 3-D-ferromagnet with easy plane in an external field.

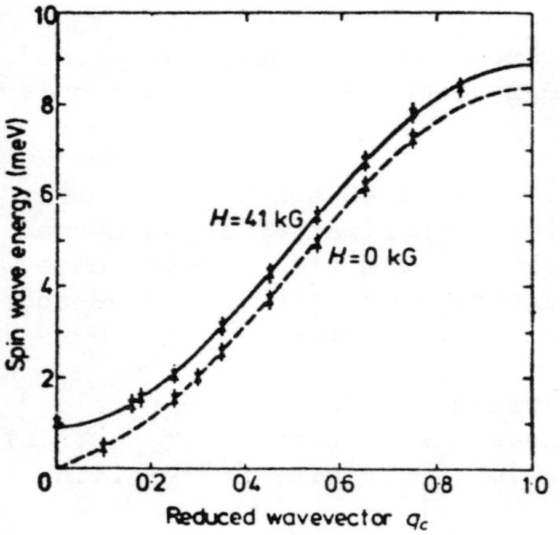

Fig. 16: Spin wave dispersion in $CsNiF_3$ at 4.9 K at H=0 and H=41 kG.

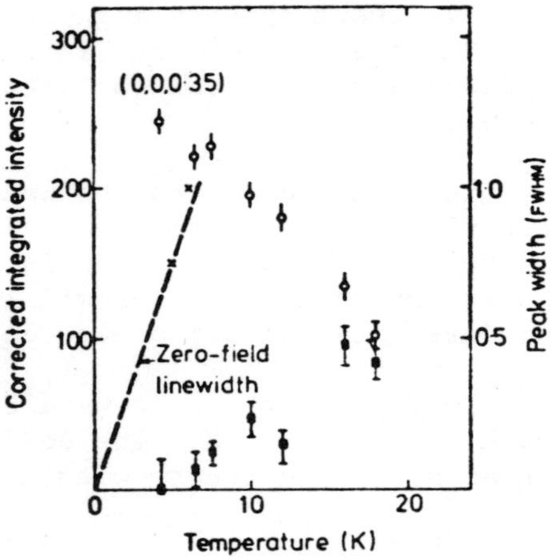

Fig. 18: Temperature dependence of the intensity ⌀ and ■ the width of the spin wave at q_c=0.35 and 41 kG.

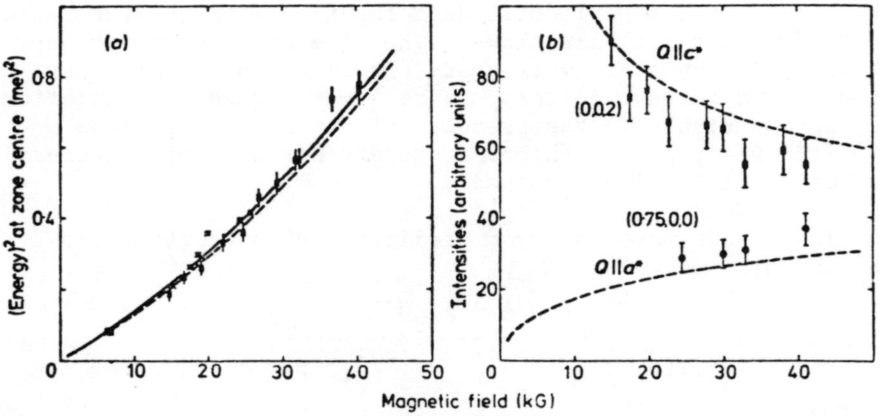

Fig. 17: Field dependence of the q_c=0 mode at 4.9 K. (a) Energy (solid line: calculated; broken line: best fit); ■ Infrared absorption data.[19] (b) Intensity for $\vec{Q}\perp c^*$ and $\vec{Q}\,||\,c^*$; broken line: calculated results.

There are however some experimental findings which do not fit easily into this picture. Figure 18 shows the temperature dependence of the spin wave line width and intensity for H = 41.0 kG. First, the spin wave line width increases much slower as in zero field as shown in Fig. 18 as well and the intensity decreases with increasing temperature whereas in zero field the intensity was constant. Since the magnetic field suppresses the fluctuations, higher temperatures than in zero field are needed to introduce enough fluctuations to affect the line width. Thus, this effect can be at least qualitatively understood in the picture we used up to now. There is however no way in explaining the change in intensity within this picture. As we will see another very different point of view is necessary to explain this, which will be discussed in the following section.

III. NONLINEAR EXCITATIONS

If we allow the excitations to have large amplitudes, then linear spin wave theory cannot work since linear theories are designed to work only for small amplitude solutions. One might expect however that large amplitude excitations are important especially in systems with strong fluctuations. It has been shown[20] that certain types of equations of motion can be solved in their full nonlinearity. Especially interesting are equations which yield stable pulse-like solutions, which are well localized and can move. One of the simplest systems, whose equations of motion are strongly nonlinear and which can be intimately related to a one-dimensional ferromagnetic chain system with planar anisotropy in a magnetic field is a set of coupled mechanically restricted pendula as sketched in Fig. 19(a), the mass of the pendula being m, the spring coupling constant being K and the overall potential is due to the gravity $\vec{G}$. As can easily be seen the equation of motion is given by

$$\frac{md^2\phi_i}{dt^2} = K(\phi_{i+1} - 2\phi_i + \phi_{i-1}) - G \sin\phi_i \tag{10}$$

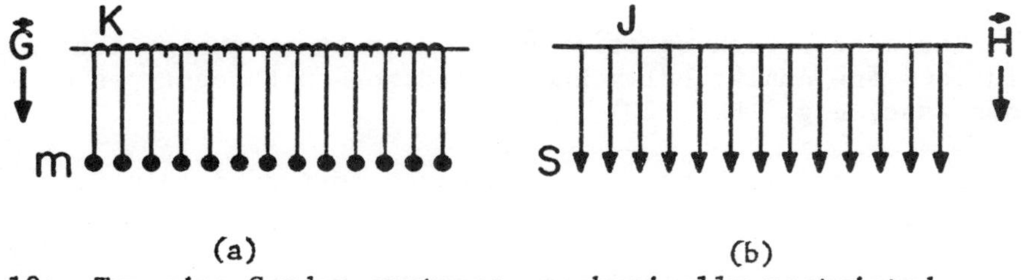

Fig. 19: Two sine-Gordon-systems: mechanically restricted coupled pendula (left) and easy plane 1-D-ferromagnet in a magnetic field.

ϕ_i is the angle the ith pendulum makes with respect to $\vec{G}$, the rotation of the pendula being restricted to a plane perpendicular to the system axis. By denoting the distance between adjacent masses by Δx we can rewrite (10)

$$m \frac{d^2\phi_i}{dt^2} = K\Delta x \left(\frac{\phi_{i+1} - \phi_i}{\Delta x} - \frac{\phi_i - \phi_{i-1}}{\Delta x} \right) - G \sin\phi_i$$

or

$$m/\Delta x \frac{d^2\phi_i}{dt^2} = K\Delta x \frac{\frac{\phi_{i+1} - \phi_i}{\Delta x} - \frac{\phi_i - \phi_{i-1}}{\Delta x}}{\Delta x} - \frac{G}{\Delta x} \sin\phi_i \tag{11}$$

(11) yields in the continuum limit the famous and ubiquitous sine-Gordon equation

$$\frac{\partial^2\phi}{\partial x^2} - \frac{m/\Delta x}{K\Delta x} \frac{\partial^2\phi}{\partial t^2} = \frac{G/\Delta x}{K\Delta x} \sin\phi \tag{12}$$

Of course, this is not only a <u>continuum</u>, but also a <u>classical</u> equation. The sine-Gordon equation is well known for a long time and is used in many different fields of physics: elementary particle physics, plasma physics, self-induced optical transmission, etc.

We can quite easily see by comparing Fig. 19(a) and (b) that the pendulum and the chain system are very similar with respect to their equation of motion at least in this phenomenological approach. In order to go from the pendulum system to the spin system we have to replace the mass m of the pendula by the value S of the spins, the spring coupling constant K by the exchange interaction J and finally the gravitational field $\vec{G}$ by the magnetic field $\vec{H}$. In order to make the analogy complete the rotation of the spins has to be confined to a plane perpendicular to the chain axis as in an easy plane system. Thus we conclude that it seems very likely that the sine-Gordon equation can be used to describe the dynamics of $CsNiF_3$ in a field.

The sine-Gordon equation has the great advantage that all solutions are analytically known: there are three types of solutions (see, e.g., Ref. 20)

linear excitations:	Spin waves (extended solutions)	
nonlinear excitations:	Solitons	} <u>See Fig. 20</u>
	Breathers (coupled soliton-antisoliton pairs)	

The nonlinear excitations have the following form:

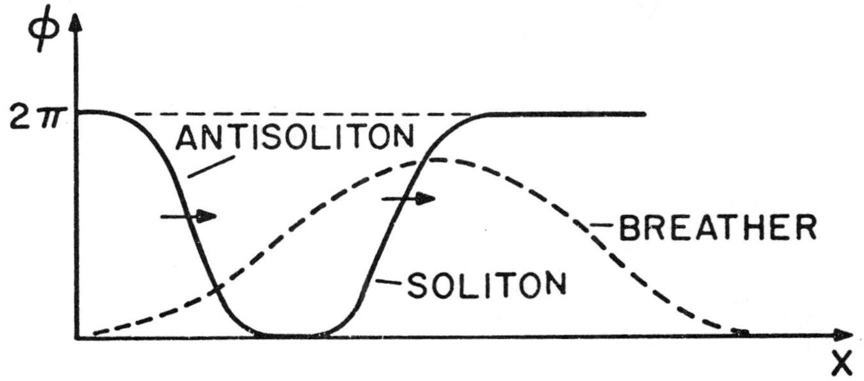

Fig. 20: The different nonlinear modes as they show up in the angle ϕ.

Soliton: $$\phi = 4 \tan^{-1} \{\exp (\pm \frac{x - vt}{d(1 - v^2/v_o^2)^{1/2}})\} \quad (13)$$

where the + and - refer to soliton and antisoliton, d is the width of the soliton and v_o is the critical velocity.

Breathers: $$\phi = 4 \tan^{-1}\{ \frac{(\frac{\omega_o^2}{\omega^2} - 1)^{1/2} \sin \{ \frac{\omega(t-vx/v_o^2)}{\sqrt{(1 - v^2/v_o^2)}} \}}{\cosh \{ \frac{(x-vt) (1 - \omega^2/\omega_o^2)^{1/2}}{d(1 - v^2/v_o^2)^{1/2}} \}} \quad (14)$$

where ω is the internal oscillation of the Breather and ω_o the frequency of the linear excitation at q = 0. From these forms for ϕ, (13) and (14), it is obvious that both solitons and breathers are localized excitations, the maximum width for the soliton being d, whereas the minimum width for the breathers is 2d. The energies of the two modes are the following:

Soliton: $$E_s = \{m_o^2c^4 + p^2c^2\}^{1/2} \qquad \text{with } E_{s,o} = m_oc^2 \quad (15)$$

m_o being the effective mass of the soliton with $m \sim 1/d$.

$$\text{Breathers:} \quad E_B = \frac{E_{B,0}}{\sqrt{1 - v^2/v_0^2}} \quad \text{with } E_{B,0} = 2E_{s,0}\left(1 - \frac{\omega^2}{\omega_0^2}\right)^{1/2} \tag{16}$$

As can be seen from (15) and (16) the soliton has a finite rest energy, whereas the breather energy can go to zero if $\omega \to \omega_0$. At the same time the breather size goes to infinity and the amplitude goes to zero. The soliton energy diverges for $v \to c$ and accordingly the size of the soliton goes to zero. Thus the two nonlinear modes have very different properties: the soliton always remains a well localized excitation whereas the breather gradually approaches the behavior of an extended mode when $\omega \to \omega_0$.

Since the sine-Gordon equation is a Lorentz invariant field equation, in Eqs. (13)-(16) "relativistic" terms like $\sqrt{1-v^2/v_0^2}$ appear. The role of the speed of light plays in this case the phase velocity v_0 of the linear excitation.

In the continuum limit the thermodynamics of a system described by the sine-Gordon equation have been worked out by several authors.[21,22] Using the fact that the nonlinear modes are reflectionless potentials for the linear excitations resulting in a phase shift of the linear excitations during an interaction with nonlinear modes, one finds the following free energy density:[23]

$$F = F_0 - k_B T n_s^{tot} - k_B T n_B$$

$$F \cong k_B T \left\{ \ell^{-1} \ln(h\omega_o \beta d/\ell) - \frac{13}{6\pi d \beta E_{s,o}} - \frac{1}{\sqrt{2\pi}} \frac{4}{d} (\beta E_{s,o})^{1/2} e^{-\beta E_{s,o}} \right\} \tag{17}$$

(17) shows that the contributions of the linear modes, the breathers and the solitons separate. This fact allows us to give a simple and physically very appealing picture of a sine-Gordon system in the low temperature limit. The nonlinear excitations behave as a gas of noninteracting thermally activated quasi-particles running along the chain. There are some doubts however whether the breathers still exist if one is going over from a continuous to a discrete system.

Let us now go over to the real system $CsNiF_3$ in an external field perpendicular to the chain direction. At least at low temperatures $T < 15$ K, when the anisotropy keeps the spin out of the z-direction, the chain direction, the sine-Gordon equation seems to be a good description of the spin dynamics. We can write down the sine-Gordon equation in the following form:[24]

$$\frac{\partial^2\phi}{\partial z^2} - \frac{1}{c^2}\frac{\partial^2\phi}{\partial t^2} = m^2 \sin\phi \qquad (18a)$$

with the following definitions:

$$c = aS(4AJ)^{1/2} \qquad \text{and} \qquad m = \{g\mu_B H^x/2JSa^2\}^{1/2} \qquad (18b)$$

The notation corresponds to that of (1) and a is the distance between the Ni^{2+} ions (a = c/2). In this notation the energy of a soliton is

$$E_s = 8m\frac{2Ja}{\sqrt{1 - u^2/c^2}}$$

u being the velocity of the soliton. In the case of the ferromagnetic chain the soliton is a magnetic particle with a length 1/m which carries a magnetic moment of approximately $1/m.gS\mu_B$ and travels with a velocity u. Since 1/m >> 1 the change in the angular moment ΔS of the chain is $\Delta S >> 1$ and thus the neutron cannot excite a soliton. In fact, always a pair of solitons and an antisoliton has to be excited for topological reasons. This does not imply that in this case $\Delta S = 0$, since only the chirality is different for the two, but sign and size of ΔS are the same for both. One therefore concludes that the only possible scattering process is the one corresponding to scattering of neutrons from a gas of real finite size particles, which gives rise to a central peak in the dynamical structure factor $S(q,\omega)$. The calculation by Mikeska[24] for the correlation-function for the solitons only yields

$$S(q,\omega) \sim \frac{\beta}{2cq}\, e^{-8\beta m}\left(\exp^{-\frac{2\beta m\omega^2}{(cq)^2}}\frac{\pi q/2m}{\sinh \pi q/2m}\right)^2 \qquad (19)$$

which is analogous to the cross section for a dilute gas.[3] Three different contributions to the cross section (19) can be identified easily:

$e^{-8\beta m}$ describes the thermal activation of the solitons and thus the temperature dependence of the intensity.

$e^{-\frac{4\beta m\omega^2}{(cq)^2}}$ this energy distribution centered around $\omega = 0$ reflects the velocity distribution of the soliton gas.

$\left(\frac{\pi q/q_m}{\sinh \pi q/q_m}\right)^2$ this term reflects the finite size of the solitons and describes the q-dependence of the intensity.

Obviously (19) is not complete since spin waves are expected to show up as usual at $\pm\,\omega_{spinwave}(q)$. Breathers have been neglected since they might not exist in a discrete lattice or chain. In order to verify the assumption that the dynamics of a 1-D-ferromagnetic chain system with planar anisotropy in an external magnetic field do indeed contain stable nonlinear excitations like solitons, a detailed measurement of the above terms in $S(q,\omega)$ is necessary. From the studies described in Section II it was clear that if nonlinear excitations were important at all, they should be most important at moderate fields and not too low temperatures. Experiments were thus performed at $q_c = 0.1$ r.l.u. ($\vec{Q} = (0, 0, 1.9)$) $3\ K < T \leq 15\ K$ and $0 \leq H \leq 15$ kG.[25] The most serious problem of these experiments and their interpretation becomes obvious in Fig. 21. At H = 5 kG the two spin wave peaks are clearly seen, whereas around zero energy a strong resolution limited peak due to incoherent scattering is observed, which obscures any weak contribution from solitons in this region. This background correction is critical and has been done in the following way: the same spectrum has been measured at low temperature, T = 3.2 K, and high field, H = 41.8 kG. By this, the spin waves have been moved to higher energies and any nonlinear effects are expected to be unmeasurably small. Such a measurement is shown in Fig. 21 as the solid line. By subtracting this background from the measurements the remaining spectra due to magnetic scattering are shown for different temperatures and H = 5 kG in Fig. 22. These spectra clearly show a three peak structure. The two peaks at $\pm\,\omega_{SW}$ are due to spin waves and their energies agree with earlier measure-

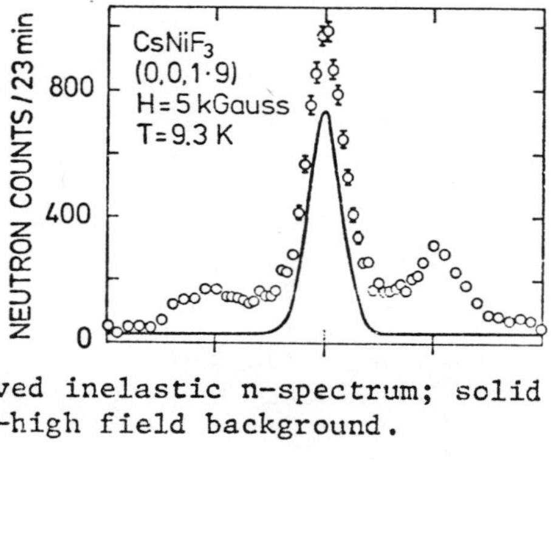

Fig. 21: Observed inelastic n-spectrum; solid line: low temperature-high field background.

ments. In order to decide upon the origin of the central peak, which obviously increases rapidly with temperature, its temperature, field and q-dependence has to be analysed. In order to do this the experimental data have been described by a dynamic structure factor $S'(q,\omega)$ of the following form: $S'(q,\omega)$ was the sum of a central component according to (19) plus two spin wave peaks at $\pm\,\omega_{SW}$ having a Lorentzian line shape all convoluted with the resolution function of the instrument. In addition, corrections for detailed balance and transmission of the instrument have been applied. By fitting this $S'(q,\omega)$ to the experimental data, which gave very good agreement between calculation and experiment, it was possible to determine the temperature and field dependence of the integrated intensity of the central peak and of the spin wave peaks, and the temperature dependence of the energy width of the central peak. The thus determined temperature dependence of the integrated intensity I and of the energy width ΔE are shown in Fig. 23. Since the only parameters, which describe the behavior of the central peak are J, A, T and H, are all well known, one can calculate $\Delta E(T)$ which contains no free parameter; the calculation of I(T) involves a scale factor since the intensity is not measured on an absolute scale. The calculations are shown in Fig. 23 as solid lines, the intensity scale being set at 14 K. The agreement between theory and experiment is obviously quite good, using the theoretical value of 34 K for the rest energy 8m. The least squares fits performed to obtain the best experimental value for 8m yielded $8m,\mathrm{exp} = 27$ K, which is slightly smaller than the theoretical one. Another crucial test of this interpretation

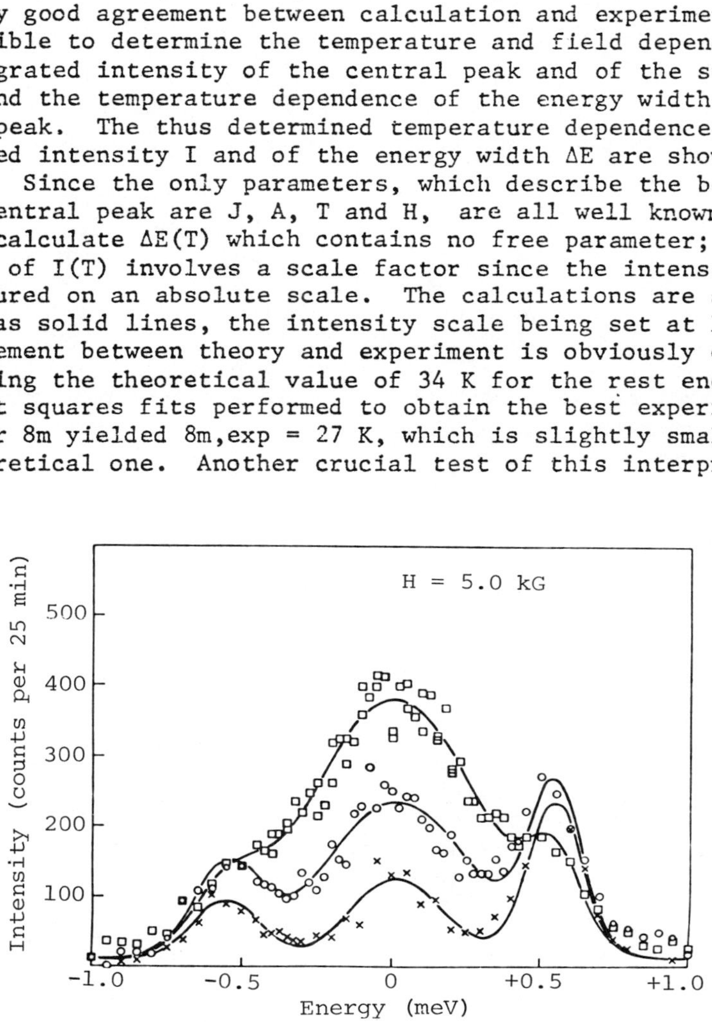

Fig. 22: The pure magnetic inelastic spectrum at x:6.3 K, 0:9.3 K and □:14 K.

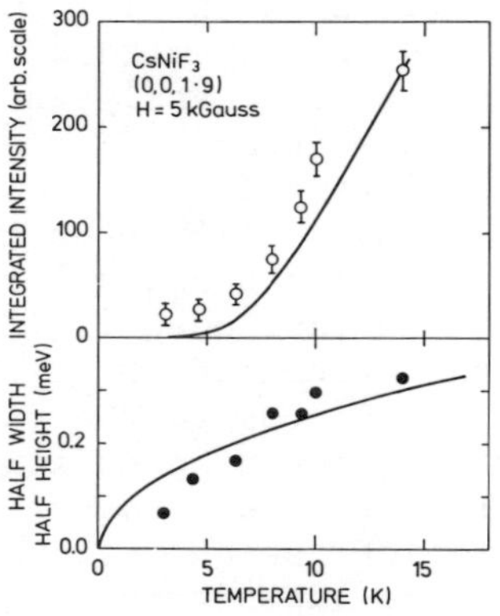

Fig. 23: Temperature dependence of the intensity and width of the central component.

is the q-dependence of ΔE and I, which has recently been measured at 10 K, 5 kG[26] yielding the following results: as the temperature dependence, the q-dependence of ΔE follows very closely the theory showing a linear q-dependence of ΔE, I(q) shows, however, only qualitative agreement with the theory yielding a too high value for the rest energy. Despite this, these experiments give very good evidence for the existence and importance of nonlinear excitations, solitons, in this system.

What are the reasons for the discrepancies between theory and experiment, and where are the weak points of the theoretical approaches used?

First of all, the principal approach was in all the theoretical calculations the use of the continuum and classical limit. There is no reason to believe that the classical approach is worse in the case of the nonlinear than in the linear excitations, where it worked very well. The discreteness of the real system seems to lead to principal questions: the continuum theory clearly shows the existance and stability of two types of nonlinear excitations, solitons and breathers, whereas in any discrete calculation the breathers do not show up as stable solutions. The solitons however are not affected as long as the length is larger than approximately 3 lattice constants. We were thus considering only solitons in the interpretation of our experimental data. In computer simulations however[27] breathers do show up and are even dominating the nonlinear part of the dynamics due to their low rest energy. Much more work is necessary to decide upon the existence and stability of the breathers on a discrete lattice.

The second important assumption made for the evaluation of the cross section (19) was that relativistic effects have been neglected. This is correct at low temperatures where the velocity distribution for a gas of relativistic quasiparticles is very similar to that of nonrelativistic ones. It turns out however when looking at the actual quantities for the system, that this assumption becomes incorrect at very moderate temperatures. The relativistic energy is given by

$$E = \frac{m_o c^2}{\sqrt{1 - v^2/c^2}} \sim m_o c^2 + \frac{1}{2} m_o v^2 + \frac{3}{4} \frac{m_o v^2}{2} \frac{v^2}{c^2} + \ldots$$

showing that as long as the third term in the expansion is small compared to the second term (the kinetic energy) the system should behave as a nonrelativistic one. From the definitions (18a) and (18b) we can calculate m_o to be 10^{-22}g at 5 kG. Using a Maxwell distribution for the velocities v the mean value for $|v|$ is found from the Maxwell distribution as $\overline{|v|} = 2.63.10^3$cm/sec at 4 K whereas c is 8.10^{+3}cm/sec, thus the correction v^2/c^2 is about 0.1; at 15 K however v^2/c^2 reaches 0.4 clearly indicating that even well below T = 15 K relativistic corrections might be necessary. How different the velocity distribution looks in the relativistic and nonrelativistic case, is shown in Fig. 24, clearly demonstrating the relativistic characters of the system. This relativistic character will

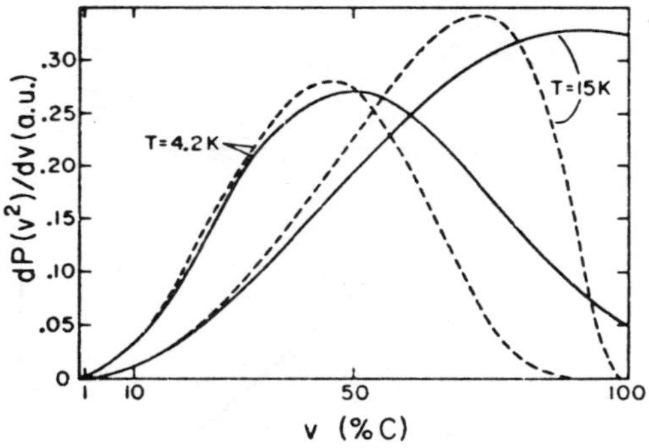

Fig. 24: Velocity distribution in the relativistic (---) and nonrelativistic(—) case at 4.2 K and 15 K.

influence the dynamic structure factor $S(q,\omega)$ as well, but it is not clear yet how it will be changed. Especially the q-dependence of the intensity should be strongly affected, which is in accordance with the strong deviations found recently.[26] Relativistic effects are believed to show up in the computer simulations as well[27] where the central peak in $S(q,\omega)$ splits at high temperatures and small q.

As a last problem we mention the spin wave intensity in the region where nonlinear and linear peaks are present in the dynamics. As shown in Fig. 25 the spin wave intensity as obtained from the procedure discussed above, decreases with increasing temperature. In principal this must be due to the effect of sharing degrees of freedoms between the different modes present in the system.[21,22,23] Since the number of states is finite in a discrete system, an increasing number of excited solitons take degrees of freedoms, which could be used by the spin waves before (each soliton takes one degree of freedom). It is, however, not necessary that the spin wave looked at must have just the inverse temperature dependence as the solitons, since the degrees of freedom are not taken in one particular ω-range, but rather from over the whole Brillouin zone.

In summarizing the section on the nonlinear excitations, it seems fair to say that the experimental evidence is very strong that in $CsNiF_3$ in an external field solitons are an important part of the dynamics and that the sine-Gordon theory describes the results reasonably. The discrepancies and weak points in this picture as outlined above demand more work in theory and experiment.

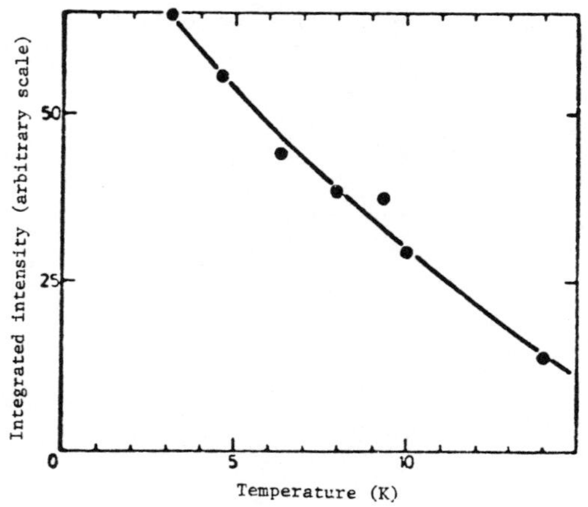

Fig. 25: Temperature dependence of the spin wave intensity measured at q_c = 0.r.l.u. and H = 5 kG.

CONCLUSION

As a general point which seems very obvious, we may note that the linear excitations, the spin waves, in one-dimensional magnets are well understood by linear spin wave theory under those conditions which ensure that nonlinear modes do not contribute. In this case the existence of spin waves is very closely coupled to the correlation-length ξ in the system. There are still open questions however in those cases where quantum effects are essential. There more theoretical and experimental work is needed.

The new and fascinating field of dynamics in 1-D-magnets, which are governed by both linear and nonlinear modes poses a lot of problems and questions which are a challenge not only for the theorists but also for the experimentalists, who have to invent methods to look more specifically at the different properties of the nonlinear and linear part of the dynamics. As for the linear excitations inelastic neutron scattering seems to be very well suited for such studies.

ACKNOWLEDGMENTS

Thanks are due to colleagues who participated in parts of the work on $CsNiF_3$: Dr. B. Dorner at ILL-Grenoble, Dr. J.K. Kjems at Risø Research Establishment. Thanks are also due to the Department of Engineering and Applied Science of Yale University, where this manuscript has been prepared.

REFERENCES

1. Steiner, M., Villain, J., Windsor, C.G., Adv. Phys. 25, 87, 1976.
2. Villain, J., these Proceedings.
3. Marshall, W., Lovesey, S.W., Theory of Thermal Neutron Scattering, Oxford Univ. Press, 1971.
4. Hutchings, M.T., Shirane, G., Birgeneau, R.F., Holt, S.L., Phys. Rev. B 5, 1999, 1972.
5. Steiner, M., Dorner, B., Villain, J., J. Phys. C 8, 165, 1975.
6. Endoh, Y., Shirane, G., Birgeneau, R.J., Richards, P.M., Holt, S.L., Phys. Rev. Lett. 32, 170, 1974.
7. Walker, L.R., Dietz, R.E., Andres, K., Darack, S., Solid State Comm. 11, 593, 1972.
8. Heilman, I.U., Birgeneau, R.J., Endoh, Y., Reiter, G., Shirane, G., Holt, S.L., BNL Report 25828.
9. Steiner, M., Zangew. Phys. 32, 116, 1971.
10. Villain, J., J. Phys. Paris, 35, 27, 1974.
11. Des Cloiseaux, J., Pearson, J.J., Phys. Rev. 128, 2131, 1962.

12. Hutchings, M.T., Windsor, C.G., J. Phys. C 10, 313, 1977.
13. Shirane, G., Birgeneau, R.J., Physica 86-88B, 639, 1977.
14. Loveluck, J., Lovesey, S.W., J. Phys. C 8, 3841, 1975.
15. Mikeska, H.J., Phys. Rev. B 12, 7, 1975; ibid B 12, 2794, 1975.
16. Müller, G., these Proceedings.
17. Steiner, M., Kjems, J.K., J. Phys. C 10, 2665, 1977.
18. Lindgård, P.Å., Kowalska, A., J. Phys. C 9, 2081, 1976.
19. Grill, R.J., Dürr, U., Weber, R., Physica 86-88B, 673, 1977.
20. Scott, A.G., Chu, F.Y.F., McLaughlin, D.W., Proc. IEEE 61, 1443, 1973.
21. Bishop, A.R., to appear in Physica Scripta (Sweden).
22. Krummhansl, J.A., Schrieffer, J.R., Phys. Rev. B 11, 3535, 1975; Gupta, N., Sutherland, B., Phys. Rev. A 14, 1790, 1976; Currie, J.F., Fogel, M.B., Palmer, F.L., Phys. Rev. A 16, 796, 1977.
23. Currie, J.F., Krummhansl, J.A., Bishop, A.R., Trullinger, S.E., to be published.
24. Mikeska, H.J., J. Phys. C 11, L29, 1978.
25. Kjems, J.K., Steiner, M., Phys. Rev. Lett. 41, 1137, 1978; Steiner, M., Kjems, J.K., J. Phys. (Paris) Letters 39, L493, 1978.
26. Kakurai, K., Kjems, J.K., Steiner, M., these Proceedings.
27. Schneider, T., Stoll, E., Phys. Rev. Lett. 41, 1429, 1928.

Q-DEPENDENCE OF THE SOLITON RESPONSE IN $CsNiF_3$ AT T = 1oK AND H = 5 kG

K. Kakurai[+], J.K. Kjems[*] and M. Steiner[+]

[+]Hahn-Meitner-Institute, Glienicker Str. 1oo
D-1ooo Berlin 39, West Germany

[*]Risø National Laboratory, DK-4ooo Roskilde, Denmark

Recently the first direct experimental evidence of solitons in the one-dimensional (I-D) ferromagnetic chain system $CsNiF_3$ with easy plane anisotropy was reported by Kjems and Steiner [1]. They observed a quasi-elastic component in the dynamic structure factor which was interpreted within the theoretical framework developed by Mikeska [4] for a I-D XY-ferromagnet in an applied field. Mikeska had showed that the spin dynamics of such systems are described by the Sine-Gordon equation and this enabled him to predict explicitly the characteristic temperature and field dependence of the scattering due to non-linear modes such as kinks or solitons. The predictions agreed very well with the results of the first neutron scattering experiment [1]. Another important prediction of the theory is the wavevector dependence of the energy width and of the integrated quasielastic intensity. This has been measured in a subsequent experiment and the results are reported here.

For detailed reviews of the properties of $CsNiF_3$ we refer to the articles by Steiner [2] and Steiner et al. [3]. The present experiment was carried out on a cold-neutron source spectrometer situated at the guide tube at the Risø DR3 reactor. The sample was the same $CsNi^{58}F_3$ crystal as was used previously. The measurements were made near the (oo2) Bragg point at T=1oK and at a field of H=5kG. The scattering vector, Q, was varied from (o, o, 1.95) to (o, o, 1.875), corresponding to a reduced wavevector variation of q_c

from 0.05 to 0.125 rlu. Some of the experimental spectra obtained after background subtraction are shown in figure 1. Also shown are the results of the least squares fit to calculated line profiles as described in the article by Steiner [2]. The resulting energy widths, ΔE (HWHM), of the quasielastic component as found by the least squares fit for different q_c's are

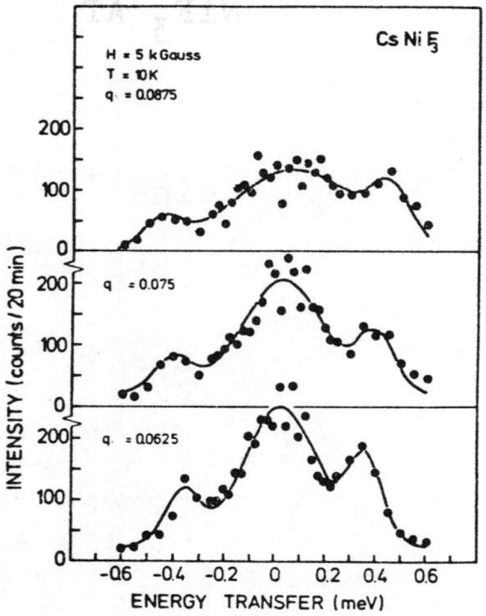

Fig. 1 Experimental spectra after background subtraction

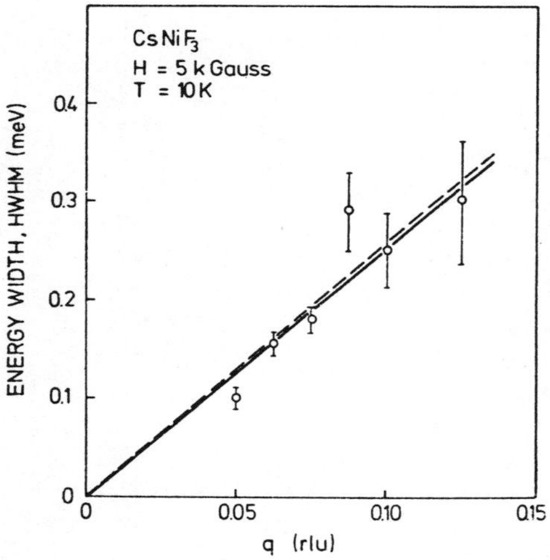

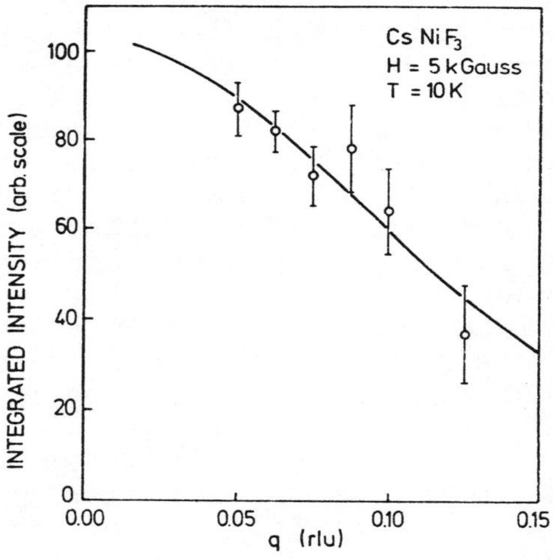

Fig. 2a Energy width, ΔE (HWHM), of the quasielastic component versus q_c.

Fig. 2b Integrated intensity of the quasielastic component versus q_c.

shown in figure 2a and the resulting integrated intensity, I, is plotted in figure 2b.

The predictions for $\Delta E\ (q_c)$ and $I\ (q_c)$ which follow from the theory by Mikeska are the following [4] :

$$\Delta E(q_c) = \sqrt{\frac{\ln 2}{4\beta m}}\ c q_c \qquad (1)$$

and

$$I(q_c) \propto \left(\frac{\frac{\pi q_c}{2m}}{\sinh \frac{\pi q_c}{2m}} \right)^2 \qquad (2)$$

where all constants are the same as in [2] and [4]. The full line in figure 2a shows the predicted linear dependence of ΔE. The best least squares fit to the data yields only a linear term and it is shown as the broken line. This result rules out a quadratic q-dependence of ΔE which would correspond to a spin diffusion process.

The least squares fit of eq.2 to the observed integrated intensities is shown by the full line in figure 2b. The resulting effective mass, 8 m, which is the only parameter, is found to be

$$8\ m/k_B = 69K$$

whereas all other experimental results can be reconciled quite well with the theoretical value of

$$8m_{theory}/k_B = 34K$$

This discrepancy may be an indication of a relativistic effect, which should appear at high temperatures [2] since Mikeska's theory is a non-relativistic approximation valid at low T only. This effect would tend to broad the response in q_c as is observed.

The present results can be summarized as follows: The observed linear q_c-dependence of the energy width of the quasielastic component confirms the interpretation of this feature in the spectral response of $CsNiF_3$ as scattering from moving kinks or solitons. It excludes an interpretation of this mode based on spin diffusion. The q_c-dependence of the integrated intensity is not in quantitative agreement with the theoretical predictions by Mikeska in contrast to most of the other experimental results obtained so far. This discrepancy and its possible connection with relativistic effects will be studied in further experiments.

References

[1] J.K. Kjems and M. Steiner, Phys. Rev. Lett. 41 (1978) 1137

[2] M. Steiner: these proceedings

[3] M. Steiner, J. Villain and C. Windsor: Adv. Phys. 25, 87 (1976)

[4] H.J. Mikeska, J. Phys. C, 11 L29 (1978)

DYNAMICS OF THE SINE-GORDON CHAIN: THE KINK-PHONON INTERACTION, SOLITON DIFFUSION AND DYNAMICAL CORRELATIONS

Nikos Theodorakopoulos

Fachbereich Physik, Universität Konstanz

7750 Konstanz, Federal Republic of Germany

I. STATEMENT OF THE PROBLEM

Recent work on the equilibrium statistical mechanics of non-linear one dimensional systems [1,2] has suggested the existence of two types of elementary excitations: linear, phonon-like oscillations about some vacuum state and intrinsically non-linear, soliton-like kinks. The consistent scheme which leads to the identification of two separate sectors in the solution space of the non-linear equation of motion carries the name of configurational phenomenology [3]. It implies the existence of non-interacting gases of elementary excitations - with appropriate self energy corrections to account for lowest order interaction effects - and provides us with a correct description of equilibrium properties. To what extent is such a conjecture, however, legitimate when applied to dynamic phenomena ? Are we justified in thinking in terms of kinks which move, more or less freely, within the lattice ? And, if not, how does one go about calculating such a quantity as the dynamical structure factor ?

I shall attempt to throw some light in these questions by considering the dynamics of the Sine-Gordon chain in the continuum limit and by going beyond standard linearisation schemes. Given the equation of motion for the displacement field $\phi(y,\tau)$ at y at time τ

$$\frac{\partial^2\phi}{\partial y^2} - \frac{\partial^2\phi}{\partial\tau^2} = \sin\phi \qquad (1)$$

with the static kink solution $\phi_k(y) = 4\tan^{-1}(\exp y)$ one may linearise (1) around $\phi_k(y)$ to obtain the excited states of kink [4],

$f_j(y)\exp(i\omega_j\tau)$. The functions $\{f_j\}$ form a complete orthogonal set and consist of the translational (Goldstone) mode with $\omega_T = 0$, $f_T = 2\,\text{sech}\,y$ and a continuum of scattering states ("phonons") with $\omega_q^2 = 1 + q^2$ and $f_q(y) = (2\pi\omega_q^2)^{-1/2}(q + i\tanh y)\exp(iqy)$. To linear order, phonons experience an asymptotic phase shift $\delta(q) = 2\arctan(1/q)$ in passing through the kink. What happens beyond this ?

II. A KINK-PHONON COLLISION

We may now describe deviations from the static kink solution by

$$\phi(y,\tau) = \phi_k(y) + \eta(y,\tau) \quad , \tag{2}$$

where

$$\eta(y,\tau) = \sum_j a_j(\tau)\, f_j(y) \tag{3}$$

and proceed to derive equations of motion for the amplitudes $a_j(\tau)$ to second order in η [5]. Here, we are interested in the description of a collision between a static kink and a propagating phonon wave packet. As discussed elsewhere [5,6] the net effect of this collision will be a spatial shift in the centers of the wave packet and the kink, by amounts $\Delta y_{ph}(q) = -\delta'(q)$ and Δy_k respectively. These spatial shifts will be related to each other via Noether's theorem for the Lorentz invariance of (1)

$$E_k^o\, \Delta y_k + E_q\, \Delta y_{ph}(q) = 0 \quad , \tag{4}$$

where $E_k^o = 8$ and E_q denote the rest energy of the kink and the energy of the phonon wave packet, respectively. Since the latter is of second order in the phonon amplitude, so will be Δy_k .

III. DIFFUSIVE MOTION OF THE KINK [6,7]

Once the physical picture of kink-phonon dynamics has been established on a microscopic scale, it is possible to go one step further and conjecture that thermal phonons will act in a stochastic manner and produce a random walk of the kink. Since phonons with different wavevectors will arrive at varying rates $\Gamma(q)$ and cause varying spatial shifts $\Delta y_k(q)$, we use an ansatz of the form

$$D = \sum_q N^2(q)\, \Gamma(q)\, \Delta y_k^2(q) \tag{5}$$

for the kink diffusion constant. For a classical phonon distribution $N(q) = (E_k^o/E_q)\bar{T}$ this yields

$$D = 0.212\ \bar{T}^2$$

where $\bar{T}$ is the reduced temperature (measured in units of E_k^0). It should be noted that the random walk described here is a process which does not involve any dissipation; thus, even to second order in the phonon amplitude, there is no mechanism which can cause a kink to change its (average) initial velocity and come to thermal equilibrium with the phonon heat bath. This is a very remarkable result because it explains, at a microscopic level, why kinks may move with unchanged average velocity ("freely") within the lattice. It should be remarked here, that the conjecture of freely moving kinks has been directly verified by recent computer simulations [8].

In summary: the influence of thermal phonons upon a kink may be described, in the latter's rest frame, in terms of the diffusion equation

$$\frac{\partial P}{\partial t} = D \frac{\partial^2 P}{\partial y^2} \tag{6}$$

where $P(y,\tau)dy$ is the probability that the kink will be centered between y and $y + dy$ at time τ.

IV. DYNAMICAL CORRELATION FUNCTIONS [10,6]

Recent neutron scattering data [9] on $CsNiF_3$, a classical one-dimensional ferromagnet with easy-plane anisotropy, has been interpreted as evidence for the existence of soliton modes in a real solid in thermal equilibrium. The spectrum of the longitudinal spin-spin correlation function

$$I(Q,\Omega) = \int dz\, dt\, \exp(iQz-i\Omega t)\langle\cos\phi(z,t)\, \cos\phi(0,0)\rangle \tag{7}$$

yields the necessary information and has been shown to contain a quasielastic contribution of Gaussian shape and width $\Gamma_Q = \bar{T}^{1/2}Q$ due to independent scattering from freely moving kinks [10]. The field ϕ denotes in this case the angle of the spin vector along the easy plane; the linear modes (magnons) are small oscillations around $\phi = 0$, a direction defined by an applied magnetic field. In the context of section II we should speak therefore of a soliton-magnon interaction.

The diffusive motion of the solitons enters the (incoherent) scattering intensity (7) via the stochastic average of a factor $\exp[iQ\Delta x(t)]$, where $\Delta x(t)$ is the kink's stochastic displacement at time t, measured in laboratory coordinates. We may transform to a frame moving with the kink's (deterministic) initial velocity, use (6) to compute the stochastic average and rewrite the result in terms of laboratory coordinates. This procedure yields

$$\langle\Delta x^2(t)\rangle = 2\, D\, t \tag{8}$$

for non-relativistic kinks, i.e. as long as $\bar{T} < 0.5$.

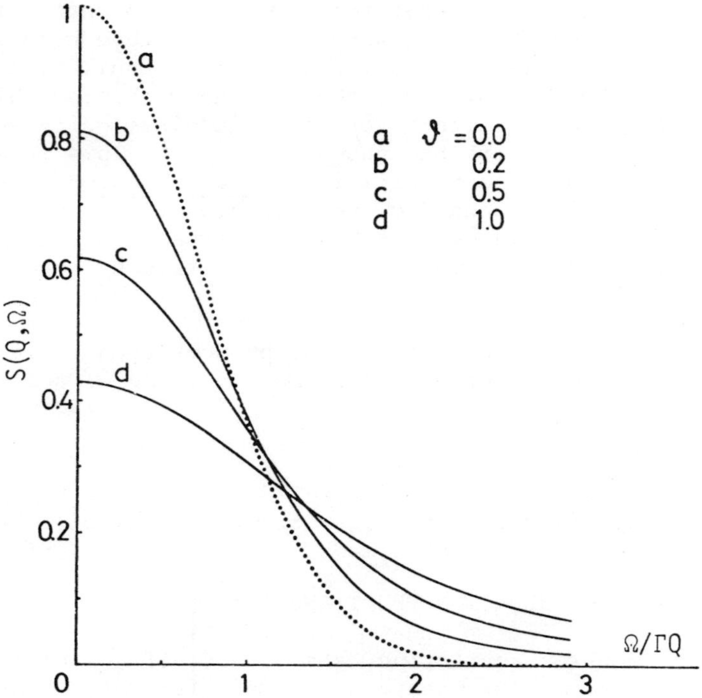

Fig. 1: The spectral intensity as a function of the frequency for various values of the parameter ϑ; the dotted curve ($\vartheta = 0$) is a Gaussian of width Γ_Q and corresponds to the limit of vanishing kink diffusion (Ref. 10). Note the decrease in peak intensity and the long-tail structure, which appear already at small values of ϑ.

The quasielastic (soliton) part of the spectrum will be now given by

$$I_{sol}(Q,\Omega) \propto n_k(\bar{T})\ F^2(Q)\ S(Q,\Omega) \tag{9}$$

where n_k is the number of kinks and antikinks present, F(Q) denotes a structure factor and the normalized spectrum is given by

$$S(Q,\Omega) = \int_{-\infty}^{\infty} dt\ e^{-i\Omega t - DQ^2|t| - \Gamma_Q^2 t^2/4} \tag{10}$$

and is plotted in Fig. 1 for various values of the dimensionless ratio $\vartheta = DQ^2/\Gamma_Q$. The effects of kink diffusion become important already for relatively small values of ϑ. The drastic decrease in the peak intensity and the long-tail structure should thus reflect the effects of soliton diffusion.

I would like to thank Professor R. Klein for helpful discussions and comments.

References

1. J.A. Krumhansl and J.R. Schrieffer, Phys. Rev. B11,3535(1975)
2. S. Aubry, J. Chem. Phys. 62, 3217(1975) and 64, 3392(1976)
T. Schneider and E. Stoll, Phys. Rev. Lett. 35, 296(1975)
3. A.R. Bishop in "Solitons and Condensed Matter", Springer-Verlag 1978 (Editors A.R. Bishop and T. Schneider)
4. M.B. Fogel, S.E. Trullinger, A.R. Bishop and J.A. Krumhansl, Phys. Rev. B15, 1578(1977)
5. W. Hasenfratz and R. Klein, Physica 89A, 191(1977)
6. N. Theodorakopoulos, Z. Physik B33 (1979)
7. Y. Wada and J.R. Schrieffer, Phys. Rev. B15, 3897(1978)
8. T. Schneider and E. Stoll, Phys. Rev. Lett. 41, 1429(1978)
9. J.K. Kjems and M. Steiner, Phys. Rev. Lett. 41, 1137(1978)
10. H.J. Mikeska, J. Phys. C11, L29(1978); This result has been previously obtained in the case of the ϕ^4 model by W. Hasenfratz, R. Klein and N. Theodorakopoulos, Solid State Comm. 18, 893(1976).

THE SPIN-WAVE CONTINUUM OF THE S=1/2 LINEAR HEISENBERG ANTIFERROMAGNET

Gerhard Müller and Harry Thomas

Institut für Physik der Universität Basel

CH - 4056 Basel, Switzerland

Hans Beck

Institut de Physique, Université de Neuchâtel

CH - 2000 Neuchâtel, Switzerland

In the S=1/2 linear Heisenberg antiferromagnet (HB AF)

$$H = J\sum_{i=1}^{N} \vec{S}_i \cdot \vec{S}_{i+1} - h\sum_{i=1}^{N} S_i^z \qquad (1)$$

- although investigated by various theoretical approaches - many important questions concerning the statics and the dynamics have remained open. Recent low-temperature neutron-scattering experiments on $CuCl_2 \cdot 2N(C_5H_5)$ (CPC), which is a good realization of an S=1/2 HB AF chain, provided new important information on the dynamics of the system, such as lineshapes and the behaviour in a magnetic field /1/. The important quantity for direct comparison with experiments of this kind is the dynamic spin-correlation function in (q,ω)-space. It is the Fourier transform of $\langle S^z(\ell,t)S^z(\ell',o)\rangle$, and for T=o it can be written as

$$G_{zz}(q,\omega) = \sum_\lambda M_\lambda \delta(\omega+E_o-E_\lambda) \ , \ M_\lambda = 2\pi|\langle o|S^z(q)|\lambda\rangle|^2 \qquad (2)$$

In a recent publication /2/ an approximate analytical expression for G_{zz} at T=o and h=o was obtained by using finite-chain calculations together with the results of two other theoretical approaches. It represents the dominant contribution to $G_{zz}(q,\omega)$ originating from a spin-wave continuum (SWC) bounded between the dispersion branches $E_1(q)=(\pi J/2)|\sin q|$ and $E_2(q)=\pi J|\sin(q/2)|$.

$$G_{zz}^{swc}(q,\omega)=2\{\omega^2-(E_1(q))^2\}^{-1/2}\theta\{\omega-E_1(q)\}\theta\{E_2(q)-\omega\}. \quad (3)$$

G_{zz} increases strongly towards the lower bound $E_1(q)$, which is the des Cloizeaux-Pearson spin-wave energy. Result (3) is in good agreement with experimental data for CPC concerning excitation energy, lineshape and integrated intensity /1,2/.

In this note we give further arguments supporting (3), which demonstrates the usefulness of finite-chain calculations for properties of the infinite system. Although we cannot give a rigorous derivation of (3), we conjecture that it represents the SWC quantatively. We have identified those excitations of the finite system which have dominant spectral weight with a special class of eigenstates in the Bethe formalism /3/, and we show that these states form, in the thermodynamic limit, a continuum exactly between the two branches $E_1(q)$ and $E_2(q)$. The Bethe Ansatz for the exact eigenfunctions consists of a linear combination $\Psi=\sum a(n_1,...,n_r)\cdot \phi(n_1,...,n_r)$ of local basis vectors with reversed spins at lattice sites $n_1,...,n_r$ with coefficients of the form

$$a(n_1,...,n_r)=\sum_p \exp\{i\sum_j k_{p_j} n_j+(1/2)i\sum_{j<\ell} \psi_{p_j p_\ell}\} \quad (4)$$

where the summation $\sum_p$ extends over all permutations of the integers $1,2,...,r$, and p_j is the image of j under the p'th permutation. The k_j and the ψ_{ij} obey the coupled equations:

$$2\cot(1/2)\psi_{j\ell}=\cot(1/2)k_j-\cot(1/2)k_\ell\ , \quad Nk_j=2\pi\lambda_j+\sum_{\ell\neq j}\psi_{j\ell}. \quad (5)$$

The integers λ_j are confined to $1\leq\lambda_j\leq N-1$, and each choice of a set $\{\lambda_j\}$ (being subject to additional restrictions) determines an eigenstate of the system. Having solved the above equations for k_j, it is straightforward to calculate wave number and energy of the corresponding eigenstate,

$$q=\sum_{j=1}^{r} k_j=(2\pi/N)\sum_{j=1}^{r}\lambda_j \quad , \quad E=-\sum_{j=1}^{r}(1-\cos k_j). \tag{6}$$

The ground state, which is a singlet (for even N), corresponds to the N/2 integers $\lambda_j=1,3,5,\ldots,(N-1)$. Des Cloizeaux and Pearson /4/ found the lowest excited states to be given by

$$\begin{array}{ll} 1,3,\ldots,(N-2n-1),(N-2n+2),\ldots,(N-2) & q>o \\ 2,4,\ldots,(2n-2),(2n+1),\ldots,(N-1) & q<o \end{array} \tag{7}$$

($q=2\pi n/N$) and calculated their energies. The result is the famous DC-P spin-wave branch $E_1(q)$. By generalization of their method we have found the sets $\{\lambda_j\}$ for all SWC states. To the highest branch $E_2(q)$, in particular, belong the sets (always for even N):

$$\begin{array}{lll} 1,3,\ldots,(N-n-2),(N-n+2),\ldots,(N-1) & n \text{ odd} & \multirow{2}{*}{q>o} \\ 1,3,\ldots,(N-n-3),(N-n),(N-n+3),\ldots,(N-1) & n \text{ even} & \\ 1,3,\ldots,(n-2),(n+2),\ldots,(N-1) & n \text{ odd} & \multirow{2}{*}{q<o} \\ 1,3,\ldots,(n-3),n,(n+3),\ldots,(N-1) & n \text{ even} & \end{array} \tag{8}$$

Using these numbers we can calculate (in the thermodynamic limit) the energies of all the excitations of the two-parameter SWC ($q>o$ for convenience):

$$E_b(q)=\pi J\left|\sin(q/2)\cos(q/2-q_b/2)\right| \tag{9}$$

where q ($o\leqslant q\leqslant\pi$) is the wave number of the excitation (now with respect to that of the ground state) and q_b ($o\leqslant q_b\leqslant q$) labels the different dispersion branches within the continuum. The lowest branch $E_1(q)$ has $q_b=o$ and the highest one has $q_b=q$ yielding $E_2(q)$. Furthermore (9) immediately provides the density of states in the SWC

$$D(q,\omega)=(N/2\pi)\ \{(E_2(q))^2-\omega^2\}^{-1/2} \tag{10}$$

According to (2) $G_{zz}^{SWC}(q,\omega)$ is the product of the density of states $D(q,\omega)$ and a spectral weight defined by the squared matrix elements between the ground state and the SWC excitations: $M(q,\omega) \equiv 2\pi|\langle o,o|S^z(q)|\omega,q\rangle|^2$, yielding

$$M(q,\omega)=(4\pi/N)\ \{(E_2(q))^2-\omega^2\}^{1/2}/\{\omega^2-(E_1(q))^2\}^{1/2} \quad (11)$$

Comparison of (11) with finite-chain matrix elements shows good agreement.

This approach to the dynamics of the S=1/2 HB AF at T=o can be extended to the h≠o case. From finite-chain calculations we have determined the excitations contributing significantly to $G_{zz}(q,\omega)$. Again we have identified this class of excitations unambiguously with a certain class of eigenstates in the Bethe formalism. The calculations to solve eq's (5) for these states are in progress. Preliminary approximate results show that $G_{zz}(q,\omega)$ is dominated by two partly overlapping continua of excitations. Fig. 1 shows the boundaries of these continua for a special value of h. Again, the spectral weight of $G_{zz}(q,\omega)$ increases strongly as the frequency is lowered towards the lower boundary of each continuum. Further we find that $G_{xx}(q,\omega)$ looks for h≠o qualitatively different from $G_{zz}(q,\omega)$. In particular, the lowest branch is inverted with respect to the axis $q=\pi/2$. Therefore we expect appropriate neutron scattering experiments to show spectra which are more complex than for h=o (having at least two dominant peaks), and which strongly depend on the relative weight of G_{xx} and G_{zz} in the scans under consideration. More details will be published elsewhere.

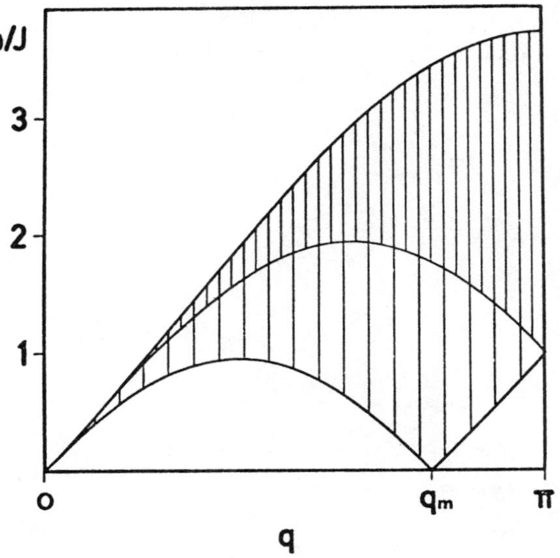

Fig. 1: The two continua of excitations dominating G_{zz} (q,ω) at T=o and $h=(1/2)h_{crit}$. In each continuum the spectral weight increases strongly towards the corresponding lower boundary. The lowest boundary corresponds approximately to the spin-wave frequency obtained numerically by Ishimura and Shiba /5/ and to the approximate analytical result by Pytte /6/. The special wave number q_m depends only on the magnetization. It is equal to π at h=o and decreases as h increases, reaching zero at the critical field h_{crit}.

References

/1/ I.U. Heilmann, G. Shirane, Y. Endoh, R.J. Birgenau, and S.L. Holt, to be published

/2/ G. Müller, H. Beck, and J.C. Bonner, to be publ.

/3/ H. Bethe, Z. Phys. 71, 205, (1931)

/4/ J. Des Cloizeaux and J.J. Pearson, Phys. Rev. 128, 2131, (1962)

/5/ N. Ishimura and H. Shiba, Prog. Theor. Phys. 57, 1862, (1977)

/6/ E. Pytte, Phys. Rev. B 10, 685, (1974)

EXCITATIONS AND PHASE TRANSITIONS IN RANDOM ANTI-FERROMAGNETS

R.A. Cowley

Department of Physics
University of Edinburgh[a]
Edinburgh EH9 3JZ
Scotland

R.J. Birgeneau

Department of Physics
Massachusetts Institute of Technology[b]
Cambridge
Mass. 02139
U.S.A.

G. Shirane

Brookhaven National Laboratory[c]
Upton
New York 11973
U.S.A.

ABSTRACT

Neutron scattering techniques can be used to study the magnetic excitations and phase transitions in the randomly mixed transition metal fluorides. The results for the excitations of samples with two different types of magnetic ions show two bands of excitations; each associated with excitations propagating largely on one type of ion. In the diluted salts the spectra show a complex line shape and greater widths. These results are in good accord with computer simulations showing that linear spin wave theory can be used, but have not been described satisfactorily using the coherent potential approximation. The phase transitions in these materials are always smeared, but it is difficult to ascertain if this smearing is due to macroscopic fluctuations in the concentration or of an intrinsic origin. Studies of these systems close to the percolation point have shown that the thermal disorder is associated with the one-dimensional weak links of the large clusters. Currently theory and experiment are in accord for the two-dimensional Ising system but features are still not understood in Heisenberg systems in both two and three dimensions.

INTRODUCTION

The transition metal fluorides are ideal systems on which to study the properties of disordered systems. Transition metal ions such as Mn^{++}, Co^{++} and Mg^{++} are chemically very similar but are magnetically very different. Consequently it is possible to grow single crystals in which magnetically very different ions are randomly distributed over the transition metal sites. It is then possible to study the excitations and phase transitions of these disordered systems in detail particularly by the use of neutron scattering techniques. The comparison of these results with theory is further facilitated because the magnetic interactions are of a relatively simple form between only near neighbours and the magnitude of these interactions is usually known from experiments on the pure materials. There are then, relatively few if any unknown parameters needed to describe the magnetic interactions in these systems and so there are few adjustable parameters which need to be introduced in confronting theory with experiment. Before 1971 most of the work on these systems had been on systems in which one of the magnetic constituents was present only in low concentration. Excellent agreement was found between the results of these experiments and theory[1]. More recently experiments have been performed on more concentrated systems and in this brief review the main experimental results will be described together with an outline of the appropriate theories.

Although all the transition metal fluorides are simple antiferromagnets, they show a wide diversity of behaviour. They have three different crystal structures; in $RbMnF_3$, the magnetic ions, $S = \frac{5}{2}$, are arranged on a simple cubic lattice and the magnetic interactions are of Heisenberg character between nearest neighbours[2]

$$H_{EX} = \frac{1}{2} J \sum_i \sum_\Delta \vec{S}_i . \vec{S}_{i+\Delta} \qquad (1)$$

where Δ is a nearest neighbour to i. MnF_2 and CoF_2 have the rutile structure with the magnetic ions arranged on a body-centred tetragonal lattice. The interaction with the body-centred ions is antiferromagnetic and there is a weaker ferromagnetic interaction with the nearest neighbours along the c-axis. In the Mn salt both these interactions are of Heisenberg character[3] but the dipolar interactions lead to a weak anisotropy field along the unique c-axis which is consequently the spin directon;

$$H_D = - H_A \sum_i (S_i^z)^2. \tag{2}$$

In CoF_2 the effect of the orbital angular momentum is that the effective exchange within the ground state doublet, $S = \frac{1}{2}$, can be written approximately[4] as

$$H_{AN} = \frac{1}{2} J \sum_{i\Delta}\sum \left[S_i^z\, S_{i+\Delta}^z + \alpha(S_i^x S_{i+\Delta}^x + S_i^y S_{i+\Delta}^y) \right], \tag{3}$$

where α is 0.81.

The third system is the structure shown in fig. 1, K_2NiF_4(S=1). The Ni ions are arranged on a square lattice with a relatively large separation and consequently little interaction between the different layers of the lattice. Although these materials order three-dimensionally, the critical scattering and magnetic excitations are characteristic of a two-dimensional system. The magnetic interactions are almost wholly between nearest neighbours and for the Mn and Ni salts of Heisenberg character[5], eqn. 1, and for the Co salt[6] the exchange is anisotropic eqn. 3 with α = 0.55. The tetragonal symmetry of these materials

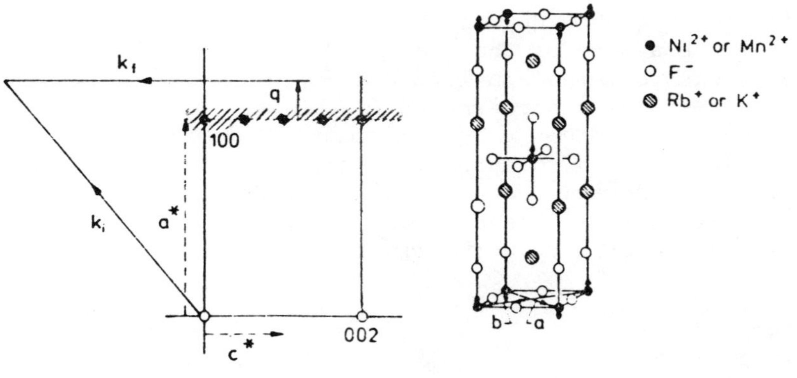

Fig. 1. The reciprocal lattice and crystal structure of K_2NiF_4. The reciprocal lattice shows the scattering geometry used to determine $S(\vec{Q})$ by taking advantage of the independence of $S(\vec{Q})$ on Q_c.

leads for the Mn and Ni salts to a weak anisotropy field which aligns the spins perpendicular to the layers. The exchange interactions and dipolar fields for these systems are listed in table I for these various different systems as determined from spin wave measurements.

Table I

Dominant Exchange Constants and Anisotropy Fields deduced from linear spin wave theory.

Material	S	J(°K)	α	H_A(°K)	Reference
$KMnF_3$	2.5	7.25	-	0	1
$KCoF_3$	0.5	112.4	-	6.7	1
$KNiF_3$	1.0	102.1	-	0	1
MnF_2	2.5	3.57	-	0.43	1
CoF_2	0.5	17.78	0.81	-	1
K_2NiF_4	1.0	104.0	-	0.87	5
K_2MnF_4	2.5	8.4	-	0.13	5
Rb_2MnF_4	2.5	7.60	-	0.15	5
Rb_2CoF_4	0.5	179.5	0.55	-	6

In the next section a brief description is given of the neutron scattering techniques. One of the first aspects studied in these systems was the nature of the excitations in very disordered systems. This aspect was reviewed[7] in 1975 and so in section 3 only a brief account is presented concentrating on the more recent developments. Attention has more recently been directed to the study of the phase transitions in disordered systems. In section 4 we describe the little experimental progress which has been made in this subject. More progress has been made in the understanding of percolation as described in the final section.

NEUTRON SCATTERING

Before any experiments can be performed single crystal specimens of the mixed systems must be grown. The starting materials are carefully purified and mixed in the desired proportions. The crystals are then grown using either Bridgman or Stockbarger techniques in a manner similar to that employed for the pure materials. One of the most difficult and frustrating aspects of this work is the determination of the concentration of the constituents in the resulting crystals, and the extent to which the concentrations are uniform throughout the crystals.

Unfortunately chemical analysis involves destroying at least parts of the crystal and even then seems to give the atomic concentration, of say a 50% concentration of Mn and Mg, to only about 3%. Lattice parameter measurements and measurements of magnetic ordering temperatures are also useful indicators of the concentration.

Measurements in many of these systems have been performed to search for evidence of any chemical ordering or clustering. In none of the materials has there been any evidence of such an ordering found.

The neutron scattering measurements have all been performed with unpolarised neutrons. The incident neutron beam with wavevector $\vec{k}_I$ is scattered to give a beam with wavevector $\vec{k}_F$, and a wavevector transfer $\vec{Q} = \vec{k}_I - \vec{k}_F$. The energy transfer to the specimen is $\hbar\omega = \frac{\hbar^2}{2m_N}(k_I^2 - k_F^2)$. The cross-section for the scattering from these systems is given by;

$$\frac{d^2\sigma}{d\Omega d\omega} = C\,\frac{k_I}{k_F}|F(\vec{Q})|^2\Big[\sin^2\Theta\, G_L(\vec{Q},\omega) + (1+\cos^2\Theta)\, G_T(\vec{Q},\omega)\Big], \qquad (4)$$

where C is a constant, $F(\vec{Q})$ the form factor of the magnetic ions, Θ the angle between $\vec{Q}$ and the z axis. The spin correlation functions are given by:

$$G(\vec{Q},\omega) = \sum_{nm} p_n \langle n|O|m\rangle\langle m|O^*|n\rangle\,\delta(\omega - \omega_{mn}),$$

where the operator O is given for $G_L(\vec{Q},\omega)$ by

$$O = \sum_i S_i^z \exp(i\,\vec{Q}.\vec{R}_i),$$

and for $G_T(\vec{Q},\omega)$ by

$$O = \sum_i S_i^x \exp(i\,\vec{Q}.\vec{R}_i),$$

while n is the initial state of the system and m the final state and $\hbar\omega_{mn}$ is their energy difference.

The excitation spectra at low temperatures are obtained from measurements of the cross-section as a function of frequency transfer for various wavevector transfers. These measurements are most conveniently performed with the aid of a triple axis crystal spectrometer which determines the energy of the incident and scattered neutrons by Bragg reflection from single crystals. The usual experimental precautions must be taken to minimize the experimental resolution while maximizing the counting rate, and to avoid the spurious scattering by the higher order neutrons reflected by the monochromators. Essentially, however, these measurements enable a quite direct measurement of $G_L(\vec{Q},\omega)$ and $G_T(\vec{Q},\omega)$.

The theory of phase transitions is most highly developed for the instantaneous properties of the system and much less highly developed for describing the time-dependent properties. In scattering terms it provides detailed predictions about the behaviour of the static correlation functions

$$G_T(\vec{Q}) = \int G_T(\vec{Q},\omega)\,d\omega, \tag{5}$$

and

$$G_L(\vec{Q}) = \int G_L(\vec{Q},\omega)\,d\omega.$$

These can be measured by measuring the differential cross-section, $G(\vec{Q},\omega)$, and numerically performing the integrals. It is, however, more convenient to measure the static correlation function directly. In a two-axis experiment the neutrons scattered in a given direction are all recorded irrespective of their energy. Unfortunately as the energy transfer varies the wavevector transfer also varies and in the integral over ω the wavevector $\vec{Q}$ is not constant for all energy transfers. The change in wavevector for a frequency transfer $\delta\omega$ from $\omega=0$ is approximately given by

$$\delta\vec{Q} = -2\vec{k}_F \frac{\delta\omega}{E_o} ,$$

where E_o is the incident energy of the neutrons. Clearly the change in the wavevector can be minimized by increasing the incident neutron energy, E_o, to be much larger than $\delta\omega$. Unfortunately increasing E_o also decreases the wavevector resolution. In the two-dimensional layer systems $G(\vec{Q},\omega)$ is largely independent of the component Q_c. Consequently[5] by performing experiments with $\vec{k}_F$ aligned along the crystallographic c axis, as shown in fig. 1, accurate measurements of $G(\vec{Q})$ can be obtained without using such high energy incident neutrons.

EXCITATIONS

In concentrated systems composed of two different magnetic

ions the excitation spectrum consists of two fairly distinct branches. This behaviour was first observed in the $KMnF_3/KCoF_3$ system[8] but has now been observed also in MnF_2/CoF_2[8,9], $KMnF_3/KNiF_3$[10] and Rb_2MnF_4/Rb_2NiF_4[11]. The results for various concentrations of $Co_cMn_{1-c}F_2$ are shown in fig. 2. For c = 0.05 there is a local mode associated with the Co ions at frequencies well above the MnF_2 host band. For c = 0.3 the Co band shows dispersion and considerable width and the Mn band has less

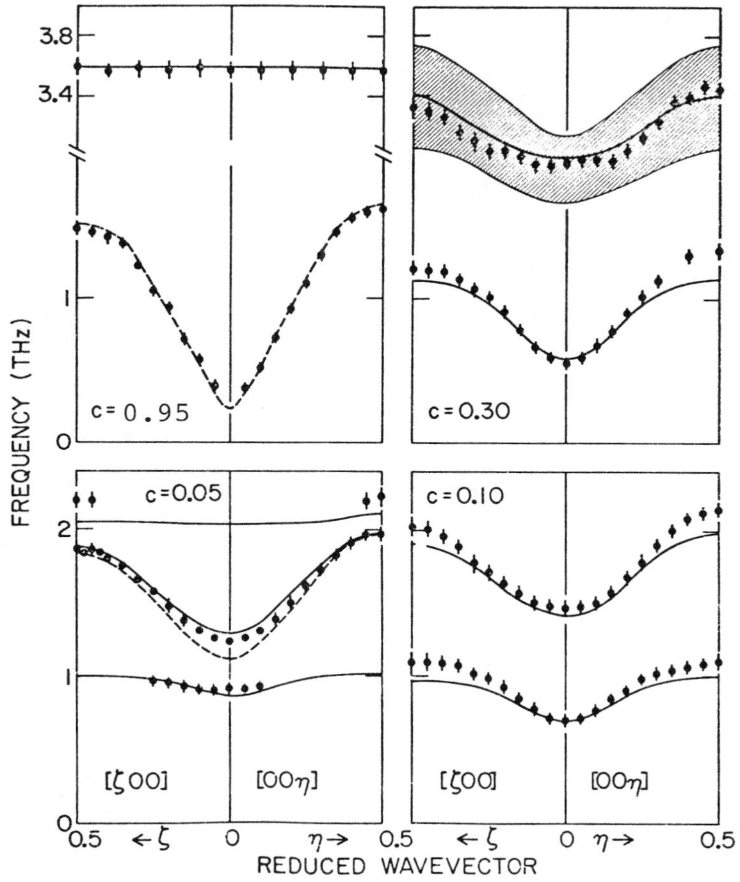

Fig. 2. Dispersion relations for the $Mn_{1-c}Co_cF_2$ system[8,9]. The solid lines are the results of CPA calculations[14] and dotted lines the dispersion relations of the pure material. The shaded region for c = 0.3 gives the width predicted by the calculations.

dispersion. This behaviour progresses steadily until for c=0.95 the Co band is very similar to that of pure CoF_2 and there is a local mode associated with the Mn ions below the CoF_2 band.

In pure materials the frequency of the magnons at the zone boundary is given by the Ising model. In the mixed systems the Ising model suggests that the energy to create a spin deviation on an ion of type λ surrounded by cz ions of type λ and (1 - c)z ions of type λ' is

$$\overline{E}_{\lambda} = cz\, J_{\lambda\lambda}\, S_{\lambda} + (1-c)zJ_{\lambda\lambda'}\, S_{\lambda'} \tag{6}$$

Experiments on systems with only a small concentration of defects[1] further suggest that

$$J_{\lambda\lambda'} = (J_{\lambda\lambda}\, J_{\lambda'\lambda'})^{\frac{1}{2}} . \tag{7}$$

Using eqn's 6 and 7 the frequencies of the mixed systems can be calculated using the results listed in table I, as shown in table II. Clearly the agreement between experiment and theory is surprisingly good for such a simple model.

A more comprehensive theory must include a calculation of the width and wave-vector dependence of the excitations. Calculations[12,13] for the $Rb_2Mn_cNi_{1-c}F_4$ system using computer simulation techniques have given a good account of the experimental results confirming that the basic model for these systems is adequate and that the excitation spectra can be obtained at low temperatures by using a linear spin wave theory.

One of the most well used ways of describing the excitations in disordered systems is the coherent potential approximation or CPA. There have been several different ways developed to apply these techniques to these systems. We believe that one of the most useful approaches is that developed by Buyers et al.[14]. It begins with the success of the Ising model discussed above. Suppose an ion λ is surrounded by r similar ions then the frequency to create a spin deviation is

$$E_{\lambda}(r) = r\, J_{\lambda\lambda}S_{\lambda} + (z-r)\, J_{\lambda\lambda'}\, S_{\lambda'} \tag{8}$$

These frequencies vary from site to site with probabilities dependent upon the concentrations of the different types of ions. If the transverse part of the exchange interactions are independent of the type of ions, then the problem is formally very similar to that of random masses in the analogous phonon problem or random single site potentials in the electron problem. For these systems

Table II

The Ising frequencies (THz) in mixed antiferromagnets from reference 7.

System	Concentration	Calculation	Experiment
MnF_2/CoF_2	0.95	3.46 1.51	3.57±0.05 1.49±0.02
	0.30	2.42 1.20	2.32±0.10 1.20±0.06
	0.10	2.05 1.05	2.02±0.08 1.09±0.08
	0.05	1.92 1.00	1.86±0.06 0.90±0.10
$KMnF_3/KCoF_3$	0.80	6.96 2.27	6.55±0.15 2.26±0.02
	0.29	7.04 2.30	6.80±0.10 2.30±0.10
$KMnF_3/KNiF_3$	0.97	8.51 2.27	7.68±0.10 2.17±0.05
	0.25	11.7 3.12	11.8 ±0.5 3.1 ±0.15
Rb_2MnF_4/Rb_2NiF_4	0.5	6.81 1.92	6.48 1.79
MnF_2/ZnF_2	0.78	1.18	1.19±0.02
	0.69	1.04	1.05±0.03
CoF_2/ZnF_2	0.86 0.69	1.60 1.28	1.59±0.02 1.12±0.05

the single site CPA describes the scattering of the magnetic excitations by the different single site frequencies. Despite this similarity there is one difference; namely that since $E_\lambda(r)$ depends on the local environment the $E_\lambda(r)$ of nearest neighbours are not completely random but have some degree of correlation. This correlation becomes most pronounced when z is low, and we are unaware of any attempt to investigate the effects of this correlation.

The approximation that the transverse parts of the exchange interactions are independent of the ions is not usually valid. By chance it is very nearly satisfied in MnF_2/CoF_2 and the results of CPA calculations for this system[14] are shown in fig. 2. Clearly the agreement between experiment and theory is quite satisfactory.

In other systems various different approximations have been employed to treat the transverse part of the exchange interactions. The most successful way is to exploit the spin scaling technique of Tonegawa[15] to minimize the randomness of the transverse exchange interactions. It is possible to eliminate the randomness entirely if $S_\lambda J_{\lambda\lambda'} = S_{\lambda'} J_{\lambda'\lambda'}$ and $S_\lambda J_{\lambda\lambda} = S_{\lambda'} J_{\lambda\lambda'}$. These relations are almost exactly satisfied in the $KMnF_3/KCoF_3$ and $KMnF_3/KNiF_3$ systems and the agreement between experiment and the CPA calculations[14,10] is very satisfactory. There are then various ad hoc ways of treating further randomness and by a judicious choice of the form reasonable agreement[16] can also be obtained for the Rb_2Ni/MnF_4 system.

The systems in which the magnetic ions are diluted with a non-magnetic ion have proved to be more difficult to explain in detail. The average zone boundary frequencies are given quite accurately for the Mn/ZnF_2[17] and Co/ZnF_2 systems[18] as shown in table 2. The CPA calculations, however, proved to be more unreliable because of the difficulty of treating the transverse parts of the exchange adequately. An alternative form of the CPA was, therefore, developed by Holcomb[19] which treats the transverse parts of the exchange interactions adequately but at the expense of treating all the Ising parts in the same average way. The frequencies are the average frequencies of eqn. 6 instead of the $E_\lambda(r)$ of eqn. 8. There is then a difference in the predictions of these two CPA theories: the theory of Holcomb predicts a smooth distribution in frequency with a single peak at each wavevector while that of Buyers et al. predicts a multi-peaked cross-section with the peaks at least loosely associated with the different arrangements of the neighbours (values of r).

The first experiments to exhibit this fine structure were performed on the Mn/ZnF_2 system[20,21]. They clearly showed the existence of fine structure in the differential scattering cross-section showing that a successful theory must treat the local environment of each ion in detail. More dramatic evidence of these effects has now been obtained using the two-dimensional layer materials. In this case z=4 and so the separation in frequency between the different $E_\lambda(r)$ is larger. In fig. 3 scattered neutron distributions[22] are shown for $Rb_2Co_{0.57}Mg_{0.43}F_4$ and they clearly show the existence of four peaks corresponding to the frequencies r = 1,2,3,4. Similar results for other concentrations in this system have been obtained by Ikeda and Shirane[23]. Fig. 3 also shows that there is much less dispersion in the frequencies than in the pure material. Computer simulations of the magnetic excitations in this system performed using the techniques developed

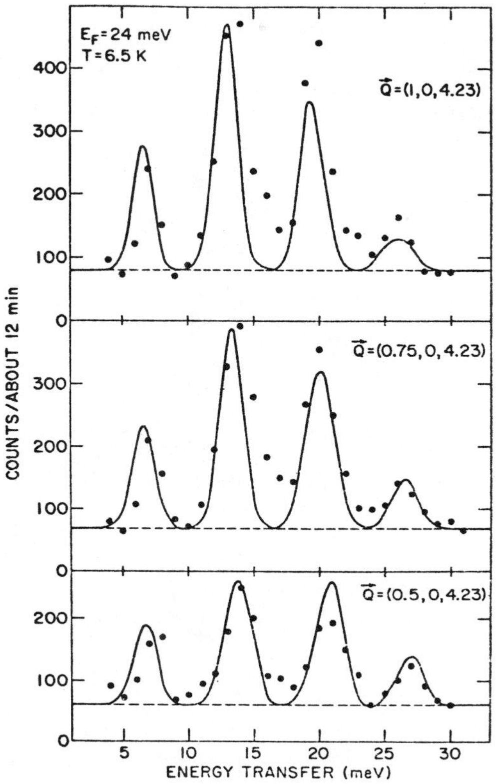

Fig. 3. Scattered neutron distributions[22] obtained from $Rb_2Co_{0.57}Mg_{0.43}F_4$ at 80 K for various wavevector transfers. The solid line is the result of a computer simulation using the techniques of Alben and Thorpe[12].

by Alben and Thorpe[12] give good agreement with the experimental results. The magnetic interactions in Rb_2Mn/MgF_4 are more nearly Heisenberg in character and consequently the fine structure is less marked, as shown in fig. 4. Nevertheless four peaks are observed corresponding to the four $E_\lambda(r)$ particularly for wavevectors close to the zone boundary[24]. Also shown in fig. 4 are the results of computer simulations which give amazingly good agreement with the experimental results. This agreement shows for both of these systems that linear spin wave theory is adequate for describing the excitations and secondly enables us to determine the exchange interaction and the concentration accurately[24].

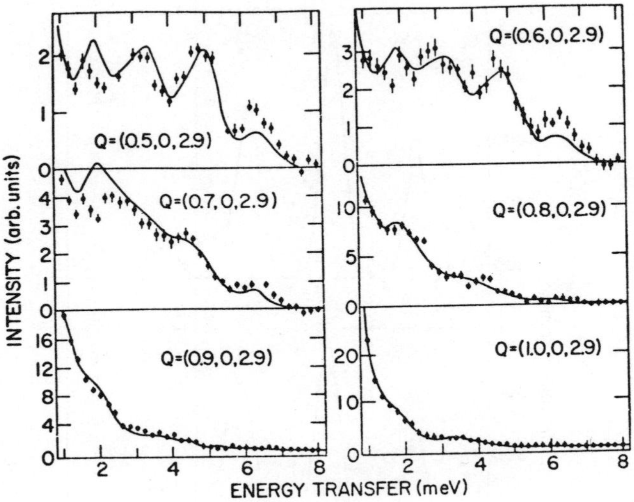

Fig. 4. Scattered neutron distributions[24] in $Rb_2Mn_{0.54}Mg_{0.46}F_4$ at 4.0 K and compared with computer simulations obtained using the techniques of Alben and Thorpe.

Calculations of these distributions using the coherent potential approximation have been performed[24]. Clearly one of the forms of theory using the $E_\lambda(r)$ must be employed if the many peaked structure is to be obtained but then there is the difficulty of treating the transverse parts of the exchange interactions. At present none of the calculations have given a good account of the results shown in fig. 4.

CRITICAL PHENOMENA AT PHASE TRANSITIONS

The study of the critical phenomena at the phase transitions of the pure antiferromagnets has provided some of the most accurate measurements of the critical exponents of different types of systems. We do not, however, review all of these results because many of the classic studies such as that[25] of $RbMnF_3$ were performed nearly 10 years ago. One result which is, however, so elegant as to be worth presenting again, is that of Ikeda and Hirakawa[26] on K_2CoF_4. The sub-lattice magnetisation is shown in fig. 5 and it agrees precisely with Onsager's exact solution of the two-dimensional Ising model.

The nature of the phase transitions in disordered systems has been the subject of considerable theoretical effort recently. The

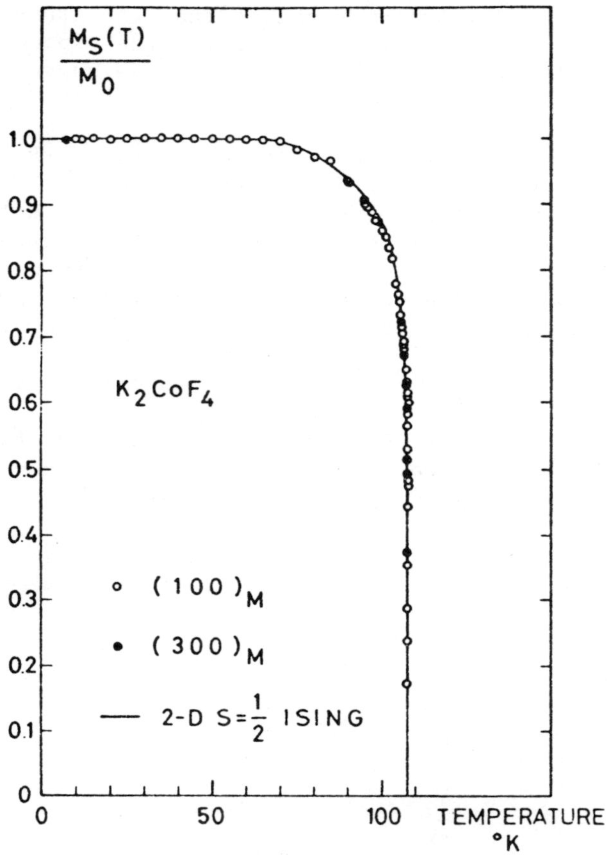

Fig. 5. The sub-lattice magnetisation, M(T), in K_2CoF_4 as a function of temperature. The solid line shows Onsager's exact solution of the two-dimensional Ising model.

work of Harris[27] suggested that if the specific heat exponent α was negative then the behaviour would be similar to that of the pure systems, but if α was positive then the concentration fluctuations would have a larger effect and might lead to a first order transition or a smeared transition or a continuous transition with different exponents from those of the pure system. Similar results[28] have now been obtained using renormalisation group theory and the ε-expansion. In the case of some systems for which α was positive renormalisation group theory suggests that the transition is nevertheless continuous but that the values of the exponents are different[29]. In contrast McCoy and Wu[30] have suggested that the phase transitions in disordered systems are smeared.

There have been two careful measurements of the critical properties of these disordered antiferromagnets published. In $Rb_2Mn_{0.5}Ni_{0.5}F_4$ the first crystal[5] showed a smeared transition with a full width at half maximum corresponding to fluctuations in concentration of 8% and a later experiment was performed on a smaller crystal which showed concentration fluctuations of only 2%. In these two-dimensional materials the critical fluctuations are extended in temperature for $\varepsilon = \frac{T - T_N}{T_N} < 0.2$ and, since the spread in concentration corresponds to $\varepsilon \sim 0.01$, it is possible to measure the properties of the critical fluctuations despite the smearing of T_c. The results were analysed by writing

$$G_T(\vec{Q}) = \frac{A_T}{K_T^2 + q^2} , \tag{9}$$

and

$$G_L(\vec{Q}) = \frac{A_L}{K_L^2 + q^2} ,$$

where $\vec{Q} = \vec{q} + \vec{\tau}$ and $\vec{\tau}$ is any reciprocal lattice vector of the magnetic lattice. Since in this material, the exchange interactions are of Heisenberg character, it might be expected that $G_T(\vec{Q}) = G_L(\vec{Q})$. This is not correct close to T_c as in the pure substances, because the presence of the anisotropy field causes $G_L(\vec{Q})$ to diverge for $\vec{Q} = \vec{\tau}$ while $G_T(\vec{Q})$ remains finite. In the analysis of the experimental results[5] it is, however, necessary to include both $G_L(\vec{Q})$ and $G_T(\vec{Q})$ to obtain the four constants A_T, A_L, K_T and K_L for each temperature. In fig. 6 we show the results for the inverse correlation length, K_L, for K_2NiF_4, K_2MnF_4 and the mixed system. Clearly the critical behaviour of all these materials is closely similar and the exponent ν is within experimental error that predicted by the exact solution of the two dimensional Ising model. These results are in accord with the theoretical predictions of renormalisation group theory ($\alpha = 0$ for the two-dimensional Ising model) provided that the smearing does result from fluctuations in the concentration.

The other study was of $Mn_{0.7}Zn_{0.3}F_2$ by Meyer and Dietrich[31]. They found a smeared transition with the smearing corresponding to

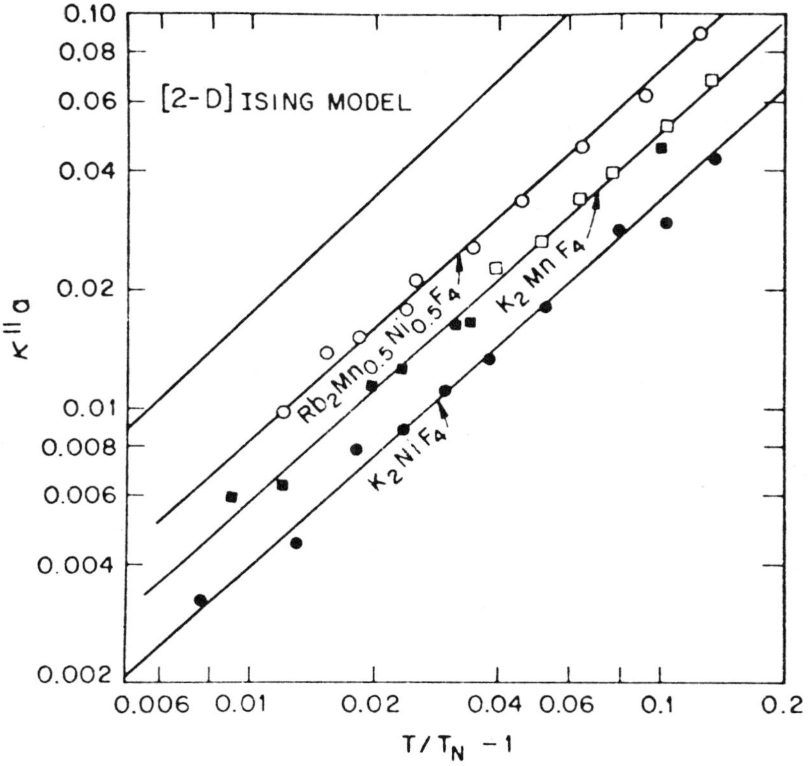

Fig. 6. The inverse correlation length, K_L, against $\varepsilon = \frac{T - T_N}{T_N}$ for K_2NiF_4, K_2MnF_4 and $Rb_2Mn_{0.5}Ni_{0.5}F_4$.

concentration fluctuations of 1%. In this three-dimensional system the range of the critical fluctuations in $\varepsilon = \frac{T - T_N}{T_N}$ is much smaller than for the two-dimensional systems. Consequently the smearing makes the extraction of the critical exponents much more difficult. The procedure followed was similar to that used for the two-dimensional system and the results yielded exponents which are not significantly different from those obtained[32] for pure MnF_2. Since the dipolar anisotropy breaks the isotropic symmetry in MnF_2, the critical behaviour is that of the three dimensional Ising model. This has $\alpha = 0.1$ and so the behaviour is expected to be modified by the concentration fluctuations. Renormalisation group theory predicts values of the new exponents[28,29] but these are sufficiently close to those of pure MnF_2 that they could not have been distinguished by the experiment.

Besides these published results similar results[33] have been found in the KMn/NiF_3 and Co/ZnF_2 systems. In both of these three-dimensional systems the smearing was sufficient to make the extraction of the exponents difficult. In conclusion, therefore, we are unable to answer what is probably the most interesting question about these phase transitions; namely whether the disorder smears the phase transition. In all the systems examined smearing has been observed but this may result from macroscopic non-uniformities in concentration and as yet it has not proved possible to decide if smearing is indeed a characteristic of uniformly random systems. If it is assumed that the smearing is due to inhomogeneities in the concentration then the exponents are consistent with those deduced using renormalisation group theory, but the accuracy of the measurements is such that this is not a very strong statement.

PERCOLATION

When the magnetic systems are diluted with non-magnetic ions the antiferromagnetic transition temperature decreases until at a critical concentration, c_p, there is no longer any long range order at T = 0. This critical concentration is known as the percolation point and is an example of a different type of phase transition where the lack of long range order is caused by geometrical disorder rather than thermal disorder. The properties of percolation considered as a problem in geometrical disorder have been studied[34] for many years. In particular for nearest neighbour interactions the critical percolation concentrations are known to be 0.59 for the square two-dimensional lattice and 0.30 for the simple cubic three-dimensional lattice. Likewise the divergence as $c \to c_p$ of the correlation functions $G_L(\vec{Q})$, have been evaluated using various techniques[35], and the results for the exponents describing the divergence of the correlation length, ν_G, and susceptibility, A_L/K_L^2 , γ_G, are listed in table III.

It is experimentally very difficult to perform experiments to test these exponents, not least because of the difficulties in controlling and accurately determining the concentrations of the specimens. Consequently it is much easier to determine the behaviour of the scattering as a function of temperature for certain fixed and hopefully known concentrations. Before these can be compared with theory it is, however, necessary to know the behaviour of the scattering as a function of temperature.

Following the suggestion of Stauffer[36], it is now generally accepted that the percolation point is an example of a multi-critical point. The behaviour as a function of temperature can then be obtained from the behaviour as a function of concentration if the scaling fields and the cross-over exponent are known. Stauffer[36] and ourselves[37] suggested that the scaling field was not temperature but the inverse correlation length, K_1, of a one-dimensional chain formed from the magnetic ions interacting as they do in the crystal at the percolation point. This suggestion was made partly as a result of computer simulations of clusters near to percolation which showed that they are very ramified (or one-dimensional), and secondly as an explanation for the experimental results discussed below.

The cross-over exponent can be calculated explicitly in the case of a one-dimensional chain where Thorpe[38] showed that as $c \to 1$ and $T \to 0$ ($c_p = 1$ for a linear chain) the inverse correlation length K is given by

$$K = K_G + K_T \tag{10}$$

where K_G is the geometrical inverse correlation length and in the linear chain is proportional to $(c - c_p)^{\nu_G}$ with $\nu_G = 1$, while K_T is the thermal contribution to K and for the linear chain is proportional to $(K_1)^{\nu_T}$ with $\nu_T = 1$. This result that $\nu_G = \nu_T$ corresponds to a cross-over exponent, $\phi = 1$. Since then Stephen and Grest[39] have shown for the Ising model that $\phi = 1$, in $6 - \varepsilon$ dimensions, by using renormalisation group theory and the argument has been extended to all orders in ε by Wallace and Young[40].

Initially we discuss the experimental results for the two-dimensional Ising system[22]; Rb_2Co/MgF_4. Four different concentrations were examined with c = 0.55, 0.575, 0.583 and 0.595, and typical measurements of the total scattering cross-section as a function of $\vec{Q}$ are shown in fig. 7. The results show a peak whose width decreased with decreasing temperature and for c = 0.595 two-dimensional long range order developed for T < 32 K. The results would be accurately accounted for by eqns. 4, 5 and 9 but with $G_T(\vec{Q})$ neglected. This is because Rb_2CoF_4 is an Ising-like material[6]. The experimental results at each concentration and temperature were fitted by varying A_L and K_L and as shown in fig. 7, good agreement was obtained. This shows, as was found in other systems[41], that the correlation function near percolation has the form of a two-dimensional Lorentzian. In fig. 8 the results for K_L are shown plotted against the temperature. The results clearly suggest that eqn. 10 is an appropriate description of these results, for all results except for T < 40 K for c = 0.595 where the result is presumably invalid due to the development of long range order. The results can be fitted to eqn. 10 with $K_T = K_1^{\nu_T}$ with K_1 calculated from the exchange constants known from the results on the excitation

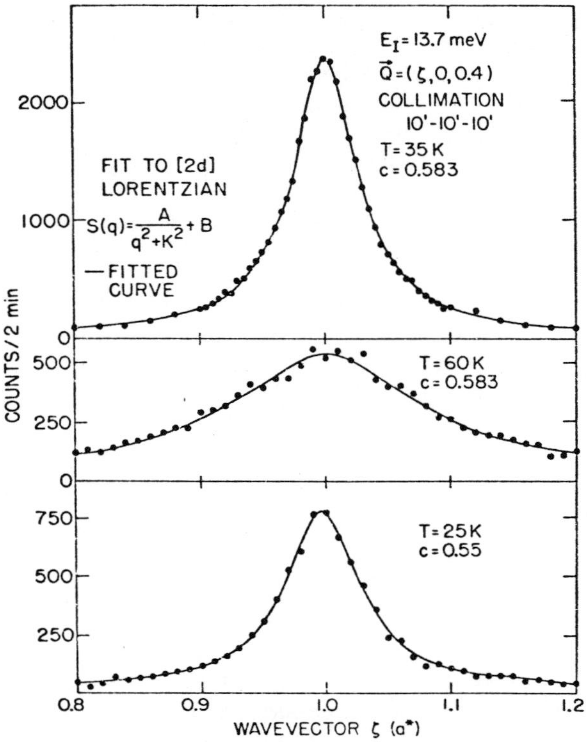

Fig. 7. Total neutron scattering intensity[22] as a function of wavevector in $Rb_2Co_cMg_{1-c}F_4$.

spectrum, fig. 3. The results shown in fig. 8 are consistent with $\nu_T = 1.33$ in agreement with the calculated value of ν_G (table III). Similar results can be obtained for the amplitude parameter A_L. Unfortunately the data analysis is complicated by a lack of knowledge of the dependence of A_L on temperature and concentration. The results suggest that the intensity $G(q^*) = \frac{A_L}{K_L^2}$ diverges at $c = c_p$ with $\gamma_T = 2.15 \pm 0.3$. As shown in table III this result is in general accord with the theoretical prediction for γ_G and a cross-over exponent, $\phi = 1$.

Similar measurements[41] were made earlier on the near Heisenberg

Table III

Exponents for Systems near Percolation

	2D Square Lattice	3D Cubic Lattice
ν_G (35)	1.35±0.01	0.83±0.02
γ_G (35)	2.43±0.03	1.68±0.05
Ising System		
ν_T (22)	1.30±0.05	-
γ_T (22)	2.15±0.30	-
Near Heisenberg		
ν_T (41,43)	0.90±0.03	0.95±0.10
γ_T (41,43)	1.50±0.10	1.84±0.20

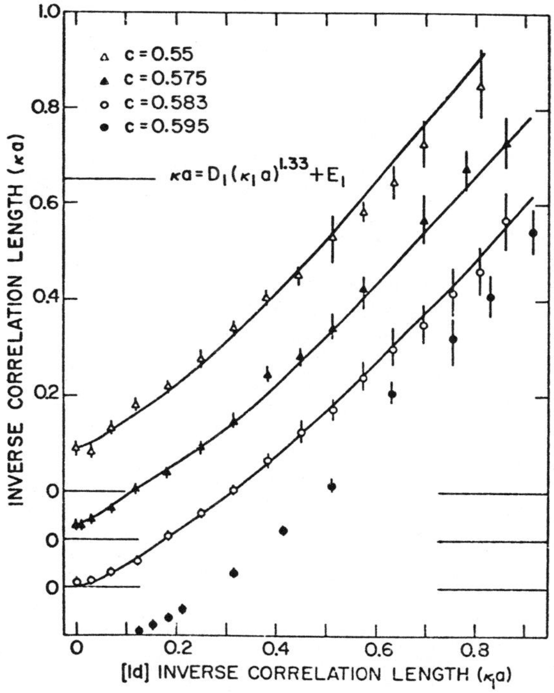

Fig. 8. The inverse correlation length, $K_L = K$, for four concentrations in $Rb_2Co_cMg_{1-c}F_4$ close to the percolation limit[22].

system $Rb_2Mn_cMg_{1-c}F_4$. In this case at high temperature $G_T(\vec{Q})$ and $G_L(\vec{Q})$ are the same but at low temperature the weak dipolar anisotropy field leads to a cross-over to Ising symmetry, and $G_L(\vec{Q})$ becomes much larger than $G_T(\vec{Q})$. This cross-over has been demonstrated experimentally[41]. The measurements can be analysed to give the correlation length at each temperature and the results are shown in fig. 9. Again they strongly suggest that K is the sum of a geometric and a thermal part. The temperature dependence of the one-dimensional correlation length, K_1, was calculated including the effects of dipolar anisotropy by using the theory and program of Blume et al.[42]. The results, fig. 9, are clearly in reasonable accord with the experimental measurements. Detailed fits suggest that ν_T for this system is 0.90±0.03, considerably different from the value for the Ising system and suggesting that the cross-over exponent for the Heisenberg system is about 1.5.

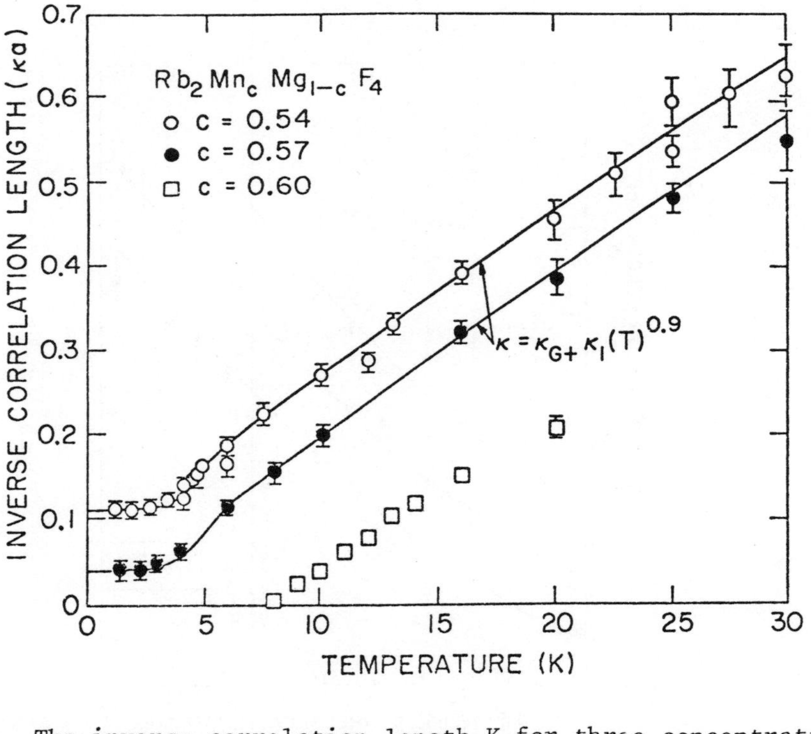

Fig. 9. The inverse correlation length K for three concentrations of $Rb_2Mn_cMg_{1-c}F_4$. The solid lines show fits to $K = K_G + (K_1)^{0.9}$.

Particularly convincing evidence of the role of the dipolar anisotropy is provided by the results of similar measurements[43] on the $Mn_cZn_{1-c}F_2$ system. The tetragonal crystal structure of this material enables measurements to be made of both $G_T(\vec{Q})$ and $G_L(\vec{Q})$. The crystal had a concentration gradient along its length so that one end ordered, whereas the other did not. Consequently by making slices of the crystal it is possible to conveniently perform experiments as a function of concentration. In fig. 10 the results for K_T and K_L are shown as a function of temperature and compared with calculations of both the transverse and longitudinal correlation functions for the appropriate linear chain. Clearly the qualitative agreement with K_L and K_T is very satisfactory showing the importance of including the dipolar anisotropy.

Finally we have not as yet discussed the development of the long range order in these systems. Samples of both Rb_2Co/MgF_4 and Rb_2Mn/MgF_4 have been examined for $c > c_p$, and measurements made of the magnetisation as a function of temperature. The long range order begins at a fairly well defined temperature T_c, at which the inverse correlation length K is close to zero. The temperature dependence of the magnetisation is very sensitive to fluctuations

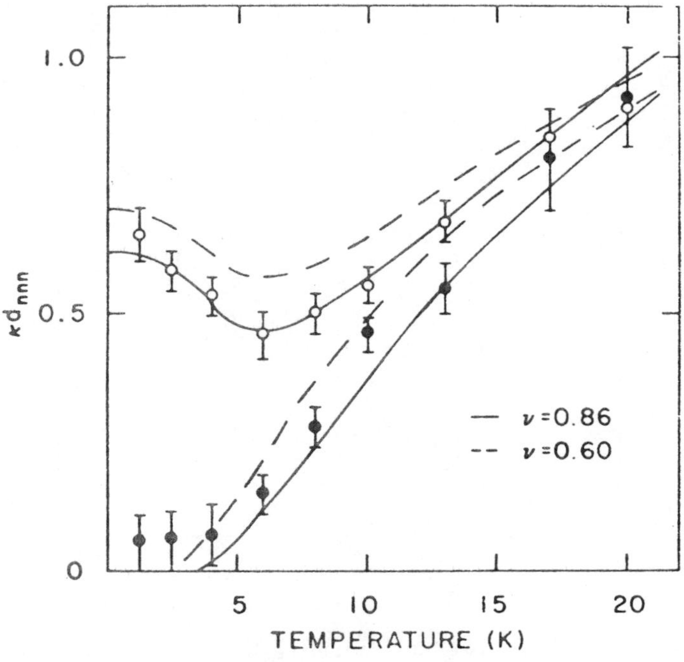

Fig. 10. Inverse correlation length K_L and K_T for $Mn_{c_p}Zn_{1-c_p}F_2$. The solid lines give the calculated inverse correlation length assuming $\phi = 1$ and $\phi = 1.4$ from ref. 43.

in the concentration and so cannot be analysed to give reliable estimates of the exponent β.

The results are less as expected in the three-dimensional systems; $Mn_cZn_{1-c}F_2$ and $KMn_cZn_{1-c}F_3$. We illustrate the behaviour in the latter system in fig. 11. There is a fairly well defined transition to an ordered structure at about 14 K, but unexpectedly the inverse correlation length is non-zero at this temperature and continues to decrease on further cooling, becoming almost zero at very low temperatures. Similar results were found in another sample with a lower concentration of Mn except that $T_c \approx 12$ K, the intensity of the Bragg reflection was less, and K steadily decreased with decreasing temperature but was non-zero at the lowest temperatures. A similar result, namely that K is non-zero at T_c, was also observed in the $Mn_cZn_{1-c}F_2$ system[43].

A further unexpected result is also shown in fig. 11; the intensity of the Bragg reflection reaches a maximum at $T \approx 6$ K and then decreases with decreasing temperature. This feature was

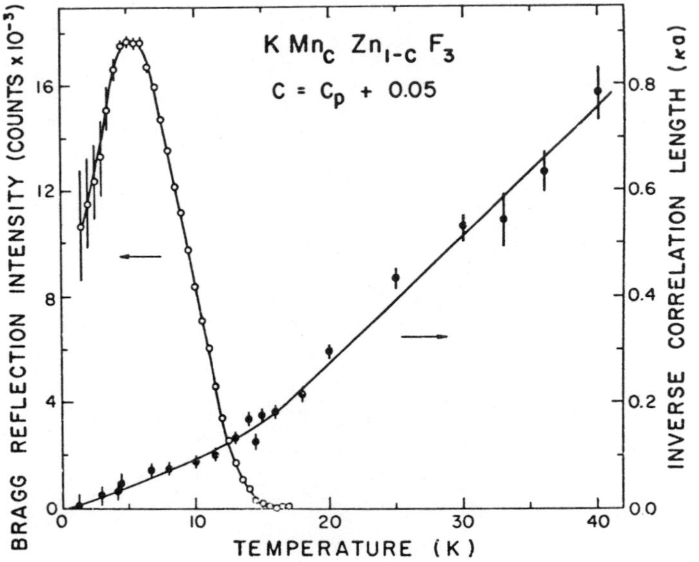

Fig. 11. Bragg scattering and correlation length in $KMn_{0.33}Zn_{0.67}F_3$ as a function of temperature[43].

not clearly established in the other crystal of this system or for the $Mn_cZn_{1-c}F_2$ system. We do not have a convincing explanation of either of these features. It is however worth remarking that the dipolar forces in the cubic perovskites will produce fields on the ions in different directions. Consequently there may be a complicated spin glass type of ordering at low temperatures.

We conclude that considerable progress has now been made in understanding percolation in these systems. The qualitative picture of the percolation point as a multicritical point with the temperature scale set by the one-dimensional correlation length is well established experimentally. The exponents of the two-dimensional Ising system are within experimental error in accord with a cross-over exponent of unity, but for the two-dimensional Heisenberg system the cross-over exponent of about 1.5 has not been obtained theoretically. The onset of long range order is reflected in the critical scattering for the two-dimensional systems but not for the three-dimensional ones. We are unaware of the reasons for this unusual behaviour.

ACKNOWLEGEMENTS

We have benefitted from the help and advice of many of our colleagues. In particular H.J. Guggenheim, E.C. Svensson, H. Ikeda, J.A. Tarvin, G.J. Coombs and K. Carneiro have collaborated with parts of this work.

REFERENCES

a Work at Edinburgh supported by the Science Research Council.

b Work at MIT supported by the National Science Foundation MRL Grant No. DMR-76-80895 Ao2.

c Work at Brookhaven supported by the Division of Basic Energy Sciences under Contract No. EY-76-c-02-0016.

1. R.A. Cowley and W.J.L. Buyers, Rev. Mod. Phys. 44, 406 (1972).
2. C.G. Windsor, G.A. Briggs and M. Kestigan, J. Phys. C 1, 940 (1968).
3. O.P. Nikotin, P.A. Lindgard and O.W. Dietrich, J. Phys. C 2, 1168 (1969).
4. P. Martel, R.A. Cowley and R.W.H. Stevenson, Can. J. Phys. 46, 1355 (1968).
5. R.J. Birgeneau, J. Als-Nielsen and G. Shirane, Phys. Rev. B 16, 280 (1977).
6. H. Ikeda and M.T. Hutchings, J. Phys. C 11, L529 (1978).
7. R.A. Cowley, A.I.P. Conf. Proc. 29, 243 (1976).
8. W.J.L. Buyers, T.M. Holden, E.C. Svensson, R.A. Cowley and R.W.H. Stevenson, Phys. Rev. Lett. 27, 1442 (1972).
9. E.C. Svensson, S.M. Kim, W.J.L. Buyers, S. Rolandson, R.A. Cowley and D.A. Jones, A.I.P. Conf. Proc. 24, 161 (1975).
10. G.J. Coombs, R.A. Cowley, D.A. Jones, G. Parisot and D. Tochetti, A.I.P. Conf. Proc. 29, 254 (1976).
11. J. Als-Nielsen, R.J. Birgeneau, H.J. Guggenheim and G. Shirane, Phys. Rev. B 12, 4963 (1975).
12. R. Alben and M.F. Thorpe, J. Phys. C 8, L275 (1975).
13. S. Kirkpatrick and A.B. Harris, Phys. Rev. B 12, 4980 (1975).
14. W.J.L. Buyers, D.E. Pepper and R.J. Elliott, J. Phys. C 6, 1933 (1973).
15. T. Tonegawa, Prog. Theor. Phys. 51, 1293 (1974).
16. G.J. Coombs and R.A. Cowley, J. Phys. C, 8, 1889 (1975).
17. G.J. Coombs, R.A. Cowley, W.J.L. Buyers, E.C. Svensson, T.M. Holden and D.A. Jones, J. Phys. C 9, 2167 (1976).
18. R.A. Cowley, O.W. Dietrich and D.A. Jones, J. Phys. C, 8, 3023 (1975).
19. W.K. Holcomb, J. Phys. C, 7, 4299 (1974).
20. O.W. Dietrich, G. Meyer, R.A. Cowley and G. Shirane, Phys. Rev. Lett., 35, 1735 (1975).
21. E.C. Svensson, W.J.L. Buyers, T.M. Holden and D.A. Jones, A.I.P. Conf. Proc. 29, 248 (1976).
22. R.A. Cowley, R.J. Birgeneau, G. Shirane, H.J. Guggenheim and H. Ikeda (unpublished).
23. H. Ikeda and G. Shirane (to be published).
24. R.A. Cowley, G. Shirane, R.J. Birgeneau and H.J. Guggenheim, Phys. Rev. B 15, 4292 (1977).
25. A. Tucciarone, H.Y. Lau, L.M. Corliss, A. Delapalme and J.M. Hastings, Phys. Rev. B4, 3206 (1971).

26. H. Ikeda and K. Hirakawa, Solid State Commun. 14, 529 (1974).
27. A.B. Harris, J. Phys. C 7, 1671 (1974).
28. G. Grinstein and A. Luther, Phys. Rev. B 13, 1329 (1976).
29. D.E. Khmelnitskii, Soviet Physics JETP 41, 981 (1975).
30. B.M. McCoy and T.T. Wu, Phys. Rev. Lett., 23, 383 (1968).
31. G.M. Meyer and O.W. Dietrich, J. Phys. C 11, 1451 (1978).
32. O.W. Dietrich, J. Phys. C 2, 2022 (1969).
M.P. Schulhof, R. Nathans, P. Heller and A. Linz, Phys. Rev. B 4, 2254 (1971).
33. R.A. Cowley and K. Carneiro (unpublished).
34. J.W. Essam, Phase Transitions and Critical Phenomena II, ed. C. Domb and M.S. Green (Academic Press, New York) (1972) p. 197.
35. P.J. Reynolds, H.E. Stanley and W. Klein, J.Phys. A11, L199 (1978).
36. D. Stauffer, Z. Phys. B 22, 161 (1975).
37. R.J. Birgeneau, R.A. Cowley, G. Shirane and H.J. Guggenheim, Phys. Rev. Lett., 37, 940 (1976).
38. M.F. Thorpe, Journal de Physique 36, 1177 (1975).
39. M.J. Stephen and G.S. Grest, Phys. Rev. Lett. 38, 567 (1977).
40. D.J. Wallace and A.P. Young, Phys. Rev., B 17, 2384 (1978).
41. R.J. Birgeneau, R.A. Cowley, G. Shirane and H.J. Guggenheim, Phys. Rev. Lett. 37, 940 (1976) and to be published.
42. M. Blume, P. Heller and N.A. Lurie, Phys. Rev. B 11, 4483 (1975).
43. R.A. Cowley, G. Shirane, R.J. Birgeneau and E.C. Svensson, Phys. Rev. Lett. 39, 894 (1977) and to be published.

EXCITATIONS OF DILUTE MAGNETS NEAR THE PERCOLATION THRESHOLD

Timothy Ziman

Department of Theoretical Physics

1 Keble Road, Oxford OX1 3NP

The random dilution of magnetic crystals with short range interactions provides for experimental investigation of percolation theories, in particular the multiscaling hypothesis for the region close to the percolation point ($p = p_c$, $T = 0$)[1,2]. Percolation has been intensively studied by series expansion, computer simulation and renormalisation group methods giving numerically well-defined exponents for quantities such as the percolation probability $P(p) \sim (p-p_c)^{\beta_p}$ and the geometric correlation length $\xi(p) \sim (p-p_c)^{-\nu_p}$[3]. Many experimentally accessible variables, however, such as correlation functions and excitation spectra cannot easily be calculated by formal renormalisation group techniques and are not reliably given by effective medium type theories. It is valuable, then, to develop scaling models consistent with known critical singularities yet simple enough to allow calculation of such experimentally determined quantities.

I shall consider geometric models of the infinite connected cluster just above the percolation concentration. Professor Cowley[1] has discussed experimental indications that at the percolation threshold local correlations can be considered to spread one dimensionally. Skal and Shklovskii[4] and de Gennes[5] introduced scaling models of the percolating cluster consistent with this; namely the infinite cluster is considered as a superlattice of nodes separated by distances of order of the correlation length $\xi(p)$ and connected by paths randomly varying in space with total length typically longer $L \sim (p-p_c)^{-\zeta}$. This defines an exponent ζ that in general is greater than ν_p but, since the paths are constrained by the overall geometry of the cluster, ζ is less than ν_p/ν_s where ν_s is the selfavoiding walk exponent for the dimension of space. Stanley et al. [6] have discussed a number of indications both theoretical and experimental that the exponent should be taken equal to the upper limit

$$\zeta_{SAW} = \nu_p/\nu_s \qquad (1)$$

We shall take this model with the Self Avoiding Walk Ansatz(1) and explore the consequences to ascertain how good a representation of the infinite cluster it is.

De Gennes[5] used the scaling model in a discussion of the conductivity of a randomly diluted resistor network just above the percolation threshold. The conductivity Σ of such a network is found to vanish with exponent $\Sigma \sim (p-p_c)^{\sigma}$. The model cluster defined above has conductivity given by

$$\Sigma = \frac{1}{L}\,(\xi(p))^{-(d-2)} \qquad (2)$$

implying the scaling relation

$$\sigma = \zeta + (d-2)\nu_p \qquad (3)$$

I shall derive a different scaling relation from (3) by exploiting the relation between the conductivity of a dilute resistor network and the spin wave stiffness D of a dilute isotropic ferromagnet[3].

$$\Sigma(p) = D(p)P(p) \qquad (4)$$

At percolation D(p) vanishes with exponent $D \sim (p-p_c)^t$ where σ and t are related by the exact result

$$\sigma = t + \beta_p \qquad (5)$$

By considering excitations of the model cluster we can derive a scaling relation for t and therefore σ by (5).

The form of long wavelength dispersion of spin waves on the model infinite cluster may be deduced by noting that for distances L' less than L the spin waves propagate by a randomly directed path travelling a Euclidean distance ℓ

$$\ell = (L')^{\nu_p/\zeta} \qquad (6)$$

and for L' greater than L a distance reduced by a constant

$$\ell = \frac{\xi(p)}{L}\,L' \qquad (7)$$

This implies a momentum variable Q conjugate to the distance L' in the equations of motion for spin waves related to the actual momentum q, as measured by neutron scattering, for example, as

$$q = \begin{cases} Q_c^{\nu_p/\zeta}\, Q/Q_c & q < q_c \sim (p-p_c)^{\nu_p} \\ Q^{\nu_p/\zeta} & q \geq q_c \end{cases} \tag{8}$$

Taking a quadratic dispersion in terms of the variable Q

$$\omega(Q) = JQ^2 \qquad Q \to 0 \tag{9}$$

leads to the scaling form for the dispersion in terms of measured momentum q

$$\omega(q) = \begin{cases} Jq^{2\zeta/\nu_p} & \\ J(p-p_c)^{2(\zeta-\nu_p)}q^2 & q \leq q_c \end{cases} \tag{10}$$

This is of the scaling form proposed by Stauffer[2] but here the dynamical exponent $z = 2\zeta/\nu_p$ is given a geometric interpretation. The spin wave stiffness vanishes as

$$D(p) \sim (p-p_c)^{2(\zeta-\nu_p)} \tag{11}$$

implying scaling relations

$$t = 2(\zeta - \nu_p) \tag{12}$$

$$\sigma = 2(\zeta - \nu_p) + \beta_p \tag{13}$$

(13) is a new scaling relation differing from relation (3). Table 1 compares numerical and mean field values of σ to predictions from relations (13) and (3) with $\zeta = \zeta_{SAW}$ and values for percolation exponents as in reference 3.

Table 1

	d=2	d=3	d>6
σ[3,7]	1.1±0.1	1.6±0.1	3
$2(\zeta_{SAW}-\nu_p)+\beta_p$ *	1.00	1.53	2
$\zeta_{SAW}+(d-2)\nu_p$ *	1.72	2.29	3

*Values of percolation exponents in [3].

It is seen that (13) works well in dimensions 2 and 3 but fails in higher dimensions where relation (3) is good. We shall now interpret this result in geometric terms. The model cluster we have taken has a total number of sites proportional to $(p-p_c)^{d\nu_p-\zeta}$. If it included all the sites of the infinite cluster this would imply a percolation probability P(p) vanishing as $(p-p_c)^{\beta_s}$ with

$$\beta_s = d\nu_p - \zeta \tag{14}$$

The exponent β_p with which P(p) vanishes is known and differs markedly from β_s; consequently the model infinite cluster has only a fraction of the number of sites in the infinite cluster. If we ignored this and took β_s in the scaling relation (13) instead of β_p we would recover de Gennes' relation (3). Thus the two scaling relations for σ differ in that in de Gennes' relation (3) the sites of the infinite cluster extra to those included in the simplest scaling model do not contribute to the conductivity whereas inclusion of β_p in (13) implies that they do. The relative success of relations (13) and (3) in low and high dimensions is interpreted geometrically by generalising the simple model cluster so far described. For low dimensions $d < d^*$ where $3 < d^* < 6$ the infinite cluster is generalised to include a diverging number of independent one-dimensional paths between each pair of nodes of the superlattice, while for dimensions $d > d^*$ the extra sites of the infinite cluster are included as branchings and dead ends and thus do not contribute to the conductivity. They do, however, affect the spin wave stiffness as the failure of relation (13) in high dimensions requires. This can be understood by considering dispersion along a regular chain with at each point a dangling side branch of length n. The equations of motion for such a system can be solved and show that the spin wave stiffness is reduced by a factor $1/(n+1)$. For the infinite cluster in high dimensions the diverging number of dead ends per site in the conducting superlattice reduces the spin wave stiffness correspondingly. It seems likely that the dimension d^* where the cluster geometry changes qualitatively should equal four since excluded volume effects are unimportant above that dimension[2].

A difficulty with the geometric models discussed here is the apparently different value of length exponent ζ_{SAW} assumed for isotropic spin systems and conductivity and the crossover exponent $\zeta_{IS} = 1$ found for discrete spin systems for all dimensions[9,10]. If the geometric model had all the essential features of the percolating cluster one would expect the exponent to be independent of the spin system concerned. The answer may lie[11] in the role of local parallelisms. Simple considerations show that anisotropic spin systems at low temperatures are much more sensitive to the existence of parallel bonds and consequently the "length" relevant to the decay of correlations in a dilute Ising system may be the number of single bonds along a chain while that in an isotropic spin system includes single or multiple bonds.

We have seen that by slight generalisation of the simplest scaling model of the infinite cluster we can reconcile known values if conductivity exponents with the Self Avoiding Walk Hypothesis. The model so developed can be applied to calculation of other physically observable quantities. In particular the form of the excitation spectra and low temperature thermodynamics can be found[11]. Application to dilute antiferromagnets is particularly relevant to experimental studies and a similar approach to that taken for a ferromagnet can be made for such systems[11]. The scaling model has also been employed to predict the combined effect of the reduced connectivity of the infinite cluster and local fluctuations in concentration in localising spin wave excitations [12].

References

1. R.A. Cowley, R.J. Birgeneau, G. Shirane 1979 (These proceedings).
2. D. Stauffer 1975 Z. Physik B22 161-71.
3. A.B. Harris and S. Kirkpatrick 1977 Phys. Rev. B16 542-76.
4. A.S. Skal and B.I. Shklovskii 1974 Fiz. Tekh. Poluprovdn. 8 1586-92 (Sov. Phys.-Semicond. 8 1029-32).
5. P.G. de Gennes 1976 J. Physique 37 L1-2.
6. H.E. Stanley, R.J. Birgeneau, P.J. Reynolds, J.F. Nicoll 1976 J.Phys.C: Solid St. Phys. 9 L553-60.
7. M.J. Stephen 1978 Phys. Rev. B17 4444-53.
8. P.G. de Gennes 1972 Physics Letters 38A 339-40.
9. M.J. Stephen and G.S. Grest 1977 Phys. Rev. Lett. 38 567-70.
10. D.J. Wallace and A.P. Young 1978 Phys. Rev. B17 2384-7.
11. T.A.L. Ziman 1979 J.Phys.C: Solid St. Phys. (in press); see also T.A.L. Ziman 1978 D.Phil. Thesis, Oxford University (Unpublished).
12. T.A.L. Ziman and R.J. Elliott 1978 J.Phys.C: Solid St. Phys. 11, L847-50.

The smaller q_z is, the less the energy gets for the S^x and S^y components whereas the anisotropy demands a relatively high energy for creating a significant S^z. Thus the amplitude in S^z gets smaller with decreasing q_z, whereas the amplitude in S^x and S^y gets larger with decreasing q_z. Finally, at $q_z = 0$ the excitation degenerates to a rotation in the easy plane without S^z components.

When increasing the temperature the OPC should disappear very rapidly since the lifetime $1/\Gamma$ of the OPC spin wave decreases rapidly due to increasing importance of thermal fluctuations in S^z. The OPC peak should however show only very small renormalization of the energy.[14] These predictions could be verified experimentally. In Fig. 12 the experimental results for the OPC peak in $CsNiF_3$ measured at Q = (0, 0, 0.435) are shown. Both the energy's and the intensity's dependence agrees well with the theoretical results.

As discussed before, the IPC peak should be strongly connected with the correlation length ξ. The evolution of the IPC peak with temperature measured at Q = (0, 0, 0.35) in $CsNiF_3$ is shown in Fig. 13. The solid lines are results of a least squares fit of a convolution of the resolution function of the instrument with the proposed Lorentzian line for the spin wave peak. The resulting temperature dependence of the line width is shown in Fig. 14. It is obvious that the line width increases proportionally to T as expected (see [4]).

In the quantum system CHAB a detailed lineshape study has not yet been done, but there is even at low temperatures a significant difference in the lineshape between classical and quantum systems. Whereas the lineshape in classical systems is quite symmetric at low temperatures and only becomes larger than Lorentzian, the quantum system shows significant asymmetry in the lineshape; this can be seen in Fig. 15.[26] This asymmetry, especially at low q_z is due to higher (S = 1) excited states which have significant weight to contribute.[21] Several theoretical approaches do predict this asymmetry.[12,17]

In summarizing the linear dynamic behaviour in zero field, some points have to be emphasized:

a) For classical (S ≥ 1) systems linear spin wave theory is able to describe all aspects of the data very satisfactorily. In quantum systems (S = 1/2), different theoretical approaches are needed to describe the behaviour of the linear excitations.

b) Planar anisotropy leads to two components in the cross section: the one is very sharp (OPC), the other is wide (IPC).

c) It appears that the correlation length ξ is the characteristic length for the spin waves as well since the width and the

CRITICAL PROPERTIES OF THE MIXED ISING FERROMAGNET

J. M. YEOMANS AND R. B. STINCHCOMBE

DEPARTMENT OF THEORETICAL PHYSICS

1 KEBLE ROAD, OXFORD OX1 3NP

A real-space rescaling method is applied to the mixed-bond, spin-$\frac{1}{2}$, Ising model. The system comprises bonds J_A, J_B with probability p, 1-p respectively. We restrict our attention to the mixed ferromagnet for which J_A, $J_B > 0$. Critical curves, T_c versus p, can be obtained for all $\alpha \equiv J_B/J_A$ and we present results for the square lattice for $\alpha = 1,2,5,10,\infty$.

The scaling transformation considered is a generalisation of that used by Bernasconi (1978) for the random resistor network problem. The square lattice is enlarged by a factor b = 2 by replacing the cluster shown in Figure 1(a) by a single bond in each

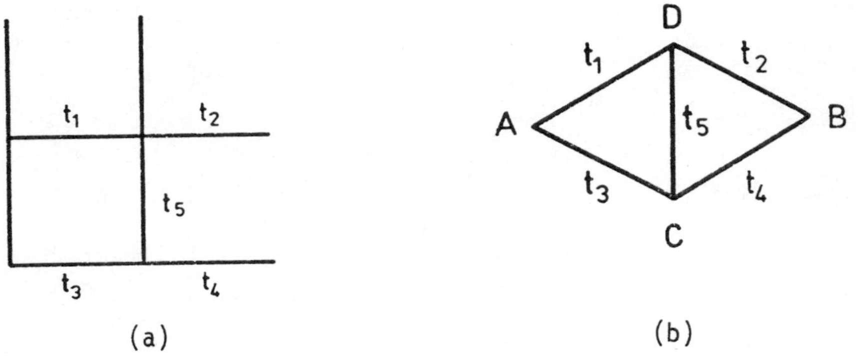

Figure 1.(a) Basic cluster which scales to a single bond in each direction under the R.G. transformation (b) "Wheatstone bridge" construction: a single horizontal bond, AB, is left after rescaling.

of the horizontal and vertical directions. In the percolation conductivity problem the renormalised horizontal bond is considered "open" if it is possible to traverse the original cluster in a horizontal direction. For the Ising problem this is equivalent to considering the cluster shown in Figure 1(b) and performing a decimation transformation (Nelson and Fisher 1975, Barber 1975, Kadanoff and Houghton 1975, Young and Stinchcombe 1976) over the spins C,D. This leads to the following relationship between the spins on the original and renormalised lattices:

$$t' = f(t_i) = \frac{t_1t_2+t_3t_4+t_5(t_1t_4+t_2t_3)}{1+t_1t_2t_3t_4+t_5(t_1t_3+t_2t_4)}, \quad t_i \equiv \tanh \beta J_i.$$

The advantage of this transformation is that it preserves the selfduality of the square lattice. Therefore the pure Ising and percolation fixed points are given exactly.

Initially each t_i is distributed according to a binary probability distribution

$$P(t_i) = p\,\delta(t_i-t_A) + (1-p)\delta(t_i-t_B).$$

Under rescaling this evolves to a new distribution

$$P'(t_i') = \int \prod_i P(t_i)dt_i\delta(t_i' - f(t_i))$$

which is, in general, considerably more complicated. Ideally this recursion equation should be iterated to give the invariant distribution from which critical properties can be obtained (Stinchcombe and Watson 1976). However, this procedure is very involved and previous work (Kirkpatrick 1977, Plischke and Zobin 1977, Jayaprakash et al. 1978, Wortis et al. 1978, Yeomans and Stinchcombe 1978, 1979(a)) has shown that for most purposes it is sufficient to approximate the renormalised distribution by another binary distribution.

$$P'_{approx}(t_i') = p'\delta(t_i' - t_A') + (1 - p')\delta(t_i' - t_B').$$

t_A', t_B', p' are then determined by matching $P'(t_i')$ and $P'_{approx}(t_i')$ as closely as possible. To determine these parameters three "matching conditions" are necessary. We consider three different possibilities:
A: matching percolation probabilities, first moments and fixing α.
B: matching first moments, second moments and fixing α.
C: matching first, second and third moments.

Results from the three approximation schemes are listed in

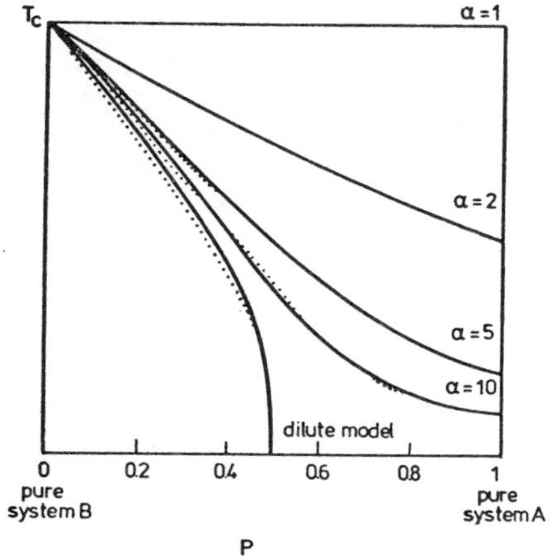

Figure 2: Critical curves for the mixed-bond Ising model on the square lattice. Results obtained using approximations A (full lines) and B (dotted lines) are shown. Approximations B and C give curves which are indistinguishable on a graph of this scale.

Table 1. For all cases fixed points ($p' = p = p^*$, $t_A' = t_A = t_A^*$, $t_B' = t_B = t_B^*$) corresponding to pure Ising systems occur at $p^*=1$, $t_A^*=0.414$ and $p^*=0$, $t_B^*=0.414$. Linearising around the fixed points gives the thermal correlation length exponent, ν_T. In the limits $\alpha \to 0,\infty$ the model reduces to the dilute Ising model and a percolation fixed point appears. Results for the percolation correlation length exponent, ν_p, and the crossover exponent, ϕ, are given in the Table.

Approximations A and B demand that $\alpha \equiv J_B/J_A$ remain invariant under the Renormalisation Group transformation. This forces the flow lines to lie in the plane α = const. which is, in general, incorrect. However, this condition makes the problem easier to handle and enables us to find the limiting slopes of the critical curves. These are listed in Table 1 and compared to the exact results (Harris 1976, Au-Yang et al. 1976, Thorpe and McGurn 1979). The good agreement suggests that the constraint on α does not have an important effect on the results obtained.

Critical curves for the three approximation schemes are compared in Figure 2. Fixing α (approximations A and B) leads to problems with the flow but has little effect on the critical curves which are seen to be very similar for the three approximations considered. Due to a duality argument (Fisch 1978) $T_c(p=\frac{1}{2})$ is known exactly. Results from the three approximation schemes are listed in Table 1 for comparison. As the decimation transformation itself preserves duality any errors must be due to the method of approximation of the integral equation. Approximation C gives $T_c(p=\frac{1}{2})$ exactly as

Table 1: Fixed points, critical exponents and limiting slopes of the critical curves of the mixed-bond Ising model on the square lattice for various values of α. Results from approximations A,B and C are compared.

$\alpha \equiv \frac{J_B}{J_A}$	1	2	5	10	∞
p=0 fixed point					
t_B^*	0.414	0.414	0.414	0.414	0.414
t_B^*(exact)[1]	0.414	0.414	0.414	0.414	0.414
t_A^*	0.414	0.217	0.088	0.044	0
ν_T	1.148	1.148	1.148	1.148	1.148
ν_T(exact)[2]	1	1	1	1	1
$\frac{1}{T}\frac{dT_c}{dp}\Big\vert_{p=0}$ (A)	0	-0.564	-0.961	-1.102	-1.246
$\frac{1}{T_c}\frac{dT_c}{dp}\Big\vert_{p=0}$ (B)	0	-0.582	-1.017	-1.177	-1.344
$\frac{1}{T_c}\frac{dT_c}{dp}\Big\vert_{p=0}$ (exact)[3]	0	-0.582	-1.010	-1.167	-1.329
p=1 fixed point					
t_A^*	0.414	0.414	0.414	0.414	
t_A^*(exact)[1]	0.414	0.414	0.414	0.414	
t_B^*	0.414	0.707	0.976	1.000	
ν_T	1.148	1.148	1.148	1.148	Percolation fixed point
ν_T(exact)[2]	1	1	1	1	$p^*=\frac{1}{2}$
$\frac{1}{T_c}\frac{dT_c}{dp}\Big\vert_{p=1}$ (A)	0	-0.753	-1.375	-1.427	p^*(exact)[5]$=\frac{1}{2}$
$\frac{1}{T_c}\frac{dT_c}{dp}\Big\vert_{p=1}$ (B)	0	-0.730	-1.297	-1.344	$\nu_p = 1.428$
$\frac{1}{T_c}\frac{dT_c}{dp}\Big\vert_{p=1}$ (exact)[3]	0	-0.727	-1.284	-1.329	ν_p(series)[6]= 1.34±0.02
					$\phi = \frac{\nu_T}{\nu_p} = 1$
					ϕ(exact)[7]=1

Table 1 (Continued)

$\alpha \equiv \frac{J_A}{J_B}$	1	2	5	10	∞
p=½					
T_c(A)	1	0.721	0.488	0.376	
T_c(B)	1	0.723	0.503	0.400	
T_c(C)	1	0.723	0.502	0.399	
T_c(exact)[4]	1	0.723	0.502	0.399	

1. Kramers & Wannier (1941)
2. Onsager (1944)
3. Thorpe & McGurn (1979)
4. Fisch (1978)
5. Sykes & Essam (1964(a))
6. Dunn, Essam & Ritchie (1975)
7. Wallace & Young (1978)

the recursion equations preserve the symmetry of the system under the transformation $(t_A, t_B, 1-p) \leftrightarrow (t_B, t_A, 1-p)$.

The good agreement with exact results and the weak dependence of the results on the method of approximation of the integral equation confirms that real space rescaling methods can give useful information about the critical properties of disordered systems. We have obtained similar results to those presented here for the mixed-site Ising model in which A and B-type sites are arranged at random. These results, together with a more detailed account of the results obtained for the bond disordered model will be presented in a forthcoming publication (Yeomans and Stinchcombe 1979(b)).

An interesting extension to the work described here would be to allow the lattice to contain both ferromagnetic and antiferromagnetic interactions as in the Edwards-Anderson model of a spin glass. However, as previous work has indicated (Kirkpatrick 1975, Young and Stinchcombe 1976, Kinzel and Fischer 1978),care must be taken in the choice of cluster and the method of approximating the integral equation to prevent spurious cancellation of interactions of different sign.

References

Au-Yang, H., Fisher, M.E. and Ferdinand, A.E., 1976 Phys. Rev.B13 1238-65.
Barber, M.N., 1975 J. Phys. C. Solid St. Phys. 8 L203-7.
Bernasconi, J., 1978, Phys. Rev. B18 2185-91.
Dunn, A.G., Essam, J.W. and Ritchie, D.S., 1975 J. Phys. C. Solid St. Phys. 8 4219-35.
Fisch, R., 1978 J. Stat. Phys. 18 111-114.
Harris, A.B., 1974 J. Phys. C. Solid St. Phys. 7 1671-92.
Jayaprakash, C., Riedel, E.K. and Wortis, M., 1978 Phys. Rev. B18 2244-55.
Kadanoff, L.P. and Houghton, A., 1975 Phys. Rev. B11 377-86.
Kinzel, W. and Fischer, K.H., 1978 J. Phys. C. Solid St. Phys. 11 2115-21.
Kirkpatrick, S., 1977 Phys. Rev. B15 1533-8.
Kramers, H.A. and Wannier, G.H., 1941 Phys. Rev. 60 252-62.
Nelson, D.R. and Fisher, M.E., 1975 Ann. Phys. 91 226-74.
Onsager, L., 1944 Phys. Rev. 65 117-49.
Plischke, M. and Zobin, D., 1977 J. Phys. C. Solid St. Phys. 10 4571-9.
Stinchcombe, R.B. and Watson, B.P., 1976 J. Phys. C. Solid St. Phys. 9 3221-47.
Sykes, M.F. and Essam, J.W., 1964 J. Math. Phys. 5 1117-27.
Thorpe, M.F. and McGurn, A.R., to be published in Phys. Rev. B.
Wallace, D.J. and Young, A.P., 1978 Phys. Rev. B5 2384-7.
Wortis, M., Jayaprakash, C. and Riedel, E.K., 1978 J. App. Phys. 49 1335-40.
Yeomans, J.M. and Stinchcombe, R.B., 1978 J. Phys. C. Solid St. Phys. 11 4095-4104.
Yeomans, J.M. and Stinchcombe, R.B., 1979(a) J. Phys. C. Solid St. Phys. 12 347-60.
Yeomans, J.M. and Stinchcombe, R.B., 1979(b) submitted to J. Phys. C. Solid St. Phys.
Young, A.P. and Stinchcombe, R.B., 1976 J. Phys. C. Solid St. Phys. 9 4419-31.

STRUCTURE AND PHASE TRANSITIONS

IN PHYSISORBED MONOLAYERS

John P. McTague, M. Nielsen, and L. Passell

University of California
Department of Chemistry
Los Angeles, CA U.S.A.

Riso National
Laboratory
Roskilde, Denmark

Brookhaven
National Lab.
Upton, NY U.S.A.

INTRODUCTION

We present here an overview of the experimental phase behavior of simple atomic and molecular adsorbates on homogeneous substrates, emphasizing recent investigations using the (001) surface of graphite where in many cases there is close correspondence between the measurements and expected 2-D behavior. Because of space limitations only a partial view of the subject is presented; in particular, the beautiful thermodynamic studies of adsorbed He done by Dash, Bretz, Vilches and others are not covered. Illustrative examples have been chosen to a large extent from the authors' own work, not because of any exaggerated sense of their relative importance and/or interest, but merely because the necessary figures were already in hand!

HISTORY AND BACKGROUND

The history of physisorption is a long and venerable one, with roots deeply imbedded in several major commercial processes. For instance, many millions of pounds of carbon black are used annually in rubber reinforcement, and particle size has been shown to be an important parameter in determining the rubber properties. Adsorption isotherms have been used for more than 50 years to characterize the surfaces of various charcoals and graphites.[1] Likewise, silica gels (amorphous SiO_2) of high surface area are common texturizers and thickeners in many cosmetics and even foodstuffs. For the latter use it is important that surfaces be smooth and rounded!

The most common technique for characterizing particle size and surface homogeneity is the adsorption isotherm, often with N_2 as the adsorbate. Typical high surface area substrates produce rather smooth, relatively unstructured isotherms, as shown in Fig. 1. It is qualitatively reasonable that, for adsorbate gases whose interaction with the substrate is sufficiently strong, the vapor pressure should increase markedly upon addition of sufficient adsorbate to cover the substrate surface with a monolayer, so parameterization of the isotherm curves with simple models, such as the B.E.T. one[2] can give a measure of the surface area. Some thirty years ago Halsey[3] postulated that the observed smooth curves obscured a number of stepwise adsorptions on a heterogeneous surface. Isolated observations of such stepwise isotherms were reported on alkali halides[4] and graphite,[5] but in 1953 Polley, Schaeffer, and Smith[6,7] made a systematic study of adsorption isotherms on graphite as a function of annealing temperature and demonstrated that increase in substrate crystalline order was accompanied by evolution of stepwise isotherms indicative of well-defined first, second, · · ·, layer formation. There had been earlier reports of submonolayer phase transitions,[8] but the availability of well-characterized, homogeneous, inert, high (001) surface area graphite accellerated study of the monolayer and submonolayer phenomena which are the concern of these lectures.

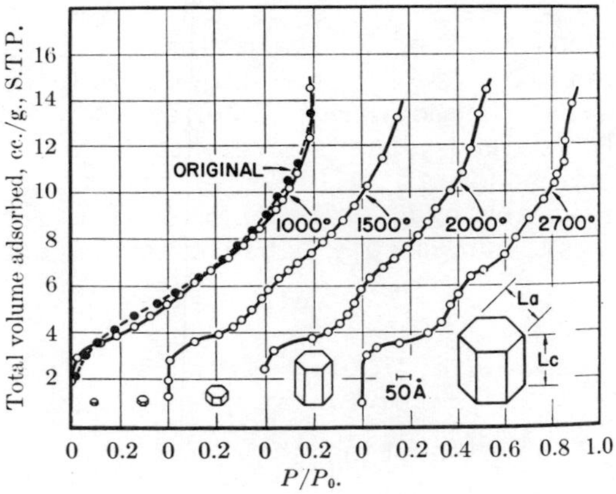

Figure 1. Argon adsorption isotherms (T = 78K) on p-33 carbon black samples annealed at various temperatures. Hexagonal inserts show X-ray determined substrate crystallite coherence lengths.

In the following sections we will briefly outline the statistical thermodynamics of adsorption and the connection with 2-dimensional properties. Following that, experimental observations of the thermodynamics and structure of monolayer films will be reviewed.

STATISTICAL THERMODYNICS OF PHYSICAL ADSORPTION

In a typical adsorption experiment a reservoir of gas adsorbate at some temperature and pressure is in thermal, mechanical, and material equilibrium with a cell containing a substrate of surface area A. The equilibrium conditions require T, p, and the chemical potential μ of the adsorbate to be independent of position $\vec{r}$. Unless there is appreciable absorption of the gas on the interior of the substrate crystals, it is usual to treat the substrate as an inert solid which is the source of an external potential $u(\vec{r})$ for the adsorbate. Sufficiently far from the surface $u(\vec{r})$ is negligible, so the chemical potential μ has the field-free value everywhere, and for a dilute system it will be equal to that of an ideal gas:

$$\mu = kT \ln (\Lambda^3 p/kT),$$

where for a monatomic gas

$$\Lambda = (h^2/2\pi mkT)^{1/2}.$$

Thermodynamic functions[9] for the system of volume V and surface area A are given by appropriate derivatives of the grand partition function Ξ :

$$\Xi = \sum_N Q_N (T,V) \exp(N\mu/kT)$$

$$= \sum_N Z_N \left(\frac{p}{kT}\right)^N/N! \equiv \sum_N Z_N \zeta^N/N!.$$

Here Q_N is the canonical partition function for an N molecule system, and Z_N is the corresponding configuration integral:

$$Z_N = \int_V d\vec{r}^N \exp\{-U(\vec{r}^N)/kT\}.$$

In particular, the average number of molecules in the system is given by:

$$\langle N\rangle = \left(\frac{\partial \ln \Xi}{\partial \ln \zeta}\right)_{T,V,A}.$$

Comparison with the same quantity $\langle N^0\rangle$ for an ideal gas in volume V but not interacting with a surface gives the surface excess

$$\langle N^{(S)}\rangle = \langle N\rangle - \langle N^0\rangle = \left(\frac{\partial \ln (\Xi/\Xi_0)}{\partial \ln \zeta}\right)_{T,V,A}.$$

$\langle N^{(S)}\rangle$ is interpretable as the number of adsorbed molecules. Analogous definitions for the adsorbed internal energy, surface spreading pressure, etc. can be constructed.

To what extent are these surface excesses to be considered as separate phases? The answer is, of course, that it matters not what one calls them so long as the statistical thermodynamics is consistent! In particular, surface phase behavior is always essentially linked to that of the bulk through the requirement of a spatially-independent chemical potential. It seems reasonable, however, to consider surface excesses as constituting separate phase behavior when the particle density $\rho(\vec{r})$ is markedly higher near the surface than ρ_0, its value in the bulk gas. A rough estimate of $\rho(\vec{r})/\rho_0$ can be obtained by neglecting adsorbate-adsorbate interactions and considering only the adsorbate-substrate term $u_s(\vec{r})$, giving $\rho(\vec{r})/\rho_0 = \exp(-u_s(\vec{r})/kT)$. For rare gases on graphite[9] u_s/k is of order 1000 - 1500 K at the adsorbate-graphite potential minimum r_0, so we have $\rho(\vec{r}_0)/\rho_0 > 2 \times 10^4$ for $T < 100$ K.

When comparing with theoretical models it is important to assess how closely such surface phases approximate 2-dimensional behavior. The out-of-plane vibrational spectrum of Ar on graphite has been determined by neutron scattering[10] (cf Fig. 8) for submonolayer coverages over a range of temperatures, and shows a dispersionless peak at $\hbar\omega \sim 5.5$ meV, indicating an r.m.s. motion perpendicular to the surface $\langle u_\perp^2\rangle^{1/2} \sim 0.05$ Å at $T \approx 60$K. Since $\langle u_\perp^2\rangle^{1/2}$ is small compared to the molecular diameter (≈ 3.8Å) most low temperature properties of such submonolayer films should be essentially 2-D in character. Classical molecular dynamics simulations[10] of adsorption have also verified that an 87% monolayer coverage of Ar is essentially 2-dimensional for $T < 80$ K, while a 98% coverage film shows noticeable 3-dimensional character for $T > 60$K.

Whenever there is true phase coexistence on the surface the chemical potential, monitored through the bulk gas pressure p ($\mu(T,p) = \mu^0(T) + kT \ln p$), must be independent of N(s), the number of adsorbed molecules.[11] This results in a vertical section for adsorption isotherms; the most celebrated case, krypton on graphite, was studied by Thomy and Duval,[12] and is shown in fig. 2.

The isotherms for $T < 87$ K point to a first-order transition between a dense and a dilute phase ending in a critical point, as well as apparent phase coexistence between two dense phases. The high quality of these isotherms has stimulated much theoretical and experimental study of submonolayer phase behavior, which is still in a period of rapid growth. Thermodynamic measurements such as adsorption isotherms and heat capacity studies are very powerful techniques for locating phase boundaries and characterizing the various transitions, but they do not usually give a direct indication of the phases involved. Without direct structural determinations the best one can do is to make inferences based on the magnitudes of heat capacities, heats of adsorption and their coverage dependences, etc., as well as reasoning by analogy with experience from the ordinary 3-D world and/or comparison with model calculations.

The striking similarity between the Kr-graphite adsorption isotherm and typical 3-D gas-liquid-solid phase diagrams led Thomy and Duval to speculate that they were observing a similar 2-D case,

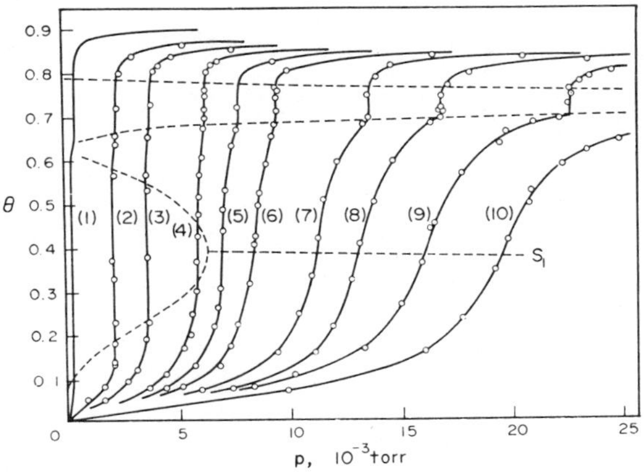

Figure 2. Isotherms for Kr on exfoliated graphite at temperatures of: (1) 77.3K; (2) 82.4K; (3) 84.1K; (4) 85.7K; (5) 86.5K; (6) 87.1K; (7) 88.3K; (8) .0K; (9) 90.1K; (10) 90.9K. Dashed lines represent early conjectures on a possible phase diagram.

with first order boundaries characterized by a gas-liquid critical point near T = 86 K and a gas-liquid-solid triple-point near T = 77 K. Subsequent constant coverage heat capacity studies by Butler, Litzinger, and Stewart[13] have given complementary information on the phase boundaries which appear to rule out the existence of the triple point. Some of their results are shown in fig. 3. Indeed, there appears to be no direct evidence for a 2-D liquid phase in the gases Ar and Kr on graphite; whether such a phase exists (i.e. a liquid-gas coexistence) remains the topic of much present investigation.

Not all systems show the sharp phase boundaries seen in Kr, although many (^3He, ^4He, D_2, H_2, N_2, CH_4) do, and indeed they have phase diagrams with much in common with Kr. Ar, on the other hand, behaves quite differently. For instance, the heat capacity anomaly is much lower and broader than in Kr and N_2, apparently indicating a continuous transition: The natures of the Kr- and Ar-like transitions have been clarified by structural studies, and are discussed in the next sections.

STRUCTURAL INVESTIGATIONS OF MONOLAYERS

Low energy electron diffraction is the most commonly used tool for the study of monolayer crystal structures in chemisorbed systems. It has the important advantage of high sensitivity; typical

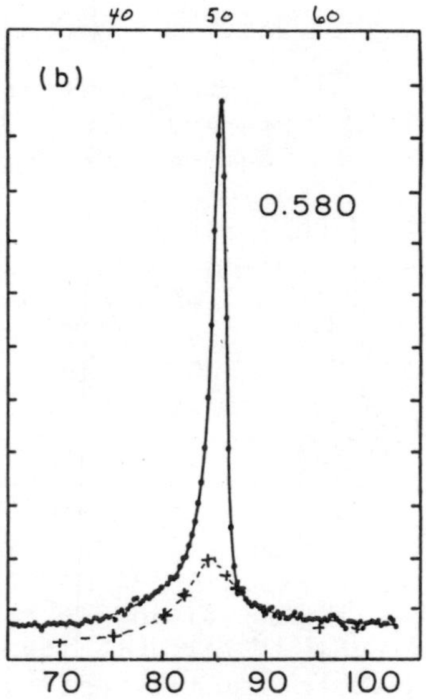

Figure 3. Submonolayer heat capacity peaks for registered Kr (.——) and incommensurate Ar (---). Temperature (K) scales are Ar (top) and Kr (bottom).

scattering cross sections are of order 10 $Å^2$ per molecule, so essentially all of the incident beam is scattered by the adsorbed monolayer or the first one or two substrate planes. However, it has some significant drawbacks for physisorbed systems: poor absolute accuracy (3-5%) for lattice parameter determination (but relative accuracy can be considerably higher), distortion of Bragg peaks by multiple scattering, and the requirement of high vacuum in the experimental chamber. It has the presently unique advantage of sufficient sensitivity to study a single crystal surface with dimensions of order 1 mm x 1 mm. Neutron and x-ray scattering cross sections are much smaller than those for LEED. Multiple scattering is therefore not a major problem, but the lack of sensitivity requires samples of at least a few cm^2 (x-rays) to ~100 m^2 (neutrons).

In 1966, Lander and Morrison[15] made the first direct determinations of the crystal structures of physisorbed monolayers. Using LEED, they investigated several substances, including Xe and Br_2 on the (001) surface of a graphite single crystal. They interpreted the Xe pattern at T = 90 K and p = 10^{-3} torr as indicating a $\sqrt{3} \times \sqrt{3}$ registered structure (fig. 4) requiring a nearest-neighbor separation of 4.26 Å, a linear compression of some 3% from that of bulk Xe. More recent measurements[16] suggest that, under these conditions, Xe is expanded slightly out of registry, but that it may register near the second layer condensation line.

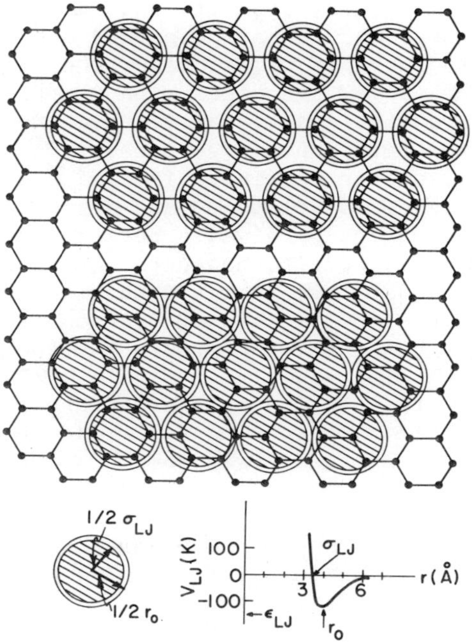

Figure 4. Schematic representation of the $\sqrt{3} \times \sqrt{3}$ registered phase (top) and of the incommensurate Ar monolayer phase (middle). The Ar-Ar Lennard-Jones potential obtained from gas phase studies is plotted at the bottom.

As with all other simple substances whose monolayer structure on graphite is known[17] (^3He, ^4He, N_2, O_2, H_2, Kr, CH_4, Ar, NO) both commensurate and incommensurate phases of Xe have a basically triangular lattice, although both O_2 and NO have phases which are significantly distorted from the equilateral triangular geometry.

Single crystal LEED studies are particularly appropriate for investigating epitaxy, and we will return to such work in a later section. To date, the majority of physisorbed crystal structure studies on graphite, and all of the dynamical ones, have utilized neutron scattering, a technique which on first glance appears ill-suited to monolayer research. Typical neutron scattering cross-sections, σ_N, are of order 10^{-7} Å^2 per nucleus, and are constant (within ± one order of magnitude for all elements and isotopes). σ_N is to be contrasted with σ, the geometric cross section or size of an atom, of order 10 Å^2. Comparison of σ_N with σ shows that only $\sim 10^{-8}$ of an incident neutron beam is scattered by the first layer of a solid sample. Clearly, neutron scattering cannot be used to study single monolayer samples: in fact, of order 10^5 monolayers are required. Fortunately, such high surface area graphites (Grafoil, Papyex, UCAR-ZYX, Spheron, Graphon, etc.) are commercially available. Although the integrated background scattering from the substrate is orders of magnitude greater than that of the monolayer, most of it usually occurs in different regions of $\vec{Q}$, ω space, making background subtraction manageable.

The first neutron study of 2D-like monolayer Bragg peaks was made by Kjems et. al.[19,20] in 1974. They observed that, at low coverages and temperatures, N_2 formed a $\sqrt{3} \times \sqrt{3}$ registered (commensurate) lattice (nearest neighbor distance a_{nn} = 4.26Å), but that upon completion of a registered monolayer, addition of more N_2 created a denser (a_{nn} = 4.04Å) incommensurate triangular phase whose exact a_{nn} was temperature and filling dependent. For fillings intermediate between registered and complete dense monolayer, a broadened peak was observed. The data did not enable one to conclude whether or not the transition was first order.

^{36}Ar has the largest coherent neutron scattering cross sections known, making it highly suited for monolayer studies. Taub et. al.[21] have examined both the structure and the dynamics of submono-

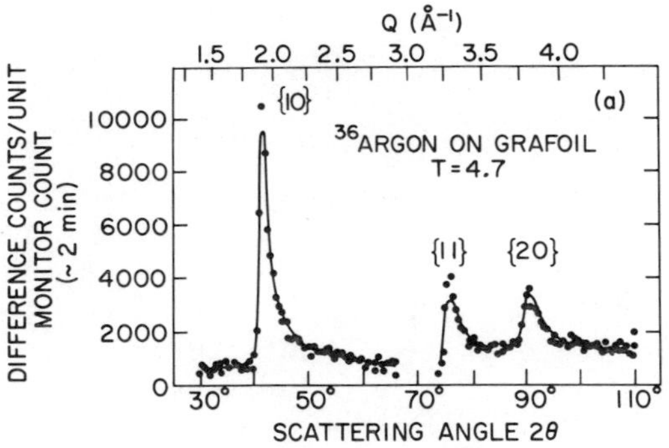

Figure 5. Diffraction pattern from ^{36}Ar monolayer showing the {10}, {11}, and {20} peaks from the film.

layer ^{36}Ar, and found behavior very different from N_2. At low temperatures they observed three Bragg peaks, indexed as the (10), (11), and (20) reflections from a 2-D triangular structure. The diffraction pattern is shown in fig. 5. Note the asymmetric shape of the peaks. This is characteristic of scattering from a system which has only 2-D periodicity. Since there is then no constraint on scattering along the third dimension, those crystallites not oriented in the scattering plane will scatter at $Q > \tau_{hk}$, the position of the hk_{th} Bragg peak predicted from the 2-D periodicity. In other words, for crystallites tilted out of the scattering plane the projected atom-atom distances are shorter, causing scattering at higher Q.

The observed peaks indicate an Ar-Ar spacing $a_{nn} = 3.86Å$, some 10% smaller than the registered one, and about 3% larger than a_{nn} for bulk Ar. The width of the leading (low Q) edge of the Bragg peak reflects the effective crystallite linear dimension, or correlation length L. For submonolayer Ar and N_2 at low temperatures L was found to be coverage independent and of order L $\sim$ 110Å on Grafoil. The coverage-independence indicates that under these conditions the adsorbate clusters into relatively large (L > 120Å) crystallites, while the constancy of L regardless of adsorbate type and lattice parameter indicates that this value of L is controlled by substrate surface imperfections. The temperature dependences of L (Ar) and L(N_2) are shown in fig. 6. Registered N_2 maintains

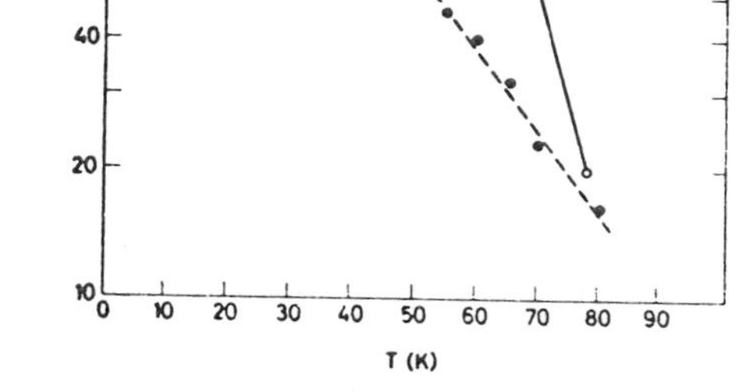

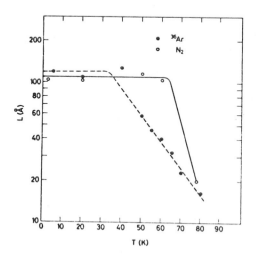

Figure 6. Temperature dependence of the correlation range (or cluster size) L in adsorbed ^{36}Ar and N_2 layers.

L ∿ 110Å up to a sharp, possibly first order, "melting" region, where L decreases to ∿20Å. In contrast, for the incommensurate Ar film a continuous decrease of L was observed over the range 40 < T < 80K. This behavior indicates that there is no qualitative difference between the "crystalline" and "fluid" states for submonolayer Ar. 2-D computer molecular dynamics (MD) simulations[10,22] are in excellent agreement with both the scattering and heat capacity measurements on Ar, indicating that such continuous behavior is a property of 2-D atomic systems. Neither the heat capacity nor the M-D calculations showed any evidence for a gas-liquid coexistence.

Inelastic scattering from the Ar film offers further insight into this melting transition. At low temperatures with the scattering vector parallel to the graphite (001) surface, two peaks centered at $\hbar_\omega$ ∿3 meV and $\hbar_\omega$ ∿ 5.5 meV are seen in the scattering groups. Model harmonic phonon calculations using the standard Ar-Ar Lennard-Jones potential are in good agreement with the observation, and indicate that the low frequency peak corresponds to the high density of states for transverse zone boundary (Z.B.) phonons, while the 5.5 meV one reflects zone boundary longitudinal ones. In fig. 7 we display the temperature dependence of the inelastic response for Q = 3.5Å^{-1} (left), and of the effective density of states obtained by dividing the observed spectra by the phonon population factor [n(ω) + 1] (right). For a 2-D harmonic solid the right hand side

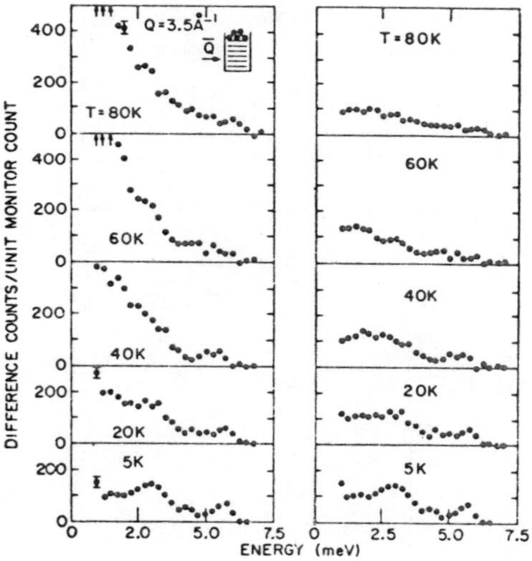

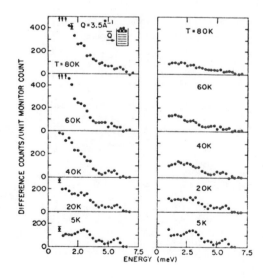

Figure 7. On the left: the temperature dependence of in-plane spectra from Ar monolayers at Q = 3.5Å^{-1}. On the right: the temperature dependence of the nominal "density of phonon states" obtained by dividing the observed spectra by the phonon population factor [n(ω) + 1].

would be temperature independent. Both of the peaks characteristic of the 2-D solid are identifiable up to T ≈ 40K; at this temperature the longitudinal Z.B. one appears to renormalize slightly, decreasing to 5.3 meV, while the transverse Z.B. peak renormalizes relatively much more, going from about 3 to 2 meV. This suggests that the melting process involves a continuous decrease in the resistance to shear, or equivalently that there is a continuous increase in fluidity concommitant with the increase in pair positional disorder, (measured through the width of the Bragg peak).

The temperature dependence of the out-of-plane dynamics is quite different from the in-plane dynamics. At low temperatures the spectral response is dominated by a Q-independent peak near $\hbar\omega$ = 5.6 meV, indicative of an Einstein-like resonant out-of-plane mode. It persists as a recognizable entity up to T = 80K, well above the region where in-plane response becomes fluid-like and overdamped.

The Ar a_{nn} is strongly temperature-dependent, with a total linear expansion of 9% up to T = 60 K. In contrast, the bulk solid has only a 2% expansion in the same range. Approximately 2/3 of the monolayer a_{nn} increase occurs in the range 40 < T < 60 K, above which the peak is too broad to estimate a_{nn}.

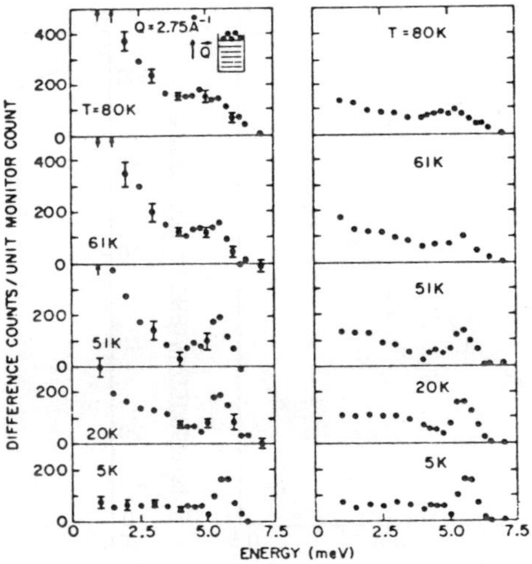

Figure 8. On the left: the temperature dependence of the out-of-plane spectra from Ar monolayers at Q = 2.75Å^{-1}. On the right: the temperature dependence of the nominal "density of states" obtained by dividing the observed spectra by the phonon population factor [n(ω) + 1].

SUBSTRATE INFLUENCES

The tendency of a monolayer to register with the substrate is influenced by the ratio of the monolayer natural packing distance, a, and the substrate repeat distance do, as well as by the ratio of absorbate-substrate periodic potential force constants to the adsorbate-adsorbate ones, and of course, the ratio of the corresponding energies to kT. When the natural misfit ratio is sufficiently small registry occurs. The ratio of the bulk a_{nn} to d_0 gives some measure of the "goodness of fit", but quantitative results would require explicit evaluation of the 2-D interactions, including the enhanced 2-D zero-point motion. The ratios of the bulk a_{nn} to 4.26Å for several systems are: Ar-0.84; Kr - 0.94; Xe - 1.02; N_2 - 0.94; O_2 - 0.77; H_2 - 0.88; D_2 - 0.84. Neutron measurements show that, at submonolayer coverages and low temperatures, Kr, N_2, H_2, and D_2 all register, while Ar and O_2 do not. LEED measurements indicate that Xe registers at high temperature and coverage. The commensurate-incommensurate solid transition has been studied in some detail for Kr and Xe by Fain,[23,24] Venables,[25] and their respective coworkers.

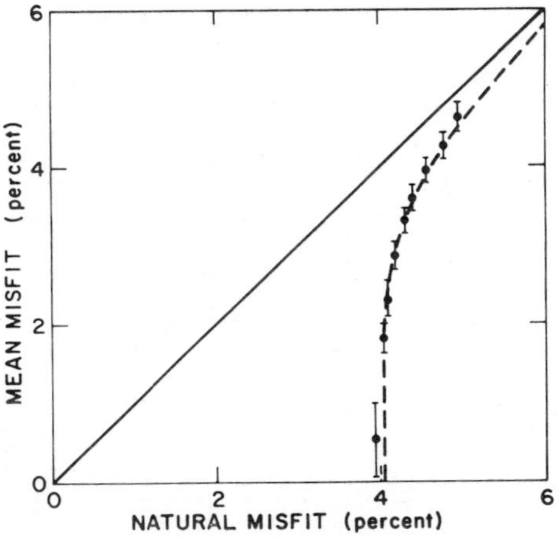

Figure 9. Mean misfit $(d_0 - d)/d_0$ vs natural misfit for Kr on graphite at T = 54 K. These and Venables' results both point to a second-order commensurate-incommensurate transition. For Kr, registry occurs when the uniform system would prefer a 4% smaller lattice parameter. Venables and Fain point out that this discrepancy can be accounted for by a model which includes coherent dislocations, or density modulations, as discussed by Villain in these proceedings.

using electron diffraction techniques (LEED and THEED). Fig. 9 shows the T = 54 K results of Chinn and Fain for Kr, plotted as the mean misfit $(d_0 - d)/d_0$ vs the natural misfit, calculated from an assumed Kr-Kr potential in the absence of the graphite lateral potential.

The melting transition in simple submonolayer registered films is, however, very sharp and appears to be first order. Fig. 10 shows the temperature dependence of the Kr (10) X-ray Bragg peak for a series of nearly monolayer films on UCAR-ZYX, a highly oriented graphite with large intrinsic surface correlation length. Horn et al.[26] have interpreted the lower coverage behavior as indicating first-order melting broadened by pressure and finite system size, while the sharp transiton at the registered filling was interpreted as a second order transition consistent with $\beta \approx 0.08$. Recently, Ostlund and Berker[27] have made a detailed renormalization group analysis of a lattice-gas model of the Kr transitions. They predict a submonolayer first-order melting region and a higher-order transition line at higher temperature and coverages. When finite system size effects are taken into account, their predictions are in striking agreement with experiments on Kr (Fig. 11) and N_2. The two phase diagrams are very similar, and neither shows two-fluid (liquid-gas) coexistence.

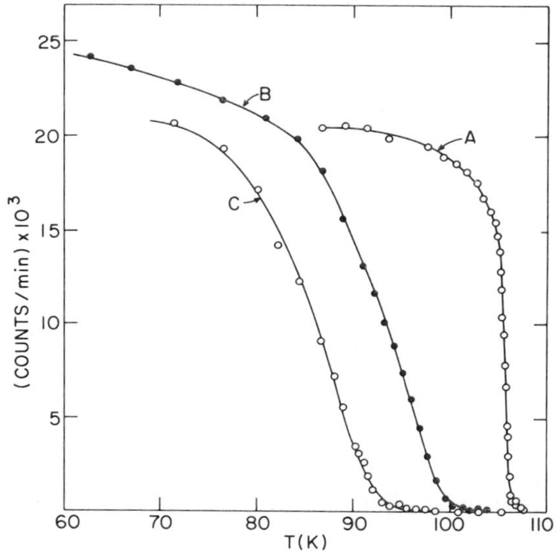

Figure 10. X-ray (10) Bragg peak intensities for Kr on UCAR-ZYX. B and C are for coverages slightly less than perfect $\sqrt{3} \times \sqrt{3}$ registry, while A is at or near the registered filling.

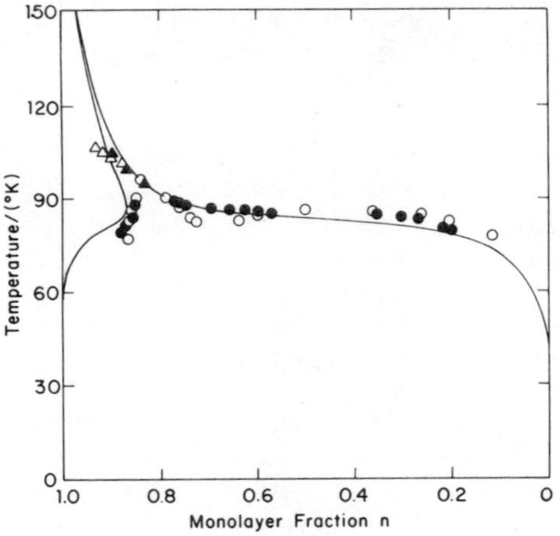

Figure 11. Predicted phase diagram for krypton on graphite, with smearing ΔT = 0.8 K. Phase boundary points from experiments are as follows: first-order transitions (circles); higher-order transition (triangles). Data points are from adsorption isotherms.

Commensurate-Incommensurate Transition and Orientational Epitaxy

Villain gives a detailed discussion of the theory of the commensurate-incommensurate transition in which deviations from perfect registered coverage are accomodated as coherent patterns of density modulations (dislocation walls) around an essentially registered structure. One can conceive of the density defects being accomodated randomly in which case they will not significantly influence the scattering patterns, (except the intensity), or as one- or two-dimensional dislocation arrays. As discussed earlier, the electron diffraction experiments of Fain and Venables[23-25] on Kr and Xe show a continuous change in the position of the Bragg maximum with coverage which begins to deviate from the registered position at a coverage different from that of perfect registry. Except possibly for unresolved peaks near this breakaway coverage, the symmetry of the served spot pattern preserved the triangular structure, suggesting a 2-D domain wall pattern. As pointed out in Villain's paper a transition from registry to a hexagonally symmetric incommensurate phase is predicted to be first-order, but there is no experimental confirmation of discontinuities. On the other hand, existence of an intermediate 1-D dislocation region breaks the symmetry and allows higher order behavior, but there is no confirmation of this 1-D modulation. Perhaps the resolution of the paradox lies either in finite experimental resolution or in finite size effects due to surface defects in the graphite substrate.

Novaco and McTague[28-30] have pointed out that the density modulations in incommensurate systems can cause the adsorbed layer to rotate off the symmetry axis of the substrate. The physical basis of this rotation is that, for symmetry orientations, density modulations involve longitudinal strains, while some of the strain is

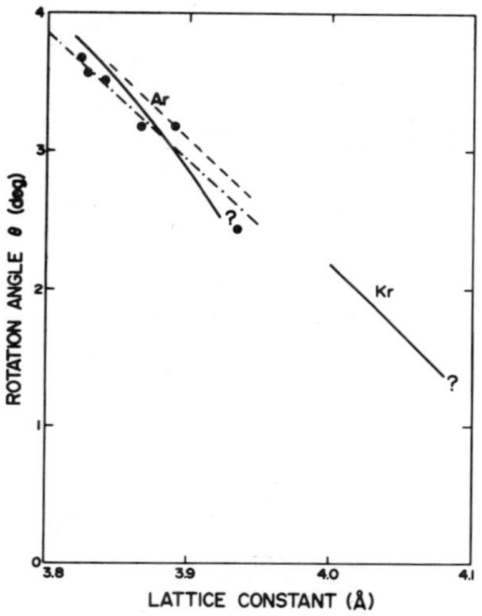

Figure 12. Rotation angle θ_{min} vs lattice constant for Ar and Kr on graphite: (——) from numerical evaluation; (----) from the analytical approximation with $C_L = \sqrt{3}\, C_T$; (-.-.-), best fit line with $C_L = 1.68\, C_T$. indicates breakdown of the linear response approximation. Experimental data points (●) are from Shaw, Fain and Chinn.

transverse for non-symmetry directions, and shear strain costs less energy. When sufficiently far from registry to allow a linear response analysis N-M show that the angle of minimum energy θ_{min} has a particularly simple functional dependence:

$$\cos\theta_{min} = \frac{1 + Z^2(1 + 2\eta)}{Z[2 + \eta(1 + Z^2)}, \quad \eta \geq 1/Z$$

$$\theta_{min} = 0, \eta < 1/Z$$

Here, $\eta = (C_L/C_T)^2 - 1$, where C_L, C_T are the longitudinal and transverse sound velocities for the monolayer film, and $Z = \tau/G$, the ratio of the adsorbate (τ) and substrate (G) reciprocal lattice vectors. For self-bound Lennard-Jones, $C_L/C_T = \sqrt{3}$. Note that, in the linear approximation, Θ_{min} depends only on geometry (τ/G) and on the sound velocities of the film, and is independent of the magnitude of the substrate lateral potential.

Shaw, Chinn, and Fain[31] have observed this orientational epitaxy in incommensurate Ar films on graphite. Their T = 54 K data as a function of lattice parameter are shown in fig. 12, and compared with the predicted values. For a given lattice parameter, the orientation angle is a sensitive function of the ratio of the sound velocities, suggesting the possibility of using this phenomenon to

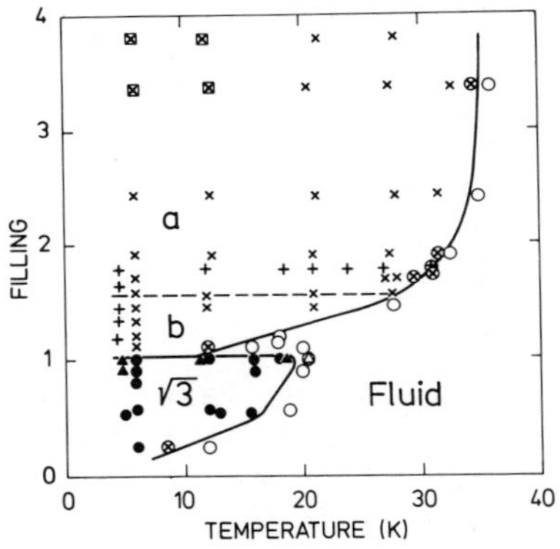

Figure 13. Phase diagram for D_2 and H_2 adsorbed on Grafoil. Filling is the amount of gas adsorbed measured in units of that gas amount which is needed to complete the commensurate $\sqrt{3}$ structure. The broken line indicates where promotion to the second layer starts. The symbols show the following: ● D_2 and ▲ H_2 layers in the $\sqrt{3}$ phase, x D_2 and + H_2 layers in the a and b phases, ○ D_2 and △ H_2 in the Fluid phase, deformed groups with high background.

estimate these properties in films where direct dynamic measurements are not feasible.

Both the structural and dynamic aspects of the commensurate-incommensurate transition in H_2, D_2, and HD films have been studied extensively by neutron scattering.[32] The phase diagram for all three species is essentially identical, and is shown in fig. 13. ^3He and ^4He have similar phase diagrams,[32] but with narrower registered regions.

At low temperatures and for coverages less than or equal to that for a complete registered monolayer ($\rho = 1$) a registered structure is observed with the (10) Bragg peak at $\tau = 1.703$ Å^{-1}. At low temperatures (T < 5K) the behavior for slight overfillings is remarkably similar to Kr (fig. 9). For $1 < \rho \leq 1.04$ a sharp peak is seen at the registered position (fig. 14), but on further filling there is a rapid increase in τ to the position corresponding to uniform incommensurate coverage. With increasing temperature the $\rho > 1$ Bragg peaks broaden and develop fine structure. There is also a marked decrease in peak height, as shown in fig. 15. Note that, at these fillings, the average neighbor distance a_{nn} is of order 4.2-4.25Å, considerably expanded from the bulk self-bound values of 3.75Å (H_2) and 3.59Å (D_2), and it may well be that, in the

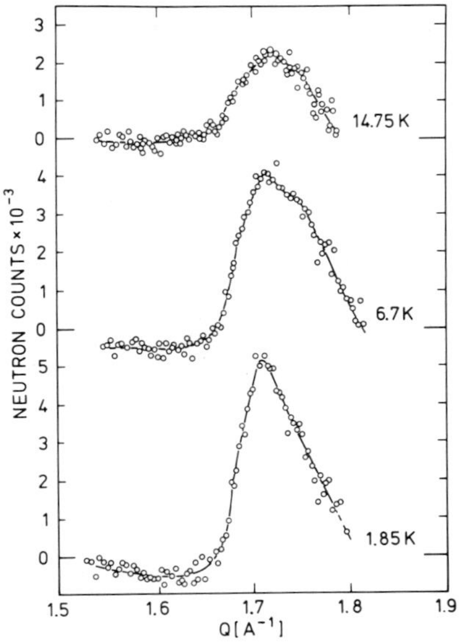

Figure 14. Neutron Scattering groups from D_2 at $\rho = 1.04$ vs T.

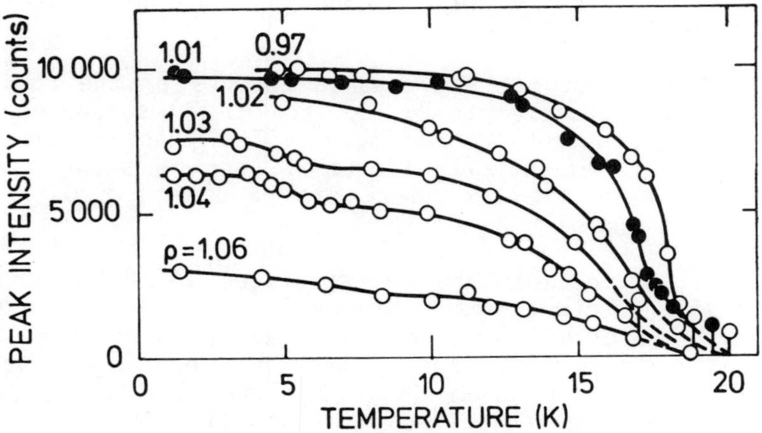

Figure 15. Intensity of the peak of the D_2 (10) reflection as a function of T near the registered coverage (ρ = 1).

absence of a substrate lateral potential, these systems could not form such an expanded solid. Indeed, it seems reasonable to view their structures in this regime as being essentially registered, with the extra filling creating fairly localized misfit dislocation arrays (domain walls), as discussed above and in Villain's paper. The melting behavior of these films is shown in fig. 15. For registered coverages and below there is a sharp melting, while the incommensurate films gradually evolve into a more "argon-like" broad transition.

Even more dramatic is the 2-D phonon density of states, as determined on an HD sample. Below the registered coverage there is a single sharp feature centered at $\hbar\omega \approx 4.2$ meV, reminiscent of an Einstein-like localized well mode. For coverages well beyond $\rho = 1$ this evolves into a more structured density of states. There is clearly much food for thought here in the commensurate-incommensurate behavior of such strongly quantum systems.

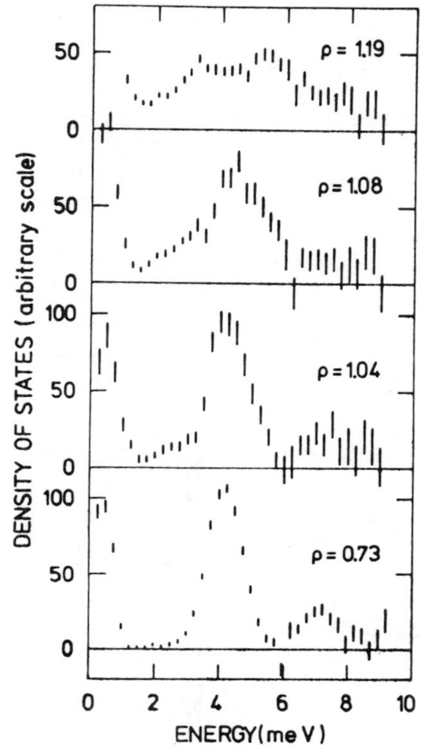

Figure 16. Phonon density of states (T = 4.2 K) for HD on Grafoil as a function of filling ρ.

Antiferromagnetism in O_2 Films

O_2 molecules have two unpaired electrons $[\pi_g^*(2p_x)\ \pi_g^*(2p_y)]$ coupled in an S = 1 triplet. The molecular ground state has the magnetic moment perpendicular to the molecular axis. In condensed phases these antibonding π^* electrons can interact via a direct overlap coupling. Two such molecules can have extra attractive interactions through virtual low-lying ionic O_2^- - O_2^+ states only if (as required by the Pauli principle) their spins $\vec{S}$ are not parallel. To lowest order then, there is an interaction between O_2 molecules with an effective Heisenberg Hamiltonian with J strongly dependent on overlap. This direct exchange mechanism causes bulk O_2

$$E_{12}(r_{12},\Omega_1,\Omega_2) = |J(r_{12},\Omega_1,\Omega_2)\ [\vec{S}_1 \cdot \vec{S}_2],$$

to order antiferromagnetically below T = 23 K. In the higher temperature paramagnetic phase the closest-packed plane has a 6 nearest neighbor triangular structure, with a_{nn} = 3.30Å. It is not possible for all 6 neighbors to align antiferromagnetically, but magnetoelastic distortions can produce 4 antiferromagnetic nearest neighbors, with the other two ferromagnetic ones moving out slightly to form the low temperature antiferromagnetic α phase. The α and β structures shown in Fig. 17

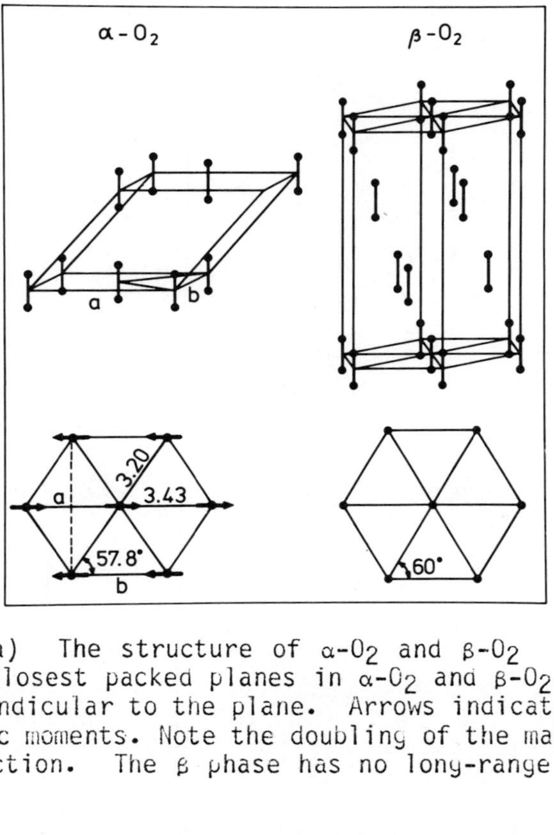

Figure 17. (a) The structure of α-O_2 and β-O_2 (b) The structures of the closest packed planes in α-O_2 and β-O_2. The molecular axes are perpendicular to the plane. Arrows indicate the directions of the magnetic moments. Note the doubling of the magnetic unit cell in the a direction. The β phase has no long-range magnetic order.

Neutron observations indicate similar behavior in the monolayer regime, showing essentially 2-D antiferromagnetic and paramagnetic phases. In all, three distinct phases have been found, all incommensurate with the graphite substrate and all having nearest-neighbor distances similar to those in the closest-packed α and β bulk phases (where the molecular axes are perpendicular to the plane, thus restricting the magnetic moments to being in the plane). The

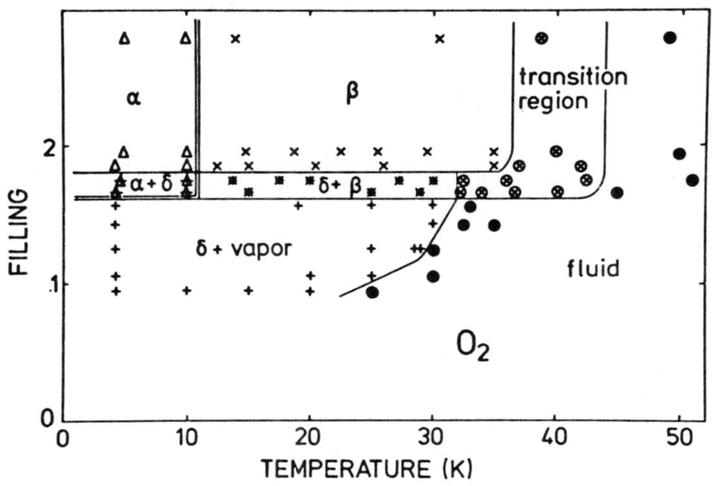

Figure 18. Phase diagram for O_2 thin films on grafoil.

neutron phase diagram is shown in fig. 18. The "filling" units are relative to a (hypothetical) registered filling, i.e. one O_2 molecule for each 3 surface graphite hexagons. Complete monolayer coverage for the α and β phases occurs at $\rho = 1.69$, while complete δ phase occurs at $\rho = 1.59$. There is α + β and β + δ phase coexistence in the intermediate regions, indicative of first order transitions. Representative T = 4.2 K diffraction profiles are shown in fig. 19. The δ-phase appears to be essentially triangular, with the reciprocal lattice vector $\tau_{10} = 2.15Å^{-1}$ at completion, but with some expansion, to $\tau_{10} \approx 2.10$ for $\rho < 1.0$. There is some excess intensity on the high Q side of the peak, and broad magnetic scattering around the superlattice position $Q \sim 1.12Å^{-1}$. These observations may indicate short-range magnetic structure damped by large molecular librational motions which become more constrained as ρ increases, but it must be said that the δ-phase is not yet understood. On the other hand, the α and β phases bear striking analogies to those of bulk O_2. In the β-phase there is an equilateral triangular lattice with $a_{nn} = 3.30Å$. The absence of a magnetic superlattice peak shows that this phase does not have long-range magnetic order. However, the α-phase shows a split (10) peak, indicative of 2 neighbors at a = 3.41Å and 4 at $a_{nn} = 3.22Å$, with a magnetic superlattice

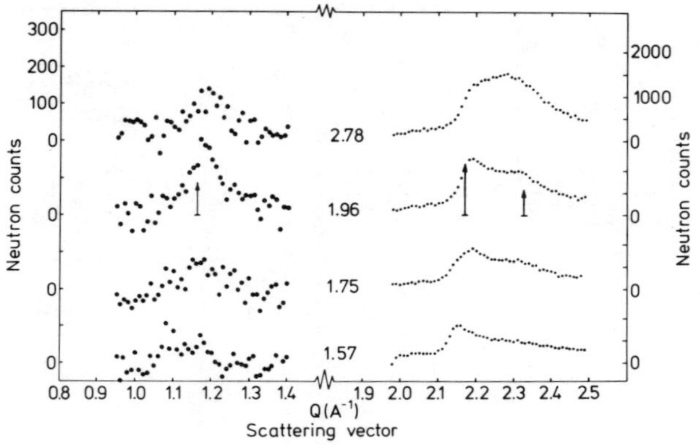

Figure 19. Neutron diffraction groups from O_2 layers adsorbed on Grafoil at T = 4.2 K. Monolayer completion corresponds to ρ = 1.69.

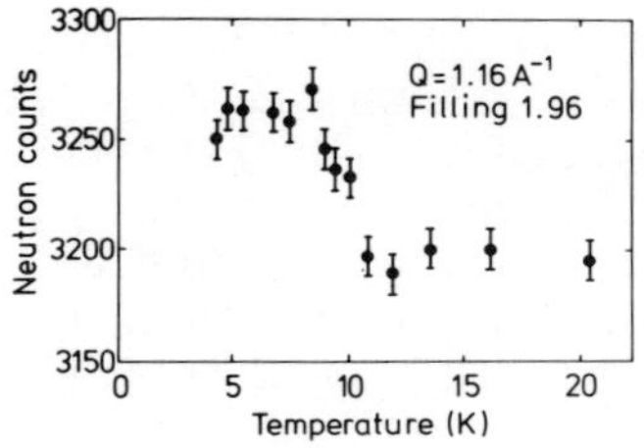

Figure 20. Scattering intensity at Q = 1.16Å^{-1}, the peak of the magnetic superlattice reflections, vs T for ρ = 1.96. The count rate 3200 represents Grafoil background.

peak corresponding to 2 a_{nn} = 6.44Å, or τ (MAGNETIC) = 1.15Å^{-1}, indicating antiferromagnetic doubling of the unit cell as in bulk α-O_2. The intensity indicates that the magnetic moments are aligned as in fig. 17. The peak intensity of this superlattice peak is shown as a function of T in fig. 20.

The magnetic transition is seen to be continuous but the weak magnetic scattering prevents quantitative analysis. On the other hand the magnetoelastic splitting τ_1 - τ_2 of the (10) structural

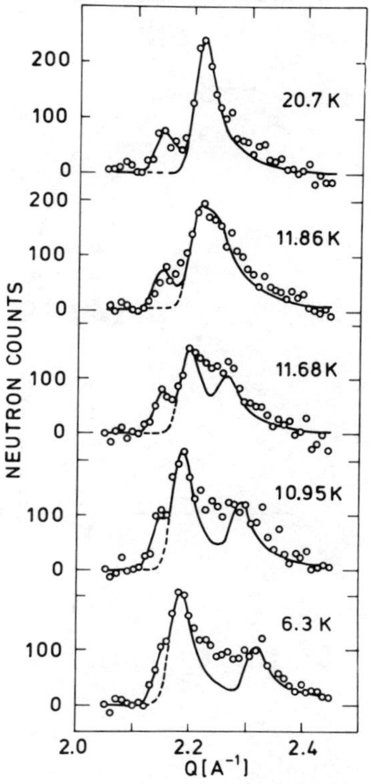

Figure 21. Neutron diffraction groups observed in the α-β transition region on ZYX graphite. At the filling used there is a slight amount of δ phase present (small low Q peak). The split α phase peak continuously evolves into the undistorted phase with increasing temperature.

peak has been studied on ZYX graphite[34] and the peak shapes are shown in fig. 21.

We interpret this splitting as a measure of the magnetic order parameter. $\tau_2 - \tau_1$ is plotted in fig. 22. The temperature variation corresponds to a continuous transition (with $\beta \sim 0.1$-0.2), but since the data go only to $\Delta\tau/\Delta\tau_{T=0} \approx 0.25$ it is not possible to deduce an accurate β from these data. The transition temperature is $T_C = 11.92 \pm 0.08$ K, and is only weakly filling-dependent in the range 1-2 monolayers.

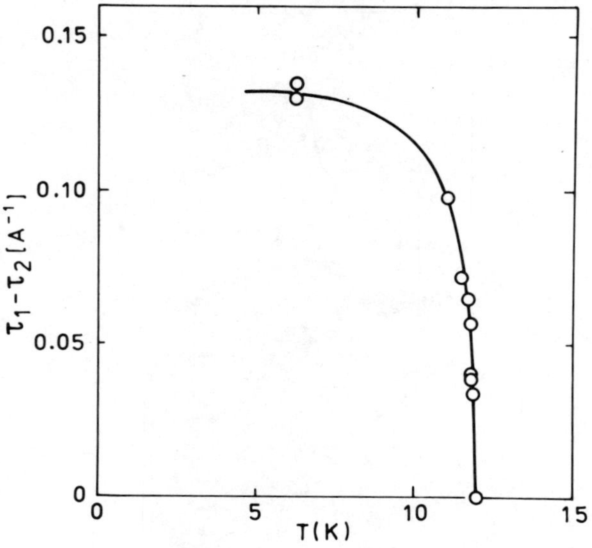

Figure 22. Splitting of the O_2 (10) Bragg peak (fig. 21) vs T.

Stoltenberg and Vilches[35] have studied the heat capacity of monolayer O_2. Fig. 23 shows a trace of their data in the region of the α-β transition, confirming its continuous nature. Domany and Riedel[36] have treated the properties of a phenomenological Landau-Ginzburg-Wilson Hamiltonian for molecules on an equilateral triangular lattice with antiferromagnetic Heisenberg coupling and cubic anisotropy, together with an elastic distortion term and a magneto-elastic coupling. For the region of parameters appropriate to O_2 the magnetic terms cause order, driving the lattice distortion. Their model then yields a single continuous transition of cubic Heisenberg character.

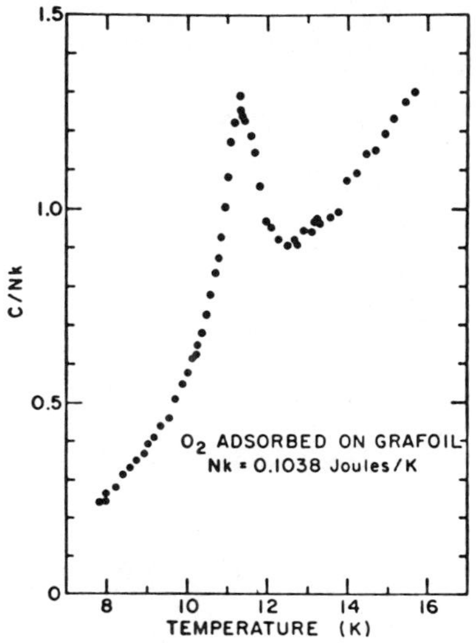

Figure 23. Heat capacity of an O_2 film (ρ = 1.9) as a function of temperature, in the region of the $\alpha \rightarrow \beta$ transition.

CONCLUSION

Complementary structural and thermodynamic studies have revealed a rich variety of essentially 2-D phase transitions in monolayer films involving 2-D melting, registered-disordered, commensurate-incommensurate, antiferromagnetic, etc., phenomena. Because of the simplicity of the systems and the relatively detailed knowledge about the micro scopic interactions involved, these systems offer fertile ground for investigating 2-D phase transition theories, especially those which include finite system size effects.

REFERENCES

1. F. Paneth and A. Radu, Chem. Ber. 57:1221 (1924).
2. S. Brunauer, P. H. Emmett and E. Teller, J. Am. Chem. Soc. 60:310 (1938).
3. G. D. Halsey, Jr., J. Chem. Phys. 16:931 (1948).
4. W. J. C. Orr, Proc. Roy. Soc. (London) A175:349 (1939).
5. E. A. Gulbrandsen and K. F. Andrew, Ind. Eng. Chem. 44:1039 (1952).
6. M. H. Polley, W. D. Schaeffer and W. R. Smith, J. Phys. Chem. 57:469 (1953).
7. W. D. Schaeffer, W. R. Smith and M. H. Polley, Ind. Eng. Chem. 45:1721 (1953).
8. B. B. Fischer and W. G. McMillan, J. Chem. Phys. 28:549 (1958)

and earlier work cited therein.
9. W. A. Steele, "The Interactions of Gases with Solid Surfaces," Pergamon Press (1974).
10. F. E. Hanson, M. J. Mandell and J. P. McTague, J. Phys. (PARIS) 38:C4-76 (1977).
11. J. G. Dash, "Films on Solid Surfaces," Academic Press (1975).
12. A. Thomy and X. Duval, J. Chim. Phys. 67:1101 (1970).
13. D. M. Butler, J. A. Litzinger and G. A. Stewart, Phys. Rev. Letters 42:1289, (1979).
14. T. T. Chung, unpublished. Quoted in ref. 18.
15. J. J. Lander and J. Morrison, Surface Sci. 6:1 (1967).
16. J. A. Venables and P. S. Schabes-Retchkiman, J. Phys. (PARIS) 38:C4-105 (1977).
17. J. P. McTague, M. Nielsen, and L. Passell, CRC Crit. Rev. Solid State Mat. Sci. 8:135 (19).
18. M. Nielsen, W. D. Ellenson and J. P. McTague, in "Neutron Inelastic Scattering" 1977, (IAEA Vienna), Vol. II., p. 433 (1978).
19. J. K. Kjems, L. Passell, H. Taub and J. G. Dash, Phys. Rev. Letters 32:724 (1974).
20. J. K. Kjems, L. Passell, H. Taub, J. G. Dash and A. D. Novaco, Phys. Rev. B 13:1446 (1976).
21. H. Taub, K. Carneiro, J. K. Kjems, L. Passell and J. P. McTague, Phys. Rev. B 16:4551 (1977).
22. D. Frenkel, F. E. Hanson and J. P. McTague, these proceedings.
23. M. D. Chinn and S. C. Fain, Jr., Phys. Rev. Letters 39:146 (1977).
24. S. C. Fain, Jr. and M. D. Chinn, J. Phys. (PARIS) 38:C4-99 (1977).
25. J. A. Venables and P. S. Schabes-Retchkiman, J. Phys. (PARIS) 38:C4-105 (1977).
26. P. M. Horn, R. J. Birgenau, P. Heiney and E. M. Hammonds, Phys. Rev. Letters 41:961 (1978).
27. S. Ostlund and A. N. Berker, Phys. Rev. Letters 42:843 (1979).
28. A. D. Novaco and J. P. McTague, Phys. Rev. Letters 38:1286 (1977).
29. A. D. Novaco and J. P. McTague, J. Phys. (PARIS) 38:C4-116 (1977).
30. J. P. McTague and A. D. Novaco, Phys. Rev. B , (1979).
31. C. G. Shaw, S. C. Fain, Jr. and M. D. Chinn, Phys. Rev. Letters 41:955 (1978).
32. M. Nielsen, J. P. McTague and W. D. Ellenson, J. Phys. (PARIS) 38:C4-10 (1977).
33. J. P. McTague and M. Nielsen, Phys. Rev. Letters 37:596 (1976).
34. M. Nielsen and J. P. McTague, Phys. Rev. B
35. J. Stoltenberg and O. Vilches, unpublished.
36. E. Domany and E. K. Riedel, J. Appl. Phys. 49:1315 (1978).

TWO-DIMENSIONAL SOLIDS AND THEIR INTERACTION WITH SUBSTRATES

Jacques Villain

Department de Recherche Fondamentale
Laboratoire de Diffraction Neutronique
Centre d'Etudes Nucléaires, 85 X, 38041 Grenoble Cedex

ABSTRACT

Adsorbed monolayers on an ideal substrate may exhibit solid phases incommensurable with the substrate. Near the transition to a commensurable solid state (C-I transition) the incommensurable phase can be described as an array of walls separating nearly registered domains.

After a review of earlier theories, the properties of the incommensurable phase are studied near the C-I transition. In the case of an anisotropic substrate, a continuous transition is predicted both at T=0 and T≠0, but with different critical behaviour. For a hexagonal substrate a first order transition is predicted except possibly at T=0.

ORGANISATION OF THESE LECTURES

These lectures deal with two-dimensional solids and their effective realisation by adsorbed layers.

In Section I, theories of ideal 2-D systems are outlined as well as experimental data on real systems. Among these systems, adsorbed layers are our main interest. The effect of a substrate is discussed in Section II. The following Sections are mainly devoted to the "incommensurable solid" phase, which is a good approximation of a 2-D solid. The methods we use are appropriate near the transition to a commensurable phase (C-1). They are based on the concept of domains and walls, which is introduced in Section III through a one-dimensional model. A two-dimensional extension appropriate for an anisotropic substrate is treated in detail in

Section IV. The hexagonal case is treated at T=0 in Section V, where preliminary results on thermal effects are also presented. Orientational epitaxy is treated in Section VI.

The substrate will always be supposed to be an infinite, ideally flat single crystal.

LIST OF ABREVIATIONS

A number of abreviations are used:

- 2-D, 3-D = two-dimensional, three-dimensional
- C-I transition: Commensurable-incommensurable transition (see §§ II,4, III-2, IV-2)
- FVdM = Franck Van der Merwe (see § III-2)
- BMVW: Bak Mukamel-Villain-Wentowska (see § V-2).

I. COLLECTIVE PHENOMENA AND PHASE TRANSITIONS IN TWO DIMENSIONS

I.1 Early Theoretical Works

Physics in two dimensions became fashionable after the year 1967, when Wegner pointed out the remarkable theoretical properties of 2-D, XY magnets, and Jancovici produced a similar theory of the 2-D, harmonic solid.

It was proved by Wegner that the 2-D, XY ferromagnet has a phase transition below which the susceptibility is infinite though the magnetization is zero - an astonishing property first conjectured by Stanley and Kaplan (1966).

Since these lectures are devoted to 2-D solids rather than magnets, a brief outline of Jancovici's theory will now be given. One starts with a set of N atoms which occupy at T=0 the sites $\vec{R}$ of a Bravais lattice. Their position at T≠0 are $\vec{R}+\vec{u}_R$ and the Hamiltonian can be written in the harmonic approximation as:

$$\mathcal{H} = \sum_k \sum_{\alpha\gamma\xi\eta=x,y} g_{\alpha\gamma}^{\xi\eta} k_\xi k_\eta u_k^\alpha u_{-k}^\gamma \qquad \text{(I-1)}$$

In this expression the kinetic energy has been omitted, $\vec{u}_k$ is the Fourier transform of $\vec{u}_R$ and the coefficients

$$g_{\alpha\gamma}^{\xi\eta}$$

have the dimension of an energy multiplied by a surface. Expression (I-1) is correct for short range forces and for long wavelengths.

The usual way of observation of crystal structures is by X-ray or neutron diffraction. The measured quantity is:

$$S(\vec{k}) = \frac{1}{N} \sum_{\vec{R},\vec{R}'} \exp i\, \vec{k}\cdot(\vec{R}-\vec{R}') \langle \exp i\, \vec{k}\cdot(\vec{u}_R-\vec{u}_{R'}) \rangle \qquad \text{(I-2)}$$

In 3-D crystals, $S(\vec{k})$ exhibit delta-function singularities ("Bragg peaks") at the reciprocal lattice vectors $\vec{k} = \vec{\tau}$ defined by $\exp i\, \vec{\tau}\cdot\vec{R} = 1\ \forall\vec{R}$. For D = 2, Jancovici proved that there is only a power law singularity:

$$S(\vec{k}) \sim |\vec{\tau} - \vec{k}|^{-(2-\theta\tau^2)} \qquad \text{(I-3)}$$

where θ is proportional to T and depends smoothly on the direction $\vec{k}/k$. A fascinating property is that the singularity strongly depends on $\vec{\tau}$: for small values of $\vec{\tau}$, $S(\vec{k})$ diverges for $\vec{k} = \vec{\tau}$, not for large values! Similar effects occur in smectic liquid crystals (see Als-Nielsen's lectures). A proof of Jancovici's theorem is summarized in Appendix A.

Although Wegner and Jancovici gave a correct description of the low temperature properties, they did not investigate the phase transitions. The transition of the 2-D, XY ferromagnet was analysed by Kosterlitz and Thouless (1974), Kosterlitz (1975) and Young (1978). It was shown by Knops (1977) that this transition is dual to the roughening transition (treated in Weeks' lectures) which will be encountered in the course of these lectures (§ IV-4 and Appendix B).

In the case of the 2-D solid, the transition analogous to the ferro-paramagnetic transition of the XY model is a transition from a solid phase to a kind of "liquid crystal" phase. Complete melting into a liquid phase might only be achieved through a second transition (Halperin, Nelson, 1978). Experimental support of these theoretical conjectures is still lacking. In the present lectures, melting will not be considered, rather interest will be focussed on solid phases.

I.2 Experimental Situation

Most experiments available on collective phenomena in 2-D systems can be related to two main classes:

a) Experiments on quasi-two-dimensional magnets like K_2CuF_4. These experiments are very precise, but crossover effects to D=3 are dramatic. Critical exponents have been accurately determined, but Wegner's predictions can hardly be checked because of crossover effects. Another problem is that Heisenberg magnets do not show the strange properties of XY magnets (Polyakov 1975, Brezin and Zinn-Justin 1976).

b) Experiments on adsorbed layers. Since the pioneering experiments of Thomy and Duval (1969) the phase diagram of rare gas mono-layers on graphite has been investigated in detail. References can be found in the book of Dash (1975) or in Mc Tague's lectures. Adsorbed layers can, in certain cases, be considered as two-dimensional solids as argued in Section II. However, experiments are very difficult and the critical behaviour is not accurately known in most cases.

One difficulty is to prepare surfaces as ideal and flat as possible. So far, theorists assume the surface to be ideal... Hopefully they will not always do so! Dirty surfaces are quite interesting for practical uses, like catalysis. But this will be left to future developments, and ideal surfaces will be assumed throughout these lectures.

Most experimental investigations of phase diagrams are concerned with adatoms interacting through Van der Waals forces with a substrate of hexagonal symmetry like graphite or lamellar halides. There is no reason to impose such limitations to theories, and lower symmetries will also be considered here. In principle, the theories developed can apply as well to interacting chemisorbed layers whose binding energy is much larger than for "physisorbed" layers like rare gases on graphite.

As a conclusion to this introduction, two kinds of experiments will be briefly mentioned though they are beyond the scope of these lectures: i) Experiments on superfluid ^{4}He films (Berthold et al. 1977) which are the best experimental check of sophisticated two-dimensional theories (Nelson, Kosterlitz 1977). ii) Interesting attempts to study magnetic surface states which have been suggested (Spanjaard et al. 1978) to occur in ^{3}He adsorbed layers (Godfrin et al. 1978).

II. EFFECT OF SUBSTRATE

II.1 Two-Dimensional Solids and Adsorbed Layers

The true 2-D systems considered in Section I are theoretical concepts. These lectures are concerned with certain properties of adsorbed monolayers (layers whose thickness is one atomic diameter). Are these layers two-dimensional objects? May they be considered as 2-D solids like the one studied in § I.1 and characterized by the Hamiltonian (I-1)? There are essentially three differences which are reviewed in the next three paragraphs. It will be seen, however, that certain phases of adsorbed layers have properties analogous to ideal, 2-D solids.

II.2 Substrate Distortion and Related Effects

A first peculiarity of the substrate is that it is an elastic medium. Adatoms produce strains which, for an anharmonic substrate can easily be eliminated and give rise to an indirect, long range interaction (Lau 1978) as briefly explained below.

According to elasticity theory (Landau and Lifshitz, 1959) a force acting at a point of the surface produces a strain proportional to the inverse distance $1/r$. However, an adatom exerts a force dipole rather than a force, so that the strain is proportional to $1/r^3$. The reason is that, if the adatom is at rest, the total force acting on it is zero and, if all forces acting on the adatom are due to the substrate the total force exerted on the substrate by the adatom is also zero. The argument neglects the weight and other external forces.

The interaction energy between two adatoms at distance ρ on the surface is proportional to

$$\int_{r,|\vec{\rho}-\vec{r}|>a} d^3r \, r^{-3} \, |\vec{r}-\vec{\rho}|^{-3} \sim 1/\rho^3$$

where a is the interatomic distance.

For nearest neighbours, the indirect interaction is about 50 times smaller for rare gas monolayers on graphite than direct Van der Waals interactions (Lau 1978) and will generally be neglected in these lectures. Its long range nature, however, may be important, and we shall come back to this point in § V.4.

Other sources of long range interactions (Lau, Kohn 1977 and references therein) are: i) indirect interaction through conduction electrons of a metallic substrate. ii) electric dipole interactions.

II.3 Chemical Potential

Another difference between adsorbed layers and usual 3-D crystals is that the vapour of the adsorbate is present in all experiments and acts as a reservoir of particles. The Hamiltonian (I-1), which corresponds to a fixed number of particles, is never sufficient and an additional free energy should be added, namely

$$F = \mu N \qquad \text{(II-1)}$$

where the chemical potential μ depends on pressure and can be varied at will. The fixed quantity is the number $\mathcal{N}$ of substrate sites, not the number N of adatoms.

In practice, one can solve the whole statistical mechanical problem for a given value of N, and minimize the free energy with respect to N at the end of the calulation. In the whole calculation a single phase may be assumed. This would not be necessarily correct if both N and $\mathcal{N}$ were really fixed, since phase coexistence would be possible.

In practice, experiments are not carried out at constant μ, but for a constant number of atoms both in the monolayers and in the vapour.

II.4 Substrate Potential

The most important and interesting effect of the substrate is to create a periodic potential, acting on adatoms, which can be represented by a Fourier series:

$$H = \sum_{Q} A_Q \cos \vec{Q}\cdot(\vec{R} + \vec{u}_R) \tag{II.2}$$

Where $\vec{Q}$ are the vectors of the reciprocal lattice of the surface of the substrate. In practice the infinite sum (II-2) is often replaced by a sum over a "star" of vectors deduced from any of them by the symmetry operations of the substrate. For a substrate of symmetry lower than square or hexagonal, the star contains only 2 vectors $\vec{Q}$ and $-\vec{Q}$ and the sum (II-2) is limited to a single term.

The main effect of the substrate is that a monolayer can have three types of phases:

a) a state characterized by an average adatom density which has the full translation symmetry of the substrate surface - and nothing more.

b) A state whose invariance group is a subgroup of the invariance group of the substrate surface.

c) A state which has none of the properties of the substrate.

In cases a) and b) the diffraction spectrum exhibits delta singularities at the points $\vec{Q}$ of the reciprocal lattice of the substrate surface. In case b) there are additional delta singularities at superstructure points. Case c) occurs at low temperature and high density, when the adatoms form a 2-D solid which is just slightly perturbed by the substrate. "Bragg singularities" of the Jancovici type, as described in § I.1 may therefore be expected. Their existence will be confirmed in §IV-3.

From now on we will be mainly concerned with phases of type c) or "incommensurable solids".

Transition form a-phase to b-phase is similar to the freezing of a lattice gas. Phases of type a) can be called "fluid" (gas or liquid according to their density) whereas phases of type b) are "commensurable solids", (Fig 4a). These "commensurable solids" have not the fascinating properties of 2-D solids studied in § I.1.

In the case of rare gas monolayers on grahite, the interaction (II-2) with the substrate is weak. Therefore, commensurable solids do not exist except if adatoms have appropriate sizes as is the case for Kr. He also has a commensurable solid phase, may be because of quantum motion. Kr monolayers undergo a commensurable - incommensurable (C-I) transition when increasing pressure. The transition is apparently continuous (Chinn and Fain 1977, Larher 1978 and references therein).

II.5 Conclusion

The main problem which will be treated in the next Sections is the competition between the interaction (I-1) and the substrate potential (II-2). Even when the latter is weak, it cannot easily be treated as a perturbation because the natural perturbation expansion involves an infinity of terms which do not decrease regularly (Novaco, Mc Tague 1977). The opposite way will be chosen here i.e. (II-2) will be treated as the dominant term. This will be found to be correct in the incommensurable phase near the transition to a commensurable solid phase.

III. WALLS AND DOMAINS

III.1 A One-Dimensional Model

The concepts of domains and walls, introduced a long time ago by F. Bloch (1932) in solid state physics, can be useful for the physics of adsorbed layers. The best way to introduce these ideas is probably a one-dimensional model.

We consider N adatoms at positions $x_1, x_2, \ldots, x_N$ on a line. They are subject to an harmonic interaction between nearest neighbours and to the periodic potential of a rigid substrate. The latter, eq. (II-2) can be limited to a single term. The Hamiltonian can be written

$$H = J \sum_{n=1}^{N} (x_{n+1}-x_n-a)^2 - \sum_{n=1}^{N} A \cos 2\pi\, x_n/b + \tilde{\mu} N \qquad \text{(III-1)}$$

where a is the interatomic distance of the absorbate in the absence of substrate, and b is the interatomic distance of the substrate. Minimization with respect to x_n for constant N yields:

$$2J \; (x_{n+1} + x_{n-1} - 2x_n) = \frac{2\pi A}{b} \; \sin 2\pi x_n/b \qquad \text{(III-2)}$$

One particular case is the commensurable (or "registered") solution of (III-2) : $x_n = npb$. Here, p is in principle any integer, but such a solution is physically acceptale only if pb is not too different from a. Thus, $p \simeq a/b$. It will always be assumed that the stability conditions of the registered structure (to be specified below) are not far from being satisfied.

III.2 The Theory of Frank and Van der Merwe (1949) [FVdM].

These authors obtained the ground state of the above 1-D model, using the additional approximation that the discrete index n can be treated as a continuum. This appears to be justified by comparison with experiment, at least in certain cases.

It is convenient to introduce the variable u_n defined by:

$$x_n = npb + u_n b/2\pi \qquad \text{(III-3)}$$

Eq. (III-2) is then transformed into the familiar sine Gordon equation:

$$\frac{d^2u}{dn^2} = 2\,\frac{\pi^2 A}{Jb^2} \; \sin u \qquad \text{(III-4)}$$

One looks for the solution of (III-4) which minimizes the continuous approximation of (III-1):

$$H = \frac{Jb^2}{4\pi^2} \int_0^N dn \left[\left(\frac{du}{dn}\right)^2 + \frac{4\pi^2 A}{Jb^2} (1-\cos u) \right] + \frac{Jb}{\pi} (pb-a)(u_N - u_1) + \mu N \qquad \text{(III-5)}$$

where μ is a shifted chemical potential:

$$\mu = \tilde{\mu} - A + J(pb-a)^2$$

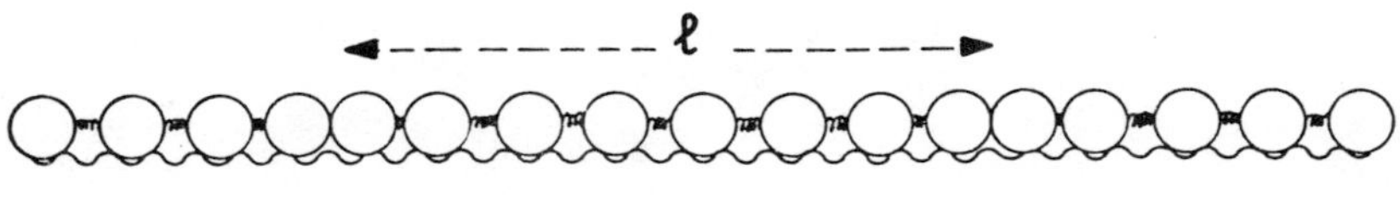

Fig. 1 The one-dimensional, FVdM model. The circles represent the adatoms, the springs the interactions between them (first term of Eq. III-1) and the wavy line the substrate potential (second term).

b) Results. With the above hypotheses, the following results are obtained. Their derivation is given below, § (c).
i) the layer is in registry with the substrate within a range of chemical potential values, $\mu_c^- < \mu < \mu_c^+$.
ii) When μ is just a little below μ_c^-, the layer is formed of large nearly registered domains (Fig. 1) of width ℓ, separated by walls of width $\lambda_0 << \ell$. The density of these walls is greater than the density of the domains, so the walls are "heavy". Similarly, when μ is just a little above μ_c^+, the layer is formed of large, nearly registered domains separated by light walls..
iii) the C-I transition is continuous and the order parameter $1/\ell$ varies according to the law:

$$|\mu-\mu_c| \simeq \text{Const.} \frac{\ell}{\lambda_0} \exp.-\frac{\ell}{\lambda_0} \qquad \text{(III-6)}$$

valid near the transition.

c) Outline of the derivation. Beside the trivial solution of (III-4), u=0, one can look for monotonic solutions satisfying the condition:

$$u_N - u_1 = \pm 2\pi$$

The general solution is well known (SCOTT et al 1973)

$$u_n = \pm f(n-n_o)$$

where

$$f(n) = 4 \text{ tg}^{-1} (\exp -\pi \frac{n}{b} \sqrt{2A/J}) \qquad \text{(III-7)}$$

This function is close to 0 or 2π, except in a region, or wall, which contains a number of atoms equal to:

$$1/H = \lambda_o \simeq \frac{b}{\pi}\sqrt{J/2A} \qquad \text{(III-8)}$$

The energy of these structures with one single wall is obtained from (III-5) after some elementary calculation, namely

$$\mathcal{H} = w_1 + \left[Jb(pb-a) - \frac{\mu}{2p}\right]\frac{u_N-u_1}{\pi} + \mu\,\mathcal{N}/p \qquad \text{(III-9)}$$

Where $\mathcal{N} = Np + (u_N-u_1)/2$ is the number of substrate sites and:

$$w_1 = \frac{4}{\pi}\, b\,\sqrt{2AJ} \qquad \text{(III-10)}$$

The quantity (u_N-u_1) is either 2π or -2π, according to whether the wall is light or heavy.

A necessary condition of stability of the registered structure ($u_n=0$) is that its energy $\mu\mathcal{N}/p$ is lower than (III-9), both for a light and a heavy wall. The result is:

$$\mu_c^- < \mu < \mu_c^+ \qquad \text{(III-11)}$$

where:

$$\mu_c^{\pm} = pw_1 \pm 2\,Jpb(pb-a) \qquad \text{(III-12)}$$

For a structure with several walls, the interaction energy between walls must be taken into account. For a pair of walls at distance ℓ, this energy is for large ℓ:

$$W_{int}(\ell) = \xi\, w_1 \exp - \ell/\lambda_o \qquad \text{(III-13)}$$

where ξ is a positive coefficient of order unity (BAK and EMERY 1976). The exponential decay results from the fact that expression (III-7) approaches its limit exponentially for large $|n|$. It can also be obtained from the linearised form of (III-4) for small u, the solution of which is obviously $\exp \pm n/\lambda_o$. The (essential) fact that ξ is positive can be obtained, for instance, from exact solutions (SCOTT et al 1973, FRANK and VAN DER MERWE 1949) or from the approximate treatment we give in the Appendix B.

For large ℓ, interactions between next - nearest - neigbour walls and further can be neglected. The structure consists of a regular succession of walls at distance ℓ (Fig. 2a). The energy is obtained from (III-9) where the first term is to be multiplied by the number $\mathcal{N}/\ell$ of walls, and (u_N-u_1) is replaced by $\pm\, 2\pi$ times the number of walls. It is also necessary to add the interaction (III-13) multiplied by the number of walls. The result is:

$$\mathcal{H} = \left[w_1 \pm 2Jb(pb-a) \mp \frac{\mu}{p} + \xi\, w_1 \exp - \frac{\ell}{\lambda_0} \right] \frac{\mathcal{N}}{\ell} + \mu \frac{\mathcal{N}}{p} \qquad \text{(III-14)}$$

Minimisation with respect to ℓ for a given value of μ yields $\ell = \infty$ is (III-11) is satisfied. Thus, this condition is not only necessary, but sufficient for the stability of the registered phase. When μ is slightly smaller than μ_c^- (or slightly larger than μ_c^+) minimisation of (III-14) yields (III-6).

In the next Section it will be useful to introduce the effective chemical potential μ which acts in walls. According to (III-14) it is a linear function of the true chemical potential μ (acting on adatoms). For heavy walls (created by increasing pressure as for Kr on graphite) the lower sign should be taken:

$$\mu^* = w_1 - 2Jb(pb-a)+\mu/p \qquad \text{(III-15)}$$

III.3 Aubry's Theory (1978)

Like the FVdM theory, Aubry's theory is a zero-temperature theory for a one-dimensional (or anisotropic) model. But it minimizes the Hamiltonian (III-1) instead of its continuous approximation (III-5). With the help of a sophisticated mathematical apparatus, Aubry was able to derive the following formula for the incommensurable ground state:

$$x_n = n\ a' + \alpha + g(na' + \alpha)$$

where α is an arbitrary phase, a' is the average interatomic distance and g is a periodic function which has the period of the substrate.

Long period structures are obviously stable in a finite range of chemical potential as in the standard registered structure. The question arises, whether true incommensurable structures can be stable at T=0 on an ideal substrate. According to Aubry, the answer is "yes"(provided the substrate potential satisfies certain differentiability conditions). Another question which arises is whether discontinuous jumps may occur between two long period structures. If the system is always in its ground state, the answer is "no", in that sense that the average interatomic distance a' (or the average distance ℓ between walls) changes continuously with the chemical potential in the thermodynamic limit. However, ℓ or a' is not a sufficient characterization of a structure: two structures with nearly equal values of ℓ, and nearly equal values of the energy, can be very distant in the phase space and separated by a potential barrier. Thus hysteresis is possible: According to Aubry, this is the explanation of the hysteresis exhibited by the large density wave system TTF-TCNQ.

So far as adsorbed monolayers are concerned, Aubry's theory might eventually find its application in chemisorbed layers like Sulphur on Iron, where a number of long period structures have been observed (HUBER, OUDAR 1 75). In the case of rare gases on graphite, no long period phases have been observed, as will be seen in Section IV-4.

III.4 Domains and Walls for Dimensions Larger than 1

The FVdM theory and the Aubry theory do apply rather generally at T=0, for any space dimensionality D, provided the system is anisotropic, so that only one wall direction is allowed. The FVdM theory even applies to anisotropic, 3-D systems at low temperatures, lower than the roughening transitions of a 2-D wall. It does not apply to 2-D solids at T≠0 because the roughening transition of a 1-D wall is at T=0, and it does not apply, even at T=0, to substrates with a square or hexagonal symmetry, as will be seen in § V-2.

However, the concept of wall, introduced to treat the FVdM model, is much more general. Walls often occur in physics as a result of competition between two types of forces: in the present case these are adatom - adatom interactions and substrate-adsorbate interactions. In magnetism (F. BLOCH 1932, L. NEEL 1944) the competing interactions are exchange and dipole forces. Other examples are charge density waves (Mc MILLAN 1976) and liquid crystals (DE GENNES 1968, LUBAN et al 1976). However, the wall concept can only be applied when the distance bewteen walls is much larger than the wall thickness, i.e. near the C-I transition. The only exception is constituted by Bloch walls in magnetism and ferroelectricity, which are useful in a very broad region.

In these lectures, we always consider incommensurable phases near the transition to a commensurable state, where the concept of walls is applicable.

This concept, introduced in this Section for a one-dimensional model, will be extended to 2-D systems in the next 2 sections. It is convenient to distinguish between anisotropic substrates (section IV) and isotropic or hexagonal substrates (section V).

Ising models with competing interactions exhibit transitions which are analogous to the C-I transition, and probably eqvivalent in the anisotropic case. This problem has been treated by various approximate methods by Liebmann, Hornreich, Schuster and Selke (to be published).

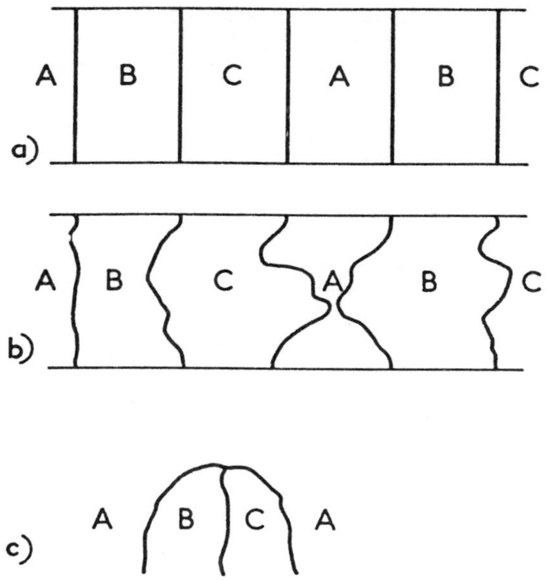

Fig. 2 The Pokrovskii-Talapov model. a) ground state. b) Typical configuration at T=0. c) A configuration discarded by the model. The lines represent walls. A, B, C represent the nature of the domains in the case of 3 sublattices (See Fig. 4)

IV - THE POKROVSKII-TALAPOV MODEL.

IV.1 Hypotheses

Pokrovskii and Talapov (1979) introduced a solvable, 2-D model which presents a C-I transition. This model is characterized by that fact that walls are roughly parallel to a given direction (say y) and cross the whole sample from the top to the bottom, without crossing each other nor coming backwards(Fig. 2). This model would be appropriate, for instance, for rare gas monolayers on graphite, if the substrate were submitted to a uniaxial stress favouring one wall direction. Strictly speaking, wall crossing as in Fig. 2c would not be forbidden in such a case, but the energy is certainly very high and it seems correct to neglect such events.

At T=0, the Pokrovskii-Talapov model is described by the FVdM theory of § III-2. At finite temperature, it behaves quite differently because the interaction between walls decays exponentially with distance. The exponential tail governs the critical behaviour at T=0, whereas at finite T, it can just be neglected, replacing the interaction by a hard core repulsion.

An additional assumption, which will be justified later, is that walls do not see the potential of the lattice, just as in the FVdM theory.

In this case, the problem depends on the chemical potential and the line tension λ of the walls. The Hamiltonian of one wall is:

$$\mathcal{H} = \gamma \sum_{y=1}^{N_y} \left[X(y+1) - X(y) \right]^2$$

$X(y)$ is the abcissa of the wall at ordinate y. In addition, one should take into account the hard core repulsion between walls and an additional free energy $\mu^* q$, where q is the number of walls, and μ^* is given by (III-15).

IV.2 Solution

As shown by de Gennes (1968), one can calculate the partition function of a system of q walls submitted to a hard core potential. The problem is first reduced to the calculation of the largest eigenvalue of the transfer matrix. Then this problem is reduced to the determination of the ground state of a system of q non-interacting Fermions. The Pauli principle accounts for the hard core interaction between the walls. The calculation is presented in the Appendix C and the resulting partition function is:

$$Z(q) = \exp \frac{1}{2\pi} N_x N_y \int_{-q\pi/N_x}^{q\pi/N_x} dk(-\beta\mu^* + 2e^{-\beta y} \cos k) \qquad \text{(IV-1)}$$

where $N_x = qp\ell$ is the number of substrate site of each row, and N_y is the number of rows. The order parameter $\ell = N_x/pq$ is obtained by minimization of this expression with respect to q:

$$4 \sin^2 \frac{\pi}{2\ell} = 2 - \beta\mu^* \exp\beta\gamma \qquad \text{(IV-2)}$$

The critical behaviour at constant temperature as a function of the chemical potential is given by:

$$\frac{\pi}{\ell} \simeq \sqrt{\beta}\, \sqrt{\mu_c^*(T) - \mu^*} \exp \frac{1}{2} \beta\gamma \qquad \text{(IV-3)}$$

where:

$$\beta\mu_c^* (T) = 2 \exp - \beta\gamma \qquad \text{(IV-4)}$$

The critical behaviour at constant potential is given as a function of temperature by:

$$\frac{\pi}{\ell} \simeq \sqrt{2(1+\beta_c\gamma)(T-T_c)/T_c} \qquad \text{(IV-5)}$$

These formulae were obtained by Pokrovskii and Talapov (1979).

IV.3 Bragg Singularities

Bragg diffraction is the best method to identify a solid. It is therefore of interest to investigate the neutron or X-ray scattering cross section near a reciprocal lattice vector. Let $\vec{\tau} = (\tau_x, \tau_y)$ be a reciprocal lattice vector of the adsorbed layer in its commensurable phase. The incommensurable phase is supposed to be still commensurable in the y direction, so that τ_y is not modified. The τ_x coordinate, however, is expected to be shifted by a relative amount $1/\ell p$ (in the case of a contraction, due to heavy walls; for light walls the shift is $-1/\ell p$).

We assume that the layer is a square lattice with inter-atomic distance ℓ in the commensurable phase. Note that different units were used in Section III. Let $(n+u_n^m/p,\ m)$ be the atomic coordinates in the incommensurable phase. The quantity which is measured by neutron scattering is:

$$N_x N_y S(\vec{k}) = \sum_{nn'mm'} e^{i k_y(m-m')+ik_x(n-n')} \left\langle \exp \frac{i}{p} k_x (u_n^m - u_{n'}^{m'}) \right\rangle \qquad \text{(IV-6)}$$

If the walls are sufficiently stiff, u_n^m depends very little on m; then

$$S(\vec{k}) \simeq \frac{1}{N_x} \sum_{\tau_y} \delta\,(k_y-\tau_y) \sum_{nn'} e^{ik_x(n-n')} \left\langle \exp \frac{i}{p}\, k_x (u_n^m - u_{n'}^m) \right\rangle \qquad \text{(IV-7)}$$

This quantity is calculated in the Appendix D by the transfer matrix technique, using the fact that $(u_n^m - u_{n'}^{m'})$ is just the number of walls which separate points (n,m) and (n', m'). The result near a Bragg point is proportional to:

$$S_x\ (k_x) \simeq \int_1^\infty dx\ \left(\frac{x}{\rho}\right)^{-\tau_x^2/2\pi^2 p^2} e^{ix\delta k_x} \qquad \text{(IV-8)}$$

where δk_x is the distance to the shifted Bragg point $\tau_x(1+1/\ell p)$. Eq. (IV-8) describes a delta function for $\tau_x = 0$. For $k_x \neq 0$, $S(k_x)$ becomes infinite for $k_x = \tau_x$ for small values of τ_x. For instance, when $p = 3$, $S(\tau_x)$ diverges for $\tau_x = 2\pi$ or 4π but not for higher values. The singularity of $S(K_x)$ is reminiscent, though slightly different, of the Jannovici singularity of harmonic, 2-D crystals mentioned in Section I.

It has been argued that the Bragg singularity of adsorbed layers can be approximated by a delta function in most experimental situations: "A 2-D crystal the size of the known Universe has a root mean square divergence of 10 Å" claimed NOVACO (J. Physique 38, C4-9, 1977). The above argument suggest it is not necessarily true near a C-I transition.

The delta factor in (IV-7) is an approximation for a very sharp singularity of the Jannovici type. The correct expression requires the evaluation of (IV-6) for $m \neq m'$. The calculation for $n = n'$, $m \neq m'$ has been given for instance by VILLAIN (1975)(*). The partial result (IV-8) is sufficient to show that Jannovici singularities arise in 2-D solids even if they are strongly anharmonic.

IV.4 The Pinning Transition

The purpose of this paragraph is to investigate the effect of the interaction between walls and substrate at finite temperature. It is indeed clear that walls "see" a periodic potential due to the substrate, which was neglected in the FVdM treatment as well as in Pokrovskii's theory. This potential clearly favours integer values of the inter-wall distance ℓ, and to a smaller extent other rational values. It will therefore be assumed that the stable structure of the adsorbate is a long period structure where ℓ is an integer, or at least that this structure is nearly stable.

Let $X_\nu(y)$ be the displacement of the ν'th wall from its normal position $\nu\ell p$. The Hamiltonian will be written as follows:

(*) Villain's statement that $(u_n^m - u_{n'}^{m'})^2$ has a finite limit when T goes to zero is wrong and results from a mistake in the evaluation of the (hopefully correct) integral (17) of this reference

$$\mathcal{H} = \gamma \sum_{\nu,y} [X_\nu(y+1) - X_\nu(y)]^2 + \eta_\ell \sum_{\nu,y} [X_{\nu+1}(y) - X_\nu(y)]^2$$

$$-w_2 \sum_{\nu,y} \cos 2\pi X_\nu(y) + g\, f(\ell) N_y \qquad \text{(IV-9)}$$

where q is the number of walls, and $\nu = 1, 2, \ldots\ldots, q$. The first term was already written in § IV-1. The last term contains a contribution of the chemical potential and the average value in the interaction between walls, which is analogous to (III-13):

$$f(\ell) = w_1 e^{-x\ell} + \mu^*$$

We have replaced ξ by 1, which is equivalent to a redefinition of w_1. The second term of (IV-9) is the part of the interaction between walls which is due to wall fluctations, and it corresponds to a harmonic approximation, valid for small fluctations. In principle, this term involves a coupling between fields X_ν (y) and $X_{\nu+1}$ (y') at different ordinates, and the oversimplified form (IV-9) is only correct if the walls are sufficiently stiff. The coefficient η_ℓ is

$$\eta_\ell = \frac{d^2 f(\ell)}{d\,\ell^2} = w_1\, \kappa^2 \exp -\kappa\ell \qquad \text{(IV-10)}$$

The 3rd term in (IV-9) is the interaction of the walls with the substrate. Its order of magnitude (FRIEDEL 1978) is:

$$w_2 \approx w_1 \exp - \lambda_o \qquad \text{(IV-11)}$$

This expression is similar to that of the Peierls potential of a dislocation (NABARRO 1967).

b) Properties of the model (IV-9)

Two factors tend to destabilise a long period structure, pressure and temperature. As discussed in § III-3, the stability condition at T = 0 is similar to (III-11) i.e. the structure is stable in a finite range of chemical potential. On the other hand, regular structures (Fig. 2a) become unstable at high temperature (for any value of the chemical potential) because the entropy of an irregular structure (Fig 2b) is too large. The stability domain of a long period structure is therefore expected to be limited by a bow (dot-dashed curves on Fig. 3). There is a special value of the chemical potential which does not contribute to destabilise the long period structure. For this special value, ℓ can be considered as a constant independent of temperature and the last term of (IV-9) can be ignored. The resulting model exhibits a roughening transition (Appendix B) which corresponds to unpinning of walls, or transition from Fig. 2 a to Fig. 2 b. The transition temperature is estimated in the Appendix E:

$$K_B T_C \approx \frac{W_o}{\mathrm{Log}\ \dfrac{K_B T_C}{\eta_\ell}}$$

where

$$W_o = \sqrt{\gamma W_2}$$

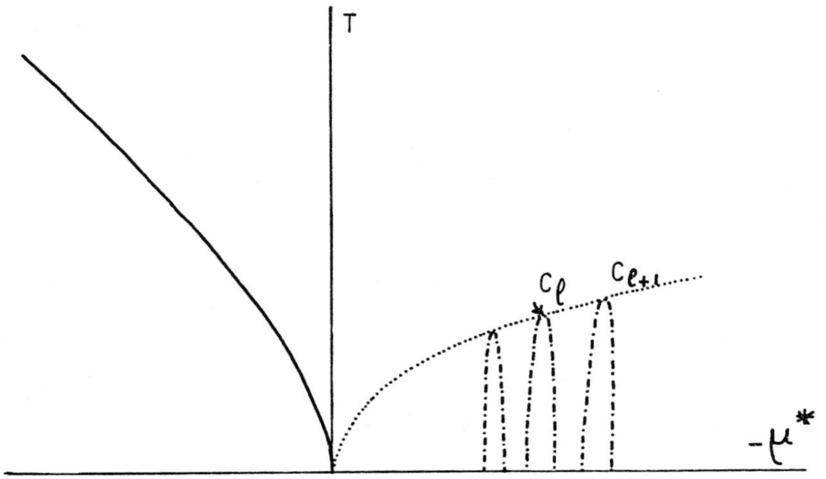

Fig. 3 Phase diagram of the Pokrovskii-Talapov model. At the left of the full curve: registered phase. Below the dot-dashed bows: long period phase, walls are pinned by the lattice. The full curve is given by Eq. (III-6).

This implicit equation can be solved by iteration. The first step is:

$$K_B T_c \approx \sqrt{\gamma w_2} \,/\, \mathrm{Log}\,(\sqrt{\gamma w_2}/\eta_\ell)$$

The value (IV-12) corresponds to the top of the curve (point C_ℓ of Fig. 3) because the transition temperature is expected to be higher when the transition is produced by temperature only than when pressure also comes into play (HALPERIN, NELSON 1978).

The locus Γ of the points C_ℓ (dotted curve of Fig. 3) is obtained by insertion into (IV-12) of expression (IV-10) for η_ℓ and (III-6) for ℓ. The result is:

$$\mu^* + \frac{w_o}{w_1} e^{-\beta w_o} + \frac{e^{-\beta w_o}}{\beta w_1} \;\mathrm{Log}\;\beta w_1 = 0 \qquad \text{(IV-13)}$$

Outside this curve Γ, the adsorbate is an incommensurable solid.

c) Discussion

The harmonic approximation used in (IV-9) for the interaction between walls is drastically different from the "contact approximation" of §§ IV-1, 2, 3. It is correct in the right hand side of Fig. 3, whereas the contact approximation is only valid in the close vicinity of the transition. A detailed calculation shows that the contact approximation is justified if $\ell > \ell_1$

$$\ell_1 \approx \lambda_o \beta\gamma + \lambda_o \;\mathrm{Log}\; \frac{\beta}{4\ell_1 \;\sinh\; \frac{1}{2}\, k}$$

Expression (IV-9) is formally identical with the Hamiltonian of an harmonic solid on a substrate. X_ν (Y) would be the abcissa of the atom of the y' th row and ν' th column, instead of being the abcissa of a wall. Comparison with (III-1) is quite suggestive. Therefore, the continuous curve and the dot-dashed bows of Fig. 3 represent in principle the same type of transition. But the parameters are quite different. For instance, in the case of rare gas monolayers on graphite the pinning potential (IV-11) may be very small because the wall thickness λ_0 is large, whereas the atomic substrate potential A in eq. (III-1) is relatively appreciable. This is probably the reason why long period structures have not been observed for rare gas monolayers on graphite.

On the other hand, the continuous curve of Fig. 3 is not expected to have a top (analogous to points C_ℓ) because a transition to a liquid phase should occur before.

V - RARE GAS MONOLAYERS ON GRAPHITE OR LAMELLAR HALIDES

V.1 Introduction

This system characterized by the hexagonal symmetry of the substrate, is of particular interest because of the amount of available experimental data. Kr on graphite exhibits a C-I transition which is apparently continuous and has roughly speaking the critical behaviour predicted by the FVdM theory, eq. (III-6) (CHINN, FAIN, 1977). No long period structures have been observed probably because the pinning energy analogous to w_2 (see eqn IV-9) is too low because of the large wall width. More information can be found in Mc Tague's lectures.

The concept of walls can be applied without difficulty (Venables et al 1977). The honeycomb cells of the substrate fall in 3 sublattices A B C (Fig. 4a) and an adatom can be said to belong to (e.g.) a A-domain if its centre lies in a A-cell. This definition applies even in the incommensurable phase far from the C-I transition when the substrate potential can be

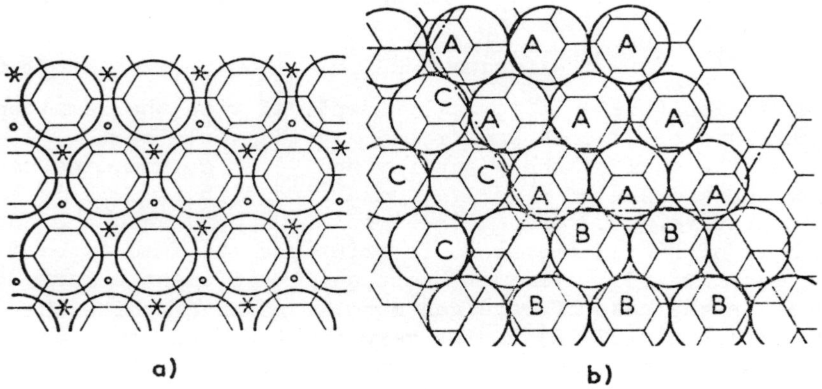

Fig. 4 Krypton monolayer on graphite a) registered phase. The centres of the adatoms (circles) occupy the A sites. Dots denote the B sites and stars are the C sites. b) incommensurable phase far from the C-1 transition.

neglected. Fig. 4b shows the network of walls obtained in this case. It is a regular honey comb network, but of course the interwall distance ℓ and the wall thickness λ are of the same order of magnitude. The wall concept can only be useful for $\ell > \lambda_0$. For rare gas monolayers on graphite we evaluate $\lambda_0 \gtrsim 5$ rare gas diameters, whereas ℓ is accessible to LEED experiments if it is smaller than 30 rare gas diameters (FAIN, private communication).

Thus, there is some overlap between the experimentally accessible region and the present theory.

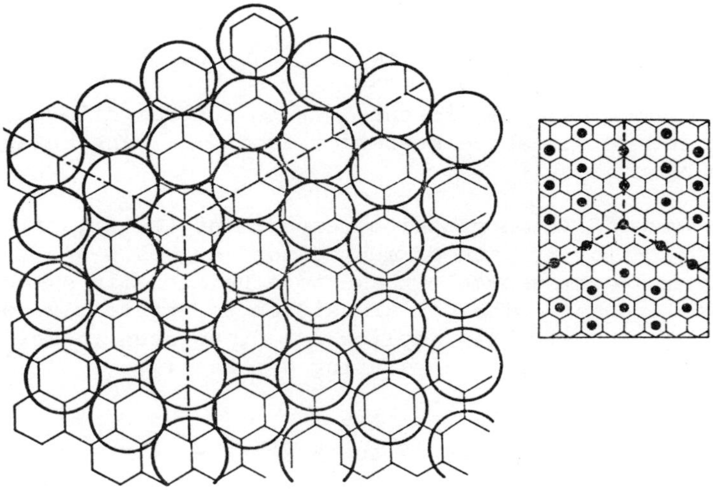

a) b)

Fig. 5 Krypton monolayer on graphite. Incommensurable phase near the C-I transition, showing 3 walls and their intersection. a) "realistic" picture. b) schematic picture for "light" walls. Krypton corresponds to "heavy" walls.

V.2 The Zero Temperature Theory of Bak, Mukamel, Villain; and Wentowska (1979) [BMVW]

It is tempting to assume this picture to remain true when the commensurable phase is approached: the wall would form a regular honeycomb network (Fig. 5) with a lattice constant ℓ which would go to infinity. Bak et al (1979) have shown that this picture is not correct. Indeed wall crossings have a finite energy Λ. If Λ is positive, the energy is lowered (for large ℓ) if the walls are reorganized into a parallel array. (Fig. 2a) as in the Pokrovskii - Talapov model. Indeed one spares a wall crossing energy proportional to Λ/ℓ^2 and one only loses an exponentially small repulsion between parallel walls (see § III-3c).

What happens now if Λ is negative? The hexagonal symmetry is probably preserved but the transition becomes first order because as soon as one wall is created, creation of a number of walls in the other directions lowers the energy. The reader is referred to the original article for the detailed derivation. The case of the square lattice is simpler. If the walls form a rectangular array with wall distance ℓ and L, the energy is given by a straightforward extension of (III-11):

$$W = C\frac{\mathcal{N}}{L\ell}\left[\Lambda+\mu^*(L+\ell)+w_1 L\, e^{-k\ell} + w_1 \ell e^{-kL}\right] + \text{Const} \qquad \text{(V-1)}$$

where C is a numerical coefficient. It is a straightforward exercise to derive the above statement from V-1. The hexagonal case is analogous.

In the case of Kr on graphite, experiment (Marti, private communication; Fain, private communication) suggests an evolution from a commensurable phase to a hexagonally symmetric, incommensurable phase. The C-1 transition is apparently continuous (Larher 1978). This implies $\Lambda > 0$, as seen above, and a sequence of 3 transitions separating 4 phases:

i) the commensurable phase, $L = \ell = \infty$
ii) an incommensurable phase of the FVdM type, $L = \infty$, $\ell \neq \infty$
iii) another phase of low symmetry incommensurable in all directions: $L \neq \ell$, both finite
iv) the hexagonally symmetric, incommensurable phase, $L = \ell \neq \infty$

The transition from (i) to (ii) is described by the FVdM equation (III-6). Phases (iii) and (iv) occur when ℓ becomes so large that the exponential repulsion between parallel walls can no more be neglected.

These features are in contradiction with experiment, which gives no evidence of more than one transition. Is this discrepancy a result of finite temperature? The following paragraph suggests that it is not.

According to FAIN (private communication) experimental data for CO on Pd, Ni and Pt (111) (CONRAD et al 1978, and references therein) can be accounted for by the BMVW theory.

V.3 Rare Gas Monolayers on Hexagonal Substrates at $T \neq 0$

The situation is very complicated in principle because walls can have many types of irregularities (Fig. 6) in addition to their lack of rigidity.

We present here a model which we believe to be realistic though not strictly correct. It rests upon the following assumptions (Fig. 6).

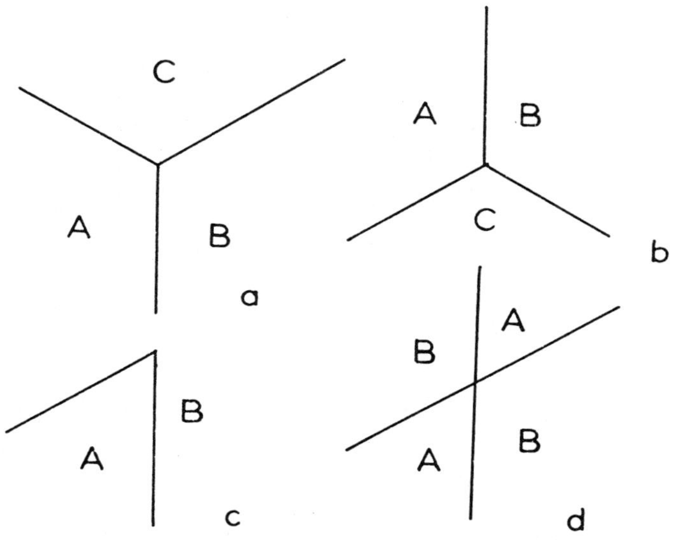

Fig. 6 Various types of wall accidents. It is probably realistic to consider (a) and (b) exclusively.

i) Strictly rigid walls parallel to 3 allowed directions.
ii) the only allowed intersections between walls are those shown on Fig. 6 a and b.
iii) all interactions between walls except the wall crossing energy are neglected.

The kinks and intersections of Fig. 6 c, d, have a high energy and their number is always small at low T, whereas intersections of type a or b have a large density far from the commensurable phase, as pointed out in § V-1. This justifies assumption (ii).

The assumption (i) of rigid walls may look surprising in comparison with the Pokrovskii-Talapov model, where this assumption would be completely incorrect. We believe it is correct on an isotropic substrate because the wall distortion between two intersections is small.

Finally, assumption (iii) has the same justification (for large ℓ and $T \neq 0$) as in the Pokrovskii-Talapov model.

At $T \neq 0$, the essential difference with respect to the BMVW theory is that the free energy contains an entropy term $-K_B T \text{ Log } g$ in addition to the energy (V-1).

The number of states g can be calculated as follows. Let the structure be built up from the top to the bottom. For instance (Fig. 7) points, A_1, A_2, A_3 ....are assumed to be chosen and points C_1 B_1 C_2 B_2 .... are to be chosen. They can be chosen from the left to the right. For instance when B_1 is known, number of possible choices of C_2 is proportional to the distance ℓ between A_1 B_1 and A_2 B_2. The total number of possible structures is therefore

$$g \approx \ell^{\frac{N_x}{\ell} \frac{N_y}{L}} \tag{V-2}$$

Where N_x and N_y is the number of atomic rows in two directions, ℓ is the average distance between vertical walls and L is the average distance between walls of other orientations. Formula (V-2) is probably correct for $L >> \ell$, otherwise the number of choices is reduced. For instance (Fig. 7) C_3 cannot be chosen too close to B_2, otherwise the intersection B_3 would not exist. However, formula (V-2) is expected to remain qualitatively correct for $L > \ell$. We propose the following generalisation:

$$-K_B T \text{ Log } g(\ell, L) = -K_B T \frac{\mathcal{N}}{\ell L} f\left(\frac{\ell}{L}\right) \text{ Log } \frac{L\ell}{L+\ell} \tag{V-3}$$

where $f(\ell/L)$ is of order unity. This term dominates the first term of (V-1) for large values of ℓ:

$$\ell \gtrsim \exp \beta\Lambda$$

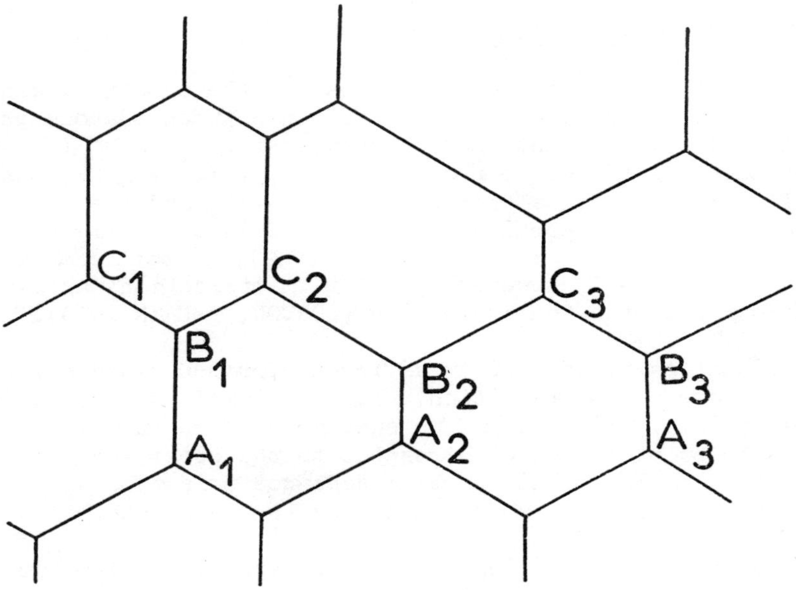

Fig. 7 Wall arrangement near the C-I transition for rare gas monolayers on graphite.

Apart from the logarithmic factor, the entropy term (V-3) which is to be added to (V-1), is rather similar to the Λ-term, but it is systematically negative. Since a negative value of Λ implies a first order transition without loss of symmetry (i.e. $\ell \approx L$) as seen in § V-2, the same features are expected at $T \neq 0$, whatever be the value of Λ.

The conclusion is that there is a disagreement between theory (which predicts a first order transition) and experiment (which observes a continuous transition). This disagreement may be due to the small size of the surface experimentally used.

V.4 Effect of Substrate Distortions

As seen in § II-2, the elastic shear of the substrate can be accounted for by an indirect interaction between adatoms proportional to $1/r^3$. The interaction between two walls of length ℓ at distance ℓ is therefore proportional to $\ell^2/\ell^3 = 1/\ell$. The number of wall segments is proportional to N/ℓ^2, and the resulting energy is proportional to N/ℓ^3. This is negligible for large ℓ with respect to the entropy (V-2), proportional to $T\,\ell^{-2}$ Log ℓ. At $T = 0$, however, the $1/r$, interaction can change the behaviour near the C-I transition (GORDON, VILAIN 1979).

The situation is completely different for 3-D systems like $TaSe_2$ which have a C-I transition. Parallel walls will be assumed for simplicity (Fig. 2a). The energy (or free energy at $T \neq 0$) has to be minimised with respect to the cross section S parallel to the walls. This problem does not arise for an adsorbed layer because S is determined by the substrate.

In the hope of clarifying this controversial problem, we shall present an oversimplified argument, assuming a phenomenological free energy of the form

$$F = K\mathcal{N}\left(1-\frac{1}{\ell}\right)\left(\frac{S}{S_o}-1\right)^2 + K\frac{\mathcal{N}}{\ell}\left(\frac{S}{S_o}-\alpha\right)^2 \qquad (V-4)$$

This expression means that the system likes to have the cross section S_o, except at the walls. Walls like to have the cross sections αS_o, where α is a constant. (V-4) is by no means the most general expression, but it is sufficient for our purpose. It can be written as:

$$F = K\mathcal{N}\left(\frac{S}{S_o}-1+\frac{1-\alpha}{\ell}\right)^2 + K\frac{\mathcal{N}}{\ell}(\alpha-1)^2 - K(\alpha-1)^2\mathcal{N}/\ell^2$$

The first term should be put equal to zero when minimizing with respect to S. The second term is a correction to the chemical potential. The last term is equivalent to an attractive interaction between walls and produces a first order transition, similarly to § V-2 (BRUCE and COWLEY 1978).

As remarked by BAK and TIMONEN (1978) indirect interactions through non-uniform strains do not necessarily produce a first order transition. But the cross section S is necessarily a uniform quantity since it is a macroscopic quantity whereas the effective field acting on it (due to walls) has a microscopic wavelength ℓ. A macroscopic modulation of S at this microscopic scale would clearly break the crystal.

VI - THE NOVACO-Mc TAGUE ORIENTATIONAL INSTABILITY

VI. 1 General Argument

The adsorbate was in the previous chapters assumed to have the same symmetry directions as the substrate. Novaco and Mc Tague (1977) discovered that the adsorbate may in fact be tilted with respect to the substrate in the incommensurable phase. They used a harmonic approximation, which is correct, though not very easy to handle, far from the C-I transition.

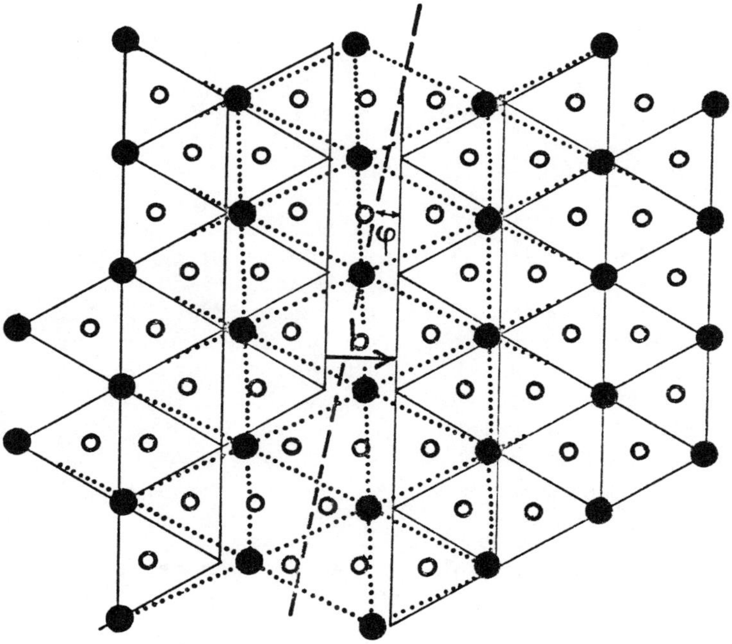

Fig. 8 A wall exhibiting the Novaco-Mc Tague distortion.

We have proposed (Villain 1978) a theory based on the concept of walls, which is only correct near the C-I transition. This theory proves that, if certain conditions to be specified below are fullfilled, the walls are tilted with respect to the symmetry directions of the substrate. There results (Fig. 8) a kink of the atomic rows of the adsorbate, at the places where they intersect a wall. When many walls are present, they produce a succession of kinks which is equivalent to a tilt of the adsorbate - the Novaco-Mc Tague effect.

The derivation is given below for T = 0. Its principle is extremely simple. Experimentally, the effect has been observed by Shaw et al (1978).

VI.2 Case of Parallel Walls

This case (Fig. 9) corresponds to the Pokrovskii-Talapov Model, or to the BMVW theory near the C-I transition (§ V-1).

Let ℓ be the distance between walls (assumed to be uniform). In the FVdM theory of Section II, the energy was found to have the following form:

$$W(\ell) = \frac{\mathcal{N}}{\ell}\left(w_1 e^{-k\ell} + \mu + A\right) \tag{VI-1}$$

where μ is an effective chemical potential and w_1 is a constant. Now, if the walls are allowed to have an arbitrary orientation defined by their angle ϕ with some given direction, A, W_1 and μ are to be replaced by functions of ϕ: A (ϕ), $W_1\mu(\phi)$, $\mu^1(\phi)$. $\mu(\phi)$ can easily be calculated if the layer is assumed to keep the topology of an ideal solid, with no vacancies nor dislocations (or if vacancies and dislocations are assumed to have no effect). $\frac{N}{\ell}\mu(\phi)$ is indeed proportional to the number of extra adatoms with respect to the commensurable phase. In the simplest cases (like rare gas monolayers on graphite, the adatoms of the n'th domain are translated by a vector $n\vec{b}$ with respect to the commensurable phase (Venables 1977, Villain 1978) as can be seen from Fig. 8. $\vec{b}$ is a given vector directed along a symmetry axis, and which is chosen to be parallel to the x axis on Fig. 8. If ϕ is the angle of the walls with the orthogonal axis y, $\frac{N}{\ell}\mu(\phi)$ is therefore proportional to the number $(N_x \cos\phi/\ell)$ of walls, and to the dimension N_y along y. Therefore, in appropriate units:

$$\mu(\phi) = \mu \cos\phi \tag{VI-2}$$

Expression (VI-1) reads

$$W(\ell) = \frac{\mathcal{N}}{\ell}\left[A(\phi) + w_1(\phi)e^{-k\ell} + \mu \cos\phi\right] \qquad \text{(VI-3)}$$

where $\mathcal{N} = N_x N_y$

Minimisation with respect to ℓ yields either $\ell = \infty$ or:

$$A(\phi) + k\ell w_1(\phi)e^{-k\ell} + w_1(\phi)e^{-k\ell} + \mu \cos\phi = 0 \qquad \text{(VI-4)}$$

It is always possible to assume

$\cos \Phi > 0$

since this only implies a suitable definition of ϕ. The Novaco-Mc Tague distortion occurs if the minimum of (VI-3) corresponds to a non-vanishing value of ϕ.

Insertion of (VI-4) into (VI-3) yields:

$$W(\ell) = -\frac{\mathcal{N}}{\ell} k\ell w_1(\phi)e^{-k\ell} \qquad \text{(VI-5)}$$

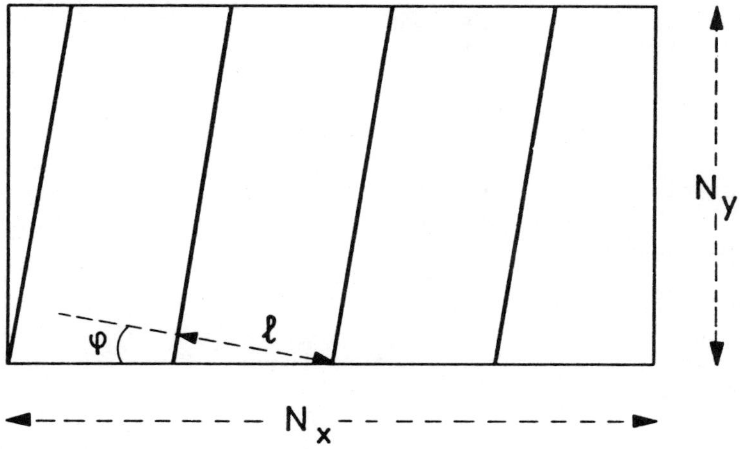

Fig. 9 A distorted array of parallel walls. The x axis is parallel to $\vec{b}$. The number of walls is $N_x \cos\phi/\ell$ and the total wall length is $\mathcal{N}/\ell$.

It will be assumed that $w_1(\phi)$ is positive, as it is in a simplified model (Villain 1978). In this case, expression (VI-5) is lower than $W(\infty) = 0$, therefore, if (VI-4) has solution, one of these solutions is the minimum minimorum of (VI-5). (VI-4) has a solution if

$$\mu < - A(\phi)/\cos\phi$$

The Novaco-Mc Tague distortion occurs near the C-I transition if $- A(\phi)/\cos\phi$ is not maximum for $\phi = 0$, or

$$A''(0) + A(0) < 0 \qquad \text{(VI-6)}$$

where $A''(\phi) = d^2A(\phi)/d\phi^2$

If $A''(0) + A(0)$ is positive, the ground state is probably not distorted, though it cannot be excluded that $-A(\phi)/\cos\phi$ has its absolute maximum for $\phi \neq 0$ though $\phi = 0$ is a relative maximum.

Thus, the crucial problem is to determine the sign of $A(0) + A''(0)$. $A(\phi)$ is the part of the wall energy which depends on the interactions of the atoms between themselves and with the substrate - but not on pressure.

The effective calculation of $A(\phi)$ is a microscopic, numerical problem which lies beyond the general arguments of these lecture. Certain speculations will be presented in § VI-5 but we shall first consider the symmetric case.

VI.3 Case of a Regular Network of Intersecting Walls

For simplicity, we shall only consider the symmetric case $L = \ell$ (in the terminology of Section V). After addition of the wall crossing energy to expression (VI-3), the energy reads:

$$W(\ell) = \frac{\mathcal{N}}{\ell}\left[A(\phi) + \frac{1}{\ell}\Lambda(\phi) + \alpha w_1(\phi)e^{-k\ell} + \gamma\mu\cos\phi\right] \qquad \text{(VI-7)}$$

where α and γ are numerical coefficients. For a square lattice $\alpha = \gamma = 2$ as in Eq. (V-1).

We shall only consider the vicinity of the first order, C-I transition which occurs when Λ is negative (§ V-2). The values ℓ_c and μ_c at the transition are given by the system $W = \frac{\partial W}{\partial \ell} = 0$. Assuming a large value of ℓ_c one obtains

$$\gamma\mu_c(\phi) \simeq - \frac{A(\phi)}{\cos\phi} + k|\Lambda(\phi)|/\mathrm{Log}\,\frac{\alpha w_1(\phi)}{k\Lambda(\phi)} \qquad \text{(VI-8)}$$

The actual value of μ_c and the value of ϕ_c of ϕ at the transition are obtained by minimisation of this expression with respect to ϕ. If ℓ_c is large, the second term can be neglected and it is again found that the Novaco-Mc Tague distortion occurs if (VI-6) is satisfied.

A detailed calculation has been given by SHIBA (1978) using a simplified microscopic model.

VI.4 Finite Temperatures

The above theory suggests that at $T \neq 0$, the C-I transition, if continuous (as it is in the Pokrovskii-Talapov model) does not lead to a Novaco-distorted incommensurable phase. The distortion can only take place through another transition, for a finite value of ℓ. A first order, C-I transition can lead to a distorted phase or not.

The reason is just that walls can have kinks. They become zig-zag chains along which the values ϕ_c and $-\phi_c$ alternate and the average value is $\bar{\phi} = 0$.

VI.5 Microscopic Theories

All existing calculations (Villain 1978, SHIBA 1979) refer to a rather unrealistic model. The adsorbate is treated as a continuum, and the substrate potential is assumed to be harmonic except on the walls - a model first proposed by Luban et al (1976). For an isotropic or hexagonal substrate $A(\phi)/A(0)$ depends only of the two Lamé coefficients, or alternatively on the correlation lengths $1/\kappa$ and $1/\kappa_T$ of compression and shear:

$$A(\phi) = C\left(\frac{1}{\kappa}\cos^2\phi + \frac{1}{\kappa_T}\sin^2\phi\right)$$

where C is a positive constant. The Novaco-Mc Tague distortion takes plase if (VI-6) is satisfied, or

$$2\kappa_T > \kappa$$

This condition can be satisfied for appropriate values of the interactions. It is not satisfied for a triangular lattice with nearest neighbour interactions (Villain 1978).

The spirit of this calculation is analogous to the spirit of the original calculation of Novaco and Mc Tague; the basic idea is that shear strains are energetically cheaper than compressions.

ACKNOWLEDGEMENTS

If there is some relation between these lectures and experimental reality, let S. Fain, Y. Larher and C. Marti be thanked for that.

If the list of References is reasonably up to date, let T. Garel's help be acknowledged.

If I was able to give a satisfactory account of Pokrovskii's theory, I could only do it after an illuminating discussion with D. Haldane.

Finally, I am grateful to R. Pynn who was a friendly but efficient Referee.

REFERENCES

S. AUBRY (1978) in "Non linear structure and dynamics in condensed matter". Springer Verlag, Berlin.
P. BAK. D. MUKAMEL, J. VILLAIN, K. WENTOWSKA (1979). Phys. Rev. B 19, 1610.
P. BAK, J. TIMONEN (1978), J. Phys. C 11 4901
P. BAK, V.J. EMERY (1976) Phys. Rev. Lett. 36, 978
J.E. BERTHOLD, D.J. BISHOP, J.D. REPPY (1977) Phys. Rev. Lett 39, 348
F. BLOCH (1932) Z. Phys. 74, 295
A.D. BRUCE, R.A. COWLEY (1978) J. Phys. C 11, 3577, 3591, 3609
E. BREZIN, J. ZINN-JUSTIN (1976) Phys. Rev. Lett. 36, 691
M.D. CHINN, S.C. FAIN (1977) Phys. Rev. Lett. 39, 146
H. CONRAD, G. ERTL, J. J. KÜPPERS (1978) Surface Science 76, 323
J.G. DASH (1975) "Films on Solid Surfaces"(Acad. Press, New York)
F.C. FRANK, J.H. VAN DER MERWE (1949) Proc. Roy Soc. London, 198, 205
J. DRIEDEL (1978) "Extended defects in Materials". Trieste lectures, to be published.
P.G. de GENNES (1968) Solid State Comm. 6, 163
H. GODFRIN, G. FROSSATI, D. THOULOUZE, M. CHAPELLIER, W.G. CLARK (1978) J. Physique 39, C6-287
P.G. de GENNES (1968) J. Chem. Phys. 48, 2257
M. GODRON, J. VILLAIN (1979) Phys. C. to be published.

B.I. HALPERIN, D.R. NELSON (1978) Phys. Rev. Lett. 41, 121
M. HUBER, J. OUDAR (1975) Surface Science 47, 605
B. JANCOVICI (1967) Phys. Rev. Lett. 19, 20
H.J.F. KNOPS (1977) Ohys Rev. Letters 39, 766
J.V. JOSE, L.P. KADANOFF, S. KIRKPATRICK. D.R. NELSON (1977) Phys. Rev. B 16, 1217
J.M. KOSTERLITZ (1974) J. Phys. C 7, 1046
J.M. KOSTERLITZ, D.J. THOULESS (1973) J. Phys. C 6, 1181
L.D. LANDAU, E.M. LIFSHITZ (1959) "Theory of Elasticity" (Pergamon, London)
Y. LAHRER (1978) J. Chem. Phys. 68, 257
K.H. LAU (1978) Solid State Comm. 28, 757
K.H. LAU, W. KOHN (1977) Surface Science 65, 607
E.H. LIEB, F.Y. WU (1972) in "Phase transitions and critical phenomena ", Vol. 1, P 361, Edited by C. DOMB and M.S. GREEN, Acad. press, London
M. LUBAN, D. MUKAMEL, S. SHTRIKMAN (1976) Phys. Rev. A 10, 360
W.L. Mc MILLAN (1976) Phys. Rev. B 14, 1496
W.L. Mc MILLAN (1977) Phys. Rev. B 16, 4655
F.R.N. NABARRO (1967) "Theory of Crystal dislocations" Clarendon Press, Oxford
L. NEEL (1944) Cah. Phys. 25, 1.

D.R. NELSON, J.M. KOSTERLITZ (1977) Phys. Rev. Lett. 39, 1201
A.D. NOVACO, J.P. Mc TAGUE (1977) Phys. Rev. Lett. 38, 1286, J. Physique 38, C4-116
V.L. POKROVSKII, A.L. TALAPOV (1979) Phys. Rev. Lett. 42, 65
A.M. POLYAKOV (1975) Phys. Letters 59 B, 79
A.C. SCOTT, F.Y.F. CHU, D.W.Mc LAUGHLIN (1973) Proc. IEEE 61, 1443
C.G. SHAW, S.C. FAIN, M.D. CHINN (1978) Phys. Rev. Lett. 41, 955
H. SHIBA (1978) Tech. Rep. I.S.S.P. 1 940 (Tokyo)
D. SPANJAARD, D.L. MILLS, M.T. BEAL-MONOD (1978) J. Physique 39, C6-293 and preprint
H.E. STANLEY, T.A. KAPLAN (1966) Phys. Rev. Lett.17,913
G. THEODOROU, T.M. RICE (1978) Phys. Rev. B 18, 2840
A. THOMY, X. DUVAL (1969) J. Chimie Phys. 66, 1966
J.A. VENABLES, P.S. SHABES-RETCHKIMAN (1977) J. Phys. 38, C4-105
J. VILLAIN (1975) J. Physique 36, L-173
J. VILLAIN (1978) Phys. Rev. Lett. 41, 36
F. WEGNER (1967) Z. Phys. 206, 465
A.P. YOUNG (1978) J. Phys. C 11, L. 453

APPENDIX A: BRAGG SINGULARITIES OF A 2-D, HARMONIC CRYSTAL (JANCOVICI 1967)

The average square fluctations is readily deduced from (I-1). It is of order:

$$\langle u_{-k}^{\alpha} u_{k}^{\gamma} \rangle \approx K_B T/gk^2 \qquad \text{(A 1)}$$

where g is the typical of $g_{\alpha\gamma}^{\xi\eta}$

The average value (1-2) can be transformed by applications of a well-known formula which uses the fact that i $\vec{k}.(\vec{u}_R - \vec{u}_{R'})$ is the sum of independent gaussian random variables:

$$S(\vec{k}) = \frac{1}{N} \sum_{RR'} \exp i\, \vec{k}.(\vec{R}-\vec{R}')\exp.-\frac{1}{2}\langle |\vec{k}.(\vec{u}_R - \vec{u}_{R'})|^2 \rangle$$

$$= \sum_{R} \exp i\, \vec{k}.\vec{R} \exp - \frac{1}{N} \sum_{q} \langle |\vec{k}.\vec{u}_q|^2 (1-e^{i\vec{q}.\vec{R}}) \rangle$$

Insertion of (A 1) yields:

$$S(\vec{k}) \approx \sum_{R} \exp i\vec{k}.\vec{R} \exp -\frac{K_B T}{4\pi^2} \int d^2q\, \frac{k^2}{gq^2} (1-e^{i\vec{q}.\vec{R}})$$

$$\approx \sum_{R} \exp i\vec{k}.\vec{R} \exp -\frac{K_B T k^2}{2\pi g} \int_{q>1/R} dq/q$$

$$= \sum_{R} \exp i\vec{k}.\vec{R} \exp - (\frac{K_B T}{2\pi g} k^2 \text{ Log } R)$$

Finally:

$$S(\vec{k}) \approx \sum_{R} R^{-\theta k^2} \exp i\vec{k}.\vec{R} \qquad \text{(A 2)}$$

Explicit calculation of (A 2) near the Bragg peaks yields (I-3).

APPENDIX B. INTERACTION BETWEEN TWO SOLUTIONS

The energy in the continuous limit:

$$W=\int dx\left[\tfrac{1}{2}\phi'^2 + V(\phi)\right] \qquad (B\ 1)$$

Minimisation with respect to ϕ yields the Euler equation:

$$\phi'' = V'(\phi) \qquad (B\ 2)$$

where

$$\phi'(x)=d\phi(x)/dx, \quad \phi''=d^2\phi/dx^2, \quad V'=dV/d\phi$$

$V(\phi)$ is
i) periodic with period 2π
ii) even in ϕ
iii) minimum for $\phi = 2\pi n$
iv) analytic.

If $V(\phi) = 1-\cos\phi$, (B 2) is the sine Gordon equation.

Far from a wall, $\phi = 2\pi n + \psi$ and (B 2) can be linearized.

$$\psi'' = V''(0)\ \psi \qquad (B3)$$

Let $\phi = f(x)$ be the solution with one wall which satisfies:

$$f(x) - \pi = -[f(-x)-\pi]$$

$$f(-\infty) = 0, \qquad f(\infty) = 2\pi$$

Let ϕ_1 and ϕ_2 be defined as:

$$\phi_1(x) = -f(-x-\ell) \qquad \phi_2(x) = f(x-\ell)$$

The function

$$\phi(x) = \phi_1(x) + \phi_2(x) \qquad (B\ 4)$$

is an approximate solution of (B 2) with two walls for large ℓ. Far from a wall, this results from the linearized expression (B 3).
Inside a wall, one of the function ϕ_1 or ϕ_2 is small and (B 2) is satisfied to order 0 with respect to this function - a poor approximation, but (B 2) is satisfied to order 1 everywhere else.

The energy corresponding to (B 4) is:

$$W(\ell) = \int_0^\infty dx[\phi_1' + \phi_2']^2 + 2\int_0^\infty dx\ V(\phi_1+\phi_2)$$

$$\simeq \int_0^\infty dx[\phi_1' + \phi_2']^2 + 2\int_0^\infty dx[V(\phi_2)+\phi_1 V'(\phi_2) +$$

$$+ \tfrac{1}{2}\phi_1{}^2 V''(\phi_2)]$$

The expression

$$\int_0^\infty dx\phi_2'^2 + \int_0^\infty dx[V(\phi_2)+\tfrac{1}{2}\phi_1{}^2 V''(\emptyset)] + \int_0^\infty dx\phi_1'^2$$

$$\simeq \int_{-\infty}^\infty dx[\phi_2'^2 + V(\phi_2)]$$

is approximately equal to the the energy W_o of a solution with one wall. The interaction energy is therefore

$$W_{int}(\ell) = W(\ell) - 2\,W_o$$

$$= \int_0^\infty 2dx[\phi_1'\phi_2' + \phi_1 V'(\phi_2)] + \int_0^\infty dx\phi_1{}^2\ [V''(\phi_2) - V''(0)]$$

or using (B 2)

$$W_{int}(\ell) = -2\phi_1(0)\phi_2'(0) + \int_0^\infty dx\phi_1{}^2[V''(\phi_2) - V''(0)] \quad (B\ 5)$$

It follows from (B 3) that

$$\phi_2(x) = f(x-\ell) \simeq C\ e^{x(x-\ell/2)} \qquad (x < -\lambda_o + \ell/2)$$

$$\phi_1(x) = -f(-x-\ell) \simeq C\ e^{-x(x+\ell/2)} \qquad (x > \lambda_o - \ell/2)$$

where λ_o is the wall width, C is a positive coefficient and:

$$\kappa^2 = V''(0)$$

The second term of (B 5) is proportional to $e^{-2\kappa\ell}$ and negligible. The first term yields:

$$W_{int}(\ell) \simeq 2\ C^2\kappa\ e^{-\kappa\ell}$$

This expression shows that the interaction between walls of the same sign is repulsive. We emphasize that this result is not limited to the sine Gordon case, $V(\phi) = 2\sin^2 \frac{1}{2}\phi$.

APPENDIX C: PARTITION FUNCTION OF THE POKROVSKII-TALAPOV MODEL NEAR THE COMMENSURABLE-INCOMMENSURABLE TRANSITION

This calculation is due to de Gennes (1968). This Appendix requires some knowledge of the "transfer matrix" method. See for instance LIEB and WU (1972) and references therein.

One considers a rectangular array of N_y atomic rows. (Fig. 2b). The i'th wall instersects the y'th row at the abcissa $x_i(y)$. The partition function for q wall is:

$$Z_q = \mathrm{Tr}\hat{\theta}_q^{\,Ny} \qquad \text{(C-1)}$$

where Θ_q is a matrix which acts on the orthonormal vectors

$$|X_1, X_2, \ldots., X_q\rangle \qquad (X_1 < X_2 < \ldots < X_q)$$

and is defined by its matrix elements:

$$\langle X_1 X_2 \ldots X_q | \Theta_q | X_1 X_2 \ldots X_q \rangle = \exp -\beta\mu^*{}_q \qquad \text{(C 2a)}$$

$$\langle X_1 \ldots X_{i-1}, X_i, X_{i+1} \ldots X_q | \theta_q | X_1 \ldots X_{i-1}, X_i \pm 1, X_{i+1} \ldots X_q \rangle$$

$$= \exp -\beta\mu^* q \exp -\beta\gamma \qquad \text{(c 2b)}$$

More generally, the transfer matrix θ_q can translate p walls by $\pm$ 1 and the corresponding matrix element is $\exp -\beta\mu^* q \exp -\beta\gamma p$. It is convient to write:

$$|X_1 X_2 \ldots X_q\rangle = c^+(X_1)c^+(X_2)\ldots c^+(Xq)|0\rangle$$

where $c^+ (X)$ is the creation operator of one Fermion at point X and $|0\rangle$ is the state with no Fermion. The Pauli principle ensures that the condition $X_1 (y) < X_2(y) < \ldots . < X_q (y)$ is satisfied for any y if it is satisfied for y = 0.

Using (C 2), the transfer matrix θ_q can be written for $\beta\gamma >> 1$ as:

$$\hat{\theta}_q = \hat{P}_q (\exp - \hat{\mathcal{H}})\hat{P}_q \qquad (C\ 3)$$

where $\hat{P}_q$ is the projection operator on the states with q Fermions and

$$-\hat{\mathcal{H}} = \beta\mu^* \sum_{i=1}^{N_x} c_i^+ c_i + e^{-\beta\gamma} \sum_{i=1}^{N_x} (c_i^+ c_{i+1} + c_i^+ c_{i-1}) \qquad (C\ 4)$$

For large values of N_y, the trace (C 1) can be replaced by the diagonal matrix element

$$Z_q = \langle q | \hat{\theta}_q^{N_y} | q \rangle = \langle q | \hat{\theta}_q | q \rangle^{N_y}$$

$$= \langle q | \exp - \mathcal{H} | q \rangle^{N_y} = \exp - N_y \langle q | \mathcal{H} | q \rangle \qquad (C\ 5)$$

where $|q\rangle$ is the q-Fermion eigen vector of $-\mathcal{H}$ which has the largest eigenvalue. This eigenvector is easily obtained since (C 4) is a free Fermion Hamiltonian:

$$|q\rangle = \prod_{k=-q\pi/N_x}^{q\pi/N_x} c_k^+ |\emptyset\rangle \qquad (C\ 6)$$

where c_k^+ is the Fourier transform of c^+ (X).

Insertion of (C 6) into (C 5) yields eq. (IV-1).

APPENDIX D: BRAGG SINGULARITIES OF THE POKROVSKII-TALAPOV MODEL NEAR THE C-I TRANSITION

Expression (IV-7) can be written as:

$$S(\vec{k}) = \sum_{\tau y} \delta(k_y - \tau_y)\; S_x(k_x)$$

where

$$S_x(k_x) = \frac{1}{N_x} \sum_{nn'} \exp ik_x(n-n' + u_n^m - u_{n'}^m /P)$$

$$\langle \exp \frac{i}{p}\, k_x(u_n^m - u_{n'}^m - \langle u_n^m - u_{n'}^m \rangle)\rangle$$

Gaussian fluctations will be assumed. One obtains:

$$S_x(k_x) = \frac{1}{N_x} \sum_{nn'} \exp i\, k_x[n-n' + u_n^m - u_{n'}^m \rangle /P]$$

$$\exp -[k_x^2 \langle (u_n^m - u_{n'}^m - \langle u_n^m - u_{n'}^m \rangle)^2 \rangle /2P^2] \qquad (D\ 1)$$

$(u_n^m - u_{n'}^m)$ is the number of walls between n and n'. Therefore:

$$\langle u_n^m - u_{n'}^m \rangle = (n - n')/\ell$$

In the notations of Appendix C, the fluctation is:

$$\Delta \equiv \langle (u_n^m - u_{n'}^m - \langle u_n^m - u_{n'}^m \rangle)^2$$

$$= \langle q | (\sum_{i=i_o}^{i_o+n'-n} c_i^+ c_i)^2 | q \rangle - \langle q | \sum_{i=i_o}^{i_o+n'-n} c_i^+ c_i | q \rangle^2$$

$$= \frac{1}{N_x^2} \sum_{i,j=i_o}^{i_o+n'-n} \exp\,[i(k-k')(i-j)]\, \langle q | c_k^+ c_{k'} c_{k'}^+ c_k | q \rangle$$

$$= (2/N_x^2) \sum_{k=0}^{k_F} \sum_{k'=k_F}^{\pi} \sin^2 \tfrac{1}{2}(k-k')(n'-n+1)/\sin^2 \tfrac{1}{2}(k-k')$$

where $k_F = \pi/\ell$. For $n'-n \gg \ell$, one finds $\Delta \simeq \mathrm{Log}(n'-n)/\ell$, and insertion into (D 1) yields expression (IV-8).

APPENDIX E: THE ROUGHENING TRANSITION

The purpose of this Appendix is the calculation, in an anisotropic case, of the order of magnitude of the roughening transition temperature for the following Hamiltonian:

$$\mathcal{H} = \sum_{n=1}^{N} \sum_{m-1}^{M} [J(u_{n+1}^{m} - u_{n}^{m})^2 + J'(u_{n}^{m+1} - u_{n}^{m})^2] - A \cos 2\pi\, u_{n}^{m}$$

The condition $J << A << J'$ will be assumed to hold.

Let J be first neglected. The rows n = Const are independent. Each row, is, at low temperature, made of domains u_m = Const, separated by walls. The energy of a wall is of order $\sqrt{AJ'}$, as results from equation (III.10). Therefore, the average size of domains is about $\exp\beta\sqrt{AJ'}$.

If the interaction J is now introduced, and if the system is assumed to be in its low temperature phase, each of these domains have an excitation energy of order:

$$J \exp \beta\sqrt{AJ'}$$

if they are not in the "correct" position. If this quantity is larger than $K_B T$, excited domains are impossible and the ordered phase is stable. Otherwise the ordered phase is expected to be unstable. The transition temperature is therefore given in order of magnitude by

$$K_B T_C \approx \sqrt{AJ'}/\text{Log}\, \frac{K_B T_C}{J}$$

THE DISLOCATION THEORY OF MELTING: HISTORY, STATUS, AND PROGNOSIS

R.M.J. Cotterill

Department of Structural Properties of Materials
The Technical University of Denmark
Building 307, DK-2800 Lyngby, Denmark

1. INTRODUCTION

Of all the physical properties of a crystal, its ultimate transition to the liquid state, at a sufficiently high temperature, is one of the most mysterious. It is surprising that no widely-accepted theory of melting has yet emerged, but the same is true of the liquid state itself, and it could well be that a full understanding of both problems will come simultaneously. The aim of this brief review is a commentary on one particular melting model: the dislocation theory of melting. According to this, melting occurs through the sudden catastrophic proliferation of dislocations. The model usually assumes that this implies acceptance of a picture of the liquid state in which the latter is essentially a crystal filled to saturation with dislocations, though it might later transpire that this is not an absolutely necessary consequence of dislocation mediated melting.

2. HISTORY

The dislocation theory of melting could be said to have its roots in the 1938 paper of Mott and Gurney [1], although their analysis was actually couched in terms of grain boundaries rather than dislocations. Since this classic paper appeared, it has become appreciated that grain boundaries can always be described by appropriate arrays of dislocations. The analysis showed that at a certain temperature the free energy of a polycrystalline array, with a grain size of a few atomic diameters, would be equal to the free energy of a defect-free crystal and that a first-order transition would thus occur. The liquid state was therefore regarded as being like a polycrystal with minimal grain size, and the high fluidity of that state was attributed to the type of intergranular motion that we now know does occur in high temperature creep.

The next significant milepost was the 1947 paper of Bragg [2], and again we have a publication whose relevance to this subject has appeared only in retrospect. This paper addressed itself to the question of the core of a dislocation, and its energy in particular. It should be recalled that the strain field around a dislocation is slowly varying and that the distortion to the crystal lattice increases with decreasing distance from the center of the defect. Inside a certain distance, the core radius, the strain actually exceeds the elastic limit. Because of this, elastic continuum theory is no longer valid, and it is not possible to derive an analytic expression for the core energy. Bragg argued that the disturbance to the core region could not be greater than would be the case if the atomic configuration had become like that of a liquid, and he calculated an upper bound for the core energy from the experimental value of the latent heat of melting. An actual calculation of the core energy, using a two-body interatomic potential became possible only with the advent of the electronic digital computer, and it was found [3] that the Bragg estimate was remarkably good. One sees that a crystal saturated with dislocation cores might thus be a good first approximation of the liquid state, and the latent heat of melting would then be the core energy multiplied by the saturation dislocation density. This view does indeed appear at a later point in the story.

The next piece of the puzzle was supplied by Shockley [4] in 1952. Using the information then extant concerning dislocation mobility, he was able to account for a liquid's fluidity, in a quantitative manner, by assuming a value for the effective dislocation concentration in the liquid state, together with certain assumptions regarding dislocation width.

The next great advance was achieved independently and simultaneously by Ookawa [5] and Mizushima [6]. It was based on the fact that dislocation energy depends on the dislocation concentration in a crystal. The more dislocations there are already present, the lower is the energy required to generate more of these defects. The free energy, F, of a crystal containing a concentration c of the defects at temperature T is found to be given by

$$F(c,T) = -(Gb^3/8\pi)\ c\ \ell n\ c\ \delta^2 - \alpha cT \qquad (1)$$

where G is the shear modulus, b is the Burgers vector, αc the dislocation entropy per atom, and δ (which is less than unity) takes account of the core contribution. The concentration c is given as the fraction of all atoms in the lattice which lie within

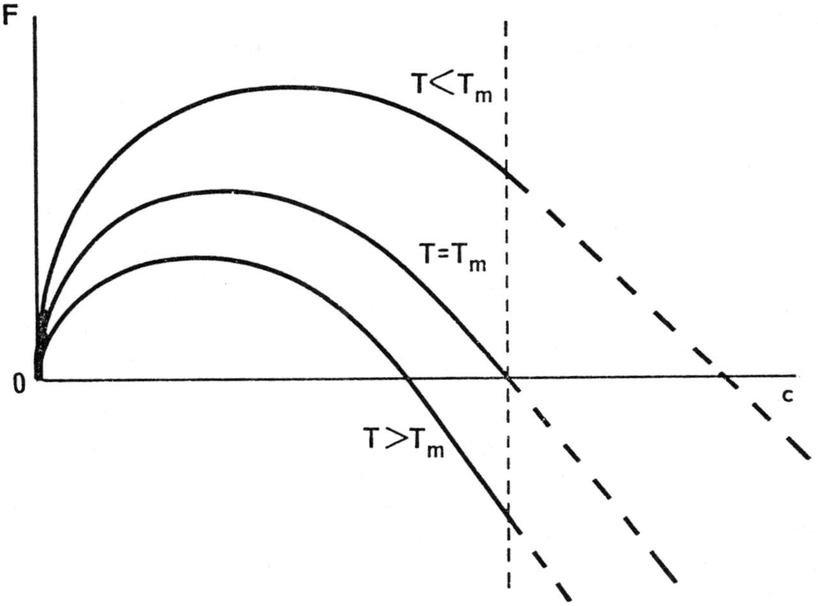

Figure 1. The free energy of a crystal containing a concentration c of dislocations, for various temperatures. The vertical dotted line denotes the saturation condition.

the core of a dislocation. One sees that Equation (1) gives a family of curves as shown in Figure 1, with the vertical dotted line denoting the saturation condition, for which every atom lies within a dislocation core. Identifying the melting isotherm, T_M, as that for which $F(0, T) = F(1, T)$ we have

$$T_M = - Gb^3 \ln \delta^2/8\alpha\pi \tag{2}$$

and it is found that for reasonable values of δ, this gives good agreement with experimental melting temperatures. In fact, the fit is even better if one divides by the factor $(1 - \nu/2)$ which allows for an equal division between edge and screw dislocations, ν being Poisson's ratio.

An important embellishment of these ideas appeared in the 1965 paper of Kuhlmann-Wilsdorf [7]. She invoked the cooperativity embodied in Equation (1) twice, by suggesting the pairing of dislocations whose Burgers vectors were of equal magnitude but opposite sign. This configuration is known as a dislocation dipole, and it has the important characteristic that the long-range strain fields of the two members of the dipole cancel each other.

If the dipole separation is comparable to the core diameter, the total energy is approximately equal to twice the core energy, and is thus favorably small. Moreover, dislocations generated in the interior of a crystal (as opposed to those that migrate inwards from the crystal surface) must, for topological reasons, be generated in such dipoles, by a shear mechanism.

Figure 1 is of course reminiscent of the behavior expected of a system undergoing a first-order phase transition, c being the order (actually disorder) parameter. It is usually assumed, however, that F ultimately rises again for values of c larger than the saturation value, so as to make the equality of free energies a unique situation. The free energy curve then osculates with the abscissa at the transition temperature, as shown in Figure 2. Two recent approaches have presented evidence that this will be the case. Edwards [8] has given a continuum argument, while the present author [9] has made computer calculations of the core energy as a function of dislocation concentration and found the expected increase of F with c even for c values comparable to, and in excess of, the saturation value. The underlying cause of this behavior is that a dislocation dilates a crystal lattice. If Figure 2 is to be valid even for the saturation condition, however, it is necessary to establish that this behavior occurs even when the continuum approach no longer holds. Here, one could invoke the

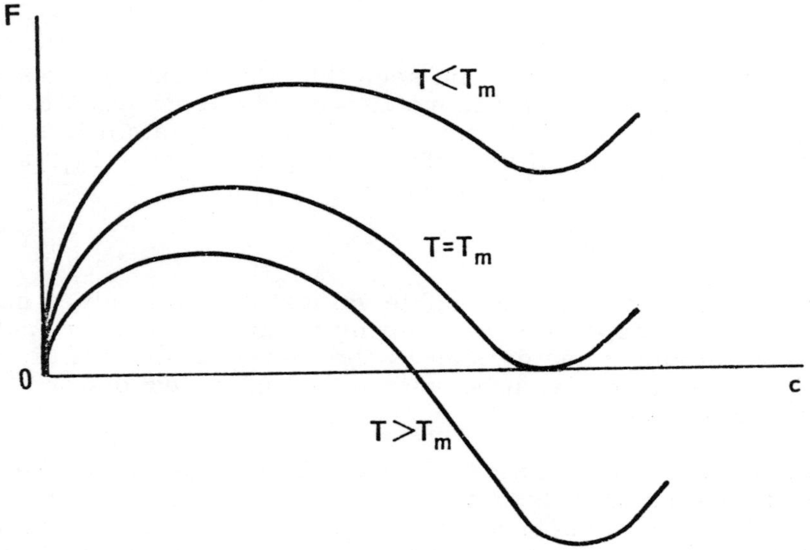

Figure 2. The free energy of a crystal containing a concentration c of dislocations, including the increase of dislocation energy caused by the dislocation-induced dilation of the lattice [8, 9].

fact that there must always be some atoms in the core for which the coordination number must be lower than for the perfect lattice. This means that the volume of the assembly must always be increased by the introduction of dislocations. Hence the stretching of interatomic bonds which causes F to ultimately rise with increasing c will always occur.

3. STATUS

In this commentary a somewhat arbitrary division will be made between the many publications which have dealt with the dislocation model. Those quoted heretofore could, however, be said to have established the model's feasibility. We turn now to papers which have described results which endorse the model, and which have thus contributed to its present status; a situation in which the dislocation theory of melting appears increasingly tenable but not, as yet, incontrovertibly verified.

The melting process is difficult to study experimentally because it occurs over such a brief interval of time. The newly-acquired technique of computer simulation by molecular dynamics is ideally suited to resolve the problem, because it presents the unique combination of a system observable at the atomic level and on a time scale comparable to the natural processes occurring in a crystal. Efforts in this direction have now been recorded for several years, and a brief enumeration of their salient results will suffice. The first simulational study was made on a two-dimensional Lennard-Jones system [10], and it revealed that melting proceeds through the spontaneous proliferation of dislocation dipoles. As far as could be discerned, the transition was sharp, and the behavior expected of a first-order phase change was observed. On the basis of this result, it was predicted [11] that melting of a three-dimensional (face-centered cubic) crystal would occur through the generation of loops of Shockley partial dislocation, and this was in fact observed [12] by the same simulational technique. An analogous study of a very thin crystal produced a depression of the melting point and also evidence that the dislocations are preferentially generated at the crystal surface [13]. This result is of interest in that crystals are normally observed to melt at their surfaces.

Although, as alluded to earlier, experiments on melting are difficult, some reports of such studies have now appeared. Crawford [14] has studied rubidium within a few millikelvin of T_M, by neutron scattering, and has found evidence of incipient dislocation generation [15]. Similarly, an electron diffraction study of aluminum [16], in which the diffraction pattern was filmed as the specimen was taken through the melting transition, produced evidence of Shockley partial dislocation generation.

The ultimate product of the melting transition is, of course, a liquid, and the instantaneous structure of that state is of obvious interest in connection with the dislocation model. Two recent studies have aimed at elucidating this structure in both two-dimensional and three-dimensional liquids. In the two-dimensional case [17], time-lapse plotting revealed that the instantaneous distribution of vibrating atoms and translating atoms was heterogeneous as shown in Figure 3. The liquid is divided, at any instant, into groups of atoms, comprising between about twenty and fifty atoms, which move with respect to one another in a manner quite reminiscent of grains in a creeping polycrystal. The similarity to the original conjecture of Mott and Gurney is striking. In a similar study of a three-dimensional liquid [18], the instantaneous distribution of free volume was probed, and it was discovered that this is quite heterogeneous. Concatenations of holes were seen (see Figures 4a and 4b), and these were assumed to be the remnant dislocation cores that had been referred to as pseudodislocations in an attempt [19] to unify melting, liquids, and the glassy

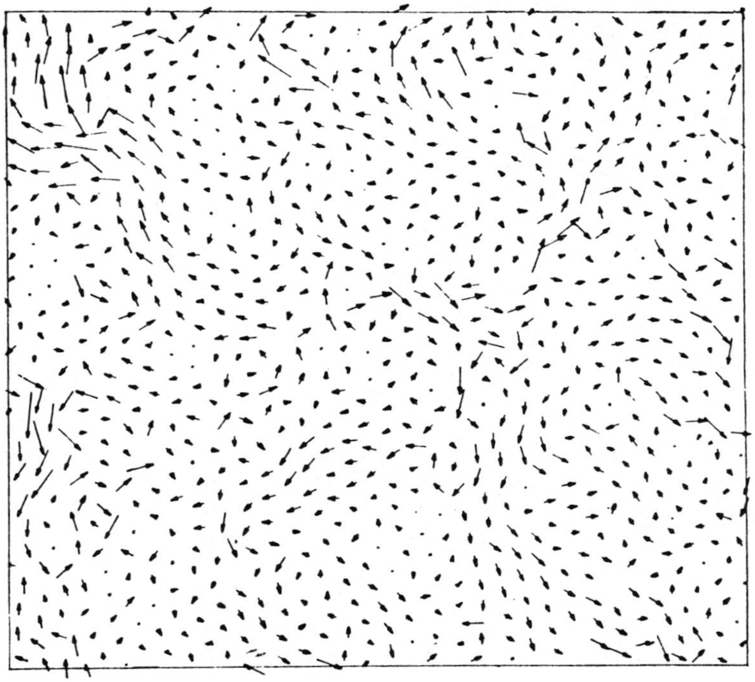

Figure 3. The net displacement of atoms in a two-dimensional Lennard-Jones liquid during a period approximately equal to five vibrational periods. Notice that the atoms appear to move in groups.

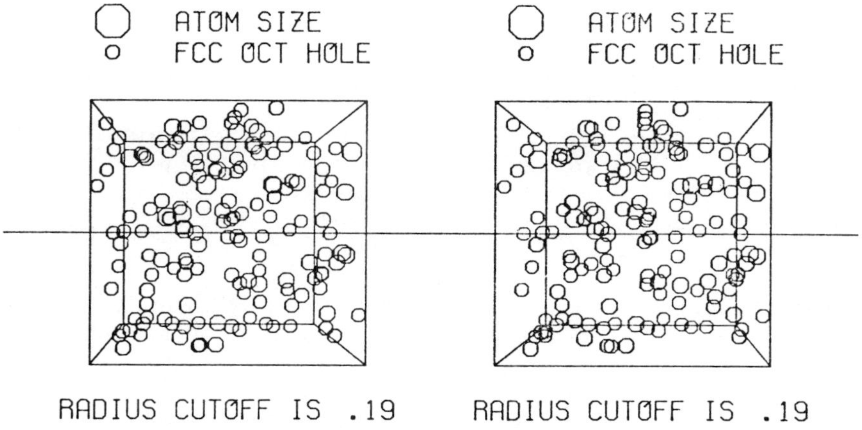

(a)

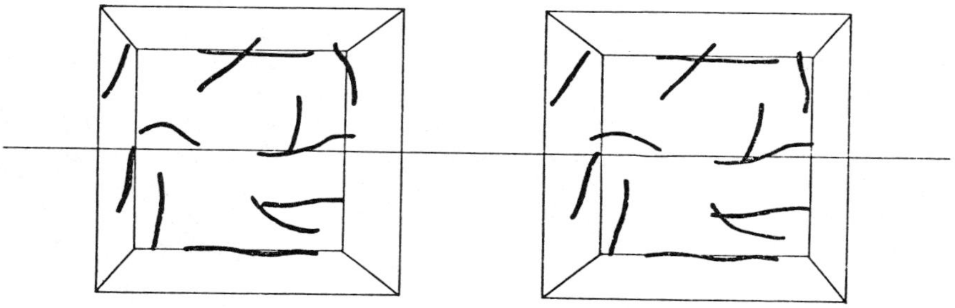

(b)

Figure 4. (a) Stereo view of the instantaneous distribution of free volume, shown as inscribed spheres, in a simulated three-dimensional liquid. The box indicates the positions of the periodic boundaries. Chains of holes, appearing as isolated segments rather than a connected network, are observed. Their positions are indicated by the lines in (b). These segments are mobile and persist for times brief compared with the vibration period.

state (an idea which has recently been further developed by Edwards and Warner [20]). Although these latest simulations are still somewhat preliminary, they do appear to constitute the first hint of dislocation-like heterogeneities in the liquid state.

4. PROGNOSIS

There is now an appreciable body of evidence implicating the dislocation in the melting transition. This has come primarily from computer simulation, but also partly from experiments on metals. There is obviously a pressing need for further experiments, and it might transpire that the newly-developed synchrotron radiation techniques, giving as they do very brief pulses, are well suited to this purpose.

One of the chief sources of resistance to the dislocation theory of melting has been its requirement that a liquid be regarded as a crystal saturated with dislocations. Not surprisingly, this is frequently regarded with anathema (see, for example, Frank and Steeds [21]). One source of misunderstanding concerns the Burgers vector. It is often argued that dislocations cannot exist in a liquid because one cannot define the Burgers vector. This is probably not so problematic as it might seem, since dislocations in a liquid should rather be regarded as being of the more general type treated by Volterra and the Italian school at the beginning of the century. Both Gilman [22] and Ashby and Logan [23] have suggested ways of circumventing the difficulty, the former by proposing that the Burgers vector might not be constant along a dislocation in the non-crystalline state. It is possible that one must even give up the idea of a definite Burgers vector at any given point on a dislocation line in a liquid and replace it by a "Burgers vector envelope" in which one instead defines the most probable Burgers vector and a modulus of spread. This envelope might become totally spread out only at the critical point, at which point the dislocation concept would finally break down.

Of related recent developments, mention must be made of Weisskopf's elegant demonstration [24] that the atoms in a liquid must move around in groups. Here again one sees independent endorsement of the original Mott-Gurney idea.

This article has been concerned primarily with three-dimensional melting. There has, however, been an interesting development in the two-dimensional case. Kosterlitz and Thouless [25] have given a renormalization group treatment of the many-dipole situation and related this to two-dimensional melting. The approach was extended by Halperin and Nelson [26], who showed that a liquid-crystal like phase, referred to as a "hexatic fluid phase", should intervene between the crystalline and isotropic liquid states. The first-order melting transition is then replaced by two second-order transitions. This prediction has now actually been verified by computer simulation [27].

In closing, it would appear that there are now two aspects of this topic which are in particular need of development. The first, and no doubt most difficult, concerns the question of dislocations in the liquid state. Efforts should be directed towards the measurement of instantaneous anisotropy in liquids. It is possible that the newly-developed technique of XAFS, with its potential for probing the distributions of bond distances, will be of some use in this respect. Finally, there is the question of the analytical treatment of a situation in which the traditional continuum approach is not valid. At the very high dislocation concentrations envisaged for the liquid state, inclusion of discrete and anharmonic effects is unavoidable. In this respect, a lead can be taken from Toda's [28] treatment of the anharmonic lattice, which revealed the importance of solitons at high vibrational amplitudes. The link with the dislocation model of melting has recently been discussed [29], the essential equivalence of solitons and dislocations being already evident in the work of Seeger and Kochendörfer [30]. It is particularly encouraging to note [29] that the soliton approach is consistent with what is probably the best known of all pieces of evidence regarding melting, namely the empirical rule of Lindemann [31].

References

[1] N.F. Mott and R.W. Gurney, Trans. Faraday Soc. 35, 364 (1939).

[2] W.L. Bragg, Symposium on Internal Stresses (Institute of Metals, London, 1947) p. 221.

[3] R.M.J. Cotterill and M. Doyama, Phys. Rev. 145, 465 (1966).

[4] W. Shockley, l'Etat Solide (Inst. International de Physique Solvay, Brussels, 1952) p. 431.

[5] A. Ookawa, J. Phys. Soc. Japan 15, 70 (1960)

[6] S. Mizushima, J. Phys. Soc. Japan 15, 70 (1960)

[7] D. Kuhlmann-Wilsdorf, Phys. Rev. 140, A1599 (1965)

[8] S.F. Edwards, Polymer 17, 933 (1976)

[9] R.M.J. Cotterill (to be published)

[10] R.M.J. Cotterill and L.B. Pedersen, Solid State Communications 10, 439 (1972)

[11] R.M.J. Cotterill, in High Temperature Materials Phenomena, ed. J.G. Rasmussen (Polyteknisk Forlag, 1972) p. 285

[12] R.M.J. Cotterill, W. Damgaard Kristensen, and E.J. Jensen, Phil. Mag. 30, 245 (1974)

[13] R.M.J. Cotterill, Phil. Mag. 32, 1283 (1975)

[14] R.K. Crawford, Bull. Amer. Phys. Soc. 24, 385 (1979)

[15] R.K. Crawford (to be published).

[16] R.M.J. Cotterill and J. Klæstrup Kristensen, Phil. Mag. 36, 453 (1977).

[17] R.M.J. Cotterill, Physica Scripta (in the press).

[18] R.M.J. Cotterill, Phys. Rev. Letters (in the press).

[19] R.M.J. Cotterill, E.J Jensen, W. Damgaard Kristensen, R. Paetsch, and P.O. Esbjørn, J. de Physique 36, C2-35 (1975).

[20] S.F. Edwards and M. Warner (to be published).

[21] F.C. Frank and J.W. Steeds, in The Physics of Metals 2, ed. P.B. Hirsch (Cambridge University Press, 1975) p. 68.

[22] J.J. Gilman, J. Appl. Phys. 44, 675 (1973).

[23] M.F. Ashby and J. Logan, Scripta Metall. 7, 513 (1973).

[24] V.F. Weisskopf, Trans. New York Acad. Sci. 38, 202 (1977).

[25] J.M. Kosterlitz and D.J. Thouless, J. Phys. C 6, 118 (1973).

[26] B.I. Halperin and D.R. Nelson, Phys. Rev. Letters 41, 121 (1978).

[27] D. Frenkel and J.P. McTague, Bull. Amer. Phys. Soc. 24, 362 (1979).

[28] M. Toda, J. Phys. Soc. Japan 22, 431 (1967).

[29] R.M.J. Cotterill, Physica Scripta 18, 37 (1978).

[30] A. Seeger and A. Kochendörfer, Z. Phys. 130, 321 (1951).

[31] F.A. Lindemann, Z. Phys. 11, 609 (1910).

THE KOSTERLITZ-THOULESS THEORY OF TWO-DIMENSIONAL MELTING

A.P. Young
Dept. of Mathematics
Imperial College
London SW7

This lecture is intended as an elementary introduction to a theory of two-dimensional melting proposed by Kosterlitz and Thouless[1]. Their theory is also applicable, with minor differences, to the two-dimensional planar spin model and since this is somewhat simpler we shall discuss it first.

In my other lectures at this school I pointed out that two-dimension rotationally invariant spin models with n component spins had no phase transition at finite temperature if $n > 2$. We shall now discuss why the case $n = 2$ is special and does have a transition at finite temperature, albeit of a rather unusual type. Devoting the angle of the spin vector relative to some reference axis by θ the Hamiltonian is given by

$$\mathcal{H} = -J \sum_{\langle ij \rangle} \cos(\theta_i - \theta_j) \tag{1}$$

where the sum is over neighbouring pairs on a square lattice, say. Since we shall mainly discuss the behaviour of this model at low temperatures where spins on neighbouring sites are strongly correlated we expand out the cosine and keep only the quadratic term. For fluctuations of wavelength much longer than a lattice spacing, a_0, one can make a continuum approximation, replacing the discrete variables θ_i by a continuous function $\theta(\vec{r})$. The Hamiltonian may then be written as

$$\mathcal{H} = \frac{J}{2} \int (\partial_\mu \theta)^2 d^2r \tag{2}$$

If, finally, one neglects the fact that $\theta + 2\pi n$ is equivalent to θ for n integer and extends the range of integration from $-\infty$ to ∞ one has

$$Z = \int_{-\infty}^{\infty} \mathcal{D}\theta \; e^{-\frac{J}{2k_BT}\int d^2r(\partial_\mu\theta)^2} \tag{3}$$

which can readily be evaluated since it only involves products of Gaussian integrals. The Hamiltonian described by (3) represents what is known as the spin wave approximation to the model and was first discussed by Wegner[2].

Of greater interest than the partition function is the correlation function $C(\vec{r})$ defined by

$$C(\vec{r}) = \langle \exp[i(\theta(\vec{r}) - \theta(0))] \rangle. \tag{4}$$

Because the Hamiltonian is quadratic in θ one can replace (4) by

$$C(\vec{r}) = \exp[-\tfrac{1}{2}\langle(\theta(\vec{r}) - \theta(0))^2\rangle] \tag{5}$$

just as for a Debye-Waller factor. From (2) it follows that

$$\langle[\theta(\vec{r}) - \theta(0)]^2\rangle = \frac{2k_BT}{J}\int \frac{d^2k}{(2\pi)^2}\frac{[1 - e^{i\vec{k}.\vec{r}}]}{k^2} \tag{6}$$

where the wavevector integral is cut off beyond $k \sim 1/a_0$ to represent the discrete lattice in the original model. For $r \gg a_0$ one finds

$$\langle[\theta(\vec{r}) - \theta(0)]^2\rangle = \frac{k_BT}{\pi J}\ln(r/a_0) \tag{7}$$

which immediately shows there is no long range order. This can, in fact, be proved rigorously[3]. Inserting (7) into (5), the spin-spin correlation function becomes

$$C(\vec{r}) = r^{-\eta} \qquad (r \gg a_0) \tag{8}$$

where

$$\eta = \frac{k_BT}{2\pi J} \tag{9}$$

Equation (8) is rather unexpected. Since there is no long range order one would naively expect the correlations to decay exponentially with a finite correlation length, as in a conventional paramagnetic phase. Here, by contrast, we have a much slower, power law decay. It is instructive to note that power law decay is characteristic of a conventional second order phase transition precisely at its critical point, fig. 1a. In this case the exponent, η, is universal for a particular class of systems. One can therefore view the spin wave approximation to the $n = d = 2$ situation as a *line* of critical points, fig. 1b, extending from $T = 0$ to $T = \infty$, each critical point characterized by its own, non-universal, exponent η.

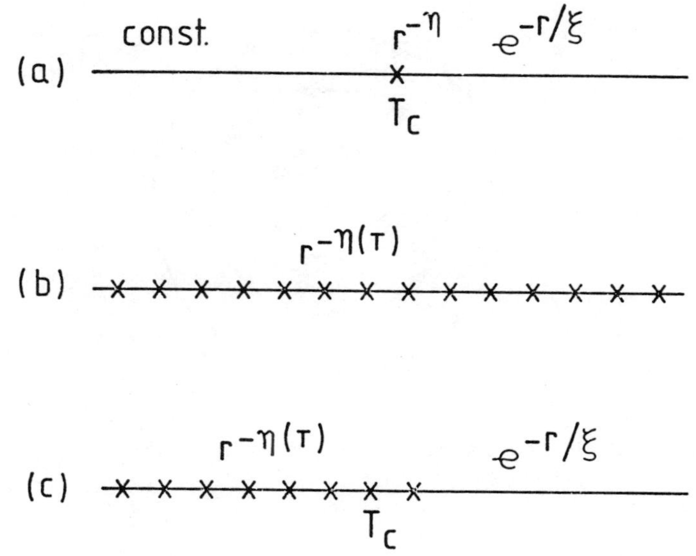

Fig. 1. Qualitative behaviour of correlations at different temperatures. The horizontal line represents temperature increasing to the right. (a) conventional second order phase change, (b) two-dimensional planar spin model in spin wave approximation where every temperature corresponds to a critical point, (c) same as (b) but in Kosterlitz-Thouless theory. The line of critical points now terminates at a finite T_c.

Although the approximations made in deriving the spin wave theory seem reasonable at low temperatures they cannot be trusted when the temperature is high and indeed, it is highly implausible that correlations only decay with a power law up to $T = \infty$. Much more likely is a *termination* of the line of critical points at a finite temperature T_c above which the correlations would decay exponentially. The 'phase transition' in the two-dimensional planar model therefore corresponds to a change from power law to exponential decay of correlations.

Kosterlitz and Thouless proposed that a finite T_c would occur if certain singular spin configurations called vortices, neglected in the spin wave approximation, were included. Because θ is a multivalued function it is possible that a line integral of the type

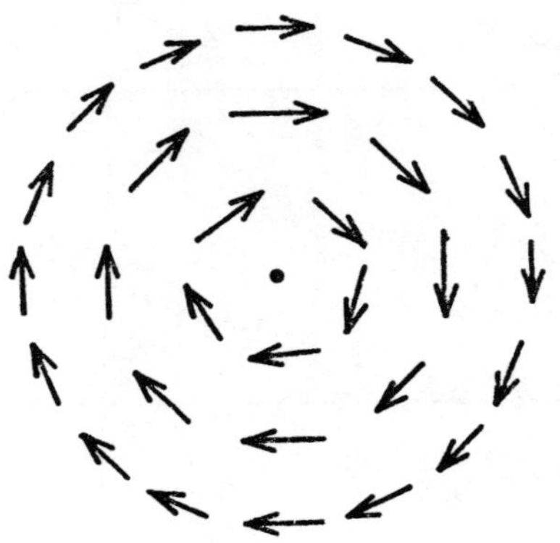

Fig. 2. A sketch of a vortex with winding number unity.

$\oint \vec{\nabla}\theta . d\vec{\ell}$ around a closed contour is not zero but is equal to $2\pi n$, where n, an integer, is called the winding number. An example is shown in fig. 2 for n = 1. A contour with $n \neq 0$ must enclose at least one singularity known as a vortex. For the configuration in fig. 2, $|\vec{\nabla}\theta| = 1/r$, so the energy of a single isolated vortex is given by

$$E_{\text{1-vortex}} = \frac{1}{2} J \int (\partial_\mu \theta)^2 d^2 r = \pi J \ln(L/a_0), \tag{10}$$

where L is the linear dimension of the system. A single vortex therefore costs an infinite amount of energy in a macroscopic system. However the energy of a pair of opposite vortices (i.e. one with n = 1, the other with n = -1) is finite and

$$E_{\text{pair}} = 2\pi J \ln(r/a_0), \tag{11}$$

where r is the separation of a pair. At low temperatures there will consequently be a finite density of tightly bound pairs of vortices but no isolated vortices. Kosterlitz and Thouless then argue that at sufficiently high temperature isolated vortices will appear because of the resulting gain in entropy and identify this temperature with the T_c which characterizes the termination of the line of critical points. The entropy of a single isolated vortex is $2k_B \ln(L/a_0)$ so its free energy is given by

$$F_{\text{1-vortex}} = (\pi J - 2k_B T) \ln(L/a_0) \tag{12}$$

which is negative for $k_BT > \pi J/2$. Later on we will see that, with minor reinterpretation, this estimate of

$$k_BT_c = \pi J/2 \tag{13}$$

is numerically correct.

We have pointed out that the phase transition is characterized by a change from power law to exponential decay of correlations. It is instructive to look at this from a somewhat different point of view. At a conventional phase transition, which occurs in *three* dimensions, for example, there is an order parameter M which vanishes as $|t|^\beta$ when the reduced temperature t tends to zero. In addition the transverse susceptibility varies as Ck^{-2} for $k \to 0$ below T_c as a consequence of a broken continuous symmetry. This is discussed for example in my other lectures here. If one defines $S = M^2C$ then S denotes the stiffness with respect to changing the angle θ of the order parameter, i.e. the change in free energy of a twist is,

$$\Delta F = \frac{1}{2} S \int (\vec{\nabla}\theta)^2 d^dr \,. \tag{14}$$

S also vanishes at T_c but with an exponent $2\beta-\eta\nu$, [4]. We have therefore two quantities S and M which vanish at T_c and remain zero above, at a conventional transition. In two dimensions rigorous results[3] prove that M = 0 but make no statement about S. Indeed, within the spin wave approximation, (3), we have S = J, despite the fact that the same approximation, through (7), indicates M = 0. The Kosterlitz-Thouless theory predicts that S is reduced from J at finite temperature because of vortex pairs and becomes zero once free vortices appear at T_c. One result of the Kosterlitz-Thouless is that equations (9) and (13) which give the exponent η and the transition temperature T_c should actually have J replaced by the stiffness S. In other words

$$S(T_c^-) = \frac{\pi}{2} k_BT_c$$

so the ratio of the stiffness as $T \to T_c^-$ to the transition temperature has the *universal*[12] value of $\pi/2$ and the exponent η as $T \to T_c^-$ is also universal and equal[11] to 1/4.

The classical planar model may be used as model of superfluid He^4, in which case θ is the phase of the condensate wave function and $S = \rho_s(\hbar/m)^2$ where m is the mass of a helium atom and ρ_s is the superfluid density, the density of fluid which flows without resistance. Thus one can have a superfluid transition in two-dimensional He^4 without long-range order. Experiments have recently been carried out on helium films[5] which support this contention and confirm a number of specific predictions of the Kosterlitz-Thouless theory.

Before discussing in more detail the theory of the transition we shall go back and relate these ideas to the problem of melting in two dimensions. The analogue of our spin wave approximation is continuum elasticity theory, according to which the Hamiltonian describing long wavelength elastic deformation of a lattice is

$$\mathcal{H} = \int d^2r[\frac{\lambda}{2}(\phi_{\alpha\alpha})^2 + \mu\phi_{\alpha\beta}\phi_{\alpha\beta}] \tag{15}$$

where μ is the shear elastic constant, λ is a Lame coefficient related to the bulk modulus, B, by $B = \lambda + \mu$, (but see the discussion on this in ref. 7), and $\phi_{\alpha\beta}$ is the conventional symmetrized strain tensor,

$$\phi_{\alpha\beta} = \frac{1}{2}(\partial_\alpha u_\beta + \partial_\beta u_\alpha) \tag{16}$$

where $\vec{u}$ is the displacement. Equation (16) is correct for an isotropic continuum and also for a triangular lattice but is not valid for a square lattice, the symmetry of which requires a third elastic constant. We shall mainly be interested in the triangular lattice because it corresponds to possible experimental situations. The density is defined by

$$\rho_{\vec{k}} = \sum_i e^{i\vec{k}.\vec{R}_i} \tag{17}$$

where $\vec{R}_i = \vec{r}_i + \vec{u}_i$ is the position of the ith atom and $\vec{r}_i$ is the position of the i-th lattice site in a state of lowest potential energy. Of interest is the structure factor $S(\vec{k})$ where

$$S(\vec{k}) = \frac{1}{N}\langle\rho_{\vec{k}}\,\rho_{-\vec{k}}\rangle$$

$$= \sum_{\vec{r}} e^{i\vec{k}.\vec{r}}\, C_k(\vec{r}) \tag{18}$$

and

$$C_k(\vec{r}) = \langle\exp[i\vec{k}.(\vec{u}(\vec{r}) - \vec{u}(0)]\rangle \tag{19}$$

is the analogue of $C(\vec{r})$ in eq. (4). The summation in (18) is over all lattice sites. Within continuum elasticity theory the average in equation (19) can be evaluated. One finds that $\langle[\vec{u}(\vec{r})-\vec{u}(0)]^2\rangle$ diverges logarithmically, as in eq. (7), and likewise demonstrates the absence of long range order, i.e.

$$\langle\rho_{\vec{G}}\rangle = 0\ . \tag{20}$$

except for $\vec{G} = 0$. This result can be proved rigorously[6] and is not simply a consequence of our approximations. $C_G(\vec{r})$ varies as $r^{-\eta_G}$ where

$$\eta_G = \frac{|\vec{G}|^2}{4\pi} \cdot k_B T \cdot \frac{3\mu + \lambda}{\mu(2\mu + \lambda)} \tag{21}$$

(see for example ref. 7).

It is clear that continuum elasticity theory is the precise analogue of the spin wave approximation. What then are the analogue of vortices? Singularities in the displacement can arise, just as for the angle field $\theta(r)$, because it is a multivalued function. Replacing $\vec{u}$ by $\vec{u} + \vec{a}$ where a is a lattice vector describes the same situation. Consequently the integral $\oint \partial_i \vec{u} d\ell_i$ around a closed loop can equal a lattice vector and is not necessarily zero. This is known as a Burger's vector, written as $\vec{b}$, and the corresponding singularity is called a dislocation. It is illustrated in fig. 3 for a square lattice and involves inserting an extra half line of atoms in a direction at right angles to the Burger's vector.

The energy of an isolated dislocation corresponding to a Burger's vector of one lattice spacing is given by eq. (10) with

$$J = \frac{1}{2\pi^2} \frac{\mu(\mu + \lambda)}{2\mu + \lambda} a_0^2 \tag{22}$$

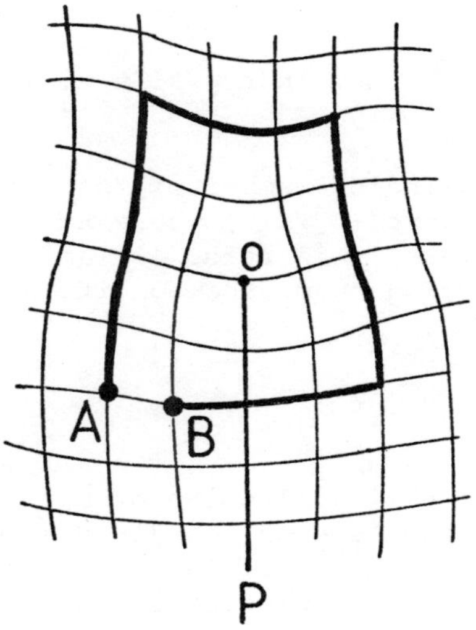

Fig. 3. A sketch of a dislocation on a square lattice. An extra half-line of atoms OP has been inserted into an otherwise perfect crystal. The solid line indicates a path which would close in a perfect crystal and the mismatch $\vec{AB}$ is called the Burger's vector.

and, at large distances, the energy of a pair with opposite Burger's vector is given by (11). Consequently the energy-entropy balance argument goes through as before and free dislocations are only expected for $T > T_c$ where $k_B T_c = \pi \tilde{J}/2$.

It was earlier pointed out in the context of the planar spin model that the Kosterlitz-Thouless transition is associated with vanishing stiffness for fluctuations in the transverse spin component. The analogue in melting is the shear modulus, which controls transverse strain fluctuations. Consequently the low temperature phase has no long range order but is 'solid' since the shear modulus is finite. The high temperature phase has a vanishing shear modulus so in this sense we can say the transition corresponds to two-dimensional melting.

After these long introductory remarks let us consider the theory of the transition in more detail. Again for simplicity we consider first of all the spin model. Kosterlitz and Thouless assume that the small oscillations (spin waves) superimposed on top of any vortex configuration have the same energy as when there are no vortices. The Hamiltonian thus breaks up into independent spin wave and vortex parts so the vortex Hamiltonian can be considered in isolation. This neglect of spin wave-vortex interactions was later shown by Villain[8] to be exact for a special type of spin interaction and his ideas have been elaborated on by José et al[9]. The vortex Hamiltonian can be written as

$$\overline{H} = \beta \mathcal{H} = \frac{2\pi J}{k_B T} \sum_{<ij>} \ln(r_{ij}/a_0) n_i n_j + \ln y \sum_i n_i^2 \qquad (23)$$

where $n_i = \pm 1$ is the winding number of the i-th vortex, $r_{ij} = |\vec{r}_i - \vec{r}_j|$ and $y = \exp(-E_c/k_B T)$ where E_c is a core energy and is discussed in some detail in refs. [1] and [9]. To avoid extra logarithmically diverging terms the condition

$$\sum_i n_i = 0 \qquad (24)$$

must be imposed.

Equation (23) is the Hamiltonian of a two-dimensional classical Coulomb gas since Poisson's equation gives a logarithmic interaction in two dimensions. The number of charges is not fixed so y is equivalent to the fugacity in a grand canonical ensemble and eq. (24) corresponds to charge neutrality.

In all problems with Coulomb forces we know that screening is important. If for example we consider a test pair of charges separated by a rather large distance r, fig. 4, then it will tend to orient smaller pairs separated by r' which come into its field of force and thereby reduce the force between the test pair from its

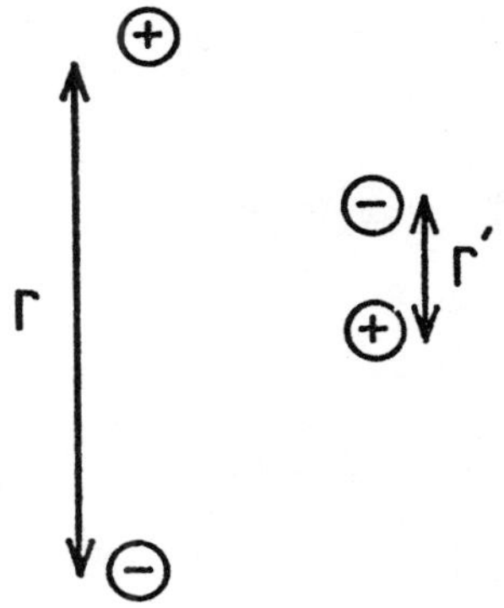

Fig. 4. A test pair of charges separated by r is screened by smaller dipoles of separation r'.

unscreened value of $2\pi J/r$, to a screened value of $2\pi J/\varepsilon r$ where ε is a dielectric constant. The number of smaller pairs which participate in the screening depends on the separation of the test pair so ε is expected to depend on r. The concept of a scale dependent dielectric constant $\varepsilon(r)$ is of crucial importance in the Kosterlitz-Thouless theory.

Kosterlitz and Thouless treat the smaller dipoles as a continuous medium so they can use dielectric theory, according to which

$$\varepsilon(r) = 1 + 4\pi\chi(r) \tag{25}$$

where

$$\chi(r) = \int_a^r n(r')\alpha(r')dr' \tag{26}$$

is the susceptibility of all pairs of separation less than r, n(r) is the number of pairs of separation r and α is their polarizability. The integral in (26) stops at r because the larger pairs polarize the smaller ones but not vice-versa.

$\alpha(r')$ is given by

$$\alpha(r') = \frac{1}{2}\frac{\pi J}{k_B T}(r')^2 \tag{27}$$

while, for $y \ll 1$, n(r') is equal to

$$n(r') = 2\pi r' y^2 e^{-V(r')/k_B T} \tag{28}$$

where $V(r')$ is the energy of the pair separated by r', and is modified from $2\pi J \ln(r'/a_0)$ by the screening effect of still smaller pairs. $V(r')$ is obtained by integrating up the force so

$$V(r') = 2\pi J \int_{\ln a_0}^{\ln r'} \frac{d \ln r''}{\varepsilon(r'')} \tag{29}$$

Combining equations (25) to (29) yields a rather complicated self-consistent integral equation for $\varepsilon(r)$. This can be cast in a more transparent form[10] by defining

$$K(\ell) = \frac{J}{k_B T \, \varepsilon(r)} \tag{30a}$$

with

$$\ell = \ln(r/a_0) \tag{30b}$$

and

$$y(\ell) = y\left(\frac{r}{a_0}\right)^2 e^{-V(r)/kT} \tag{30c}$$

so $k_B T \times K(\ell)$ is the effective stiffness including the screening effects of all vortex pairs of separation less than $a_0 e^{\ell}$. One then obtains the following pair of coupled differential equations

$$\frac{d}{d\ell} K^{-1}(\ell) = 4\pi^3 y^2(\ell) \tag{31a}$$

$$\frac{dy(\ell)}{d\ell} = [2 - \pi K(\ell)] y(\ell) \tag{31b}$$

first obtained by Kosterlitz[11] using a renormalization group method. These equations have to be solved for $K(\ell=\infty)$, which is ($1/k_B T$ times) the renormalized stiffness including all vortex effects, subject to the boundary condition that $y(\ell=0) = y$, $K(\ell=0) = J/k_B T$. Some trajectories are shown schematically in fig. 5 together with a possible locus of starting values. Of great importance is the straight line PC which represents a trajectory terminating at C ($K^{-1}(\infty) = \pi/2$, $y(\infty) = 0$). P is where this trajectory crosses the locus of starting values. If the starting point is to the left of P the trajectory ends on the horizontal axis and the renormalized stiffness is finite but reduced somewhat from its spin wave value of J by polarization of vortex pairs. However if the starting point is to the right of P then K goes to zero as $\ell \to \infty$ so the renormalized stiffness vanishes. We can therefore identify P with the critical point of the model. Notice that $K(\ell=\infty)$ is equal to the universal[12] value of $2/\pi$ at $T = T_c^-$ but vanishes above T_c. Furthermore for T just below T_c it is not difficult to show that

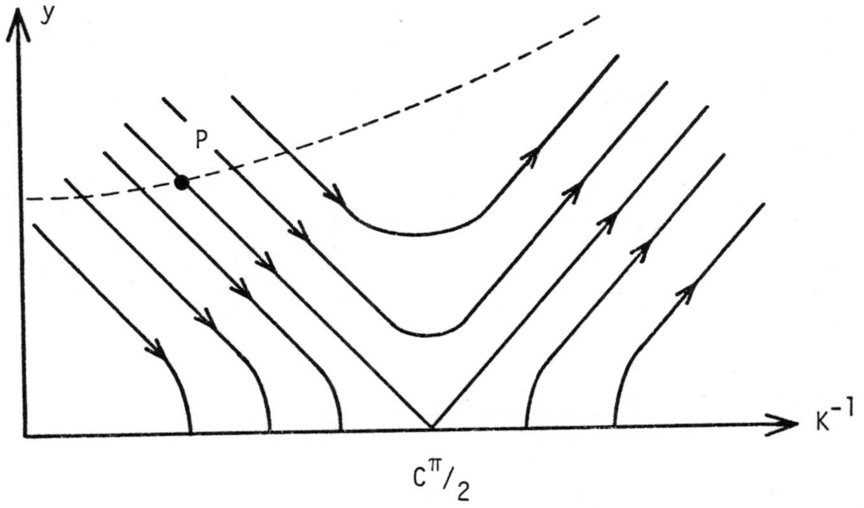

Fig. 5. Sketch of the solutions of equations (31) in the vicinity of $y = 0$, $K^{-1} = \pi/2$. The dashed line is a typical locus of initial values.

$$K(\infty) = \frac{2}{\pi} + C|t|^{\bar{\nu}} \tag{32}$$

where

$$\bar{\nu} = \frac{1}{2} \tag{33}$$

and C is a non-universal constant (it depends on y). The stiffness therefore varies close to T_c as shown in fig. 6. Some experimental verification of this has now been obtained from experiments[5] on He^4 films.

Other predictions of the theory are (i) only an essential singularity $\sim e^{-A\pm/|t|^{\bar{\nu}}}$ in the specific heat and (ii) an exponentially increasing correlation length above T_c, i.e. $\xi \sim e^{B/t^{\bar{\nu}}}$. Point (i) is not inconsistent with the jump in stiffness at T_c because this only occurs on infinite length scales. Energy, involving local correlations, is given by an integral of fluctuations over all length scales which leads to a much weaker singularity.

The application of this theory to melting does not involve many great changes. Some of the differences are

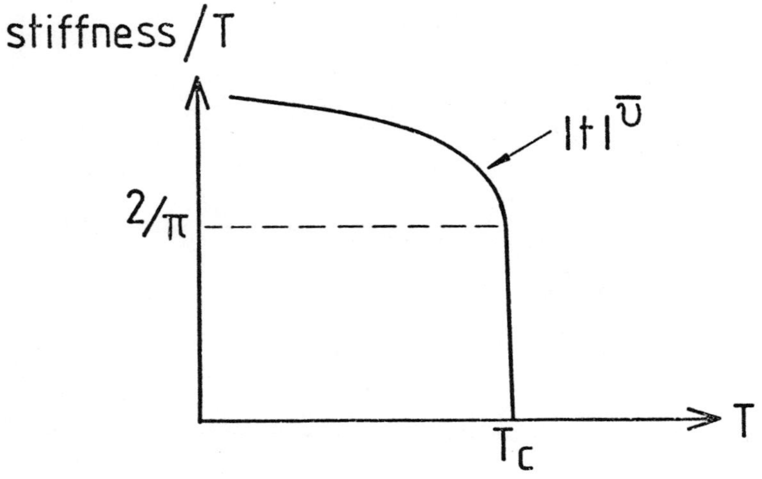

Fig. 6. Sketch of the renormalized stiffness (divided by k_BT) obtained from equations (31) of the text, in the region close to T_c.

(i) The shear modulus has a jump at T_c but this is non-universal because J in eq. (22) involves λ as well as μ.

(ii) The exponent $\bar{\nu}$ is equal[7,13] to 0.3696 ... for a triangular lattice but is still probably 1/2 for a square lattice. The difference arises from the possibility of combining three Burgers vectors to satisfy the neutrality condition, $\Sigma\vec{b} = 0$, on the triangular lattice. This leads to another term in equations (31).

(iii) If one allows for a substrate potential which gives a preferred orientation[7] to the adsorbate but does not pin the atomic positions then $\bar{\nu}$ is non-universal and depends on the strength of the orientational bias. The effect is, however, rather small[13]. Substrate interactions will be discussed in detail in the lectures by Villain.

(iv) The phase above T_c is actually not a liquid[7] but more like a liquid crystal with power law decay of *angular* correlations. A second transition, of similar type, involving unbinding of disinclinations is required to complete the melting process.

Equations (31) represent the first term in an expansion in y and so should be satisfactory if the density of pairs is very small at T_c. If this condition is not met one cannot rule out some other type of behaviour such as a first order transition. Experiments and computer simulations would be most welcome in sorting this out (see the seminar by D. Frenkel at this school).

REFERENCES

1. J.M. Kosterlitz and D.J. Thouless, J. Phys. C 6, 1181 (1973).
2. F.J. Wegner, Z. Phys. 206, 465 (1967).
3. N.D. Mermin, Phys. Rev. 176, 250 (1968).
4. B.D. Josephson, Phys. Letters 21, 608 (1966).
5. D.J. Bishop and J.D. Reppy, Phys. Rev. Lett. 40, 1727 (1978).
6. N.D. Mermin, J. Math. Phys. 8, 1061 (1967).
7. B.I. Halperin and D.R. Nelson, Phys. Rev. Lett. 41, 121; ibid. p.519; also D.R. Nelson and B.I. Halperin, Phys. Rev. B (to be published).
8. J. Villain, J. Phys. (Paris) 36, 581 (1975).
9. J.V. Jose, L.P. Kadanoff, S. Kirkpatrick and D.R. Nelson, Phys. Rev. B16, 1217 (1977).
10. A.P. Young, J. Phys. C 11, L453 (1978).
11. J.M. Kosterlitz, J. Phys. C 7, 1046 (1974).
12. D.R. Nelson and J.M. Kosterlitz, Phys. Rev. Lett. 39, 1201 (1977).
13. A.P. Young, Phys. Rev. B (to be published).

PHASE TRANSITIONS AND ORIENTATIONAL ORDER IN A TWO DIMENSIONAL LENNARD-JONES SYSTEM

Daan Frenkel, Frank E. Hanson and John P. McTague

University of California
Department of Chemistry
Los Angeles, CA 90024

It has long been suspected that the solid-fluid transition in two dimensions might be rather different than its 3-D counterpart, because of the lack of translational order in 2-D solids. Recently, a detailed theory of 2-D melting has been put forward by Halperin and Nelson.[1] This theory provides a picture of 2-D melting that is indeed very different from what is observed in the three dimensional world. In particular, Halperin and Nelson (henceforth referred to as HN) make the intriguing prediction that, if 2-D melting is not a first order transition, then two second order transitions are required to go from the solid to the isotropic fluid phase. The solid and isotropic fluid phases will be separated by a peculiar liquid crystal-like phase which exhibits short range translational order but long range "orientational" order.

Experiments on the melting of adsorbed monolayers[2,3] strongly suggest that for some of the systems studied the solid-fluid transition is in fact continuous. Unfortunately, direct experimental verification of the predictions based on the HN theory is non trivial. This is so because the quantities that are probed most readily in experiments are expected to exhibit only subtle (and for all practical purposes, unobservable) changes at the phase transitions. On the other hand, physical properties which, according to the HN theory, should change markedly at the phase transitions, turn out to be very hard to measure experimentally.

In order to gain insight into the behavior of the different quantities that play a central role in the HN theory, we chose to perform a computer "experiment" (molecular Dynamics) on a collection of two dimensional Lennard-Jones (12-6) atoms. This particular system seemed a suitable candidate for detailed investigation because the results of earlier calculations by Hanson and McTague[4] indicated that the temperature dependence of the thermodynamic and structural properties of 2-D Lennard-Jonesium closely reproduce those of argon on graphite. This latter system appears to have a continuous melting transition. Although the 2-D Lennard-Jones system is quite similar to argon on graphite, it differs in some respects from the model for 2-D matter used in the HN theory. In the first place, the HN theory describes two dimensional matter as an elastic continuum with embedded dislocations, whereas the calculations are performed on a collection of interacting particles. Secondly, periodic boundary conditions are used in the MD calculations. As a consequence, all fluctuations with wavevector $k<2\pi/L$ (L = boxlength) are excluded. In contrast, long wavelength fluctuations play an essential role in the HN renormalization group treatment of the two dimensional phase transitions. The box length used in the MD calculations is, however, of the same order of magnitude as the effective size of 2-D argon crystallites grown on exfoliated graphite surfaces.

Molecular Dynamics calculations were performed on a 256 particle Lennard-Jones system at a reduced density $\rho^* = \rho\sigma^2 = 0.8$ and at reduced temperatures ranging from $T^* = kT/\varepsilon = 0.25$ to 1.25. The duration of most runs was approximately 100 τ ($\tau = \sigma(m/\varepsilon)1/2$; 100 τ = 20,000 time steps), though runs at least twice as long were done to obtain transverse current correlation functions. Each run was preceded by an equilibration period of 10 to 15 τ. The ρ^* = 0.8 isochore was traversed in both directions to test for possible hysteresis effects that tend to accompany discontinuous phase transitions.

Fig. 1 shows the temperature dependence of the energy⁺, pressure and heat capacity along the $\rho^* = 0.8$ isochore. The important thing to note about fig. 1 is that hysteresis seems to be virtually absent. This behavior is in marked contrast to what has been observed in 3-D, where freezing occurs through nucleation only upon significant under cooling.[5] A two dimensional system that shows first order melting is, for instance, the registered phase of Kr on graphite. Results of computer simulations on this latter system[6] are also shown in figure 1 for the sake of comparison. In this case the solid and fluid phases correspond to distinct branches in the E vs T plot. In contrast, the thermodynamic properties of the flat L.J. system appear to be continuous functions of temperature. Closer inspection of fig. 1 shows that both E and P change slope around $T_1^* \approx 0.36$ and $T_2^* \approx 0.57$. This fact, in itself, is no indication of the occurrence of 2 higher order phase transitions.

⁺In fact, the function plotted is $E^* - 2T^*$, i.e. the energy per particle minus the energy per particle in a harmonic lattice at the same temperature.

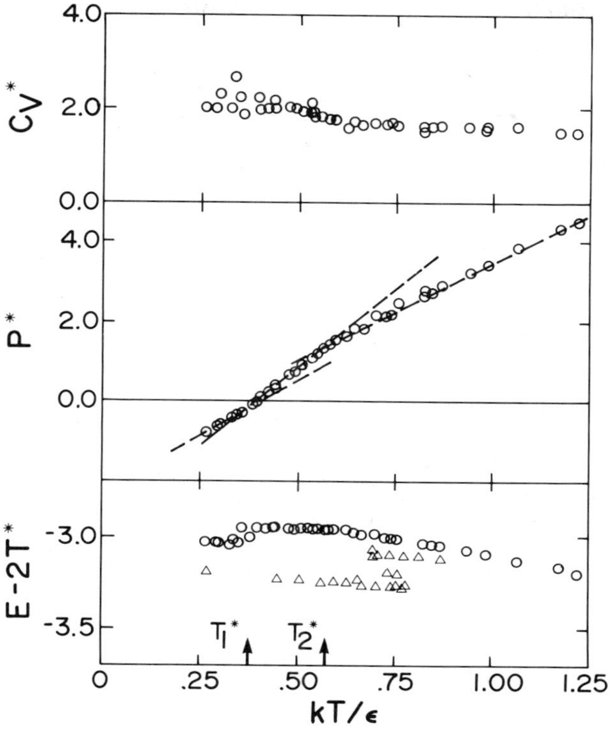

Figure 1. From top to bottom: Heat capacity, virial pressure and energy[+] of a flat 2-D Lennard Jones system at $\rho^* = 0.8$ (○). T_1^* and T_2^* indicate the approximate temperatures where E and P change slope. To guide the eye, straight line segments have been drawn through the P vs T points; these line segments do not however have any theoretical significance. In particular, no discontinuous changes in slope of P at T_1^* and T_2^* are implied. Typical first order behavior is shown in the lower figure (△). The points correspond to "Kr on graphite" at $\rho^* = 0.825$ (from ref. 6).

In fact, in an infinite system, one would expect to observe something similar if a first order phase transition at constant density is pressure broadened; the region $T_1^* < T^* < T_2^*$ would then be a two phase region. In a finite system, however, the creation of interfaces in general requires a non-negligible amount of free energy (this is in fact one of the reasons why hysteresis occurs in 3-D). The fact then, that no hysteresis is observed seems to indicate that, if the transition were first order, the free energy of interface formation is negligible. But that is equivalent to the statement that the system shows critical behavior. The MD calculations therefore seem compatible with the interpretation of T_1^* and T_2^* as 2nd order phase transitions. The temperature dependence of the structure factor obtained from the machine calculations provides additional evidence that the melting transition is continuous. In particular, the width of the first Bragg peak is a smooth function of temperature. (The actual values agree almost quantitatively with the experimental data for argon on graphite.)

An intriguing prediction of the HN theory concerns the temperature dependence of the correlation function of the orientational order parameter. The orientational order parameter that has the symmetry properties of a triangular lattice is defined as:

$$\psi(\vec{r}) = \exp(6\ i\ \theta(\vec{r})) \tag{1}$$

where $\theta(r)$ is the angle between some fixed axis and the line joining the centers of mass of two neighboring atoms. In the continuum description used by HN, $\theta(r)$ can be expressed in terms of the displacement field. Although no infinite range translational order exists even in a harmonic, infinite 2-D crystal, orientational order is long range, i.e., $\langle\psi^*(o)\ \psi(r)\rangle \to c \neq 0$ for $r \to \infty$.[7] Halperin and Nelson predict that if melting in 2-D is not first order, $\langle\psi^*(o)\ \psi(r)\rangle$ will decay algebraically in the intermediate (hexatic) phase and exponentially in the high temperature isotropic fluid phase. For computational purposes it is more convenient to define the orientational order parameter in the following way:

$$\psi_6(\vec{r}) = \frac{1}{N} \sum_{i=1}^{N} \delta(\vec{r} - \vec{r}_i) \left\{ \frac{1}{6} \sum_{j=1}^{6} \exp(6\ i\ \theta_{ij}) \right\} \tag{2}$$

where j = 1 to 6 are the 6 nearest neighbors of atom i. Because $\psi_6(r)$ is only defined at the site of an atom, the correlation function $\langle\psi_6^*(o)\ \ \psi_6(r)\rangle = g_6(r)$ exhibits oscillations (see fig. 2). Partly this effect is due to oscillations in the radial distribution function, $g(r)$. But even after dividing $g_6(r)$ by $g(r)$ (which has been done in fig. 2), oscillations remain. These oscillations can be understood by considering a regular triangular lattice with a few interstitial atoms. If both r_i and r_k are lattice sites, $\psi_6^*(r_i)\ \psi_6^*(r_k) = 1$ but if r_k is the site of an interstitial atom, $\psi_6^*(r_i)$

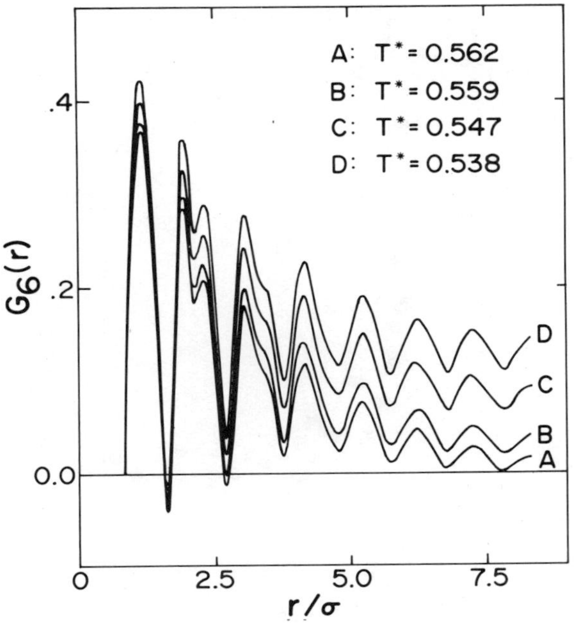

Figure 2. Correlation function of the orientational order parameter (as defined in text) in the vicinity of T_2^*. The function plotted is $\langle\psi_6^*(o)\ \psi_6(r)\rangle/g(r)$. Note that at the lower temperatures $g_6(r)$ does no longer die out within half a box length.

$\psi_6(r_k) = -1$. For comparison with the continuum theory it is only meaningful to speak about the envelope of $g_6(r)$. In the high temperature fluid phase this envelope is found to be very nearly exponential for all but the shortest distances. We denote the correlation length of this exponential by $\xi_6(T)$. The HN theory predicts that this correlation length will diverge very strongly (in fact, as $\exp(b/(T - T_i)^{1/2})$) as the hexatic-isotropic fluid transition (T_i) is approached from above. The MD results for the temperature dependence of $\xi_6(T)$ are shown in fig. 3. Clearly, $\xi_6(T)$ increases dramatically around $T^* = T_2^*$, in sharp contrast to the rather unspectacular behavior of the thermodynamic properties. We have not plotted $\xi_6(T)$ down to lower temperatures, because once $\xi_6(T)$ be-

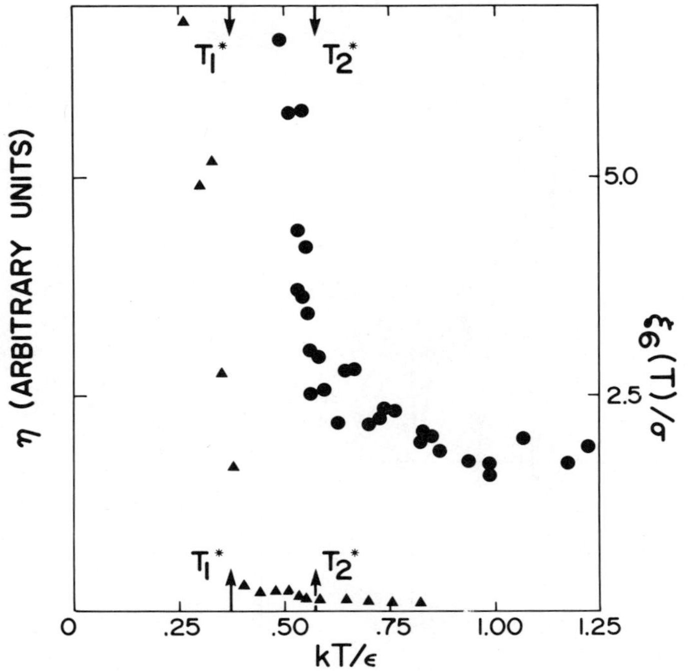

Figure 3. Correlation length of orientational order parameter vs. temperature (●). Shear viscosity, as obtained from the transverse current spectrum; (▲).

comes of the order of half a boxlength ($L/2 \approx 7.5\sigma$) the periodic boundary conditions start to dominate its behavior (see fig. 2). According to the HN theory, $g_6(r)$ will decay algebraically in the hexatic phase. We did not attempt to verify this prediction, let alone determine power law exponents, because the periodic boundary conditions will necessarily distort such slowly decaying correlation functions beyond recognition.

If we assume for the moment that the behavior of $\xi_6(T)$ indicates that there is indeed a hexatic-isotropic fluid transition around T_2^*, we should expect a solid-hexatic fluid transition at lower temperatures. Although the thermodynamic properties of the system are expected to be quite insensitive to this transition, the

HN theory predicts a rapid and dramatic drop in the resistance to shear as the system goes from the solid to the hexatic phase. The temperature dependence of the resistance to shear of the L.J. system is followed most conveniently by studying the decay of transverse current fluctuations. In the hydrodynamic limit, transverse currents decay as $\exp(-k^2\eta t/\rho m)$; hence the total area under the transverse current correlation function (i.e. the $\omega = 0$ component of the power spectrum $G_T(k;\omega)$) is proportional to $(k^2\eta)^{-1}$. Fig. 3 shows the temperature dependence of η as obtained from $[k^2 G_T(k;0)]^{-1}$ (after averaging over 3 long wavelength transverse phonons). Clearly there is a very dramatic drop in viscosity around $T^* = T_1^*$. Close to T_2^* the viscosity decreases by about 50%. At higher temperatures the viscosity appears to be more or less constant.

In summary, it then appears that, to the extent to which our calculations can be compared with the continuum theory, they seem to support its predictions. We tentatively identify the phase between T_1^* and T_2^* as a hexatic liquid bordered at lower and higher temperatures respectively by a solid and an isotropic fluid phase. It should be noted that in the HN theory the melting transition is driven by the dissociation of dislocation pairs. Our results therefore reaffirm indirectly the importance of dislocations in 2-D melting. Of course, the earlier work of Cotterill and coworkers[8] provides more direct qualitative evidence for the relation between dislocations in 2-D melting. It is not possible to construct a phase diagram on basis of the limited data available at present. It appears that the solid-fluid transition becomes first order at higher temperatures and densities.[9,10] Several authors have reported results of machine calculations that seemed to suggest a first order liquid-gas transition at lower densities.[9,11] We have tried to reproduce those results by performing constant N, P, T Monte-Carlo calculations (256 particles, 4.10^6 configurations/run) in the relevant region of the phase diagram ($T^* = 0.5$, $P^* = 0.05$ to 0.1). Apart from the observation that the system is very sluggish and shows large density fluctuations, we failed to find any evidence for 2-phase behavior; in particular, the average density was found to be almost proportional to the pressure.

We are grateful to Professor David Nelson for several helpful conversations and suggestions. This work was supported in part by NSF grants CHE76-21293, CHE77-15387, and GP32304.

REFERENCES

1. B. I. Halperin and D. R. Nelson, Phys. Rev. Lett., 41:121 (1978).

2. J. P. McTague, M. Nielsen and L. Passell, CRC Critical Reviews in Solid State and Materials Sciences, 8:125 (1979) and references therein.

3. T. T. Chung, unpublished, as quoted in ref. 4.

4. F. E. Hanson, M. J. Mandell and J. P. McTague, J. Phys. (PARIS), C-4:76 (1977).

5. M. J. Mandell, J. P. McTague and A. Rahman, J. Chem. Phys., 64:3699 (1976); 66:70 (1977).

6. F. E. Hanson and J. P. McTague, to be published.

7. N. D. Mermin, Phys. Rev., 176:250 (1968).

8. R. M. Cotterill, E. J. Jensen and W. D. Kristensen, in: "Anharmonic Lattices, Structural Transitions and Melting," Ed. T. Riste, Noordhoff, Leiden, 1974.

9. F. Tsien and J. P. Valleau, Mol. Phys., 27:177 (1974).

10. S. Toxvaerd, preprint.

11. D. Henderson, Mol. Phys., 34:1 (1977).

THE ROUGHENING TRANSITION

John D. Weeks

Bell Laboratories
Murray Hill, N.J. 07974

I. INTRODUCTION

The idea that there could be a "roughening" of the interface of a crystal in equilibrium with its vapor at a particular temperature T_R was first suggested by Burton and Cabrera (1949) and further developed in a now classic article by Burton, Cabrera and Frank (BCF) (1951). Representing the crystal surface by a two-dimensional (2D) Ising model they suggested that there would be large fluctations in the surface structure at the Ising model's critical temperature $T_C(2D)$ and a disappearance of the nucleation barrier to crystal growth. Jackson (1958, 1967) further developed and extended these ideas to the case of melt growth and showed that the morphology and growth mechanism of a wide class of crystals could be understood by assuming they were grown above or below the appropriate surface roughening temperature.

Although these ideas were well known to most material scientists and workers in the field of crystal growth, it is only fairly recently that their importance and relevance has been appreciated by condensed matter physicists. The roughening transition is of interest today not only because of its implications for surface physics but also because of its relationship to phase transitions in a number of different systems, several of which are discussed at this Institute.

In these lectures, we will give a brief introduction to the crystal growth models and ideas that lead BCF to suggest the possibility of surface roughening, followed by a review of the modern work relating the roughening transition to phase transitions in a number of 2D systems, including the planar (XY) model, the F model and the coulomb gas. We describe the application of the Kosterlitz (1974) renormalization group theory to describe the statics and dynamics of the roughening transition and the results of Monte Carlo calculations which seem in good accord with the theory. More complete discussions and a guide to the literature of various aspects of these lectures can be found in review articles by Weeks and Gilmer (1979), Gilmer and Jackson (1977), Müller-Krumbhaar (1977), and Leamy et al. (1975).

II. THE SOLID-ON-SOLID MODEL

Consider for simplicity the case of a (001) face of an impurity-free simple cubic crystal in equilibrium with its vapor. We model this situation using a restricted version of the usual lattice gas (Ising model) in which every site is either vacant or occupied by a single atom whose interaction with another atom in a nearest neighbor site is ϕ. If we further require that every occupied site be directly above another occupied site (thus excluding "overhangs") we obtain a "solid-on-solid" (SOS) model. A SOS model can thus be thought of as an array of interacting columns of varying integer heights. The surface configuration is represented by the 2D array of integers specifying the number of atoms in each column perpendicular to the (001) face, or equivalently by the height of the column relative to the flat T = 0 reference surface. Growth or evaporation of the crystal involves the "surface atoms" at the tops of their columns. As shown in Fig. 1, complex surfaces with steps and other kinds of disorder can be represented using the column model.

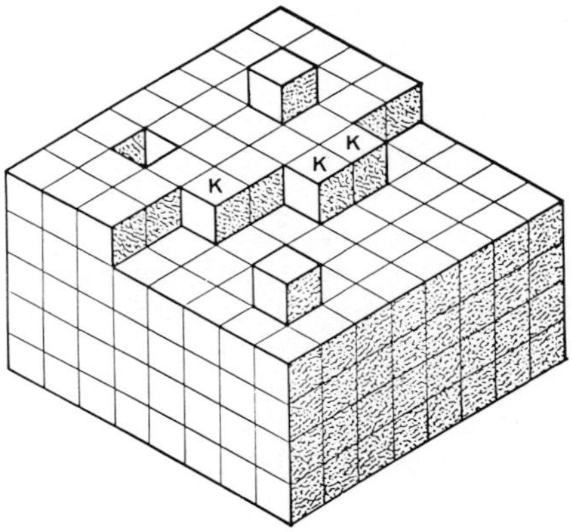

Fig.1 Atoms on a (001) face of a simple cubic crystal. Surface atoms may have up to four lateral neighbors. An atom in a kink site, indicated by a k in the figure, has two lateral neighbors.

For the restricted lattice gas, the energy between neighboring columns of heights h_j and $h_{j+\delta}$ is determined by counting the number of "broken bonds" and hence is proportional to $\phi|h_j - h_{j+\delta}|$. The equilibrium properties of this particular model (denoted ASOS herein) can be determined from the partition function

$$Z_{ASOS} = \sum_{\{h_j\}=-\infty}^{\infty} \exp\left[-\frac{J}{2kT}\sum_{j,\delta}|h_j - h_{j+\delta}|\right] \tag{1}$$

where $J \equiv \phi/2$ and the summation is over all integer heights for each column. Note that the number of vertical broken bonds is conserved in the excitations permitted in the ASOS model. Hence we can arrive at (1) formally by considering the interface in an anisotropic lattice gas with a vertical bond strength $\phi_\perp$ which tends to infinity (Weeks et al., 1973).

The ASOS model is an accurate approximation to the interface in the unrestricted lattice gas at low T, since the overhanging configurations suppressed are of higher energy. Further we can consider a wider variety of SOS models in which the interaction energy between columns is some increasing function of the height difference $V(h_j - h_{j+\delta})$. At low T, the higher energy multiple height jumps between neighboring columns will be infrequent and the properties of all such models will be very similar. (As will become apparent later, an essential feature of all these models is that the heights can range over all integers $-\infty < h < \infty$.)

Dynamics is introduced into the model by creating or annihilating atoms at random positions on the surface. This process simulates the molecular exchange between the solid and vapor phases. Thus is it reasonable to assume that the rate of creation (deposition) of atoms per site at the surface, denoted k^+, is independent of the neighboring surface configuations. However, the annihilation (evaporation) of a surface atom is an activated process requiring the breaking of nearest neighbor bonds. We assume the evaporation rate of an atom with m lateral neighbors ($0 \leq m \leq 4$ in a cubic lattice) is

$$k_m^- = \nu \exp[-m\phi/kT]$$

where ν is the evaporation rate of an isolated adatom at the surface. It is easy to see that this choice of transition probabilities obeys detailed balance (Gilmer and Bennema, 1972). The equilibrium state, described by Eq (1), is reached when the creation rate, k_{eq}^+, equals the evaporation rate of a kink site, $\nu \exp[-2\phi/kT]$. This must be true since a layer can, in principle, grow (or be removed) by successive creations (or annihilations) of atoms only at kink sites.

We emphasize that the above is a stochastic model for the statics and dynamics of the crystal growth process. Important limitations of the model include the neglect of strain fields, and the need to assume in advance a particular lattice structure. However, the model does give a consistent and physically reasonable description of the cooperative interactions among clusters of atoms that are critical to the crystal growth process. Properly interpreted it thus provides a useful compromise between mathematical simplicity and physical reality.

III. THE BCF ARGUMENT

To gain a qualitative feel for the statics of the roughening transition we make use of the analogy between a lattice gas and a ferromagnetic Ising model, where an occupied site is represented by an "up" spin and a vacant site by a "down" spin. The configuration at $T = 0$ is described by successive 2D layers of up spins representing occupied sites in the crystal followed by layers of down spins representing the vapor. The final (surface) layer of up spins is effectively isolated since the layers above and below are magnetized in opposite directions. Thus following BCF, we might expect the surface layer to behave like a 2D Ising model with large spin fluctuations (i.e., large regions of surface vacancies and adatoms) and thermodynamic singularities near the 2D critical temperature $kT/\phi \cong 0.57$. Note that the cancellation argument holds equally well for the anisotropic lattice gas with $\phi_\perp >> \phi$. This shows that the roughening transition is not related to the bulk (3D) critical temperature which scales with $\phi_\perp$ (and indeed is infinite for the ASOS model), but rather is a transition unique to the interface.

The BCF picture implies that each crystal face has a distinct roughening temperature, the more loosely packed faces having the lower T_R. Indeed some faces, e.g., the (011) face of a simple cubic crystal, have no connected 2D net of nearest neighbor bonds and their roughening temperature is zero, the result for a 1D Ising model. For most crystal growth applications, the most important faces are the slow-growing close-packed faces.

As one might expect, the critical-like fluctuations occurring at the interface near T_R have an important effect on the crystal growth rate. Crystal growth on a relatively flat surface well below T_R is a difficult process, requiring the formation of a critical nucleus cluster. If a surface is at its roughening temperature, then BCF reasoned that the critical fluctuations produce clusters of arbitrarily large size and hence the nucleation barrier to crystal growth disappears. Another implication is that the crystal grown with $T < T_R$ is faceted with very anisotropic growth rates for the different faces, the close-packed faces growing in a layer-by-layer fashion. Above T_R, essentially isotropic growth should occur (Jackson, 1967).

It is easy to find fault with this crude argument. There can be an exact cancellation of the interactions from the layers above and below the surface layer only at T = 0. Indeed, van Beijeren (1975) used this observation to prove rigorously that $T_C(2D)$ is a lower bound to T_R. Further the restriction of the excitations to only one layer is unrealistic and must be removed for a more exact description. Still, the argument is physically very suggestive and it stimulated experimental work, some of which is described in the next section, which seems in good accord with their physical picture.

IV. EXPERIMENTAL RESULTS

There were initially few attempts to experimentally verify the BCF ideas on roughening because they estimated that a crystal in equilibrium with its vapor would melt before the closest-packed face roughened. However, recent experiments by Jackson and Miller (1977) suggest that for simple van der Waals crystals, the roughening point is well below the melting point. They studied the plastic crystals C_2Cl_6 and NH_4Cl and found dramatic changes in the morphology (faceted to essentially isotropic) of crystals grown for temperatures differing by less than five degrees. These experiments are the only ones we know of in which a crystal in equilibrium with its vapor is taken from below to above its roughening temperature.

Earlier experimental corroboration of the roughening picture involved comparison of a given crystal's structure to predictions arising from estimates of the roughening temperature. Most of this work was for melt growth, and the temperature range over which the crystal growth can be observed experimentally is very small. However, as shown by Jackson (1958, 1967), it is possible to understand both the growth mechanisms (nucleated or continuous) and crystal structure (faceted or isotropic) of a very wide variety of materials by determining whether the crystal as grown was below or above its surface roughening temperature.

These experimental results, and all others we know of, have indirectly observed the roughening transition by its effect on crystal growth. An experimental study of the equilibrium properties of the crystal-vapor interface seems called for. Then one can test a number of the detailed predictions that arise from the new developments in the theory of the roughening transition. These results are discussed in the next part of these lectures.

V. MONTE CARLO CALCULATIONS: QUALITATIVE FEATURES

The BCF one layer model is obviously inadequate in several important respects. To gain a better physical feeling for the roughening transition it is useful to consider the results of Monte Carlo simulations on the ASOS model. Fig. 2 give typical equilibrium surface configurations generated by the MC method at various values of kT/ϕ. At the lower temperature distinct adatom and vacancy clusters are visible but at the highest temperature

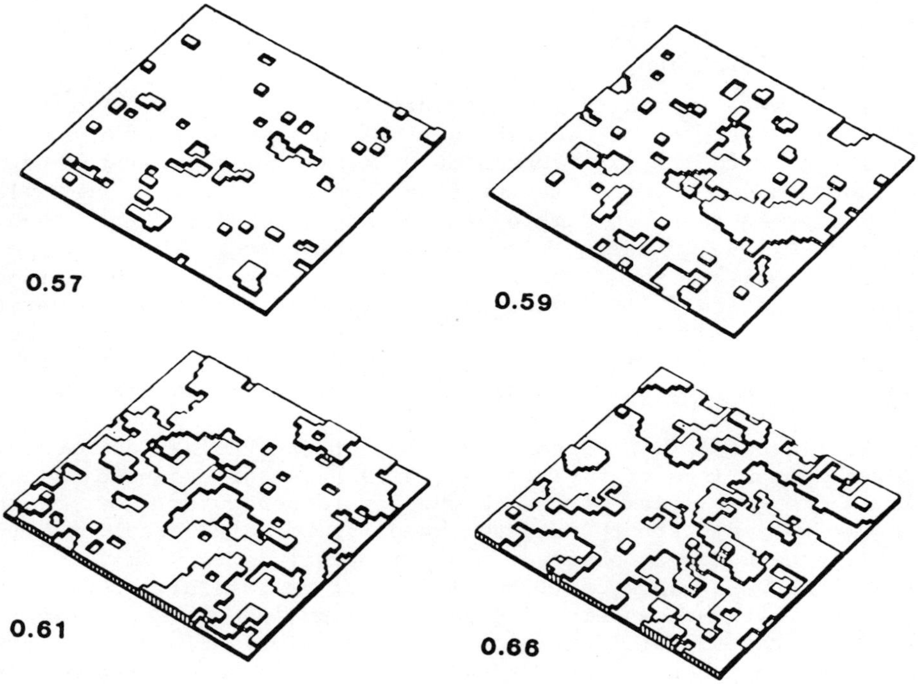

Fig. 2. Computer drawings of typical surfaces generated by the MC method at the indicated values of kT/ϕ.

the clusters have grown and merged together to such an extent that arbitrarily large clusters are present and indeed the original reference level of the surface is not apparent (Leamy et al., 1975).

Thus the BCF picture of a rapid increase in surface roughness near the 2D Ising model's critical temperature $kT_C(2D)/\phi \cong .57$ appears confirmed. However an important additional feature becomes clear: the excitations extend over many layers - evidentally arbitrarily many in the limit of an infinite system. It is clear that once a large cluster has formed it is just as easy to form another cluster on top of the given cluster as it is to form it in the original layer.

This buildup of large clusters on top of other clusters can be thought of as a long-wavelength distortion of the local position of the solid-vapor interface. Note that there are relatively few double jumps between nearest neighbor columns even at the highest temperature shown. Thus it is the long-wavelength distortions involving a single jump from one large cluster to another that dominate the physics of roughening. The essential idea of BCF is correct: there is a class of low energy excitations possible at the crystal-vapor interface. However these same excitations cause the local position of the interface to wander arbitrarily far from the original reference level.

These considerations suggest there are several equivalent ways of characterizing the roughening transition. Clusters of arbitrarily large size can be found at and above T_R. The formation of these arbitrarily large ridges also implies that the edge free energy and edge energy (per unit length) required to form a step on the crystal surface should vanish at T_R (Leamy and Gilmer, 1974). Since large clusters of adatoms and vacancies are equally probable at T_R, the average density of the surface layer should be 1/2 at and above the roughening temperature. The formation of arbitrarily large clusters in one layer implies a high probability of finding similar large clusters in adjacent layers and the loss of the original reference level. Thus the interface width should diverge at T_R in an infinite system (Weeks et al., 1973). The disappearance of the nucleation barrier implies that the susceptibility (the partial derivative of the average height with respect to an infinitesimal driving force) should diverge at and above T_R. The motion of the interface at and above T_R can be thought of as similar to that of a drumhead, whose normal modes of vibration correspond to the formation of large clusters of adatoms or vacancies on the surface.

VI. STATIC CRITICAL BEHAVIOR

We now make a more formal analysis of the properties of the roughening transition. A literal interpretation of the BCF one layer model suggests that the roughening transition lies in the same universality class as the 2D Ising model. In fact, because excitations are not restricted to a single layer, it lies in a very different universality class. This was first demonstrated in the work of Chui and Weeks (1976).

Chui and Weeks introduced the discrete Gaussian (DG) model in which the interactions between nearest neighbor columns is quadratic in the height difference:

$$H_{DG} = \frac{J}{2} \sum_{j,\delta} \left(h_j - h_{j+\delta}\right)^2 \equiv \frac{J}{2} \sum_{jj'} h_j G_1^{-1}(jj')h_{j'}, \tag{2}$$

$$= \frac{J}{2} \sum_{q} |h_q|^2 \, G_1^{-1}(q) \tag{3}$$

where

$$G_1^{-1}(q) = 4 - 2\left(\cos q_x + \cos q_y\right) \tag{4}$$

and

$$h_q = \frac{1}{\sqrt{N}} \sum_{j} h_j e^{iqj} \tag{5}$$

is the Fourier transform of the height variable h_j. The Fourier transform of the matrix $G_1^{-1}(jj')$ is explicitly given in Eq. (4). As argued before, at low T multiple jumps are unimportant and we expect Eq. (2) to give the same critical behavior as the ASOS model in Eq. (1). More generally, the roughening transition involves long wavelength fluctations in the position of different parts of the interface. Changes in the interaction energy between columns that affect only short wavelength properties should be irrelevant at the roughening point. Furthermore the Gaussian interaction is in a sense the most fundamental. Note that in Eq. (3) the interaction energy for small q goes as $q^2|h_q|^2$. This is characteristic of a surface tension (Buff et al., 1965). We expect that at high temperatures the long wavelength properties of the interface arising from virtually <u>any</u> reasonable microscopic interaction can be described using a surface tension. Thus a wide class of microscopic column hamiltonians should transform under renormalization group equations to the basic Gaussian interaction as in Eq. (3).

The DG partition function can be written

$$Z_{DG} = \int d\{h_j\} \prod_j W(h_j) \exp\left[-\frac{1}{kT} H_{DG}\right] \tag{6}$$

where

$$W(h_j) = \sum_{n_j=-\infty}^{\infty} \delta(h_j - n_j) \tag{7}$$

$$= \sum_{k_j=-\infty}^{\infty} \exp\left[ik_j h_j\right] \tag{8}$$

The weighting function $W(h_j)$ in Eq. (7) restricts the integration in Eq. (6) so that only integer values of h_j contribute. In Eq. (8) we have reexpressed $W(h_j)$ in a more convenient way using a well-known identity (see, e.g., Lighthill, 1959) which is essentially the Poisson summation formula. Here $k_j = 2\pi n$ for integer n. Substituting Eq. (8) into Eq. (6) we have

$$Z_C \equiv \frac{Z_{DG}}{Z_0} = \sum_{\{k_j\}=-\infty}^{\infty} < \exp\left(i \sum_j k_j h_j\right) >_0 \tag{9}$$

Here Z_0 is the unweighted Gaussian model's partition function [Eq. (6) with $W(h_j) \equiv 1$], which can be evaluated exactly. The angular brackets indicate an ensemble average in the unweighted Gaussian ensemble.

In Eq. (9) we note the characteristic function for the Gaussian distribution. Hence the $\{k_j\}$ also have a Gaussian distribution given by the inverse matrix to G_1^{-1} (see, e.g. Cramer, 1946) and Eq. (9) becomes

$$Z_C = \sum_{\{k_j\}=-\infty}^{\infty} \exp\left[-\frac{kT}{2J} \sum_{jj'} k_j\, G_1\,(jj')\, k_{j'}\right] \tag{10}$$

where, from Eqs. (2)-(4), the inverse matrix $G_1(jj')$ is

$$G_1\,(j,j') = \frac{1}{2N} \sum_q \frac{e^{iq(j-j')}}{G_1^{-1}\,(q)} \tag{11}$$

Eq. (10) is in fact the partition function for a neutral 2D lattice Coulomb gas (see Chui and Weeks, 1976 for further details) in which the k_j represents the charges. Note the q^{-2} dependence at small q in Eq. (11) which characterizes the Coulomb interaction.

The reduced temperature kT/J has been inverted in going from the DG model in Eq. (6) to the Coulomb gas in Eq. (10). The fact that the Coulomb gas appears is really no mystery: the matrix $G_1^{-1}(jj')$ in Eq. (2) is the lattice analogue of the Laplacian operator and hence its matrix inverse, $G_1(jj')$ in Eq. (11), is the 2D lattice Green's function, i.e., the 2D Coulomb potential.

Since Z_0 is analytic, the singularities in the DG partition function Z_D are identical with those in Z_C. These had already been discussed by Kosterlitz and Thouless (1973) and Kosterlitz (1974) in connection with their analysis of the planar (XY) model and a dislocation model for 2D melting. They established that the Coulomb gas has a phase transition from a low temperature dielectric phase with opposite charges tightly bound together in "diatomic molecules" to a high temperature metallic phase. The free charges in the metallic phase come from the now disassociated "molecules" and give the usual Debye screening. The properties of this transition can thus be directly related to those of the roughening transition and differ greatly from those of the 2D Ising model.

The most dramatic differences show up in the behavior of the correlation length ξ. Define the height-difference correlation function for two columns separated by a distance r:

$$G(r) = \langle (h_0 - h_r)^2 \rangle \tag{12}$$

where the angular brackets indicate an ensemble average in the SOS system. G(r) gives a measure of the average fluctuations in height between different regions of the interface separated by a distance r, and the square of the <u>interface width</u> is the $r \to \infty$ limit of G(r). The <u>correlation length</u> ξ is proportional to the distance r' at which G(r') is approximately equal to its asympotatic value. The results of Kosterlitz (1974) then imply that below T_R the interface width is finite with a finite correlation length ξ. At all temperatures above T_R, however, G(r) is proportional to ln r. Thus the interface width diverges logarithmically and the correlation length ξ is infinite. It is as if there were a line of "critical points" for all $T \geq T_R$ where ξ is infinite.

The renormalization group (RG) method of Kosterlitz (1974) further showed that the correlation length diverges very rapidly as $T \to T_R$ from below:

$$\xi \propto \exp\left[c/(T_R - T)^{\frac{1}{2}}\right] \tag{13}$$

and of course ξ remains infinite for $T > T_R$. This behavior is very different from that of the 2D Ising model where ξ diverges by a power law only at T_C. Further, the singular part of the free energy has a similar form near T_R:

$$F \propto \exp\left[-C'/\left(|T-T_R|^{\frac{1}{2}}\right)\right] \tag{14}$$

Note the square root dependence on $T-T_R$ in Eqs. (13) and (14). The free energy is non-analytic at T_R but the singularity is a very weak one with all temperature derivatives of the singular part vanishing at T_R. In particular there is no specific heat anomaly at T_R, in contrast to the Ising model. We will discuss the Kosterlitz RG theory and the derivation of some of these results in the next section.

This kind of behavior should apply to a wide class of interfacial models with different interaction energies between columns. Furthermore the periodic delta function weighting in Eq. (7) can be replaced by other periodic weighting which favor integer positions. A particularly interesting case was analyzed by Ohta and Kawasaki (1978), who took

$$W(h_i) = 1 + 2\, y_0 \cos 2\pi h_i \ . \tag{15}$$

A Coulomb gas partition function like Eq. (10) again results but the charges k_j now have only the values 0 and $\pm\, 2\pi$. Using the Kosterlitz RG method, they found critical behavior identical to that described above for the DG model and give further implications for the roughening transition. We discuss in the next section the dynamics of a very similar model.

Another model which has similar critical behavior is the planar model, which Kosterlitz (1974) analyzed by relating it to a 2D coulomb gas. Jose et al., (1977) and Knops (1977) have made the connection between the planar and SOS models quite explicit mathematically by showing they are related by an exact "duality" transformation which we now discuss. Consider a general SOS partition function

$$Z = \sum_{\{h_j\}} \exp\left[-\sum_{j,\delta} V\left(h_j - h_{j+\delta}\right)\right] \tag{16}$$

where $V(h_j - h_{j+\delta})$ is an arbitrary increasing function of the height difference between neighboring columns. If the 2N "bond" variable

$$n_{<ij>} \equiv h_i - h_j \tag{17}$$

(where column i is a nearest neighbor to column j having the smaller x or y coordinate) were all independent then Eq. (16) could be represented as a product of single variable partition functions. Of course there are actually N constraints which the $n_{<ij>}$ variables must obey: around each square of four columns we must have (see Fig. 3)

$$(h_1 - h_2) + (h_2 - h_3) + (h_3 - h_4) + (h_4 - h_1) = 0 \qquad (18)$$

or

$$n_{<12>} + n_{<23>} - n_{<43>} - n_{<14>} = 0 \qquad (19)$$

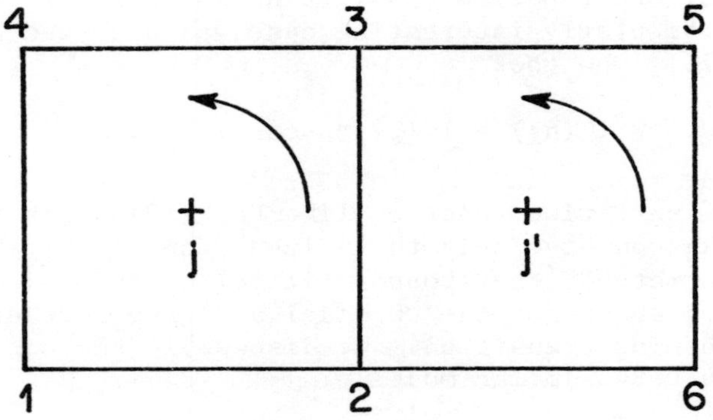

Fig. 3. Squares of four columns 1, 2, 3, 4 and 2, 6, 5, 3 and the dual lattice points j and j' around which the angles ϕ and ϕ' are measured.

We can still introduce the $n_{<ij>}$ variables in Eq. (16) and treat them as independent if we take care of the constraints (19) using the Kronicker δ function:

$$\delta_{0,\Sigma n_{<ij>}} = \int_0^{2\pi} d\phi_j \exp\left\{ i\phi_j \left[n_{<12>} + n_{<23>} - n_{<43>} - n_{<14>} \right] \right\} \qquad (20)$$

The angle ϕ_j can be associated with the center of the jth square of columns, and in general with a point on the dual lattice. We can also index the bond variables using the dual lattice points. For example in Fig. 3, we rename $n_{<23>}$ as $n_{<jj'>}$. Eq. (16) then becomes

$$Z = \sum_{\{n_{<ij>}\}} \int_0^{2\pi} d\{\phi_j\} \exp\left[-\sum_{<jj'>} \left\{V\left(n_{<jj'>}\right) + in_{<jj'>}\left(\phi_j-\phi_{j'}\right)\right\}\right] \tag{21}$$

or

$$Z = \int_0^{2\pi} d\,\{\phi_j\} \exp\left[-\sum_{<jj'>} \tilde{V}\left(\phi_j-\phi_{j'}\right)\right] \tag{22}$$

where

$$-\tilde{V}(\phi_j) = \ln\left\{\sum_n \exp\left[-V(n) + in\phi_j\right]\right\} \tag{23}$$

Eq. (22) is the partition function for a generalized planar model with a 2π periodic angular interaction between neighboring "spins" given by Eq. (23). Just as for the DG to Coulomb gas transformation discussed earlier, the weak and strong coupling limits are interchanged in the transform in Eq. (23). In particular, the DG model transforms exactly into the planar model of Villain (1975) and the usual planar model with $\tilde{V}(\theta) = \frac{J}{kT}\cos\theta$ transforms to a new SOS model with

$$V(n) = -\ln I_n\left(\frac{J}{kT}\right) \tag{24}$$

Near T_R, the energy for multiple height jumps given in Eq. (24) is bracketed by that given by the ASOS and DG models while at low (planar model) temperatures (i.e., high SOS temperatures), Eq. (24) **reduces to the Gaussian model.** If these general SOS models are in the same universality class as we have argued, then the corresponding planar models are also.

Another model which almost certainly is in the same universality class is the F model, for which an exact solution is known (Lieb, 1967). We will discuss this model later, after a study of the dynamics of the roughening transition using the Kosterlitz RG method.

VII. ROUGHENING DYNAMICS AND THE KOSTERLITZ RENORMALIZATION GROUP METHOD

In this section we review a theory of crystal growth dynamics near the roughening point introduced by Chui and Weeks (1978). We are thus dealing with the interesting transition between sub-linear (nucleated) growth below T_R and continuous growth above. We assume some familiarity with recent developments in the theory of dynamic critical phenomena. (See, e.g., Hohenberg and Halperin, 1977.)

The fundamental idea in developing a tractable theory for dynamics at the roughening point is that of dynamic universality (Hohenberg and Halperin, 1977). It is postulated that in addition to all the properties that affect the static roughening behavior, one need consider in addition only the (hydrodynamic) conservation laws and coupling between the conserved variables. Details of the dynamics which do not affect conservation laws are irrelevant for a description of the long-wavelength low-frequency behavior of the system at the roughening point. For example, systems with and without surface diffusion should exhibit similar behavior at their respective roughening transitions.

Our model for crystal growth is particularly simple since there are no conserved quantities such as the energy or momentum density to consider. We have postulated from the first a stochastic and purely relaxational model of crystal growth. Assuming dynamic universality we can thus study, for example, a simple relaxational Langevin model kinetic equation ("Model A") and obtain information about all members of this universality class.

We consider the following generalized SOS model Hamiltonian for the crystal-vapor system

$$H = \frac{J}{2} \sum_{j,\delta} \left(h_j - h_{j+\delta}\right)^2 + Jg^2 \sum_j h_j^2 - \sum_j \Delta\mu_j h_j - 2y_0 J \sum_j \cos\,(2\pi h_j) \tag{25}$$

The first term gives the interaction energy between a column at site j (and height h_j) and its nearest neighbors at sites $j + \delta$, while the second gives the interaction with a dimensionless "stabilizing field" g^2 which tends to localize the interface near $\langle h \rangle = 0$. Usually we consider the limit $g^2 \to 0^+$. The third term gives the interaction with "applied fields" $\Delta\mu_j$ which for generality can be different for different lattice sites. We will later associate $\Delta\mu_j$ with the chemical potential driving force for crystal growth. The last term, parameterized by the dimensionless quantity y_0 is a weighting function which energetically favors integer values of the h_j. For small y_0 it reduces to the Ohta and Kawasaki (1976) function, Eq. (15). Models similar to Eq. (25) have also been studied by Muller-Krümbhaar (1977), Saito (1978), and Zittartz (1978).

We introduce dynamics through the Langevin equation (See, e.g., Ma, 1976)

$$\frac{\partial h_j}{\partial t} = - \frac{\Gamma}{T}\frac{\delta H}{\delta h_j} + \eta_j$$

$$= - \Gamma K^{-1} \sum_{\delta} \left(h_j - h_{j+\delta}\right) - \Gamma K^{-1} g^2 h_j + \Gamma\left(\Delta\mu_j/T\right)$$

$$- 2\pi K^{-1}\Gamma y_0 \sin 2\pi h_j + \eta_j \tag{26}$$

Here $K^{-1} \equiv 2J/T$. (We set Boltzmann's constant equal to unity in this section.) The η_j are Gaussian fluctuating white noises which satisfy

$$< \eta_j(t) > = 0$$

$$< \eta_j(t)\eta_{j'}(t') > = 2\Gamma\delta_{jj'}\delta(t-t') \tag{27}$$

where the angular brackets indicate an ensemble average. The parameter Γ can be identified with the equilibrium (kink-site) evaporation rate (Weeks and Gilmer, 1979). We assume that the system starts from equilibrium at $t = -\infty$ and allow the applied fields $\Delta\mu_j$ to be time dependent.

If $y_0 = 0$, the Eq. (26) is a linear equation and can be solved exactly by Fourier transform methods in terms of a lattice Green's function which in the long wavelength limit has the form (de Gennes, 1971)

$$G(q,\omega) = \left[K^{-1}\left(q^2 + g^2\right) - i(\omega/\Gamma)\right]^{-1} \tag{28}$$

In the limit $g^2 \to 0^+$, which we consider hereafter, G is the Green's function for 2d diffusion. This, of course, is not surprising since when $y_0 = 0$, Eq. (26) is a finite difference analog of the diffusion equation. $G(q,\omega = 0)$ is proportional to the static Green's function in Eq. (11).

For non-zero y_0, Eq. (26) can be rewritten after taking Fourier transforms as

$$h(q,\omega) = G(q,\omega)\left[\Delta u(q,\omega) + \eta(q,\omega)/\Gamma - 2\pi K^{-1} y_0 \; F\{\sin 2\pi h(s,t)\}\right] \quad (29)$$

Here s is a dimensionless 2d lattice vector (the unit of length being the lattice spacing) locating the center of a column, and F{ } indicates a Fourier transform in space and time.

We will analyze Eq. (29) using linear response theory, assuming that the driving force $\Delta\mu$ is infinitesimally small. Hence we will try to predict the limiting slope of the growth rate curve as the driving force tends to zero. In addition, the linear response analysis gives valuable information about spatial and temporal correlations of the interface at equilibrium when $\Delta\mu = 0$ (Hohenberg and Halperin, 1977).

Expanding the solution of Eq. (27) in powers of $\Delta\mu/T$

$$h(q,\omega) = h_0(q,\omega) + h_1(q,\omega)\Delta u(q,\omega)/T + 0\left(\frac{\Delta u}{T}\right)^2 \quad (30)$$

the linear response function $\chi(q,\omega)$ is given by the ensemble average over the noise

$$\chi(q,\omega) = \langle h_1(q,\omega) \rangle \quad (31)$$

and using Eqs. (29)-(31), the unperturbed ($y_0 = 0$) response function explicitly is

$$\chi_0(q,\omega) = G(q,\omega) = \left[K^{-1}\left(q^2 + g^2\right) - i(\omega/\Gamma)\right]^{-1} \quad (32)$$

The effect of a non-zero y_0 is conveniently expressed in terms of a self-energy $\Sigma(q,\omega)$, defined as

$$\chi^{-1}(q,\omega) = \chi_0^{-1}(q,\omega) + \Sigma(q,\omega) \quad (33)$$

Substituting Eq. (30) into Eq. (29) we find after some simple manipulation a formally exact expression for Σ given by

$$\Sigma(q,\omega) = \frac{4\pi^2 y_0 K^{-1} F\left\{\langle \cos[2\pi h_0(st)]h_1(st,s't') \rangle\right\}}{\langle h_1(q,\omega) \rangle} \quad (34)$$

Note that the term transformed is a function only of the differences s-s' and t-t' since the noise ensemble is stationary.

The behavior of Σ in the limit of very low temperatures is easy to analyze. The equilibrium fluctuations of h_0 are very small at low temperatures and the weighting function localizes the interface very near $h_0 = 0$. Linearizing the sine term in Eq. (6.2) then gives a constant value for Σ of

$$\Sigma(q,\omega) \cong 4\pi^2 y_0 K^{-1} \tag{35}$$

Thus from Eq. (33) there is a finite response even in the $q,\omega \to 0$ limits at low temperature.

At high temperatures ($T > T_R$) the situation is very different. Here the weighting function has little effect on the system. Thermal fluctuations are large enough that the interface wanders arbitrarily far from its $T = 0$ location (this delocalization characterizes the roughened phase). When $y_0 = 0$, the weighting function vanishes altogether and the response function can be calculated exactly. This divergent response function [Eq. (32)] presumably gives the limiting high temperature behavior of a system with a finite y_0.

These qualitative arguments can be put on a much firmer basis by using the renormalization group method of Kosterlitz (1974) and José et al. (1977). We consider an expansion of the inverse linear response function $\chi^{-1}(q,\omega)$ in powers of y_0. Similar expansions have proved very useful in the static limit. The zeroth order term $[\chi_0^{-1}(q,\omega)]$ gives the limiting ($T \to \infty$) behavior and the higher order terms give corrections arising from a non-zero weighting function. We will use this expansion to derive differential recursion relations which relate the response in the original system with parameters K, Γ and y_0 to that in a system with renormalized parameters K', Γ' and y_0'. Integration of the recursion relations will, in fact, provide a connection for all $T \geq T_R$ between the original system and the exactly solvable system with $y_0 = 0$.

Expanding h_0, h_1 and Σ in powers of y_0, we find, using Eqs. (29)-(34), after some straightforward but tedious algebra [much of which can be found in an article by de Gennes (1971)] that Eq. (33) can be written to lowest order in q and ω as

$$\chi^{-1}(q,\omega) = \left[K^{-1} + \pi^3 K^{-2} y^2 \int_1^\infty ds\ s^{3-2\pi K}\right] q^2$$

$$-i\omega\left[\Gamma^{-1} + \Gamma^{-1}\frac{\pi^4 y^2}{(\pi K-1)} \int_1^\infty ds\ s^{3-2\pi K}\right] + 0(y^4) \qquad (36)$$

where $y \equiv y_0 \exp[-Kc]$ and c is a constant approximately equal to $\frac{1}{2}\pi^2$. We can obtain renormalization group equations from (36) by eliminating the short wavelength parts of the integrals. Divide the range of integration of each integral in Eq. (36) into two parts: 1 to b and b to ∞, with $0 < \ln b << 1$ (i.e., b is very close to unity). The small s (short wavelength) parts of the integration can be combined with the original constant term (either K^{-1} or Γ^{-1}) to yield a new parameter value and the large s part of the integration rescaled so that the integrals again run from 1 to ∞. The scale factor is absorbed in a redefined y variable. Eq. (36) can thus be rewritten in exactly the same functional form with K, y, and Γ replaced by K(l), y(l) and Γ(l), with $l \equiv \ln b$. This equivalence implies the differential recursion relations

$$\frac{dK(l)}{dl} = -\ \pi^3 y^2(l) \qquad (37)$$

$$\frac{1}{2}\frac{dy^2(l)}{dl} = -\ [\pi K(l)-2]y^2(l) \qquad (38)$$

$$\frac{d\ \ln\Gamma(l)}{dl} = -\ \frac{\pi^4 y^2(l)}{\pi K(l)-1} \qquad (39)$$

subject to the boundary conditions $K(l = 0) = K$, etc.

The first two equations are essentially identical with the static recursion relations found by Jose et al. (1977) and Nelson and Kosterlitz (1977) in their analysis of the planar model and the 2D coulomb gas. We will study them further before discussing the dynamical implications contained in Eq. (39). Defining the variable $x(l) \equiv \pi K(l)-2$, Eq. (37) can be rewritten

$$\frac{1}{2}\frac{dx^2(l)}{dl} = -\ \pi^4 x(l) y^2(l) \qquad (40)$$

and comparing with Eq. (38), we see there is a conserved quantity

$$x^2(l) - \pi^4 y^2(l) = \text{const} = x^2(0) - \pi^4 y^2(0) \tag{41}$$

As long as $x(l) > 0$, Eq. (38) drives $y(l)$ to zero as $l \to \infty$. This provides a justification for the original expansion in powers of y in this temperature regime. The roughening point can be thought of as the low temperature end point of this line of "critical" points with $y(\infty) = 0$ and at this end point we must have $x(\infty) = 0$ or $K(\infty) = \frac{2}{\pi}$. This value is <u>universal</u> [i.e. independent of the initial value of y and a number of other modifications in the initial hamiltonian that could be envisioned (Nelson and Kosterlitz, 1978)] and should hold for all roughening models. When applied to other systems, this prediction implies a universal jump in the superfluid density of ^4He films as T_C is approached from below and the universal value $\eta = \frac{1}{4}$ for the critical exponent describing the decay of correlations at T_C in the planar model (Nelson and Kosterlitz, 1978).

Another universal feature comes from Eq. (41) when we evaluate it at $l = \infty$ for temperatures greater than T_R. Then $y(\infty) = 0$ and

$$x^2(\infty) = \left[x^2(0) - \pi^4 y^2(0)\right] \qquad T \geq T_R \tag{42}$$

Very near T_R, We can expand the right hand side in a power series about $T-T_R$, noting that the constant term vanishes since at T_R, $x(\infty) = 0$. We get to lowest order

$$x(\infty) = [A(T - T_R)]^{\frac{1}{2}} \qquad |T - T_R| << 1 \tag{43}$$

where A is a nonuniversal constant, but the square root cusp is again universal. It has already shown up in Eqs. (13) and (14). Finally a (nonuniversal) estimate of T_R can be found from Eq. (42) when we set $x(\infty) = 0$. Recalling that $y = y_0 \exp[-\frac{1}{2}\pi^2 K]$ [see Eq. (36)] we have the equation

$$K = \frac{2}{\pi} + \pi^2 y_0 \exp[-\tfrac{1}{2}\pi^2 K] \qquad (T = T_R) \tag{44}$$

Setting $y_0 = 1$ to approximate the DG model, we solve Eq. (44) by iteration and find $kT_R/J \cong 1.45$. As we will see this value is in excellent agreement with Monte Carlo estimates.

Further analysis of the static equations is possible but we now examine the behavior of the dynamical parameter Γ in Eq. (39). Eliminating $y^2(l)$ between Eqs. (37) and (39) and integrating we have

$$\frac{\Gamma(\infty)}{\Gamma} = \frac{\pi K(\infty) - 1}{\pi K - 1} \tag{45}$$

Hence Γ effectively scales with K, whose behavior we have discussed above. Eq. (45) shows that the renormalized Γ is reduced from its bare value, but does not vanish along the entire fixed line of critical points which characterizes the roughened phase including the end point at T_R. Using the language of Hohenberg and Halperin, (1977), the dynamics is thus conventional. However, the mutual scaling of K and Γ represents an interesting and somewhat unconventional feature of the model. The calculations given above show that $K(\infty)$ has a square root cusp as $T \to T_R$; thus it should be possible to observe a similar anomaly in $\Gamma(\infty)$.

These results have several immediate consequences for the static and dynamic behavior of the crystal-vapor interface. For example, the average growth rate R of a crystal is related to the response to a spatially and temporally uniform driving force when the stabilizing field $g^2 = 0$. To first order in $\Delta\mu$ it is given by

$$R = \lim_{\omega \to 0} - i\omega\chi(q=0,\omega)\frac{\Delta\mu}{T} \tag{46}$$

$$= \Gamma(\infty)\,\frac{\Delta\mu}{T} \qquad (T \geq T_R) \tag{47}$$

Thus the theory predicts linear growth at and above T_R in agreement with conventional theories of crystal growth.

Below T_R the situation is very different. Approaching the roughening temperature from below, the response function has the limiting form

$$\chi(q,\omega) = \left[K'\left(q^2 + \xi^{-2}\right) - i(\omega/\Gamma')\right]^{-1} \tag{48}$$

with a finite correlation length ξ and renormalized coefficients K' and Γ'. Eq. (46) then predicts a zero growth rate for $T < T_R$ to first order in $\Delta\mu/T$. This result is consistent with the fact that growth at low temperatures occurs by a nucleation mechanism. Nucleation theory gives the result $R \propto \exp(-c/\Delta\mu)$, so in fact below T_R all terms in a power series about $\Delta\mu = 0$ should vanish.

This change in growth mechanisms is directly related to the change in the equilibrium spatial and temporal correlations between different parts of the interface. The height-height correlation function can be immediately calculated from the fluctuation-dissipation theorem (see, e.g., Ma, 1976)

$$< |h_0(q,\omega)|^2 > = \frac{2}{\omega} \, \mathrm{Im}[\chi(q,\omega)] \tag{49}$$

where Im [ ] denotes the imaginary part. In particular, for $T \gtrsim T_R$ and large s or large t,

$$< [h_0(s,t) = h_0(0,0)]^2 > \cong \frac{K(\infty)}{2\pi} \ln \left\{ \max \left[s^2, \frac{4\Gamma(\infty)t}{K(\infty)} \right] \right\} \tag{50}$$

where we have used some results of de Gennes (1971). Thus there are logarithmically diverging correlations in space and time above T_R. Note that the coefficient of the logarithm involves only the <u>renormalized</u> coupling constant $K(\infty)$. As discussed before, the RG theory predicts that $K(\infty)$ takes on the universal value $\frac{2}{\pi}$ and from Eq. (43) that there is a square root cusp near T_R. Thus an accurate determination of the equal time correlation function G(r) in Eq. (12) [i.e., Eq. (50) with t = 0] provides a direct test of these universal predictions.

The large distance limiting value of the equal time correlation function gives a measure of the interface width. Eq. (50) shows that the interface width diverges logarithmically for $T \gtrsim T_R$. Similar remarks apply to the temporal correlations. Eq. (50) also implies that the correlation length ξ is infinite for all $T \gtrsim T_R$.

Below T_R, Eq. (48) holds and the correlation functions reach <u>finite</u> asymptotic values exponentially fast. In particular, the interface width is finite below T_R and there is a finite correlation length. There are many other interesting features of the roughening point which follow from a more careful analysis of the renormalization group equations. (See, e.g., Ohta and Kawasaki, 1978). We will instead discuss an exactly solvable model where the RG predictions can be checked as well as the results of computer simulations, both of which are in accord with the Kosterlitz RG theory.

VIII. THE FSOS MODEL AND MC CALCULATIONS

Van Beijeren (1977) showed that there is a particular roughening model which is isomorphic to the exactly solvable F model. (The F model is a special case of the symmetric six-vertex model in which the two vertices with no net polarization are given the lowest energy, and hence it describes an antiferroelectric system).

Consider a SOS model for the (001) face of a face-centered-cubic crystal with nearest-neighbor interactions between the atoms. At T = 0 nearest neighbor columns in the x and y directions differ by one atom since half the columns terminate in the layer directly below the outermost surface layer. Now constrain the system such that at all temperatures these nearest neighbor columns can differ by at most $\pm$ 1 atom. Thus we are completely suppressing the higher energy multiple height jumps between neighboring columns. As argued before, this should have no effect on the critical behavior and we expect that this model, which we call the FSOS model, is in the same universality class as the other SOS models (ASOS, DG, XY, ...) we have been discussing.

Van Beijeren (1977) showed by a simple argument (which we will not reproduce here since it is very clearly presented in the original work) that the allowed column configurations in the FSOS model can be placed in exact correspondence with the vertex configurations of the F model, and hence the two systems are isomorphic. (Van Beijeren actually considered a bcc crystal with next nearest neighbor forces but his argument applies equally well to the nearest neighbor fcc model, which seems more physically realistic.) Thus one can make use of the results for the exact solution of the F model (Lieb, 1967, and Lieb and Wu, 1972) to test the predictions of the RG theory given in the last section.

Van Beijeren showed that there is indeed a roughening transition in the FSOS model and that at T_R the free energy to form a step vanishes. There is no divergence in the specific heat at T_R but the free energy has an essential singularity of exactly the form [Eq. (14)] predicted by the RG theory.

Furthermore we can make use of the very recent results of Youngblood, Axe and McKoy (1979 as discussed in this Institute) to analyze the exact behavior of the height-height correlation function G(r) [see Eq. (12)] in the FSOS model. They find for all $T \geq T_R$, the exact result for large separation r:

$$G(r) \sim \frac{A(T)}{\pi} \ln r \tag{51}$$

where at T_R, $A(T_R) = \frac{2}{\pi}$ and there is a square root cusp as A(T) approaches its value $\frac{2}{\pi}$ at T_R. These exact results are in precise agreement with the universality predictions of the RG theory as discussed after Eq. (50) and provide a dramatic confirmation of the Kosterlitz RG approach. Unlike most other applications of RG methods, where approximate results are obtained for the system of interest from, say, an ε expansion, the Kosterlitz RG method appears exact for the 2D systems it was designed to treat.

The success of the theory for the FSOS model suggests that one can use the theory to help predict the transition temperature in other SOS models. Shugard, Weeks and Gilmer (1978) performed MC calculations for G(r) in the FSOS model where the G(r) is known exactly and showed that the MC method using a 60 × 60 system gave a very accurate representation of G(r) until about r = 12 where finite size effects became significant. Their data at T_R could be very accurately fit as in Eq. (51) with $A(T_R) = \frac{2}{\pi}$. Thus accurate MC calculations of G(r) are possible despite the problems of finite system sizes and finite run times.

Shugard et al. then calculated G(r) for the ASOS and DG models and determinined T_R by finding the temperature where the best fit to the curve using Eq. (51) gave an $A(T) = 2/\pi$. They found $kT_R/J = 1.24$ for the ASOS model, in good agreement with the series expansion estimates of Weeks et al. (1973). For the DG model, they estimate $kT_R/J = 1.46$, considerably above the unrenormalized value $4/\pi = 1.28$ predicted by the theory of Zittartz (1978), but in excellent agreement with the KT estimate given after Eq. (44). We believe these values are much more accurate than previous estimates by Swendsen (1977), which were based on an assumed divergence in the specific heat. As discussed before, the KT theory predicts no divergence in the specific heat.

Shugard et al., also studied the planar model by simulating the dual SOS model given in Eq. (24). They find $kT_C/J = 0.90$ which agrees fairly well with the series expansion estimates of 0.95 given by Lambreth and Stanley (1975). The data definitely rules out the value $1.1 \sim 1.2$ given by Miyashita et al., (1978), on the basis of direct MC simulations of the planar model and shows the advantage of a simulation using the dual SOS model with its discrete excitations.

Finally we mention that simulations of time-dependent correlation functions give results in excellent agreement with the theory of Chui and Weeks (1978), Eq. (50). In particular, the diffusion-like s^2-t scaling holds for all $T \geq T_R$.

IX. FINAL REMARKS

Since the time of Burton, Cabrera and Frank (1951) there has been considerable progress in our understanding of the nature of the roughening transition. It is in the same universality class as the phase transition in the planar model and the theory of Kosterlitz and Thouless (1973, 1974) provides precise predictions for a number of experimentally accessible properties. A quantitative experimental study of the roughening transition could provide a crucial test of these important theoretical ideas.

There remain some interesting problems for the theorist. Little work has been done on the roughening transition for multicomponent systems. Under certain conditions this may be in a different universality class (Knops, 1979, private communication). Also the analysis of the crystal growth rate has only been done using linear response theory. This is inadequate to uncover the details of the disappearance of the nucleation barrier as $T \to T_R$ from below. A treatment accurate to all orders in the driving force $\Delta\mu$ could give additional insight into nucleation theory. Finally the study of roughening in models more general than the SOS model would be instructive. For example, we believe very strongly that the roughening transition for the interface in the unrestricted 3D lattice gas lies in the same universality class as the restricted SOS models, but there is no rigorous proof. A related question which the existence of a roughening transition brings up is the degree to which interfacial properties such as the interface width can be thought of as intrinsic (independent of system size and external field strength). Some preliminary thoughts on this subject have been given by Widom (1972), and Weeks (1977), but no rigorous analysis has been done.

Acknowledgments

I am very grateful to S. T. Chui, G. H. Gilmer, H. J. Leamy and W. J. Shugard, without whose collaborative efforts much of the work described herein could not have been done. I am also indebted to P. C. Hohenberg, K. A. Jackson, and F. H. Stillinger for a number of stimulating discussions.

References

Burton, W. K., and Cabrera, N., 1949, Disc. Faraday Soc. 5, 33.
Burton, W. K., Cabrera, N., and Frank, F. C., 1951, Phil. Trans Roy. Soc. London 243A, 299.
Buff, F. P., Lovett, R. A., and Stillinger, F. H., 1965, Phys. Rev. Lett. 15, 621.
Chui, S. T., and Weeks, J. D., 1976, Phys. Rev. B14, 4978.
Chui, S. T., and Weeks, J. D., 1978, Phys. Rev. Lett. 40, 733.
Cramer, H., 1946, Mathematical Methods of Statistics, Princeton University Press, Chap. 24.
de Gennes, P. G., 1971, Faraday Symposium #5 on Liquid Crystals, London, p. 16.
Gilmer, G. H., and Bennema, P., 1972, J. Appl. Phys. 43, 1347.
Gilmer, G. H., and Jackson, K. A., 1977, in Crystal Growth and Materials, North Holland, New York, p. 79.
Hohenberg, P. C., and Halperin, B. I., 1977, Rev. Mod. Phys. 49, 435.
Jackson, K. A., 1958, in Liquid Metals and Solidification, ASM, Cleveland, p. 174.

Jackson, K. A., 1967, in _Prog. in Solid State Chemistry_, ed. by H. Reiss, Pergamon Press, New York, vol. 4, p. 53.
Jackson, K. A., and Miller, C. E., 1977, J. Crystal Growth _40_, 169.
Jose, J. V., Kadanoff, L. P., Kirkpatrick, S., and Nelson, D. R., 1977, Phys. Rev. _B16_, 1217.
Knops, H. J. F., 1977, Phys. Rev. Lett _39_, 776.
Kosterlitz, J. M., and Thouless, D. J., 1973, J. Phys. _C6_, 1181.
Kosterlitz, J. M., 1974, J. Phys. _C7_, 1046.
Lambeth, D. N., and Stanley, H. E., 1975, Phys. Rev. _B12_, 5302.
Leamy, H. J., Gilmer, G. H., and Jackson, K. A., 1975, in _Surface Physics of Materials_, Academic, New York, Vol. 1, p. 121.
Leamy, H. J., and Gilmer, G. H., 1974, J. Crystal Growth _24_, 499.
Lieb, E. H., 1967, Phys. Rev. Lett. _18_, 1046.
Lieb, E. H., and Wu, F. Y., 1972, in _Phase Transitions and Critical Phenomena_, edited by C. Domb, Academic, London, Vol. 1.
Lighthill, M. J., 1959, _Introduction to Fourier Analysis and Generalized Functions_, Cambridge University Press, Cambridge, p. 68.
Ma, S. K., 1976, _Modern Theory of Critical Phenomena_, Benjamin, W.A., Reading.
Miyashita, S. H., Nishimori, H., Kuroda, A., and Suzuki, M., 1978, Prog. Theoret. Physics _60_, 1669.
Müller-Krumbhaar, H., 1977, in _1976 Crystal Growth and Materials_, North Holland, New York, p. 79.
Nelson, D. R., and Kosterlitz, J. M., 1977, Phys. Rev. Lett. 39, 120.
Ohta, T., and Kawasaki, K., 1978, Prog. Theor. Phys. _60_, 365.
Saito, Y., 1978, Z. Phys. _B32_, 75.
Shugard, W. J., Weeks, J. D., and Gilmer, G. H., 1978, Phys. Rev. Lett. _31_, 549.
Swendsen, R. H., 1977, Phys. Rev. _B15_, 5421.
van Beijeren, H., 1975, Commun. Math. Phys. _40_, 1.
van Beijeren, H., 1977, Phys. Rev. Lett. _38_, 993.
Villain, J., 1975, J. Phys. (Paris) _36_, 581.
Weeks, J. D., Gilmer, G. H., and Leamy, H. J., 1973, Phys. Rev. Lett. _31_, 549.
Weeks, J. D., 1977, J. Chem. Phys. _67_, 3106.
Weeks, J. D., and Gilmer, G. H., 1979, in _Advances in Chemical Physics_, edited by I. Prigogine and S. A. Rice, vol. 40, p. 157.
Zittartz, J., 1978, Z. Phys. _B31_, 63, 79, 89.

STATICS AND DYNAMICS OF THE ROUGHENING TRANSITION: A SELF-CONSISTENT CALCULATION

Yukio Saito

IFF, KFA Jülich
Postfach 1913
D-5170 Jülich, Fed. Rep. of Germany

I. INTRODUCTION

Crystal growth is strongly influenced by the configuration of the interface between the crystal and the vapour phases.[1] The relation between the growth rate and the interfacial configuration was investigated by Chui and Weeks[1,2] by means of linear-response theory and by using the Kosterlitz renormalization-group (RG) method. Above the roughening temperature T_R, where the interface is rough, the growth rate R is found to be proportional to the chemical potential difference $\Delta\mu$, whereas below T_R, where the interface is flat, the crystal is found not to grow. A puzzling feature of their results is that the ratio $R/\Delta\mu$ for $\Delta\mu \to 0$ is finite on approaching T_R from above, whereas it remains zero on approaching T_R from below. Interesting unanswered questions concern the extent of the non-growing region (in $\Delta\mu$) below T_R, and the nature of the growth for larger $\Delta\mu$.

In this paper the discontinuity of the ratio $R/\Delta\mu$ at T_R for small $\Delta\mu$ and the growth rate R for large $\Delta\mu$ are investigated. The range of $\Delta\mu$ considered extends beyond the linear-response region. Since the crystal grows, we have to consider states far from equilibrium. Because of the nonlinear and nonequilibrium complications, we use a simpler method than the RG method, namely the self-consistent approximation (SCA),[3] for a unified investigation of both the statics and the dynamics of the roughening transition.

By using the duality relation[4,5,6] between crystal growth models and magnetic systems in two-dimensions (2d), we also investigate the critical behavior of the 2d planar spin model in the SCA.

II. ROUGHENING TRANSITION

The interface of the crystal is characterized by height variables $\{h_i\}$ on the 2d lattice sites. The h_i should be integers, but for mathematical conveniences we consider heights varying continuously from $-\infty$ to $+\infty$ with a weight which favors integer values. The Hamiltonian is given by

$$\mathcal{H}= J/2 \sum_i \sum_\delta (h_i-h_{i+\delta})^2 + y \sum_i (1-\cos 2\pi h_i)-\Delta\mu \sum_i h_i \ , \quad (1)$$

where the periodic potential represents the weight factor mentioned above. The ground states of Eq. (1) with $y>0$ are the same as those of the discrete Gaussian (DG) model, where the height is restricted to have the integer values. The system is kept in contact with a heat bath with temperature T, and evolves according to the following Fokker-Planck equation for the probability $P(\{h_i\},t)$:

$$\tau \frac{P(\{h_i\},t)}{\partial t} = \sum_i \frac{\delta}{\delta h_i}\left(\frac{\delta \mathcal{H}}{\delta h_i} + k_B T \frac{\delta}{\delta h_i}\right) P(\{h_i\},t) \ . \quad (2)$$

The most important physical quantities are the first and the second cumulants of the height. The average height h(t) gives the growth rate R=dh/dt and characterizes the dynamics, whereas the second cumulant $G_{ij}=\langle h_i h_j\rangle-\langle h_i\rangle\langle h_j\rangle$ measures the roughness of the interfacial configuration and characterizes the statics. Due to the non-linearity of the Hamiltonian (1) these cumulants are coupled with higher order cumulants in the equation of motion. In order to eliminate the hierarchy, we introduce a Gaussian approximation for the probability distribution

$$P(\{h_i\},t)=\exp\left[-\frac{1}{2} \sum_i \sum_j (h_i-h(t))\ G^{-1}_{ij}(t)\ (h_j-h(t))\right] \quad (3)$$

which is characterized only by the average height h(t) and the correlation $G_{ij}(t)$. Then h(t) and the Fourier-transformed height correlation G(q,t) satisfy the equations of motion

$$\tau dh(t)/dt =\Delta\mu-\Delta\mu_c \sin 2\pi h(t) \ , \quad (4)$$

$$\tau dG(q,t)/dt=4JG(q,t)[q^2+\xi^{-2} \cos 2\pi h(t)-K^{-1} G(q,t)], \quad (5)$$

where

$$\Delta\mu_c=2\pi y \exp[-2\pi^2 N^{-1} \sum_q G(q,t)]=(J/\pi)\xi^{-2} \ , \quad (6)$$

$$\xi^{-2}=(2\pi)^2 (y/2J) \exp[-2\pi^2 N^{-1} \sum_q G(q,t)] \ , \quad (7)$$

$$K =k_B T/2J \ . \quad (8)$$

According to Eq. (4), the crystal growth rate is proportional to the external driving force $\Delta\mu$, but due to the periodic potential the rate is reduced. Note that the coefficient of the reduction $\Delta\mu_c$ is related to the interfacial configuration through G(q). Thus the static roughening transition, which changes the configuration of the interface, affects the dynamics of the crystal growth. The evolution (5) of G(q,t) for small wave vector q consists of a deterministic part due to the Hamiltonian and a stochastic part due to the thermal fluctuation with dimensionless temperature K. The effect of integer preference for the heights is modified by the coefficient ξ^{-2}, which is related to the interfacial configuration through G(q) in Eq. (7).

We now summarize the static and dynamic results. When $\Delta\mu=0$, the system has a static solution. The average height takes integer values, and the correlation G(q) has the Ornstein-Zernike form;

$$G_{eq}(q) = K/(q^2+\xi^{-2}) \quad . \tag{9}$$

Here the correlation length ξ is determined self-consistently by Eq. (7). ξ is infinite at high temperatures ($K \geq K_R = 2/\pi$) and remains finite, $\ln \xi \sim (K-K_R)^{-1}$, at low temperatures ($K < K_R$). The dimensionless roughening temperature $K_R=k_BT_R/2J$ is $2/\pi$. From Eq. (9) we can calculate the height difference correlation $\langle(h_i-h_j)^2\rangle$. Above K_R it diverges logarithmically for large separation, $\langle(h_i-h_j)^2\rangle \approx K/\pi \ln r_{ij}$, indicating the uncorrelated heights and the rough interface. Below K_R, on the other hand, the correlation saturates for large separation ($r_{ij} \gg \xi$) as

$$\langle(h_i-h_j)^2\rangle \approx K/\pi \ln \xi - K\sqrt{\xi/2\pi r_{ij}} \exp(-r_{ij}/\xi) \tag{10}$$

and the saturation value shows the critical behavior as $\ln\xi\sim(K-K_R)^{-1}$. This saturation indicates the strong correlation between the heights and the interface is flat.

When a finite $\Delta\mu$ is applied, the crystal grows. Since our main concern is the growth rate, we assume that the interfacial configuration is the same as the equilibrium one, i.e. $G(q,t)=G_{eq}(q)$. Then the average height evolves according to Eq. (4), with the time-independent coefficient $\Delta\mu_c$ given by Eq. (6). At high temperatures where ξ is infinite, $\Delta\mu_c$ vanishes and the growth rate is proportional to the external driving force, i.e. $R=\Delta\mu/\tau$. At low temperatures, $\Delta\mu_c$ remains finite and non-zero, and the growth is reduced for $\Delta\mu$ smaller than $\Delta\mu_c$. Only after $\Delta\mu$ exceeds the spinodal value $\Delta\mu_c$, does the crystal begin to grow. This growth occurs with oscillating R. The period of oscillation is the time necessary for the interface to proceed a unit height. The oscillation reflects the flatness of the growing interface. Solving Eq. (4) exactly, we find the averaged growth rate $\bar{R}=\sqrt{(\Delta\mu)^2-(\Delta\mu_c)^2}/\tau$.

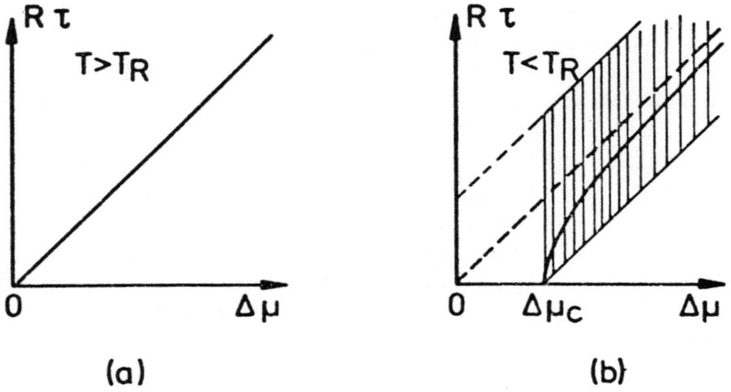

Fig. 1. The growth rate R versus chemical potential difference $\Delta\mu$ (a) at a high temperature and (b) at a low temperature. The broken line in (b) represents the growth rate for the rough interface, and the hatched region represents the region where the rate oscillates.

For sufficiently large $\Delta\mu$ the rate approaches the rough phase value $\Delta\mu/\tau$. The result is summarized in Fig. 1. When the roughening temperature is approached from below, the spinodal value $\Delta\mu_c$ decreases, the non-growth region shrinks and finally at K_R unhampered growth takes place.

III. TWO-DIMENSIONAL PLANAR MODEL

The duality relation between crystal growth models and planar magnets in 2d is well known.[4,5] For example, the magnetic correlation function g(r) of the Villain model is related to the step free energy F(r) of a unit height step of the DG model[5,6] by the formula

$$g(r_{ij}) \equiv \langle \exp i(\Theta_i - \Theta_j) \rangle = \exp[-F(r_{ij})/k_B T]. \tag{11}$$

Here Θ_i is the spin angle, and the temperature T of the DG model is inversely proportional to that of the Villain model T^V.

Approximating F(r) of the DG model by that of the modified model (1), one finds[7] that g(r) follows a power law, $g(r) \sim r^{-\eta}$ with $\eta = 1/2\pi K$ for low temperatures $T^V < T_c^V$ (or $K \geq K_R$), and an exponential law, $g(r) \sim \xi^{-\eta} \exp(-r/\xi)$ for high temperatures $T^V > T_c^V$ (or $K < K_R$). The power-law decay of the correlation function g(r) gives an infinite susceptibility χ below T_c^V, whereas χ remains finite above T_c^V. The exponent η depends on temperature, and at T_c^V (or $K=K_R$) it takes the value 1/4, which agrees with the result of the Kosterlitz RG method.

IV. CONCLUSIONS

Using a self-consistent approximation, a unified treatment was given for the statics and dynamics of the roughening transition. Qualitative features of the statics are found to be the same with those obtained by the more complex RG method, although some critical exponents differ.[1,7,8] The spinodal value $\Delta\mu_c$ of the chemical potential difference, which characterizes the growth of the crystal, is related to the correlation length which characterizes the interfacial configuration. The growth rate for large $\Delta\mu$ below roughening is obtained analytically, and a smooth change of the growth rate is found at the roughening temperature.

REFERENCES

1. J.D. Weeks, "Roughening Transition" (in this volume) and references cited.
2. D.L. Chui and J.D. Weeks, Phys. Rev. Lett. 40, (1978) 733.
3. Y. Saito, Z. Phys. B32, (1978) 75.
4. H.J.F. Knops, Phys. Rev. Lett. 39, (1977) 776.
5. J.V. Jose, L.P. Kadanoff, S. Kirkpatrick, and D.R. Nelson, Phys. Rev. B16, (1977) 1217.
6. R.H. Swendsen, Phys. Rev. B17, (1978) 3710.
7. Y. Saito, submitted to Prog. Theor. Phys.
8. T. Ohta and K. Kawasaki, Prog. Theor. Phys. 60, (1978) 365.

FLUCTUATIONS IN TWO-DIMENSIONAL SIX-VERTEX SYSTEMS

R. W. Youngblood, J. D. Axe and B. M. McCoy*

Physics Department
Brookhaven National Laboratory
Upton, N.Y. 11973

*Institute of Theoretical Physics
State University of New York at Stony Brook
Stony Brook, N.Y. 11794

ABSTRACT

The character of polarization correlations in six-vertex systems will be discussed. Making use of a connection between the 1-d Heisenberg-Ising chain and the six-vertex problem, we draw upon existing results for the chain correlations to obtain information about long-wavelength polarization correlations in six-vertex models. These results are compared with a neutron scattering study of 2-d polarization correlations in the layered compound copper formate tetrahydrate. Because the six-vertex model is equivalent to a particular roughening model, these results also explicitly predict the critical behavior of that roughening model just above its roughening temperature. The results correspond to the predictions of Kosterlitz and Thouless for the phase transition in the 2-d Coulomb gas.

Although quite a bit is known about the statistical mechanics of 2-d six-vertex systems [1], much remains to be said about the fluctuations in these models. Here, we will present a short discussion of those aspects of the fluctuations that seem particularly germane to the subject of this particular conference. In particular, we will discuss analytic expressions for the long wavelength polarization correlation functions for these models which are asymptotically exact at large distance [2]. We will apply these results to neutron scattering experiments on a quasi-2-d hydrogen-bonded system, copper formate tetrahydrate (CFT) [3]. We will also show the relevance of these results to models of the solid-on-solid interfacial roughening transformation [4], and comment on the relation of Kosterlitz-Thouless theory [5] to six-vertex systems.

The models under discussion have been introduced by Dr. Weeks in his lectures at this conference [4]. The allowed vertex configurations are prescribed by the ice rules. The vertex weighting scheme is shown in Figure 1. (In adopting these weights, we are tacitly ruling out applied fields.) If vertices 1 and 2 are favored energetically, the ground state is ferroelectric along $\pm$ y; if 3 and 4 are favored, the ground state is ferroelectric along $\pm$ x; if 5 and 6 are favored, the ground state is antiferroelectric. The parameter Δ, defined in Fig. 1, is a measure of how close the system is to being ferro- or antiferroelectric. η, also defined in Fig. 1, measures the anisotropy in the polar configurations. Baxter [5] showed that singularities in the free energy corresponding to ferroelectric and antiferroelectric transformations occur at $\Delta = +1$ and -1, respectively. The relation between the three phases (antiferroelectric, disordered, and ferroelectric) is summarized in Fig. 1. For the weighting scheme we employ, no single physical system displays the full range of behavior shown in Fig. 1; the point $\Delta = 1/2$ corresponds to $T = \infty$. But the polarization correlations we discuss are a property of the entire disordered regime, and it is natural to discuss them as a function of Δ rather than T. We will also have occasion to mention the so-called FSOS roughening model [4] and the 1-d Heisenberg-Ising chain; certain relevant attributes which arise from equivalences within these models are summarized in Fig. 1.

Our primary interest is in long-wavelength polarization fluctuations. For this purpose, it is convenient to define a coarse-grained polarization. Consider the correlation function between two parallel arrows for the isotropic case, a = b. Introduce the variable $\sigma_j(m,n)$ to denote the sense of the arrow at site (m,n) (the subscript j = 1,2 denotes the arrow direction -- right or left sloping in Fig. 1). For $\Delta = 0$, Sutherland used an equivalence to a free fermion system to show that for large r

$$\langle\sigma_j(0)\sigma_j(m,n)\rangle \sim \frac{2}{\pi^2 r^2}\left\{(-1)^{m+n} - \frac{(m^2-n^2)}{(m^2+n^2)}\right\} \tag{1}$$

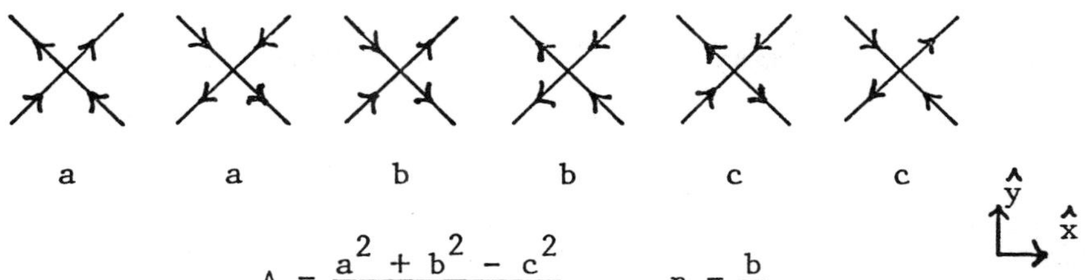

$$\Delta = \frac{a^2 + b^2 - c^2}{2ab} \qquad \eta = \frac{b}{a}$$

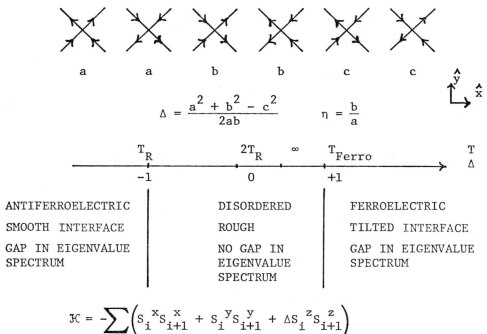

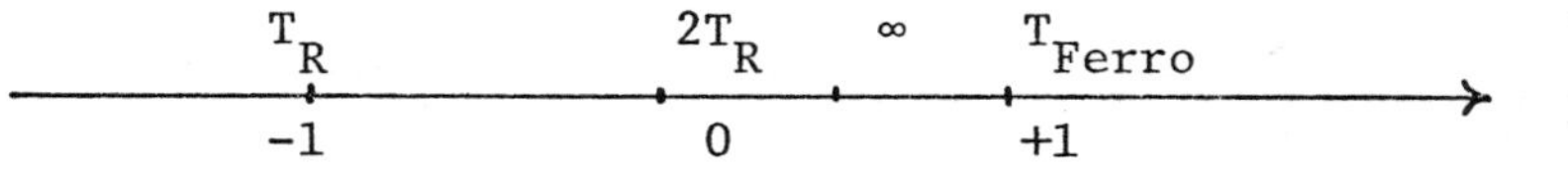

ANTIFERROELECTRIC	DISORDERED	FERROELECTRIC
SMOOTH INTERFACE	ROUGH	TILTED INTERFACE
GAP IN EIGENVALUE SPECTRUM	NO GAP IN EIGENVALUE SPECTRUM	GAP IN EIGENVALUE SPECTRUM

$$\mathcal{H} = -\sum\left(S_i^x S_{i+1}^x + S_i^y S_{i+1}^y + \Delta S_i^z S_{i+1}^z\right)$$

Figure 1.

where m,n are integer coordinates and $r = \sqrt{m^2+n^2}$. We see that there is an antiferroelectric contribution which oscillates rapidly, and a ferroelectric component which does not. We therefore form the "coarse-grained" ferroelectric polarization

$$P_j(\vec{r}) = \sigma_j(m,n) + \sigma_j(m+1,n).$$

It is easy to see that the leading contribution to $\langle P_j(0)P_j(\vec{r})\rangle$ is the smooth contribution to $\langle\sigma_j(0)\sigma_j(m,n)\rangle$; the other contributions cancel, to leading order in 1/r. In fact, making use of a connection between this model and the dimer model [6], we have shown [2] that leading contributions to the coarse-grained polarization correlations at Δ = 0 have the form

$$\langle P_y(0)P_y(\vec{r})\rangle \sim -A\,\frac{(x^2-y^2)}{(x^2+y^2)^2} \tag{2}$$

$$\langle P_y(0)P_x(\vec{r})\rangle \sim A\,\frac{2yx}{(x^2+y^2)^2} \tag{3}$$

$$\langle P_x(0)P_x(\vec{r})\rangle \sim A\,\frac{(x^2-y^2)}{(x^2+y^2)^2} \tag{4}$$

where $A = 2/\pi^2$. These quantities also depend on η; for simplicity, we have temporarily set $\eta = 1$. In these expressions and in the following, the symbol $\sim$ means "is asymptotically equal to in the limit of large r."

The simple form of these functions reflects an underlying simplicity in the system. The six-vertex condition (Fig. 1) is the condition that polarization be locally "divergenceless" at each lattice site. This condition is largely responsible for the form of Eq. (2-4). The ice rules (Fig. 1) state that the polarization is "divergenceless" at each lattice site, which in turn means that $\vec{P}(\vec{r})$ is a solenoidal vector field, $\nabla\cdot\vec{P}(\vec{r}) = 0$. This guarantees that $\vec{P}(\vec{r})$ is the curl of a mathematically simpler vector field, $\vec{h}(\vec{r})$. Fortunately, there is a physical as well as a mathematical motivation for introducing $\vec{h}(\vec{r})$. We know from the work of van Beijeren [8] that $\vec{h}(\vec{r})$ is simply related to the height variable of a surface roughening model.

Dr Weeks' lecture in this volume [4] discusses the relation of a particular roughening model (which he denotes FSOS) with the six-vertex model which we are discussing. In particular, Fig. 3 of his third lecture shows on a discrete lattice how to associate a spin variable with the local gradient of the column height h. With the appropriate choice or coordinate system, this is just

$$\sigma_1(m,n) = -[h(m,n+\tfrac{1}{2}) - h(m,n-\tfrac{1}{2})] \tag{5}$$

$$\sigma_2(m,n) = [h(m+\tfrac{1}{2},n) - h(m-\tfrac{1}{2},n)] \tag{6}$$

Thus, for example (Fig. 2a), an ordered polar six-vertex configuration pointing along (say) y, is equivalent to a surface with a monotonically increasing height, $\partial h/\partial x$ = constant. The local polarization conservation $\nabla\cdot\vec{P} = 0$ completely suppresses longitudinal polarization fluctuations, which in the language of the height variable translate into a discontinuous "tear" on the crystal surface, as shown in Fig. 2b. Such configurations correspond to unacceptably large step sizes in the roughening model.

It is clear from the foregoing paragraph that the proper choice of vector potential $\vec{h}(\vec{r}) = h(\vec{r})\hat{z}$, so that

$$\vec{P}(\vec{r}) = (P_x,P_y) = \left(-\frac{\partial h}{\partial y}, \frac{\partial h}{\partial x}\right) = -\nabla\times\vec{h}(\vec{r}). \tag{7}$$

Then, if we define a height-height correlation function

$$\Psi(\vec{r}-\vec{r}_o) = \langle h(\vec{r})h(\vec{r}_o)\rangle \sim -A\ln(|\vec{r}-\vec{r}_o|) \quad ,$$

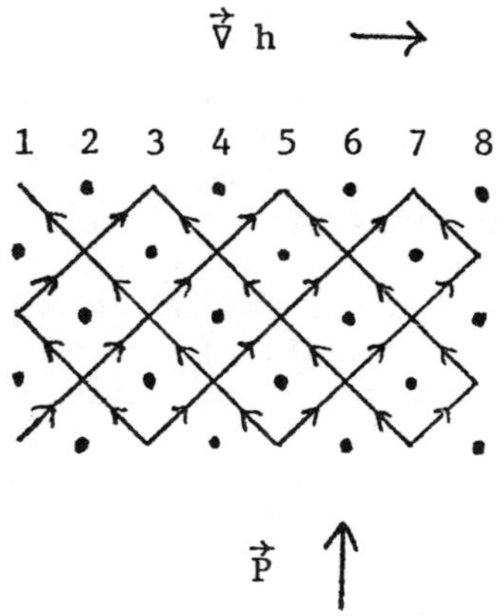

Fig. 2a. The completely polarized state of the six-vertex lattice corresponds to a monotonically sloping surface. (Compare vertex 3 of Fig. 3 of Weeks' third lecture.) The numbers give the heights of the lattice sites above which they appear.

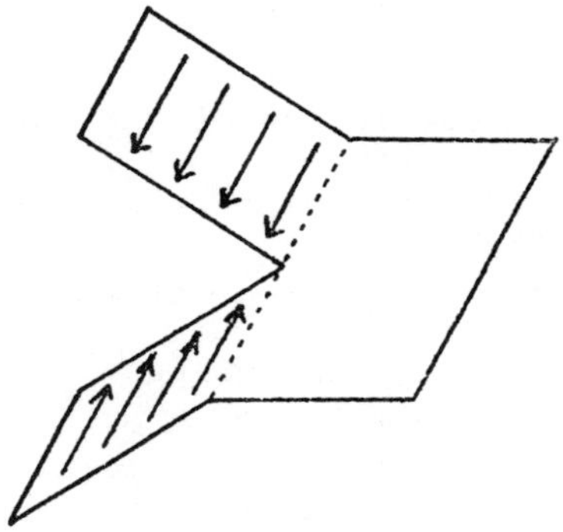

Fig. 2b. A longitudinal polarization fluctuation (violating $\nabla\cdot\vec{P} = 0$) corresponds to a tear in the surface (violating the step size constraint in the roughening model).

Eqs. (2-4) can be restated very concisely in terms of the height-height correlation function, $\Psi(\vec{r}-\vec{r}_o)$.

$$<P_y(\vec{r}_o)P_y(\vec{r})> = \frac{\partial}{\partial x_o}\frac{\partial}{\partial x}\Psi(\vec{r}-\vec{r}_o) \tag{8}$$

$$<P_y(\vec{r}_o)P_x(\vec{r})> = -\frac{\partial}{\partial x_o}\frac{\partial}{\partial y}\Psi(\vec{r}-\vec{r}_o) \tag{9}$$

$$<P_x(\vec{r}_o)P_x(\vec{r})> = \frac{\partial}{\partial y_o}\frac{\partial}{\partial y}\Psi(\vec{r}-\vec{r}_o) \tag{10}$$

Thus far, all of our results have been restricted to $\Delta = 0$, where the problem is especially simple. However, we can now proceed to extend the results throughout the regime $-1<\Delta<1$, by making contact with previous work on the 1-d Heisenberg-Ising chain. McCoy and Wu [9] showed that the transfer matrix of the six-vertex problem commutes with the Heisenberg-Ising Hamiltonian, shown in Fig. 1. Therefore, there is a connection between the correlation functions in the two problems. Recent work by Luther and Peschel [10] and Fogedby [11] gives the leading asymptotic contribution to $\langle S^z(0,0)S^z(x,t=0)\rangle$, which (by virtue of the above-mentioned commutation relation) has the same general form as $\langle\sigma_j(0,0)\sigma_j(m,n=0)\rangle$ (equation 1), including a prefactor A which is a known function of Δ. This result effectively prescribes the coarse-grained correlation function along one axis. Since $\nabla\cdot\vec{P} = 0$, there is still a generating vector potential which can be analytically continued from that axis to cover the entire x-y plane. Thus we arrive at the following result, valid for $-1<\Delta<1$.

$$\Psi(\vec{r}) \sim -A\ln r + \text{constant} \tag{11}$$

where $A = (\pi^2\theta)^{-1}$ (Luther and Peschel, [10], Fogedby [11])

and $\theta = \frac{1}{2} - \frac{1}{\pi}\sin^{-1}\Delta$ (Johnson, Krinsky and McCoy [12])

The formulation of the correlation function $\Psi(r)$ given in Eq. (11), together with Eq. (8-10) (which generate the asymptotic polarization correlation functions), constitute the principal results of this paper. We now turn to a discussion of their significance in two different areas.

Since Chui and Weeks [13] have shown that the discrete Gaussian roughening model maps onto the 2-d Coulomb gas problem, and since van Beijeren [8] has explicitly demonstrated that a similar rougening model maps exactly onto the F model, it is natural to suppose (along with Shugard et al. [14] and others [4]), that there is a close connection between the critical behavior of the Kosterlitz-Thouless transition and the critical behavior of the six-vertex transition at $\Delta = -1$. In particular, the prefactor K_∞ appearing [14] in

$$G(r) = 2\left(\langle h(0)^2\rangle - \langle h(0)h(r)\rangle\right) \sim \frac{K_\infty(T)}{\pi}\ln r + c \tag{12}$$

is to be compared with $A(\Delta)$ appearing in Eq. (11). If we can identify $K_\infty(T)$ with $2\pi A(\Delta)$, we can expand $A(\Delta)$ to obtain

$$K_\infty(T \to T_R^+) = \frac{2}{\pi} + (\text{nonuniversal constant}).\ (T-T_R)^{1/2} + .. \tag{13}$$

Both the value $K_\infty(T_R) = \frac{2}{\pi}$ and the leading square root behavior are predicted by Kosterlitz-Thouless theory for the unbinding of vortices. Thus, the present results for Ψ explicitly support the idea that the six-vertex model (together with its various equivalents) is in the same universality class as the 2-d Coulomb gas (together with its equivalents).

Now we turn to polarization fluctuations in CFT. A full description of this work is contained in Refs. [2] and [3]. Crystals of CFT contain 2-d layers of water molecules interleaved with 2-d layers of copper formate. Above $T = T_o = 248$ K (in the deuterated compound), there is icelike disorder in the hydrogen-bond network. Below T_o, the layers become ferroelectric, the direction of polarization alternating between +b and -b from one layer to the next. The in-plane longitudinal momentum coordinate is K; the transverse momentum coordinate is H. Figure 3 shows some results of a diffuse neutron scattering study of the long-wavelength polarization correlations. ("Polarization" here means that of the hydrogen atom positions, measured from the centers of their respective bonds.) The quantity plotted is the measured intensity, $I(\vec{Q})$.

$$I(\vec{Q}) = \overline{|F(\vec{Q})|^2}\; S_{yy}(\vec{q}). \tag{14}$$

$F(\vec{Q})$ is a geometrical structure factor, which is a rather slowly varying function of $\vec{Q} = \vec{G} + \vec{q}$, where $\vec{G}$ is a reciprocal lattice

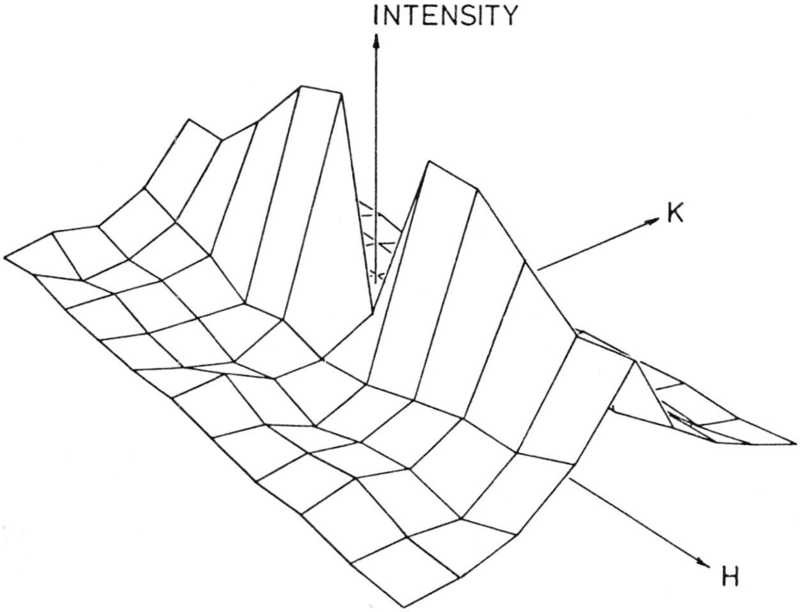

Fig. 3. Shown here are results of a neutron scattering study of polarization fluctuations in the 2-d hydrogen-bond network of copper formate tetrahydrate (Ref. [3]). The origin of the indicated coordinate system corresponds to a reduced intralayer momentum transfer $\vec{q}$ of zero. Data are shown for essentially an entire (intralayer) Brillouin zone.

vector. $S_{yy}(\vec{q})$ is the Fourier transform of $\langle P_y(0)P_y(r)\rangle$, and the bar denotes an average over instrumental resolution. The important feature in this plot is the notch at $\vec{q} = 0$. Typically (for example in Ising-type systems), pair correlations near T_c give rise to scattering which peaks at $\vec{q} = 0$; as T_c is approached, the diffuse peak associated with order parameter fluctuations is seen to grow. Here, instead of peaking, the intensity dips to near zero, in spite of the incipient transition to an ordered state in which there is a Bragg peak at $\vec{q} = 0$. (In calling the origin of Fig. 3 $\vec{q} = 0$, we have suppressed the L momentum coordinate, upon which the scattering is only weakly dependent. This is the expected quasi-2-d behavior. However, it is important to note that the data of Fig. 3 lie in a plane in which L is half-integral, in which the antiferroelectric Bragg peaks occur below T_o.)

In the CFT problem, $\eta \neq 1$, and the hitherto suppressed η dependence of the correlations must be taken into account. It is shown in Ref. [2] that this can be done by replacing r in Eq. (11) with

$$\rho(\vec{r}) = \sqrt{x^2 + \lambda^2 y^2} \tag{15}$$

For $\Delta = 0$, we have $\lambda = \eta$. Polarization correlations are still given by Eqs. (8-10). The Fourier transform of $\langle P_y(0)P_y(\vec{r})\rangle$ is given by

$$S_{yy}(\vec{q}) = \pi A\lambda \frac{h^2}{\lambda^2 h^2 + k^2} \quad \text{at small q.} \tag{16}$$

This function is plotted in Fig. 4 in units of πA, with λ chosen to correspond roughly to the experimental observations. Note that there is a strong formal resemblance between the singularity in this function and that occurring in dipolar-coupled systems (see Dr. Als-Nielsen's notes on $LiTbF_4$). However, we stress that the calculated effect is due entirely to six-vertex ineractions. The absence of longitudinal fluctuations ($S_{yy}(h,k=0)=0$) is a direct consequence of the ice rule restrictions, $\nabla\cdot\vec{P} = 0$. In real scattering experiments, the cross section one observes is somewhat smeared by finite instrumental resolution. This effect is partially taken into account in Fig. 5. There is a strong resemblance between Figs. 3 and 5; thus, the small-q regime (the notch) of Fig. 3 is evidence that the pair correlations in CFT are qualitatively obeying Eq. (2). It would be of some interest to perform further measurements with much higher resolution to test Eq. (8) quantitatively.

In summary, we see that in the disordered phase of a particular class of six-vertex systems, long-wavelength polarization correlations are governed by a logarithmic (2d-Coulomb-like) potential. The scattering cross section of hydrogen-bonded systems of this type depends on particular spatial derivatives of this potential;

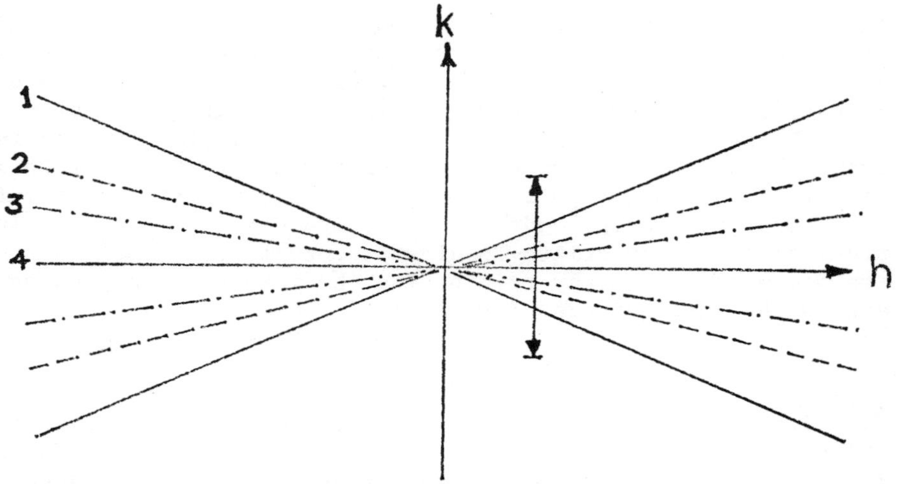

Fig. 4. This is a contour plot of $S_{yy}(q)$ (see Eq. (16)). The contours are at integer multiples of $A\pi$, for the case $\lambda = \frac{1}{4}$. The arrow indicates a longitudinal resolution width (see Fig. 5).

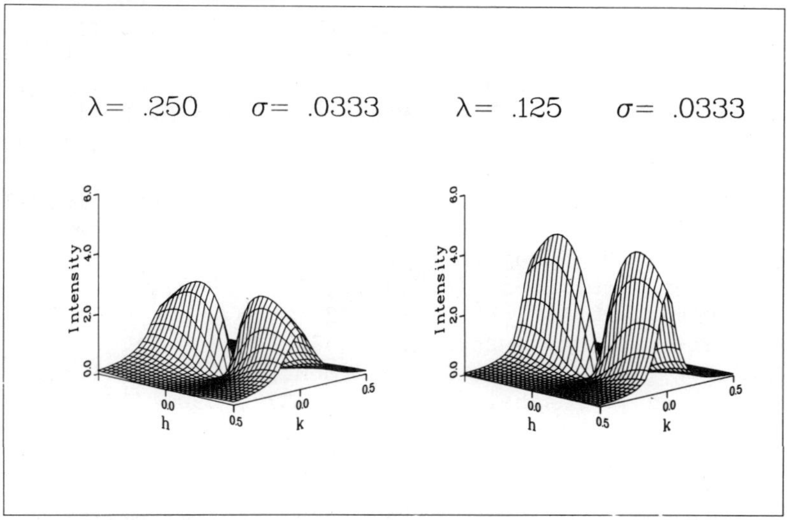

Fig. 5. These are plots of I(Q) calculated from Eq. (14). S_{yy} is given by Eq. (16), with A held constant. F(Q) corresponding to the data shown in Fig. 3 is included. σ is the halfwidth along k of a Gaussian instrumental resolution function (the resolution along h is assumed to be perfect). σ, h, and k are given in reciprocal lattice units.

the observed cross section from such correlations is quite distinctive. Given certain rigorous connections between the Heisenberg-Ising problem and the six-vertex problem, we can draw upon existing results to obtain new information about the six-vertex problem away from the free fermion limit, $\Delta = 0$. In particular, we can infer the temperature dependence of the prefactor of the logarithmic potential. In addition to making an explicit statement about a particular roughening model, the results correspond gratifyingly to predictions of Kosterlitz-Thouless theory for the vortex-unbinding transformation, thereby reinforcing the idea that roughening models are in the same universality class.

ACKNOWLEDGMENT

Two of us (RWY and JDA) have profited from conversations held at this Institute with J. D. Weeks, on the connection between roughening models and the six-vertex problem, and from several conversations with V. J. Emery.

REFERENCES

1. For a review, see the article by E. H. Lieb and F. Y. Wu in "Phase Transitions and Critical Phenomena," ed. C. Domb and M. S. Green, Academic, New York (1977).
2. R. Youngblood, J. D. Axe. and B. McCoy, submitted to Phys. Rev.
3. R. Youngblood and J. D. Axe, Phys. Rev. B17, 3639 (1978).
4. For reviews, see J. D. Weeks, this volume; J. D. Weeks and G. H. Gilmer in "Advances in Chemical Physics," ed. I. Prigogine and S. A. Rice, Vol. 40 (1979); H. Müller-Kvumbhaar in "1976 Crystal Growth and Materials," North Holland, New York (1977).
5. J. M. Kosterlitz and D. J. Thouless, J. Phys. C6, 1181 (1973); J. M. Kosterlitz, J. Phys. C7, 1046 (1974). A review is given by A. P. Young in this volume.
6. R. J. Baxter, Ann. Phys. (N.Y.) 70, 193 (1972).
7. B. Sutherland, Phys. Lett. 26A, 532 (1968).
8. H. van Beijeren, Phys. Rev. Lett. 38, 993 (1977).
9. B. M. McCoy and T. T. Wu, Nuovo Cimento 56B, 311 (1968).
10. A. Luther and I. Peschel, Phys. Rev. B12, 3908 (1975).
11. H. C.Fogedby, J. Phys. C11, 4767 (1978).
12. J. D. Johnson, S. Krinsky, and B. M. McCoy, Phys. Rev. A8, 2526 (1973).
13. S. T. Chui and J. D. Weeks, Phys. Rev. B14, 4978 (1978).
14. W. J. Shugard, J. D. Weeks and G. H. Gilmer, Phys. Rev. Lett. 41, 1399 (1978).

LIGHT SCATTERING STUDIES OF THE TWO-DIMENSIONAL PHASE TRANSITION IN SQUARIC ACID

Erling Fjær, Jorunn Grip and Emil J. Samuelsen

Institutt for Almen Fysikk, University of Trondheim
Norwegian Institute of Technology
N-7034 Trondheim-NTH, Norway

1. INTRODUCTION

Crystalline squaric acid, $H_2C_4O_4$,

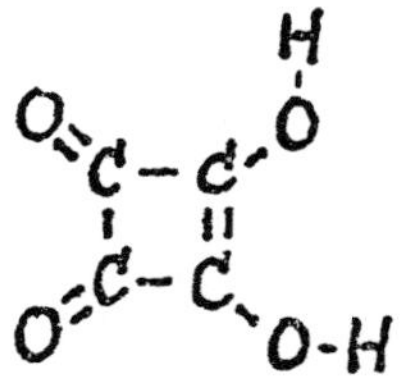

has attracted considerable interest among physicists due to its simple planar geometry [1] and the existence of a structural phase transition at about 100°C [2]. Since proton ordering in hydrogen bonds seems to play a key role in the phase change, it was soon realized that one had at hand a material which might show interesting two-dimensional Ising-type behaviour [3]. Information gained from birefringence studies [2] and neutron scattering [3] did in fact support such a view.

2. LIGHT SCATTERING STUDIES

Previously published light scattering studies of squaric acid [4, 5, 6] have been aimed at classifying the symmetry types of the various modes, rather than at elucidating the nature of the phase transition.

We shall here report on a detailed study of the 83 cm^{-1} Raman line. This line is assigned to the inplane transverse motion of the rigid squaric acid molecules [5, 7]. Referred to the Brillouin zone of the high-temperature b.c. tetragonal unit cell, 83 cm^{-1} corresponds to the zone boundary value of the (degenerate) TA modes along the unique direction normal to the molecular layers [7]. In the low temperature monoclinic phase the unit cell is doubled along the unique b-axis, making this mode Raman active below T_c but invisible (zone boundary) above. At very low temperatures (10K) a slight line splitting is observed [5, 8] due to the monoclinic structure (a and c axes distinct).

3. ORDER PARAMETER

One expects that the intensity of the 83 cm^{-1} line goes with the order parameter $\eta(T)$ squared [7], where $\eta(T)$ measures the positional deviations from the average high temperature positions. Some profiles of the line is shown in figure 1 at various temperatures, and integrated intensities versus temperature is shown in figure 2. The full line of figure 2 is the measured superreflection intensity of elastic neutron scattering [3]. It is seen that the two intensities in fact follow each other. The curve has a temperature dependence

$$I \sim \varepsilon^{2\beta} \tag{1}$$

with

$$\varepsilon = T_c - T \tag{2}$$

and with β = 0,137, as found from the more accurate neutron data [3]. Thus the light scattering data corroborates with the previous conclusion that the order parameter critical index β is quite close to the theoretical two-dimensional Ising value 0.125.

4. ORDER FLUCTUATIONS

It is seen from both figure 1 and 2 that the 83 cm^{-1} line does not vanish entirely at T_c, but remains as a broad, diffuse and unsymmetrical structure at about 78 cm^{-1}. In fact the asymmetry at the base of the peak is evident far below T_c as well, being, however most pronounced near and above T_c. - A "Rayleigh component" of scattering is furthermore observed, centred at zero frequency but with wings out to > 80 cm^{-1}. This component increases strongly at T_c. It is strongly polarised.

The tetragonal symmetry above T_c would predict a total vanishing of the line at 83 cm^{-1}. We interprete the existence of the remnant intensity above T_c in terms of local monoclinic symmetry of the squaric acid molecules being retained into the high temperature phase. This view corroborates well with the diffuse neutron scattering data [3], which indicated fairly long range (> 25 unit cells) two-dimensional correlations within the molecular layers,

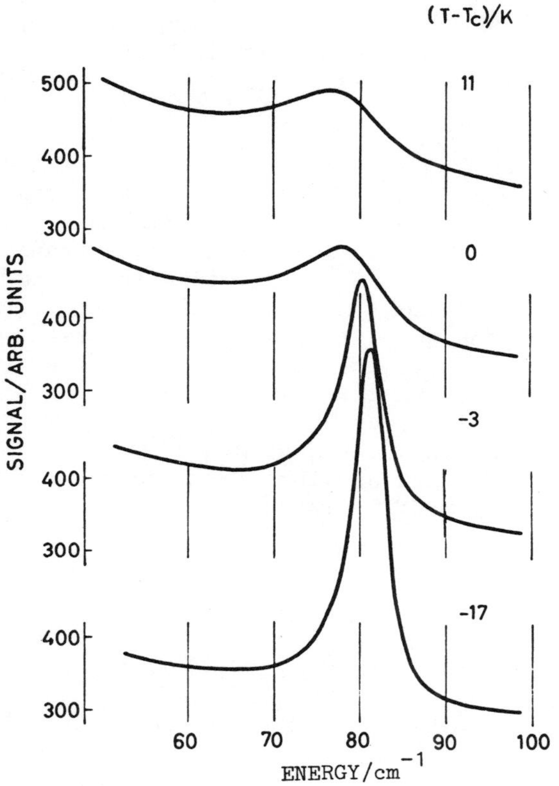

Figure 1.

Smoothed profiles at various temperatures near T_c of the 83 cm^{-1} line. Notice pronounced structure above T_c.

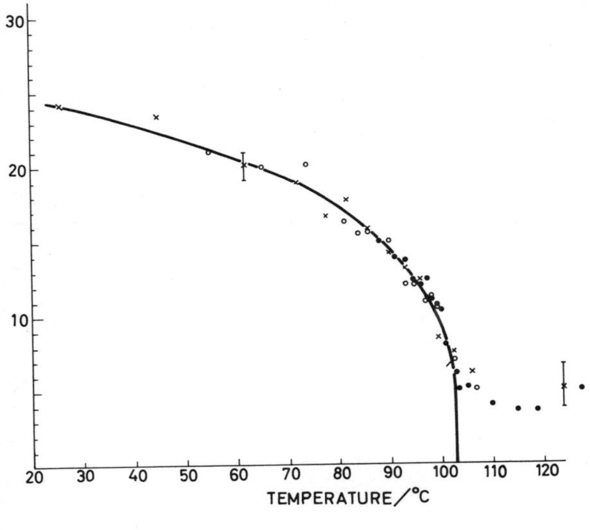

Figure 2.

Integrated intensity of the 83 cm^{-1} line versus temperature from three different series.
The full line is $(T_c - T)^{2\beta}$ with $\beta = 0.137$.

whereas individual layers are correlated only to first or second neighbours (along the unique b-axis). This is an independent indication of the two-dimensionality of the phase transition.

5. PEAK SHAPE AND WIDTH

All Raman lines of squaric acid are observed to show increased line width as temperature is increased, in the normal way. In addition an extra width increase associated with the phase transition is observed. For instance the 83 cm^{-1} line width is < 2 cm^{-1} at 10K, about 2 cm^{-1} at 300K and 12 cm^{-1} at T_c. Unlike other lines the 83 cm^{-1} line shows also a pronounced line shape asymmetry as already described.

Finite line widths are normally accounted for by a phenomenological temperature dependent damping Γ, the origin of which being anharmonic interactions among the various vibrational modes. In the limit of $\hbar\omega \ll kT$ the intensity profile is then given by

$$I(\omega) \sim \frac{kT \cdot \Gamma}{(\omega_{\vec{q}}^2 - \omega^2)^2 + \omega^2\Gamma^2} \cdot \eta^2(T) \qquad (4)$$

where $\omega_{\vec{q}}$ is the undamped excitation frequency. $\omega_{\vec{q}}$ may also be slightly temperature dependent, but no "soft mode" has ever been observed in squaric acid. For the 83 cm^{-1} mode the order parameter term $\eta^2(T)$ has been included in (4).

However, (4) can not explain the persistence of the 83 cm^{-1} line above T_c. This component, and possibly part of the "Rayleigh-like" scattering, may be accounted for by molecular disorder, to be described in the following.

6. DISORDER INDUCED SCATTERING

In this model it is assumed that the molecules retain their low-temperature monoclinic structure at all temperatures, in accordance with the existence of remnant scattering above T_c. The actual orientation of the molecules may be specified by a pseudospin variable σ_i, and one may apply a formalism analogous to that used by Geisel & Keller [9] for order-disorder of NH_4^+ molecules in NH_4Br. Then there will be three important terms

$$P^{\alpha\beta} = \sum_{r,j} P^{\alpha\beta}_{r,j} u^r_j + \sum_{j'} R^{\alpha\beta}_{j'} \sigma_{j'} + \sum_{r,jj'} S^{\alpha\beta}_{rjj'} u^r_j \cdot \sigma_{j'} \qquad (5)$$

in the expansion of the polarizability tensor, where u^r_j is the r-component of the displacement of jth molecule (rigid molecules assumed) and P, R, S are coupling terms.

The first term is the normal first order lattice dynamical term which vanishes by symmetry for the 83 cm^{-1} line. The second term is due to reorientational motion of molecules, and it might give rise to a Rayleigh-type scattering near T_c.

The intensity of 83 cm^{-1} is wholly due to the third term of (5). It would lead too far to discuss all the details of the further derivation. Following Geisel and Keller the assumption is made that the reorientational dynamics of σ is slow, compared to the lattice dynamics of u. Then well below T_c the intensity is proportional with $\langle\sigma\rangle^2 = \eta^2(T)$, as already used in (4). To get a semiqualitative picture of how the scattering may behave, we shall assume with Geisel and Keller that the coupling $S_{jj'}$ is nearest neighbour only. This gives an intensity proportional to

$$\sum_{\vec{q}} \frac{n(\omega_{\vec{q}}) + \frac{1}{2} \pm \frac{1}{2}}{\omega_{\vec{q}}} \langle|\sigma(\vec{q})|^2\rangle \tag{6}$$

We have taken a specially simplified model in order to calculate $\langle|\sigma|^2\rangle$, namely each layer to be long range ordered with an order parameter η, and the correlation between layers to decrease with distance as

$$(-\eta)^{|n_y|} \tag{7}$$

$n_y(= \pm 1, \pm 2, \ldots\ldots \pm \frac{N_y}{2},)$ numbering layers from some central one. Such a model gives sharp δ-functions in q_x and q_z, and gives for the scattered intensity a factor

$$M(q_y) = \frac{1-\eta^2}{(1+\eta)^2 - 4\eta \cdot \left(\frac{\omega_{q_y}}{\omega_m}\right)^2} \tag{8}$$

ω_m (83 cm^{-1}) is the maximum frequency of the branch. The resulting function has to be folded with the resolution of the instrument, and the q-sum be converted into energy integration, giving

$$I(\omega)=C\cdot\int_0^{\omega_m} d\omega' \frac{1-\eta^2}{(1+\eta)^2-4\eta\cdot\left(\frac{\omega'}{\omega_m}\right)^2} \frac{kT}{\omega'^2} \cdot\rho(\omega')\cdot e^{-2\ln 2\left(\frac{\omega-\omega'}{\Gamma}\right)^{\frac{1}{2}}} \tag{9}$$

The one dimensional density of state is taken to be

$$\rho(\omega') = \frac{1}{(1 - (\frac{\omega'}{\omega_m})^2)^{\frac{1}{2}}} \tag{10}$$

The calculated frequency dependent scattering based on eq. (9) is shown in figures 3 and 4. Gaussian instrumental resolution function is assumed, and intrinsic excitation width is allowed for by letting Γ be somewhat temperature dependent (FWHM 12 cm^{-1} at T_c).

The peak asymmetry is resonably well reproduced by these calculations. As is seen, (9) involves a weighted density of state whose divergency at $\omega' = \omega_m$ is smoothed by the instrumental width. The $M(q_y)$ term also shows strong enhancement at $\omega' \approx \omega_m$ for η close to unity, in fact giving strong weight to the $q_y = \frac{\pi}{b}$ contribution. Near T_c $\eta \simeq 0$ and the strong peaking at $\omega = \omega_m$ disappears, giving rise to the smooth curve of figure 4. We notice the presence of a peak even above T_c.

(9) also describes the increased scattering at low ω, reproducing the Rayleigh-like contribution centred at $\omega = 0$. In this picture the Rayleigh component is directly related to static order fluctuation and is dynamic in origin only from the phonon contribution. The details are, however, dependent upon the assumption of S being of short range.

7. CONCLUSIONS

The present light scattering study confirms the two-dimensional character of the phase transition of squaric acid, with an order parameter critical index $\beta \simeq 0,14$. It is concluded that the molecular layers retain their low-temperature symmetry above T_c. This fact gives rise to a peak asymmetry, reflecting the density of states of the TA phonons along the unique axis.

A strong line width anomaly for most Raman lines is observed at T_c. This is probably due to interaction with the disordered layers. A (seemingly) quasi elastic (Rayleigh-like) critical component is observed at T_c, part of which may be explained by static order fluctuations.

Work is in progress to study the effect of uniaxial pressure on squaric acid.

Acknowledgements:

Thanks are due to Dr. Dag Semmingsen for the loan of several single crystals of squaric acid.

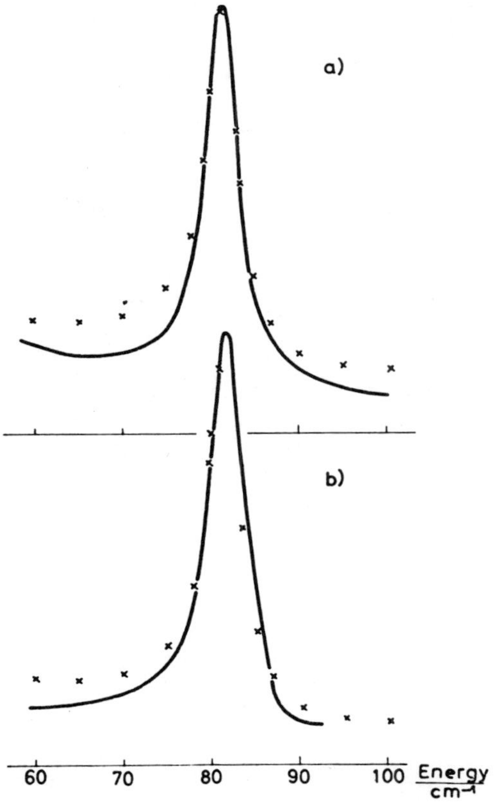

Figure 3.

Observed (x) intensity profile of the 83 cm^{-1} line at 80°C

Calculated profile based on two models:

a) Lorentzian peak eq. (4) on a Lorentzian background peak width 5 cm^{-1}.

b) Disorder model eq. (9), width 5 cm^{-1}, η = 0,75, on a constant background.

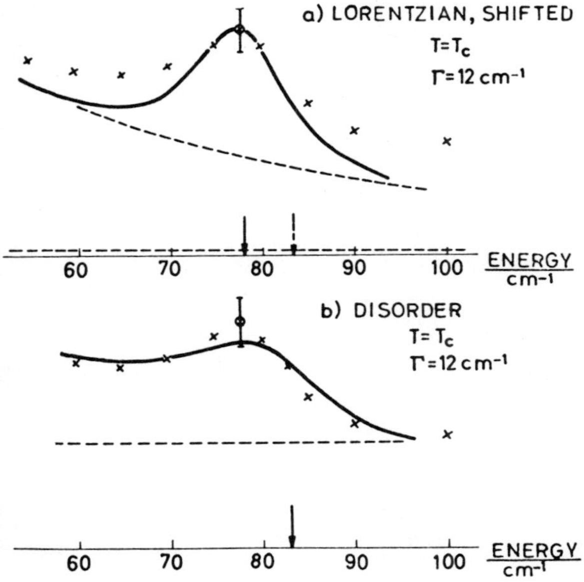

Figure 4.

Observed (x) intensity profile of 83 cm^{-1} line at T_c.

a) Lorentzian peak eq. (4) on a Lorentzian background (---) fitted at 40 cm^{-1}. Peak width 12 cm^{-1}.

b) Disorder model eq. (9), width 12 cm^{-1} η = 0, on a constant background

REFERENCES

[1] Semmingsen D., Acta Chem. Scand. 27 3961 (1973)

[2] Semmingsen D. and Feder J., Solid State Commun. 15 1369 (1974)

[3] Samuelsen E.J. and Semmingsen D., J. Phys. Chem. Solids 38 1275 (1977)

[4] Baglin F.G. and Rose C.B., Spectrochim. Acta. 26A 2293 (1970)

[5] Nakashima S. and Balkanski M., Solid State Commun. 19 1225 (1976)

[6] Bougeard D. and Novak A., Solid State Commun. 27 453 (1978)

[7] Samuelsen E.J., Fjær E. and Semmingsen D., J. Phys. C. (Solid State Physics) (to be published (1979))

[8] Fjær E., Grip J. and Samuelsen E.J., to be published

[9] Geisel T. and Keller J., J. Chem. Phys. 62 3777 (1975)

MONTE CARLO SIMULATION OF DILUTE SYSTEMS AND OF TWO-DIMENSIONAL SYSTEMS

K. Binder

IFF, KFA Jülich
Postfach 1913
D-5170 Jülich, Fed. Rep. Germany

I. INTRODUCTION

These lectures are concerned with several types of systems: ferromagnets diluted with impurities; quasi-twodimensional (2d) magnets; monolayers of adsorbate atoms at surfaces and their registered structures. Emphasis will be on static properties, as studied by Monte Carlo (MC) methods. It will be shown how such numerical results elucidate theoretical questions as well as experiments. Of course, only a few examples mainly selected from the work of the author's group can be given here; a detailed bibliography on other studies, as well as a description of the theory of the MC method and its technical aspects is found in a recent book /1/.

II. FERROMAGNETS DILUTED WITH NONMAGNETIC IMPURITIES AND RELATED SYSTEMS

Consider a magnet where a fraction (1-x) of the magnetic ions is randomly replaced by nonmagnetic ones. These defects may be quenched (i.e., fixed in their position) or annealed (i.e. mobile due to thermal fluctuations). Only the quenched case is treated here. Two questions may be asked: (i) how are local magnetic properties influenced at sites close to the defects? (ii) how are global properties influenced? Both questions can conveniently be studied by MC methods.

The simplest model has nearest-neighbor exchange J between magnetic atoms and no interaction between a magnetic atom and the impurity. Fig. 1 shows results for a classical Heisenberg magnet

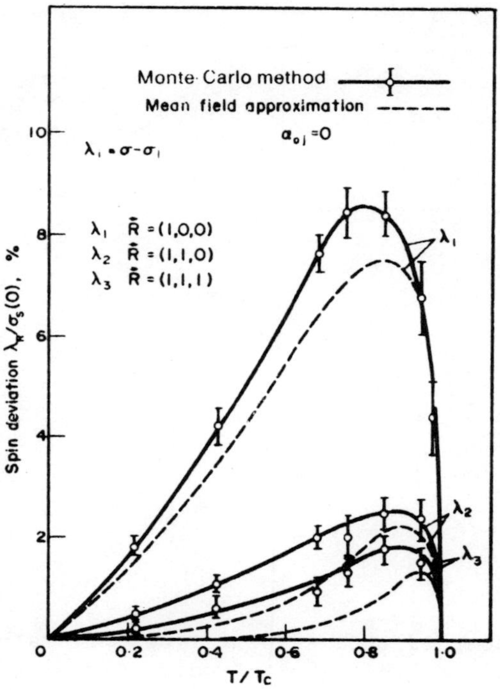

Fig. 1

Reduction of magnetization close to a defect plotted vs. temperature /2/

with one such defect at $\vec{R}=0$ /2/. Close to it the spins are much stronger fluctuating, and hence the magnetization is reduced by an amount $\lambda_{\vec{R}}$. Qualitatively this effect can be seen already in molecular field approximation (MFA), which, however, is inaccurate both in the spin wave regime and near T_c. Experimentally this local enhancement of fluctuations is important if critical exponents are measured with resonance techniques (NMR etc.) using impurity atoms (see /1/ for a discussion of such experiments on Ni, where indeed effective order parameter exponents slightly enhanced over bulk values were found).

Next we consider fcc classical Heisenberg ferromagnets with interactions J_{nn}, $J_{nnn}=-J_{nn}/2$ and a finite fraction of impurities /3/ (this is a reasonable model of $(Eu_xSr_{1-x})S$ /4/).

The decrease of critical temperature with impurity concentration (Fig. 2) as found from the simulation agrees well with experiment /4/. The peak of the specific heat at T_c, which is quite pronounced in the pure case, is quickly washed out and only

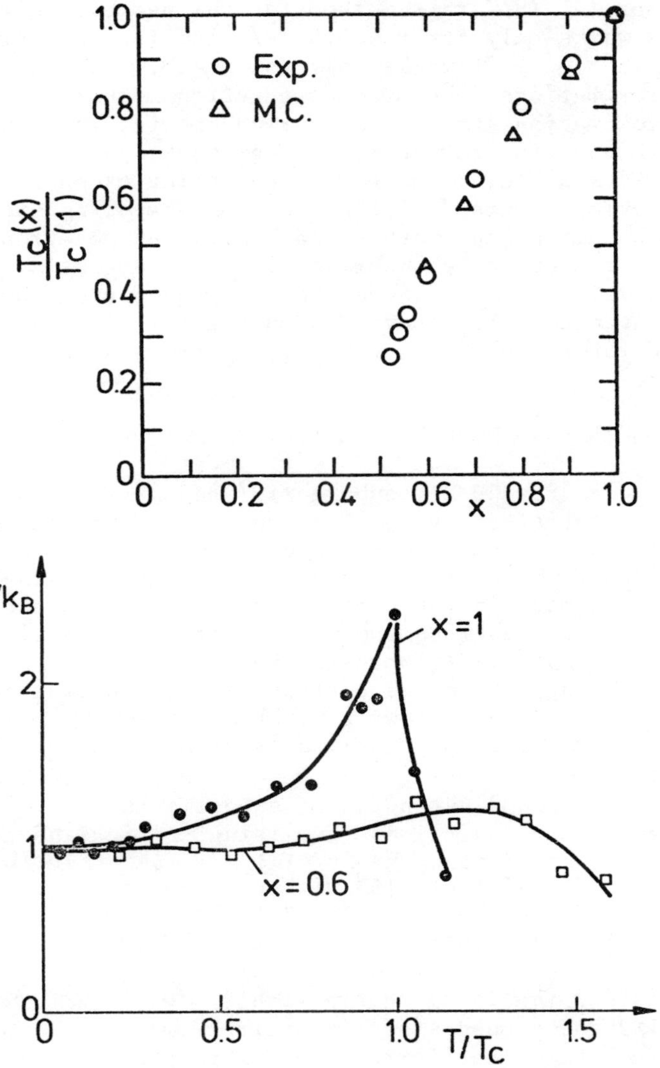

(A)

(B)

Fig. 2. A) Critical temperature plotted vs. impurity concentration
B) Specific heat plotted vs. temperature /3/

a broad Schottky-like anomaly with maximum somewhat above T_c remains. The same effect was seen in studies on diluted 2d- /5/ and 3d- /6/ Ising models, real-space renormalization group calculations /7/, and experiments /8/. While $T_c(x)$ for the Ising model can be found fairly reliable from the extrapolation of expansions in x or 1/T /9/, this method for the nearest-neighbor Heisenberg model works only for $x \gtrsim 0.5$ /9/, and for competing exchange as shown in Fig. 2 it works only for $x \gtrsim 0.7$ /3/. The accuracy of approximations like Green's function methods + CPA /10/, Bethe approximation etc. are rather uncertain, and hence MC methods clearly are useful for such inhomogeneous systems. Fig. 2 shows no results for $x \lesssim 0.5$: due to the competing exchange there "spin glass"-behavior occurs /3,4,11/. For non-competing exchange this does not happen, and one rather reaches the percolation threshold x_c. MC studies at T=0 where x is varied across x_c are standard /1/ and will not be reviewed here. Unfortunately not much has as yet been done to study the "multicritical" behavior when T is varied for x close to x_c, where experiments exist /8, 12/.

III. MODELS FOR QUASI-TWO-DIMENSIONAL (2D) MAGNETS

While MC studies for 2d Heisenberg /13/ and XY-magnets /14/ did not yield very satisfactory results, due to strong fluctuations, metastable vortex configurations etc., work was more successful for models whose critical behavior belongs to the Ising universality class. In the nearest-neighbor Ising case comparisons with exact solutions and with finite size scaling theories were performed /1,15/. More interesting in the context of strongly fluctuating systems is a study /16/ of the weakly anisotropic 2d Heisenberg model, with hamiltonian $\mathcal{H} = -J\Sigma[(1-\Delta)(S_i^x S_j^x + S_i^y S_j^y) + S_i^z S_j^z]$. Even for very small anisotropy Δ the critical behavior is still Ising-like (Fig. 3). Results of this type have contributed to a better understanding of experiments on K_2NiF_4 /17/, where first strong deviations from Ising critical behavior were suggested. In the relation $\langle M\rangle \approx B(\Delta)[1-T/T_c(\Delta)]^{1/8}$ both $B(\Delta)$ and $T_c(\Delta)$ were found consistent /16/ with a variation $\propto|\ln(1/\Delta)|^{-1}$, as expected theoretically /18/, although a nonzero "Stanley-Kaplan" /19/ - T_c could also not be ruled out.

Next we consider the Ising square with $J_{nn}<0$, $J_{nnn}<0$ /21,22/. For $R \equiv J_{nnn}/J_{nn} < 1/2$ the ground state is antiferromagnetic, while for $R > 1/2$ it is a layered antiferromagnet. This phase has a two-component order parameter and thus is of great theoretical interest: it belongs to the universality class of the XY-model with cubic anisotropy /20/. MC studies of this model /22/ indicated that the exponents depend on R (similar to the Baxter model which belongs to the same universality class /23/). A different behavior was found for R=1/2, where $T_c=0$: the correlation length seems to diverge exponentially strong as $T \to 0$ /21/, as in the 1d Ising system. Fig. 4 shows phase diagrams for several R in nonzero field H. For strong enough H the ground state is ferromagnetic, while for intermediate H it is highly degenerate: ferromagnetic rows alternate with antiferromagnetic rows. Each of the latter can start either with ↓ or ↑ and hence contributes a factor 2 to the

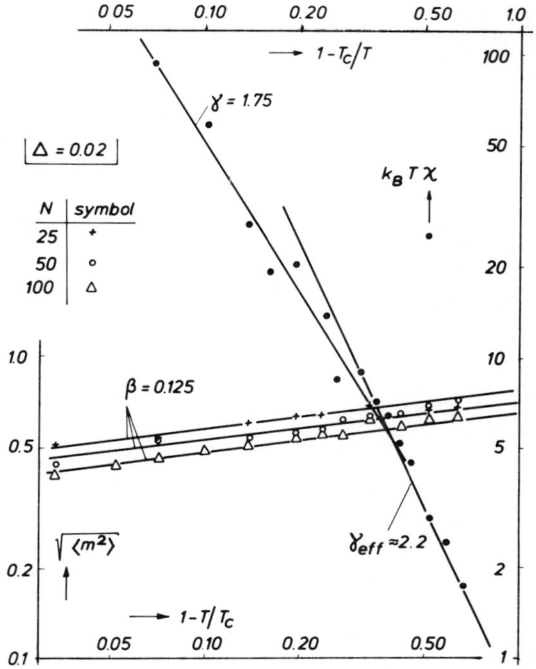

Fig. 3

Log-log plot of magnetization and susceptibility of the 2d anisotropic Heisenberg magnet /16/

degeneracy. Thus the sublattices forming the antiferromagnetic rows exhibit only 1d-order. These rows become disordered at $T \to 0$: there one has again a two-component layered structure, where now ferromagnetic rows alternate with disordered ones. Since this "superantiferromagnetic" (SAF) structure has the same symmetry as the structure where ferromagnetically aligned rows ↓, ↑ alternate, the "transition" at $T>0$ for $R>1/2$ between these structures is completely continuous, in contrast to the MFA which predicts a 1st order transition /22/. For $R<1/2$ one has both antiferromagnetic and "superantiferromagnetic" structures. At the critical field H_{c3} not only the ground state energies of the two phases become equal, but also their interface energy vanishes /22/: Therefore again the multicritical point between these orderings occurs at $T=0$. MFA again yields the phase diagram topology wrong: instead of SAF phase it yields a (2x2) structure, with a multicritical point between it and the antiferromagnetic phase at H_{c4} rather than H_{c3}. So far this model does not have a direct experimental application, but it contributes to our understanding of multicritical behavior and "universality classes" in 2d-systems /24/. A particular interesting study considered an Ising square including various two- and four-spin couplings to investigate the vicinity of the Baxter multicritical point /25/.

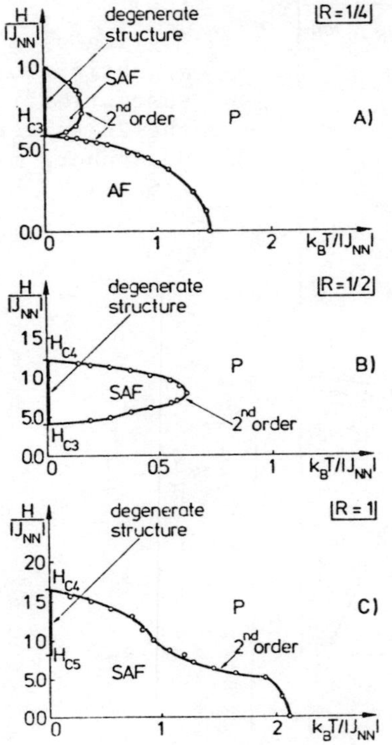

Fig. 4

Phase diagrams of the square Ising antiferromagnet in a field for various R.

IV. LATTICE GAS MODELS FOR ADSORBED MONOLAYERS AT SURFACES

Adsorbed monolayers at surfaces occur in a variety of liquid and solid phases /26/. Here only "registered" solid phases are considered (lattice structures commensurate with the substrate surface). The substrate surface is modelled as a lattice of preferred sites on which ions can be adsorbed. One is interested in the phase diagram as a function of T and coverage Θ of the adsorbate. With an occupation variable $c_i=(0,1)$ we have $\Theta=\langle c_i \rangle$. The hamiltonian is expressed in terms of the adatom-adatom interaction ϕ and the binding energy ε between adatom and substrate,

$$\mathcal{H} - \mu N_a = -\frac{1}{2}\sum_{i,j} c_i c_j \phi(\vec{r}_i - \vec{r}_j) - (\varepsilon+\mu)\sum_i c_i + \mathcal{H}_o \tag{1}$$

N_a is the number of adsorbed atoms, μ the chemical potential, and $\mathcal{H}_o$ describes other degrees of freedom (vibrations etc.). As is well-known, Eq. (1) is equivalent to an Ising magnet $[c_i=(1-S_i)/2]$

$$\mathcal{H} = -\frac{1}{2} \sum_{i,j} J_{ij} S_i S_j - H \Sigma S_i + \mathcal{H}'_o \;,\; J_{ij} = \phi/4, \; H = -\left[\varepsilon + \mu + \frac{1}{2} \sum_j \phi\right]/2 \qquad (2)$$

Assuming a square lattice {appropriate for monolayers at (100) surfaces of transition metals, for which many experimental examples exist /27/}, we have a model of the type considered in Sec. III: only now the magnetization M $[M=1-2\Theta]$ rather than the field is an independent variable. The magnetization process corresponds to the adsorption isotherms $[\mu_c \equiv -\varepsilon - \frac{1}{2} \sum_j \phi(\vec{r}_i - \vec{r}_j)]$

$$\mu/k_BT = (1/Nk_BT)\;(\partial F/\partial\Theta)_T = -2H/k_BT + \mu_c/k_BT \qquad (3)$$

Here we will only treat repulsive ϕ_{nn} (see /28/ for a brief discussion of the behavior for attractive forces) Fig. 5 summarizes

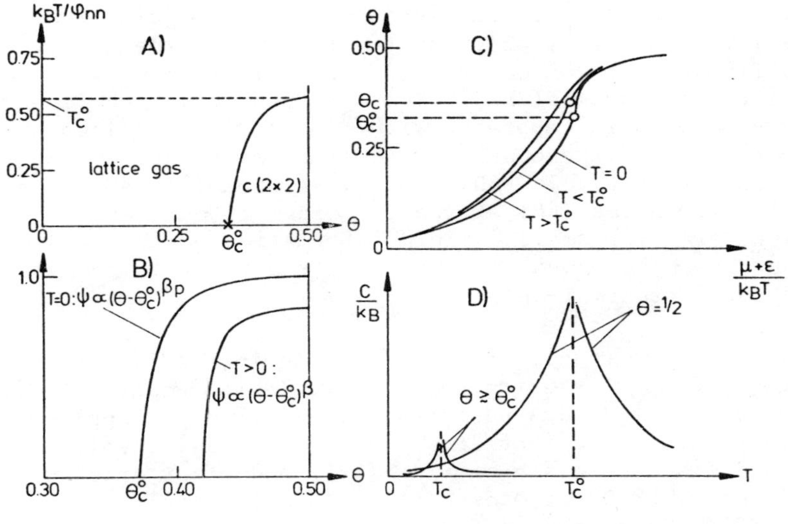

Fig. 5. Lattice gas with nearest neighbor repulsion: Phase diagram (A), Order parameter (B), adsorption isotherms (C), heat capacities (D), (B-D schematic) /28/

the behavior if interactions with more distant neighbors are zero. A transition occurs from the lattice gas to the c(2x2) structure (A) [corresponding to the above antiferromagnet]. The transition coverage Θ_c depends on $k_BT/|\phi_{nn}|$ and reaches the limit $\Theta_c^0 \cong 0.37$ /29/ for $k_BT/|\phi_{nn}| \rightarrow 0$. This "hard core lattice gas" /29/ behaves differently: energy and specific heat (C) are zero identically, entropy only determines the transition. There the adsorption isotherm (D) has an infinite slope, in contrast to nonzero $k_BT/|\phi_{nn}|$ where only its curvature diverges (the slope corresponds to the ferromagnetic susceptibility of the Ising antiferromagnet). Also the order parameter exponent may be different (B), the transition in the hard-core limit should belong to a universality class different from the Ising class. We suggest /22,28/ that it is a type of percolation transition: rather than randomly "paving" lattice cells (of size a^2) with particles of the same size we use particles of size $2a^2$, which then block both the occupied site and its nearest neighbors for further occupation. At Θ_c^0 an infinite "connected" structure appears, while for $\Theta < \Theta_c^0$ only finite clusters occur. The exponents should thus be that of 2d percolation, $\beta_p \cong 0.138$, $\delta_p \cong 18$ /30/. For small nonzero $k_BT/|\phi_{nn}|$ one should see crossover from these exponents to the Ising ones. The point T=0, $\Theta = \Theta_c^0$ is a multicritical point, just as the percolation threshold in dilute magnets /31/. This limiting case may be physically realized when atoms are adsorbed whose diameter exceeds the lattice spacing of the preferred sites.

Figs. 6,7 describe the behavior of the model of Sec. III with nonzero ϕ_{nnn}. For ϕ_{nnn} attractive (Fig. 6A) the point T=0, $\Theta=\Theta_c^0$ changes into a tricritical point (Θ_t, T_t), where the transition changes from 2nd to 1st order /32/. Below T_t we have a mixed phase region {lattice gas + c(2x2)}. For ϕ_{nnn} repulsive the degenerate structure of Sec. III appears, with two multicritical points Θ_1^*, Θ_2^* of percolation-like character (6C). As mentioned above, MFA fails even qualitatively (6D). Only weak anomalies are seen in the adsorption isotherms (6B) at the 2nd order transitions. Note that these isotherms for small Θ yield information on the binding energy ε (which here was absorbed into the abscissa to make this part of the isotherm T-independent).

Fig. 7 shows results for R=1/2,1. Again MFA is inadequate (it never can yield information on the percolation points Θ_1^*, Θ_2^*, for instance). Our study has no immediate experimental applications and rather should elucidate the general problems involved. A Monte Carlo analysis to explain similar (2x1) and other ordered structures found by LEED for O on W (110) /33/ looks very promising, however /34/. There the preferred sites form a lattice of space group C2mm rather then square. For (111) surfaces of these metals (and adsorption on grafoil) the preferred sites form a triangular lattice, which was studied also with MC /24,35/. Fig. 8 shows an example for R=-1 /35/. Ordered phases appear at $\Theta=1/3$ and $\Theta=2/3$, the system decomposes into 1(2) filled and 2(1) empty triangular sublattices (which has been seen e.g. for He on grafoil /26/). Due to the 3-fold symmetry of these structures their order-disorder transition belongs /36/ to the universality class of the 3-state Potts model /37/. Fig. 9 shows preliminary MC results for this model. Series expansion estimates for the specific heat exponent range from $\alpha=0.05$ /38/ to $\alpha=0.42$ /39/, while experiment yields /40/ $\alpha \approx 0.36$. The MC data indicating a distinctly higher value may be not close enough to T_c. Even for static and dynamic exponents of the 2d kinetic Ising model only a 10% accuracy could be reached /1/: while MC studies hence yield valuable information on phase diagrams, for estimating 2d-exponents a combination of MC and real space renormalization looks most promising /41/.

Acknowledgements: I am grateful to W. Kinzel and D.P. Landau for collaborating on parts of the work described here.

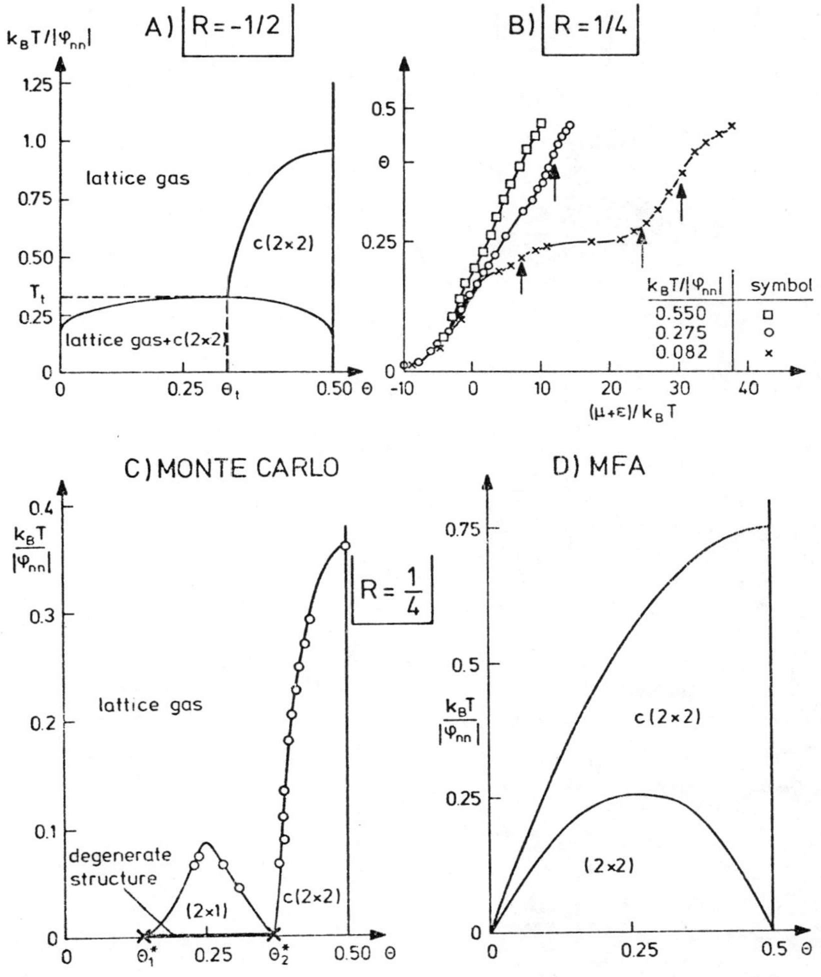

Fig. 6. Lattice gas with nonzero next nearest neighbor interaction: MC phase diagrams for 2 values of R(A,C), adsorption isotherm (B), MFA phase diagram (D) /22/

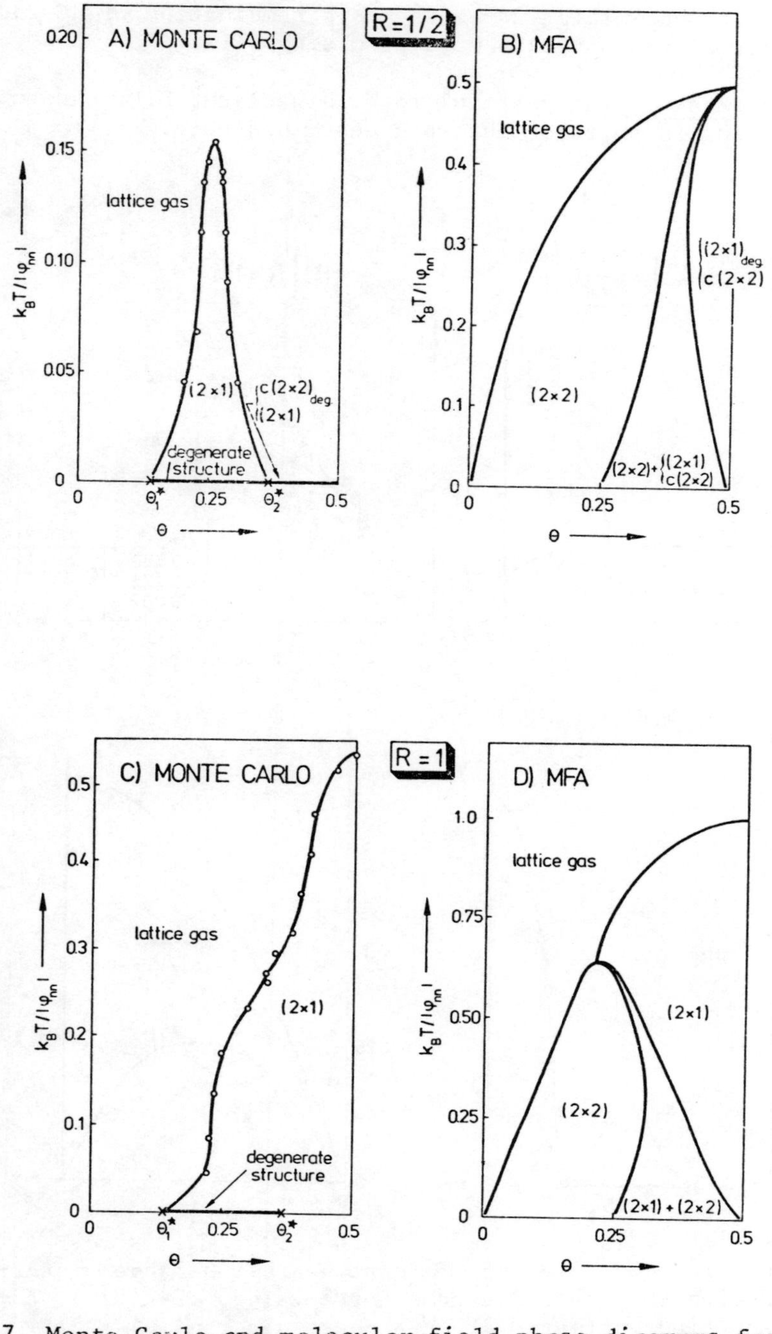

Fig. 7. Monte Carlo and molecular field phase diagrams for two values of R /22/

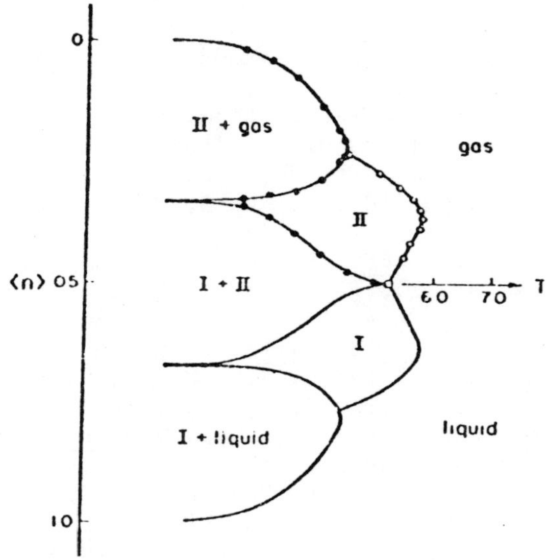

Fig. 8. Monte Carlo phase diagram for R=-1, triangular lattice /35/

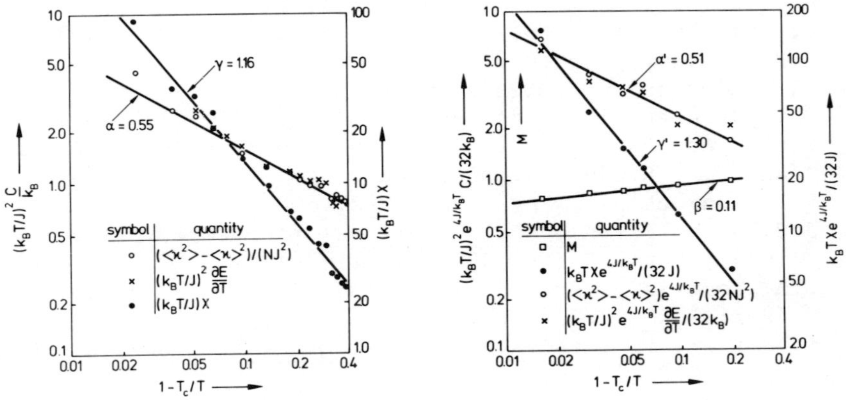

Fig. 9. Log-log plot of specific heat, susceptibility and order parameter of the 3-state Potts model on the square lattice above T_c (left) and below T_c (right side) /28/

References

1. K. Binder (ed.) Monte Carlo Methods in Statistical Physics. Springer, Berlin 1979.
2. V. Wildpaner et al., J. Phys. Chem. Sol. 34, 925 (1973).
3. K. Binder, W. Kinzel and D. Stauffer, to be published.
4. H. Maletta and P. Convert, Phys. Rev. Lett. 42, 108 (1979).
5. See e.g. W.Y. Ching and D.L. Huber, Phys. Rev. B13, 2962 (1976).
6. D.P. Landau, Physica B86-88, 731 (1979)
7. C. Jayaprakash et al., Phys. Rev. B17, 2244 (1978).
8. H.A. Algra et al., Phys. Rev. Lett. 42, 606 (1979).
9. G.S. Rushbrooke, in Critical Phenomena in Alloys, Magnets, and Superconductors (eds.: R.E. Mills, E. Ascher and R.I. Jaffee), p. 155 (McGraw Hill, New York 1971).
10. H. Dvey-Aharon and M. Fibich, Phys. Rev. B18, 3491 (1978).
11. See also lectures on Spin Glasses by K. Binder in this book.
12. R.J. Birgeneau et al., Phys. Rev. Lett. 37, 940 (1976). See also lectures by R. Cowley in this book.
13. R.E. Watson et al., Phys. Rev. B2, 684 (1970).
14. C. Kawabata and K. Binder, Solid State Comm. 22, 705 (1977).
15. D.P. Landau, Phys. Rev. B13, 2997 (1976), and references therein.
16. K. Binder and D.P. Landau, Phys. Rev. B13, 1140 (1976).
17. R.J. Birgeneau et al., Phys. Rev. B8, 1736 (1971); Phys. Rev. B16, 280 (1977).
18. D.R. Nelson and R.A. Pelcovits, Phys. Rev. B16, 2191 (1977).
19. H.E. Stanley and T.A. Kaplan, Phys. Rev. Lett. 17, 913 (1966).
20. S. Krinsky and D. Mukamel, Phys. Rev. B16, 2313 (1977).
21. D.P. Landau, preprint.
22. K. Binder and D.P. Landau,preprint.
23. J.V. José et al., Phys. Rev. B16, 1277 (1977).
24. For a general discussion of these concepts see E. Domany and E.K. Riedel, J. Appl. Phys. 49, 1315 (1978).
25. E. Domany et al., Phys. Rev. B12, 5025 (1975).
26. J.G. Dash, Films on Solid Surfaces (Academic Press, New York 1975); Phys. Repts. 38C, 177 (1978).
27. G.A. Somorjai, Surf. Sci. 34, 156 (1973); K. Masuda, J. Phys. (Paris) 40, 299 (1979).
28. K. Binder, in Trends in Physics, Proc. IVth Gen. EPS Conf., York 1978, in press.
29. L.K. Runnels, in Phase Transitions and Critical Phenomena Vol. 2 (C. Domb and M.S. Green, eds.) Chap. 8 (Academic Press, New York 1972).
30. H.E. Stanley, J. Phys. A10, L211 (1977) and references therein.
31. D. Stauffer, Z. Phys. B25, 391 (1975).
32. D.P. Landau, Phys. Rev. Lett. 28, 449 (1972); K. Binder and D.P. Landau, Surf. Sci. 61, 577 (1976).
33. T.M. Lu et al., Phys. Rev. Lett. 39, 411 (1977).

34. W.Y. Ching et al., to be published.

35. B. Mihura and D.P. Landau, Phys. Rev. Lett. 38, 977 (1976).
36. S. Alexander, Phys. Lett. 54A, 353 (1975).
37. R.B. Potts, Proc. Cambridge Phil. Soc. 48, 106 (1952).
38. J.P. Straley and M.E. Fisher, J. Phys. A6, 1310 (1973).
39. T. de Neef and I.G. Enting, J. Phys. A10, 801 (1977).
40. M. Bretz, Phys. Rev. Lett. 38, 501 (1977).
41. R.J. Swendsen, preprint, and references contained in /1/.

ORDER AND FLUCTUATIONS IN SMECTIC LIQUID CRYSTALS

J.D. Litster, R.J. Birgeneau, M. Kaplan and C.R. Safinya

Dept. of Physics and Center for Materials Science and Engineering, Massachusetts Institute of Technology
Cambridge, Massachusetts, 02139, U.S.A.

J. Als-Nielsen
Risø National Laboratory
DK-4000, Roskilde, Denmark

I. INTRODUCTION

Liquid crystals are among the most interesting condensed states of matter; they are interesting in their own right and we also expect that insights into their properties will help to understand other condensed phases that exist in nature. In all condensed phases there are thermally excited fluctuations. Depending on the spatial dimensionality along with the symmetry and range of interparticle interactions, these fluctuations can play a role of varying significance in determining properties of matter. In some materials the fluctuations play an unimportant role and the statistical mechanical calculations to understand which phases occur and their properties can be carried out simply by a mean field approximation. In other materials the fluctuations profoundly alter the material properties in the vicinity of phase transitions, and only recently have we learned how to make approximate calculations with sufficient accuracy to explain the thermodynamic behavior of these materials. This is the class of phase changes known as critical phenomena. Finally, in certain cases, the fluctuations are so important as to prevent the establishment of phases which interactions between the molecules would otherwise favor. This understanding of the importance of symmetries and geometry in determining the phases in which condensed matter can exist, has come about through the use of scattering spectroscopy to probe directly the nature of the relevant thermal fluctuations. This spectroscopy has been carried out using thermal neutrons, x-rays, visible light, and to some extent, electrons.

In these notes we shall discuss what we know about order and fluctuations in the smectic phases of liquid crystals; our theoretical models at this stage are largely phenomenological, and most of our knowledge of the important fluctuations comes from scattering of electromagnetic radiation (visible light or x-rays) and thermal measurements.

To begin, we remind you that most thermotropic liquid crystals are roughly cigar shaped organic molecules with quite anisotropic properties such as dielectric polarizability and diamagnetic susceptibility. A typical liquid crystal molecule is octyloxycyanobiphenyl (8OCB) sketched below.

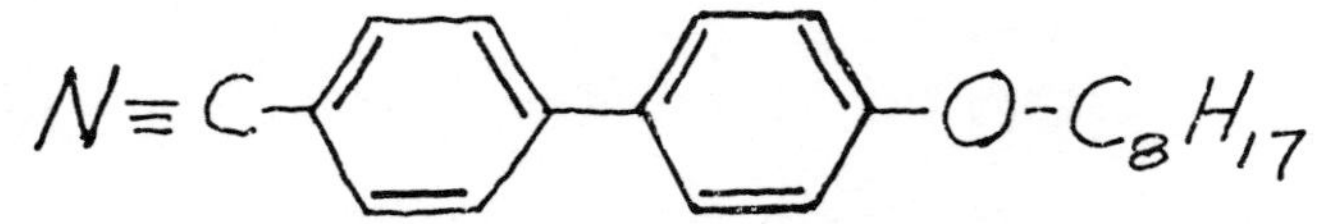

The various liquid crystal phases have orientational and translational order of the molecules intermediate between that of crystalline solids and isotropic (disordered) liquids. Three important liquid crystals phases are sketched below with the molecules represented by lines.

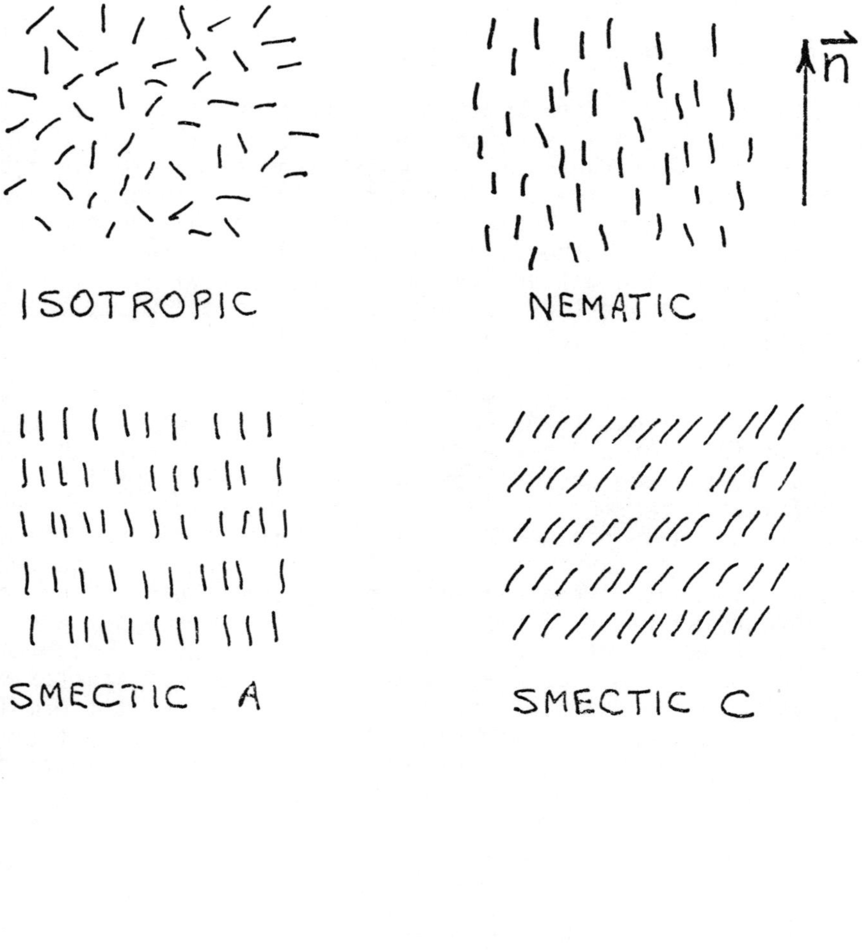

The nematic phase (N) has only orientational long range order of the molecules with no translational order beyond that of normal liquids. The smectic A (SmA) and smectic C (SmC) phases have the orientational order of the nematic and in addition have translational order in one dimension, i.e. a one dimensional density wave. The wave vector of the wave can be parallel to the molecular orientation (SmA) or at an angle (SmC). One commonly speaks of a layered structure for smectic phases, although the density wave is a rather more accurate description. In the SmA phase the molecules are normal to the smectic "layers" and neither smectic phase has translational order within the layers. Thus SmA and SmC phases may also be regarded as stacked up layers of two dimensional liquids. There are also a large number of smectic phases, called B, D, E, H, etc., which are well ordered compared to SmA and SmC phases and have varying kinds of translational order within the layers. I will discuss presently a model for these well-ordered phases.

II. THE NEMATIC PHASE

In order to discuss smectics properly we should first review the properties of the nematic phase. The books by de Gennes [1] and Chandrasekhar [2] are excellent sources for this material. In addition to the usual degrees of freedom and hydrodynamic modes of a liquid, the nematic phase has long lived excitations (Goldstone modes) as a result of the orientational long range order. These are analogous to spin waves in a ferromagnet and are usually referred to as director fluctuations, the director $\vec{n}$ being the local optic axis in the liquid crystal. They are treated quantitatively by a phenomenological model proposed by Oseen and Frank [3]. Since $\vec{n}$ is a unit vector field it can have only three types of distortion, and an elastic free energy per unit volume for director distortions that is consistent with the uniaxial symmetry may be written as

$$\Phi_N = \frac{1}{2}\left\{K_1(\vec{\nabla}\cdot\vec{n})^2 + K_2(\vec{n}\cdot\text{curl }\vec{n})^2 + K_3(\vec{n}\times\text{curl }\vec{n})^2\right\} \qquad (1)$$

The three elastic constants and corresponding types of distortion are K_1 (splay), K_2 (twist), and K_3 (bend). All have values about 10^{-6} dynes.

Fluctuations in $\vec{n}$ also cause changes in the dielectric constant tensor and intense scattering of light. Thus the Goldstone modes of nematic order can readily be studied by light scattering. An extension of the Oseen-Frank picture to include dynamics was carried out by the Orsay group [4]. The viscosities of liquid crystals are large (>10 cp) and the elastic constants are weak, thus the director modes are overdamped and do not propagate. Taking the Fourier transform of (1) one finds two normal modes whose autocorrelation functions can be calculated from the Orsay model. If $\vec{n}$ is taken to lie along z and $\vec{q}$ is in the x-z plane ($\vec{q} = q_\perp \hat{x} + \vec{q}_\parallel \hat{z}$) one finds for the fluctuations

$$\langle n_x^*(\vec{q},0)\, n_x(\vec{q},\tau)\rangle = k_B T\left[K_1 q_\perp^{\,2} + K_3 q_z^{\,2}\right]^{-1} e^{-\Gamma_1(\vec{q})\tau} \tag{2a}$$

$$\langle n_y^*(\vec{q},0)\, n_y(\vec{q},\tau)\rangle = k_B T\left[K_2 q_\perp^{\,2} + K_3 q_z^{\,2}\right]^{-1} e^{-\Gamma_2(\vec{q})\tau} \quad . \tag{2b}$$

Thus the intensity of the scattered light can be used to measure the splay, twist, and bend elastic constants K_1, K_2 and K_3. The spectrum of the scattered light contains information on the decay constants, which have the form

$$\Gamma_i(\vec{q}) = (K_i q_\perp^{\,2} + K_3 q_z^{\,2})/\eta(\vec{q}) \quad . \tag{3}$$

The viscosity $\eta(\vec{q})$ is a combination of five tensor components and is discussed in detail by de Gennes [1].

III. THE NEMATIC-SMECTIC A TRANSITION AND THE SMECTIC A PHASE

The SmA phase is usually described as being composed of equidistant planes of molecules (interlayer separation = a), each molecule being on the average perpendicular to the planes (i.e. the system is uniaxial). The equilibrium director, $\vec{n}_0$, is thus perpendicular to the planes and this direction is usually taken as the z-axis. One immediate consequence for light scattering of this layered structure is that bend and twist fluctuations are excluded in smectics. To see this, we consider a perfect (dislocation free) smectic sample. If we form the quantity

$$\frac{1}{a}\int_A^B \vec{n}\cdot d\vec{l}$$

where A and B are any two points in the sample and a is the layer spacing (which is assumed constant), this line integral simply counts the number of layers crossed in going from A to B. If we take a closed loop as the integration path then we must have (since there are no dislocations):

$$0 = \frac{1}{a}\oint \vec{n}\cdot dl$$

or, equivalently,

$$\vec{\nabla} \times \vec{n} = 0$$

This means that if you try to impose a bend or twist distortion ($\vec{\nabla} \times \vec{n} \neq 0$) on a uniform smectic sample it will relax the strain by breaking layers (i.e. creating dislocations) very near the surface rather than admitting bend or twist into the bulk. In other words, the elastic constants K_2 and K_3 are very large in the smectic phase so that as one passes from nematic to smectic one should see a divergence of these two quantities. Since it is the director modes which give a nematic its turbid appearance one expects to see a large decrease in scattering intensity in the SmA phase and this is indeed the case. Smectics A look like conventional liquids.

The divergence one expects depends on the nature of the phase transition. Most N-SmA transitions are first order, but McMillan [5] and Kobayashi [6] pointed out the transition could be first or second order depending on the sign of the fourth order term in the free energy expansion. Physically, a second order transition will occur if the nematic order parameter S is nearly saturated near T_{AN}; the layered structure of the SmA can then be created with only a slight increase in S to make the molecules perpendicular to the layers. But in a sample with a narrow nematic range a large jump in S is needed to create the SmA phase and the transition is first order. The second order transitions, if not near a tricritical point, should follow mean field behavior for spatial dimensionality $d > d^* = 4$. It has also been argued [7] that a different physical mechanism, coupling of the Sm order to nematic director fluctuations, will cause a cubic term in the free energy and make the transition weakly first order. Experimentally, liquid crystals have been found which appear to have (within the resolution limits of the experiment) a second order transition. It is these second order or nearly second order liquid crystals which have engendered the most activity not only because a theoretical framework exists to help to understand their behavior but also because this understanding can be applied to other systems which undergo phase transitions in an effort to unify the physics of condensed matter.

Both de Gennes [8] and McMillan [5] have developed Landau theories of the N-SmA transition. In what follows we shall summarize the de Gennes model and its implications for light scattering. We start by defining an order parameter for the smectic phase. The order parameter, ψ, is defined as the amplitude of a one dimensional density wave whose wave vector, q_0, is parallel to the nematic director (the z-axis):

$$\rho(\vec{r}) = \rho_0\{1 + \mathrm{Re}[\psi e^{iq_0 z}]\} \tag{4}$$

where $a = 2\pi/q_0$ is the layer spacing and $\psi(\vec{r}) = |\psi|\ e^{iq_0 u}$. Here u is the displacement of the layers in the z direction away from their equilibrium position ($u = u(x,y,z)$).

The complex order parameter, ψ, is analogous to the two component scalar order parameters found in both the superconducting transition in a metal and the superfluid transition in liquid helium. With this in mind, de Gennes [8] proposed a free energy density of the Landau-Ginzburg form:

$$\Phi = \Phi_0 + \alpha|\psi|^2 + \frac{1}{2}\beta|\psi|^4 + \frac{1}{2M_v}|\partial_z\psi|^2 + \frac{1}{2M_t}|(\partial_x + iq_0 n_x)\psi|^2$$
$$+ \frac{1}{2M_t}|(\partial_y + iq_0 n_y)\psi|^2 + \Phi_N \tag{5}$$

where Φ_N is given by Eq. (1). M_v, M_t are components of a "mass tensor" along and perpendicular to the unperturbed director $\vec{n}_0$ (which lies along z), K_1, K_2, K_3 are the Frank elastic constants and $\partial_z = \partial/\partial_z$, etc. In the SmA phase, $\langle|\psi|\rangle = \psi_0 = (-\alpha/\beta)^{1/2}$; if we assume that $|\psi|$ is spatially uniform then

$$\Phi = \Phi_0 + \frac{1}{2}\left\{-(\alpha^2/\beta) + B(\partial_z u)^2 + D[(n_x+\partial_x u)^2 + (n_y+\partial_y u)^2]\right\} + \Phi_N \tag{6}$$

where $B = \psi_0^2 q_0^2/M_v$ and $D = \psi_0^2 q_0^2/M_t$ give, respectively, the restoring forces for fluctuations in the phase of ψ (that is, in layer thickness) and for fluctuations of the molecular orientation (that is, the director) away from the normal to the layers. Following the same procedure as in the nematic, we take the Fourier transform of Φ and apply the equipartition theorem to extract the amplitude of the mean square fluctuations of the director. If we choose the wave vector $\vec{q}$ to lie in the x-z plane then we find:

<u>Mode 1</u>

$$\langle n_x^2(\vec{q})\rangle = \langle n_1^2(\vec{q})\rangle = \frac{k_B T\left[1 + \frac{B}{D}\left(\frac{q_z}{q_\perp}\right)^2\right]}{K_1 q_\perp^2 + K_3 q_z^2 + \frac{B}{D}(D + K_1 q_\perp^2 + K_3 q_z^2)\left(\frac{q_z}{q_\perp}\right)^2} \tag{7}$$

<u>Mode 2</u>

$$\langle n_y^2(\vec{q})\rangle = \langle n_2^2(\vec{q})\rangle = \frac{k_B T}{D + K_2 q_\perp^2 + K_3 q_z^2} \tag{8}$$

With $B = D = 0$ we regain the expressions for the nematic case (Eq. 2). In actual experiments these expressions simplify greatly, for instance by choosing either $q_\perp$ or q_z to be zero.

In the nematic phase $\langle\psi\rangle = 0$ but there are fluctuations in ψ analogous to fluctuation diamagnetism in superconductors. The fluctuations in ψ can be considered as little islands of of smectic continuously forming and disappearing in a sea of nematic. Since these smectic regions will not accept twist or bend, the nematic will appear stiffer as K_2 and K_3 get larger. Just as $\vec{B} = \vec{\nabla} \times \vec{A}$ is expelled from the bulk at the superconducting transition so bend and twist ($\propto \vec{\nabla} \times \vec{n}$) are expelled at the N-SmA transition. The form of the divergence can be carried over directly from Schmid's analysis of fluctuation diamagnetism [9] to give

$$K_2 = K_2{}^0 + \frac{k_BT}{24\pi} q_0{}^2 \frac{\xi_\perp{}^2}{\xi_{\parallel}} = K_2{}^0 + \tilde{K}_2 \tag{9a}$$

$$K_3 = K_3{}^0 + \frac{k_BT}{24\pi} q_0{}^2 \xi_{\parallel} = K_3{}^0 + \tilde{K}_3 \tag{9b}$$

where $K_2{}^0$, $K_3{}^0$ are the background nematic phase values of the elastic constants, $\tilde{K}_2$, $\tilde{K}_3$ are the diverging parts and $\xi_{\parallel} = (1/2\alpha M_V)^{1/2}$ and $\xi_\perp = (1/2\alpha M_t)^{1/2}$, respectively, are the longitudinal and transverse coherence lengths for fluctuations of $|\psi|$ in the nematic phase.

The expected temperature dependence of $\xi_{\parallel}$, $\xi_\perp$ depends on the model used. In the mean field model of McMillan [5], $\xi_{\parallel}$, $\xi_\perp \sim (T-T_c)^{-1/2}$ (this follows from letting $\alpha = \alpha_0(T-T_c)$ and letting M_V, M_t be constant). These mean field results appear to be correct for the superconducting transition even though d(=3) is less than d*(=4). The reason for this will be discussed below. On the other hand, de Gennes [8], using Wilson's [10] calculation of critical exponents for a two component scalar order parameter in three dimensions, expects that $\xi_{\parallel}$, $\xi_\perp \sim (T-T_c)^{-\gamma/2-\eta} = (T-T_c)^{-\nu}$ where γ, η, and ν are the usual critical exponents for the susceptibility, correlation function and correlation length, respectively [11]. The same result can be got by modifying the temperature dependence of the coefficients in the free energy $(\alpha(T) = \alpha_0(T-T_c)^\gamma;\ 1/M_V,\ 1/M_t \sim \xi^\eta)$. With Wilson's values [10] $\gamma = 1.30$, $\eta = 0.04$ we expect $\xi \sim (T-T_c)^{-0.66}$. The critical properties of helium obey this relation rather than the mean field, as one would expect for d* = 4.

The difference in behavior of two systems with apparently isomorphic order parameters, free energies and the same d* can be ascribed to the different interactions in the systems. In a superconductor the interaction range, even far from T_c, is very long so that one must get extremely close to T_c before the fluctuations of the order parameter begin to become important. Experimentally, one cannot as yet get this close to T_c so that mean field behavior is observed. In helium, the interaction range is rather short so that fluctuations start to dominate farther from T_c, i.e. the critical range is wider, and one therefore observes non-classical behavior at an experimentally accessible temperature difference, $T-T_c$. The liquid crystal interaction is also short range so that one could expect critical behavior at the N-SmA transition analogous to that of superfluid He^4.

Actually, the situation may be more complicated in the superconductor and in the smectic A liquid crystal. A renormalization group calculation [7] predicts that fluctuations will make the transition weakly first order. This effect is certainly too small to be observed in superconductors, but the original theoretical estimate of the discontinuity for smectics A is larger than has been observed in recent experiments. Recalculation as suggested by Chen and Lubensky may lead to a smaller estimate for the first order SmA-N discontinuity [12]. Since there is yet no experimental evidence of the predicted first order jump, we shall continue our discussion in terms of the helium analogue.

On the smectic side of the transition we expect the elastic constants B and D to go to zero at T_c since there are no longer any layers. Since $B = \psi_0^2 q_0^2/M_v$ and $D = \psi_0^2 q_0^2/M_t$ and both M_v and M_t are expected to have the same critical behavior in the helium analogy ($1/M_v$, $1/M_t \sim \xi^\eta$), both B and D should go to zero with the same exponent. Using scaling relations between critical exponents one can easily show [13] that ψ_0^2/M_t, $\psi_0^2/M_v \sim 1/\xi$. Thus, in the helium analogy we should see B, $D \sim (T_c-T)^{0.66}$. Mean field behavior would require M_t and M_v to be constant while $\psi_0^2 = -\alpha/\beta \propto (T_c-T)^{1.0}$ since is constant. Therefore in mean field theory we expect B, $D \sim (T_c-T)^1$.

Let us turn to the experimental results to see if this picture of the phase transition is correct; we first discuss pre-transition behavior in the nematic phase. The short range order associated with the SmA density wave can be directly probed by x-ray scattering spectroscopy. The apparatus used at M.I.T. [14] is schematized in Fig. 1 below.

The scattered x-ray intensity will be proportional to

$$<\delta\psi^2(\vec{q})> = \frac{kT\,\chi_s}{1 + \xi_\parallel^{\,2}(q_\parallel - q_0)^2 + \xi_\perp^{\,2} q_\perp^{\,2}} \tag{10}$$

as can readily be calculated from the free energy, Eq. (5). Here χ_s is a generalized susceptibility ($\sim 1/\alpha$) for the smectic order parameter. Thus by measuring the intensity and width of the x-ray scattering peak one may determine χ_s, $\xi_\parallel$, and $\xi_\perp$. A typical scan through the peak is shown in Fig. 2. An unexpected result is that the term $\xi_\perp^2 q_\perp^2$ in (10) must be replaced by $\xi_\perp^2 q_\perp^2 (1 + c\xi_\perp^2 q_\perp^2)$ in order to fit the profile of the $q_\perp$ scan - that is, the $q_\perp$ scans fall off more rapidly than a Lorentzian. The χ^2 for the fit is shown with and without the extra term. The relative importance of this term increases as $t = (T/T_c-1) \to 0$ and $\xi_\perp$ diverges. Our

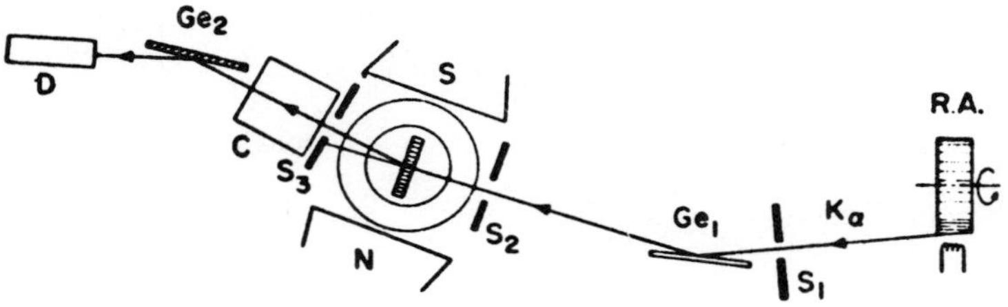

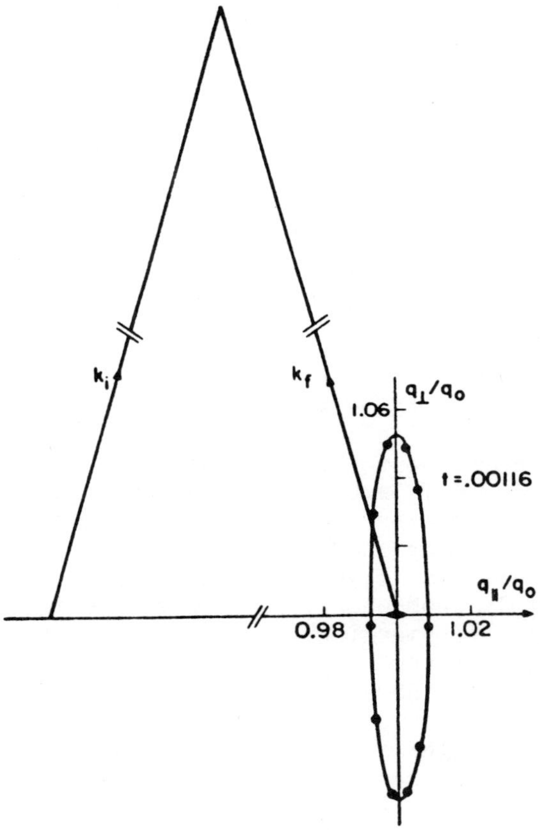

Fig. 1. Upper panel: schematic diagram of x-ray spectrometer with S_i = slits, Ge_i = perfect germanium crystals, C = horizontal Soller slits, D = sodium iodide detector. Lower panel: scattering diagram in reciprocal space showing critical scattering half-intensity contour at t = 0.00116; the solid ellipse in the center is the instrumental resolution function.

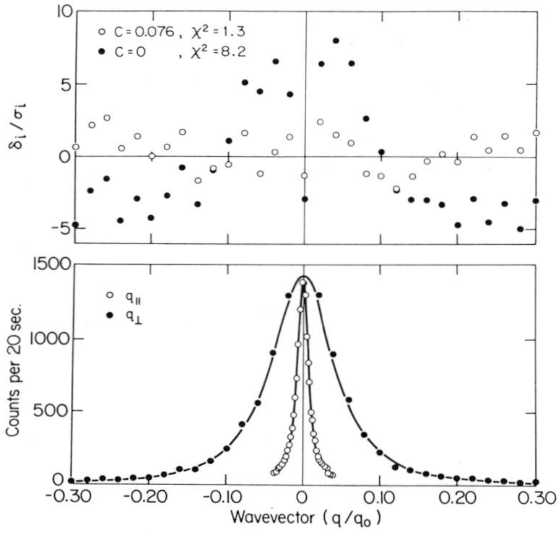

Fig. 2. Lower panel: typical transverse and longitudinal x-ray scattering profiles for 8CB. Upper panel: deviations of the fit to the transverse scan with and without the $c\xi_{\perp}^4 q_{\perp}^4$ term.

belief is that this term is a manifestation of divergent fluctuations in the phase of ψ which prevent the SmA phase from having true long range order; more about that presently. The experimental results of a series of M.I.T. experiments on pre-transitional SmA behavior in the nematic phase are contained in articles soon to appear [15, 16]. They are summarized below.

Three materials, cyanobenzylidene-octyloxyaniline (CBOOA), 8OCB, and octylcyanobiphenyl (8CB) were studied. These are all so-called "bilayer smectis", the value of q_0 (0.179, 0.197, 0.198Å^{-1}, respectively) corresponding to slightly less than twice the molecular length, indicating there is probably some antiferroelectric short range order of the molecules. When the x-ray scattering results are analyzed to determine power law singularities for the critical divergences one finds the critical exponents given in the table below ($5 \times 10^{-5} < t < 2 \times 10^{-2}$)

Exponent	CBOOA	8OCB	8CB
γ	1.30±0.06	1.32±0.06	1.26±0.06
$\nu_{\parallel}$	0.70±0.04	0.71±0.04	0.67±0.02
$\nu_{\perp}$	0.62±0.05	0.58±0.04	0.51±0.04

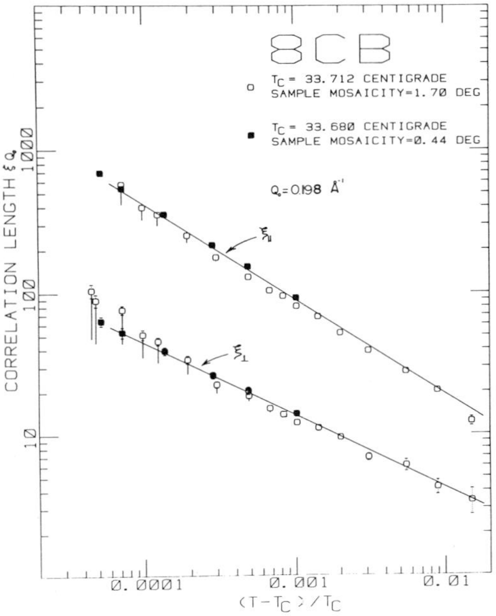

Fig. 3. The correlation lengths for short range SmA order fluctuations in the N phase of 8CB, as determined by x-ray scattering.

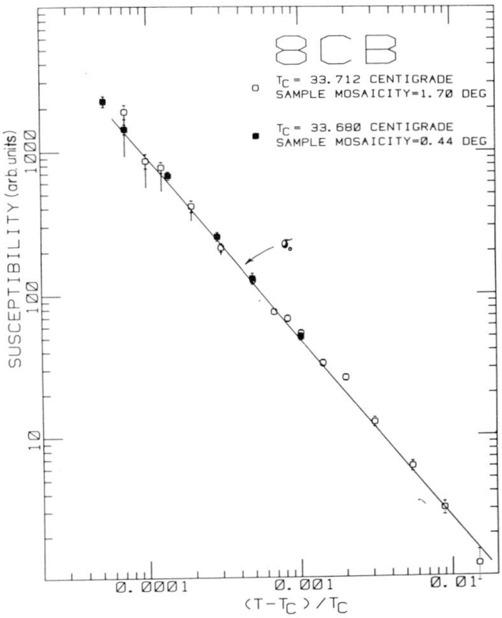

Fig. 4. Divergence of the susceptibility for SmA fluctuations in the N phase of 8CB, as determined by x-ray scattering.

These are the effective exponents obtained if one assumes a single power law divergence. The figures 3 and 4 show the data for 8CB. The resolution along $q_\parallel$ is determined by the perfect germanium crystals of the spectrometer and is quite good (HWHM 4.2×10^{-4} Å^{-1}). The resolution along $q_\perp$ is determined by the mosaic spread (variations in the director orientation) of the sample and varied from 1.5×10^{-3} Å^{-1} to 6×10^{-3} Å^{-1}, depending on sample mosaicity. The mosaicity was determined from the limiting width of $q_\perp$ scans as $t \to 0$ and was assumed to be Gaussian and temperature independent. The corrections necessary for mosaic spread in deconvoluting the data are shown as vertical lines on the figures. The mosaic correction is negligible for the sample with 0.44 deg. spread and in agreement with the corrected data for 1.7 deg. spread for X_S and $\xi_\parallel$ which lends confidence to the correction procedure. The agreement is less good for $\xi_\perp$, where the mosaicity contributes to first order, rather than to second order as in $\xi_\parallel$ measurements. The exponents obtained for γ and $\nu_\parallel$ are in satisfactory agreement with theoretical values (d=3, n=2, γ=1.316, ν=0.669) and support de Gennes' helium analogy. However, the situation is less clear with the results for $\xi_\perp$. The greatest difficulty appears to be with 8CB. In figure 5

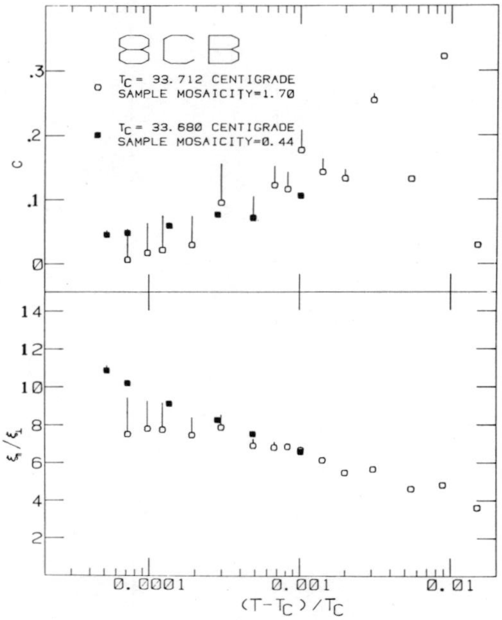

Fig. 5. Variation of the $\xi_\perp^4$ coefficient c and the ratio $\xi_\parallel/\xi_\perp$ with reduced temperature for 8CB.

are the results for the $q_\perp^4$ term c and the ratio $\xi_\parallel/\xi_\perp$ for 8CB. It should be emphased that the different exponents correspond only to a rather small evolution in the ratio $\xi_\parallel/\xi_\perp$ over three decades of t. The results are summarized in the table below.

Material	$\xi_\parallel/\xi_\perp$ at $t=10^{-2}$	$\xi_\parallel/\xi_\perp$ at $t=10^{-4}$
CBOOA	5.5±0.5	8±1
8OCB	5±1	9±1
8CB	4±1	9±1

In the worst case, 8CB, the ratio $\xi_\parallel/\xi_\perp$ changes by a factor 2.6 over $t = 2 \times 10^{-2}$ to $t = 5 \times 10^{-5}$; if the exponents were truly different this ratio must go to ∞ in the remaining 15 mK to T_C. Our experiments are not able to tell us if this is indeed the case.

As can be seen from Eq. (9), light scattering experiments can be used to measure the divergence in elastic constants and thus indirectly the correlation lengths. Experiments have been done at M.I.T. on all three of these compounds. For CBOOA the result was $\nu_\parallel = 0.74 \pm 0.04$. Measurements of K_2 and K_3 for 8OCB are shown in

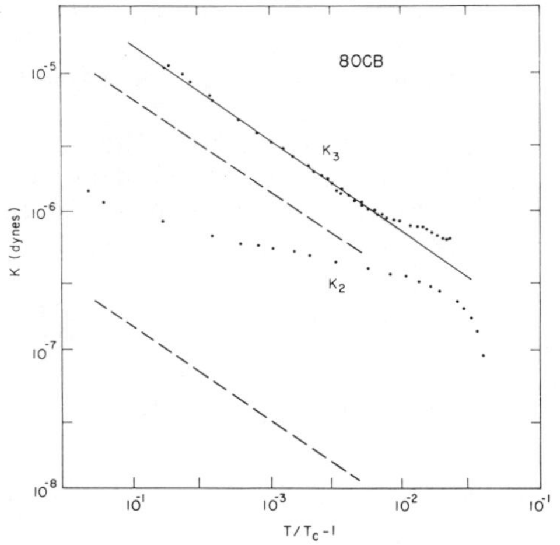

Fig. 6. Temperature dependence of elastic constants K_3 and K_2 near the N-SmA transition of 8OCB.

figure 6. The dashed lines in the figure are the predictions of Eq. (9) using q_0, $\xi_\parallel$, and $\xi_\perp$ obatined from x-ray measurements. It is apparent for K_2 (because the ratio $\xi_\parallel{}^2/\xi_\perp{}^2$ is so small) that the divergent part cannot be reliably separated from the background value $K_2{}^0$. The solid line through the K_3 data is a least squares fit and gives $\nu_\parallel = 0.66 \pm 0.04$. Thus the x-ray and light scattering data are in satisfactory agreement except that the coefficient $kTq_0{}^2/24\pi$, a mean field result, is too small by a factor four. Figure 7 shows light scattering data for K_3 in 8CB; here the data

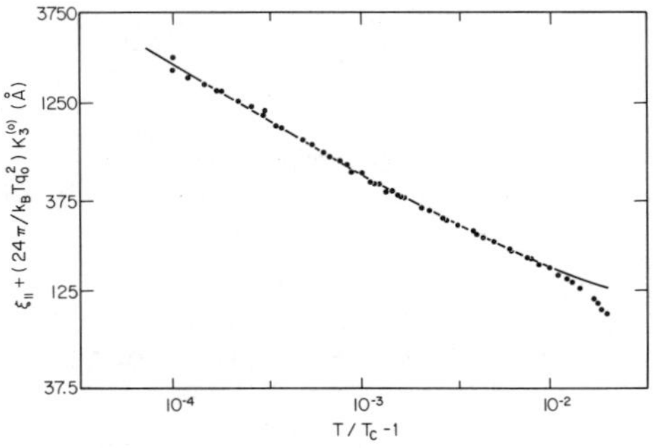

Fig. 7. Divergence of $\xi_\parallel$ near the SmA-N transition of 8CB as determined by light scattering and Eq. (9)

are normalized to agree with the x-ray result at $t = 10^{-4}$ and a temperature independent background term (50 Å) due to K_3^0 has been included. The exponent $\nu_{\parallel} = 0.62 \pm 0.03$ is in satisfactory agreement with the x-ray result.

To summarize nematic phase studies then, the results support the helium analogue (with upper marginal dimensionality $d^* = 4$) with two possible exceptions; there appears to be an evolution in the ratio $\xi_{\parallel}/\xi_{\perp}$, and the scattering cross section requires a $(\xi_{\parallel} q_{\perp})^4$ term.

We should now like to discuss the properties of the smectic phase. The Goldstone mode that arises from the SmA ordering corresponds to fluctuations in the layer displacement u (i.e. the phase of ψ). The dispersion relation for these modes, calculated from Eq. (7), is

$$\hbar\omega = K_1 q_{\perp}^4 + B q_{\parallel}^2 \tag{11}$$

The unusual nature of this arises because when $q_{\parallel} \neq 0$, $u(\vec{q})$ involves compression of the smectic layers, but if $q_{\parallel} = 0$ the only restoring force to $u(\vec{q})$ comes from curvature of the layers and involves the nematic splay elastic constant K_1. This has a profound effect on the properties of the SmA phase. If we calculate the mean squared fluctuations in u, the result is

$$\langle u^2(\vec{r})\rangle = \frac{kT}{(2\pi)^3} \int \frac{d^3q}{B q_{\parallel}^2 + K_1 q_{\perp}^2} \tag{12}$$

The limits of integration can be chosen as $(2\pi/L) \leq |\vec{q}| \leq q_0$ to obtain an approximate answer. (L is the size of the sample.) One readily obtains

$$\langle u^2(\vec{r})\rangle \simeq \frac{kT}{4\pi(BK_1)^{1/2}} \ln(q_0 L) \quad . \tag{13}$$

Thus the SmA phase in three dimensions shows the same logarithmic singularity from long wavelength Goldstone modes as one finds for solids in two dimensions; one expects the SmA phase does not exhibit true long range order, but rather the topological long range order predicted by Kosterlitz and Thouless. The lower marginal dimensionality, d^0, is three for SmA (and also for SmC) phases of liquid crystals.

If we wished to study this by x-ray scattering, we recall that the x-rays measure the Fourier transform of the pair correlation function

$$G(\vec{r}) = \langle e^{iq_o[u(\vec{r})-u(0)]}\rangle \tag{14}$$

This can be calculated from Eq. (11) in the harmonic approximation [17] to be proportional to

$$G(\vec{r}) \sim \frac{1}{(x^2+y^2)^{\eta}} e^{-\eta E_1\left(\frac{x^2+y^2}{4\lambda z}\right)} \tag{15}$$

where $\eta = (kTq_0^2/8\pi\lambda B)$, $\lambda = (K_1/B)^{1/2}$ is the analogue of the penetration depth in a type I superconductor, and E_1 is the exponential integral. From the properties of E_1 it is readily seen that $G(\vec{r})$ does not extend to infinity (as it would for long range order) but has an anisotropic power law (algebraic) decay

$$G(\vec{r}) \sim \frac{1}{r_\perp^{2\eta}} \qquad r_\perp \gg r_\parallel \tag{16a}$$

$$\sim \frac{1}{r_\parallel^{\eta}} \qquad r_\parallel \gg r_\perp \tag{16b}$$

This means the scattering from the SmA density wave in the Sm phase is not a Bragg peak, but a power law singularity, which has been observed in experiments you will hear a lot more about from Jens Als-Nielsen. A typical result for 8OCB is shown in Fig. 8 below. (The solid line is a convolution of the theoretical cross section from Eq. (15) with the instrumental resolution function, shown dashed, and gives $\eta = 0.17$, $\lambda q_0 = 3.9$.) These experiments also show the second harmonic (scattering at $q_\parallel = 2q_0$) is $\lesssim 10^4$ weaker than at q_0. The SmA density wave is very nearly a perfect sine wave, a result which can be quantitatively understood using the phase fluctuations in ψ.

Now, we turn to some other properties of the SmA phase, the elastic constants B and D, which can be measured by light scattering from the director modes. From Eq. (7), we see how to measure K_1 and B if the geometry is chosen to scatter from mode n_1, while scattering from mode n_2 enables determination of D. Figure 9 below shows both the intensity of scattering from mode 1 and decay constant Γ_1 in the SmA phase of CBOOA [18]. (The angle Φ is between the wave vector $\vec{q}$ and the smectic layers.) From Eq. (7), the minimum in the parabolic curves gives K_1, and the curvature gives the ratio K_1/B. The results [18] for several different samples of CBOOA are shown in Fig. 10.

If the He analogue correctly describes the SmA phase, then one would expect $B \sim 1/\xi_\parallel \sim -t^{0.66}$. However the solid cruve in the figure has the equation $-t^{0.33}$! (The figure also shows that K_1 divided by the splay viscosity has no anomalous behavior, as is expected.) An additional strange result is that the coefficient D vanishes as $-t^{0.5}$; not only different from the expected $1/\xi_\perp$ result, but also different from B. Additional experiments [15]

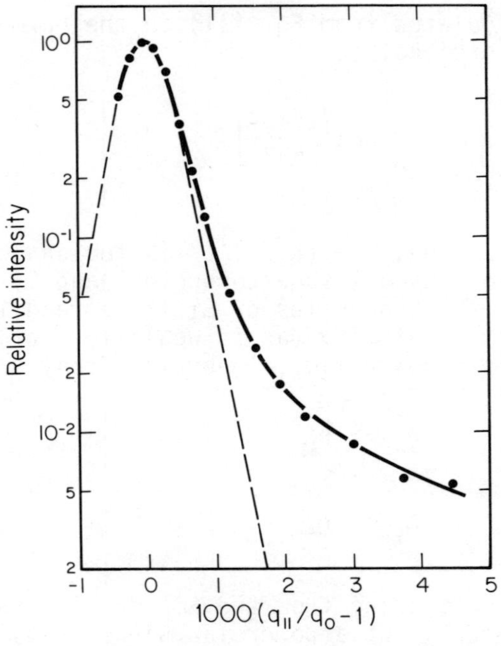

Fig. 8. X-ray structure factor for the SmA density wave in the SmA phase of 8OCB at a reduced temperature t = 9 x 10-4.

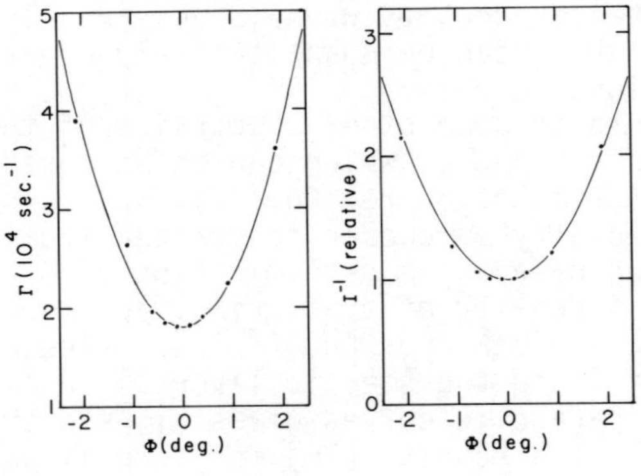

Fig. 9. The intensity and decay constant for light scattered from director mode 1 in the SmA phase of CBOOA.

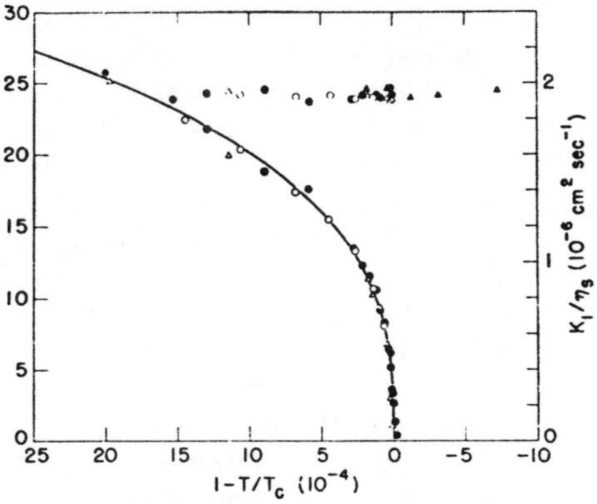

Fig. 10. Temperature dependence of K_1/η_s and B/K_1 near the SmA-N transition of CBOOA, from light scattering measurements.

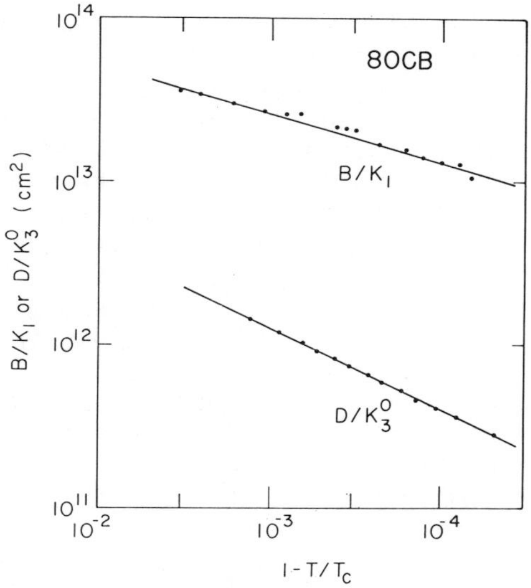

Fig. 11. Light scattering results for the temperature dependence of B/K_1 and $D/K_3^{(0)}$ in the SmA phase of 8OCB.

[15] carried out at M.I.T. show similar results in 8OCB and 8CB. The 8OCB data show $B \sim -t^{0.30}$ and $D \sim -t^{0.50}$ (see Fig. 11) while for 8CB [16], the exponent for B is 0.26±0.06.

We would now like to summarize our present understanding of the SmA phase. The N phase experiments are consistent with de Gennes' He analogue with an upper marginal dimensionality d* = 4, but we have to understand the evolution in $\xi_{\parallel}/\xi_{\perp}$. It is possible that there are indeed two lengths in the problem, but it may simply be that this evolution represents the influence of anisotropic critical behavior (because of the anisotropy of K_i) predicted by Lubensky and Chen [12]. To answer this question a calculation of the crossover function with the appropriate parameters for 8CB is needed. The term $q_{\perp}^4 \xi_{\perp}^4$ observed in the x-ray cross section may be due to phase fluctuations in the short range order - a pretransitional indication of the $q_{\perp}^4$ cross section predicted by the SmA phase by Eq. (16a). If so, this represents the only indication in the N phase of the lack of true long range order in the SmA phase. Since the phase fluctuations in the A phase raise d^0 from 2 to 3, it seems that they must ultimately raise d* from 4. However, the effects of these fluctuations are subtle and should not manifest themselves in the short range order of the N phase unless $q_0 \xi >> 1$; thus the He analogue suffices to describe short range order.

The SmA x-ray scattering appears to confirm $d^0 = 3$ and in fact represents the first direct experimental evidence of the algebraic decay of correlation functions predicted at lower marginal dimensionality. The elastic constants B and D are probably sensitive to the lack of true long range order and their anomalous critical behavior is probably associated with the divergence of long wavelength fluctuations in u. Since B is associated with layer compression ($\partial u/\partial z$) while D measures the force keeping the molecules normal to the layers (and is associated with their orientational order, which is truly long range, as well) it is perhaps not surprising the two constants have different critical behavior. These ideas have yet to be tested by quantitative calculations, however. The specific heat is determined by fluctuations in short range order and careful experiments at Kent State University [18] give results nearly in agreement with the He analogue. The divergence in C_p may be slightly stronger than that of He, but the experiments are complicated by a weak first order jump and 40 mK coexistence region, probably due to some impurities in the 8OCB sample. Experiments at M.I.T. [20] show a stronger divergence ($\alpha = 0.27 \pm 0.03$) on the N side which may be consistent with anisotropic scaling [12] if the exponents $\xi_{\parallel}$ and $\xi_{\perp}$ are really different. Kent State experiments [21] on 8S5 (pentylphenylthiol-octyloxybenzoate) show a close to logarithmic divergence of C_p. Thus the He analogue provides a nearly correct description of short range order effects near the SmA-N transition; slight discrepancies exist and it is not yet clear if they can be explained by corrections to scaling or anisotropic scaling. The analogue fails to describe properly behavior associated with long range order since the phase fluctuations in ψ prevent the establishment of true long range order. A quantitative theory, analogous to the Kosterlitz-Thouless picture for two dimensional crystals, has yet to be developed.

IV. THE SMECTIC C PHASE AND SmC-SmA TRANSITION

The SmC phase is like the SmA, except that the molecules are tilted with respect to the layers; thus director modes which do not change the tilt angle of the molecules with respect to the layers cost little free energy and are of an amplitude comparable to the nematic phase. We shall not go into everything we know about the SmC phase (which is much less than about SmA) but shall discuss some experiments relevant to our interest here: the effect of thermal fluctuations on the nature and properties of the phases that can exist in condensed matter. The SmC layers are 2D liquids just like the SmA phase and the Landau-Peierls instability should also mean that the SmC phase lacks long range translational order, although this has not been verified experimentally. In many materials, 8S5 for example, there is a transition from a SmA to SmC which appears to be second order. The SmC order parameter has two degrees of freedom: the tilt angle (magnitude) and azimuthal direction (phase). Thus it is an $n = 2$, $d = 3$ phase like superfluid He or a superconductor; the upper marginal dimensionality d^* should therefore be four. A number of experiments have been carried out on the SmA-SmC transition, many of which probably do not reliably obtain the asymptotic behavior near the transition. The experiments which appear to be reliable [22,23,24,25] seem to support mean field behavior as if $d^* = 3$. Just recently x-ray scattering experiments have been carried out at M.I.T. on 8S5 [26] which show rather clearly behavior at the SmC-SmA transition. The experiments were done with the director (molecular) orientation fixed by an external field. The smectic density wave then shows up as a peak in reciprocal space while in the SmA phase. However when the molecules tilt with respect to the layers, then the normal to the layers falls on a cone of tilt angle Φ with respect to the field. As a result, the density wave peak becomes a ring in reciprocal space from which the tilt angle can be directly measured. Several scans through this ring are shown in Fig. 12. The layer spacing can also be precisely measured by θ-2θ scans to determine the magnitude of the momentum transfer. Thus both the tilt angle and the layer spacing can be measured simultaneously. The results of the experiments are shown in Fig. 13. The primary order parameter is Φ the tilt angle, and follows mean field asymptotic behavior. The layer thickness, measured by $(1-q_A/q_C)$ is a secondary order parameter (proportional to Φ^2). Although $(1/\Phi)\arccos(q_A/q_C)$ is independent of temperature, it is less than unity ($\simeq 0.8$) indicating there may be some conformational changes in the aliphatic tails of the molecules.

We must explain the mean field exponent associated with the transition; the answer seems to lie in consideration of the Ginzburg criterion which predicts mean field behavior for [27]

$$|t| = |1-T/T_c| > \frac{k^2}{32\pi\, \xi_0^{\;6}\, \Delta C^2} \tag{17}$$

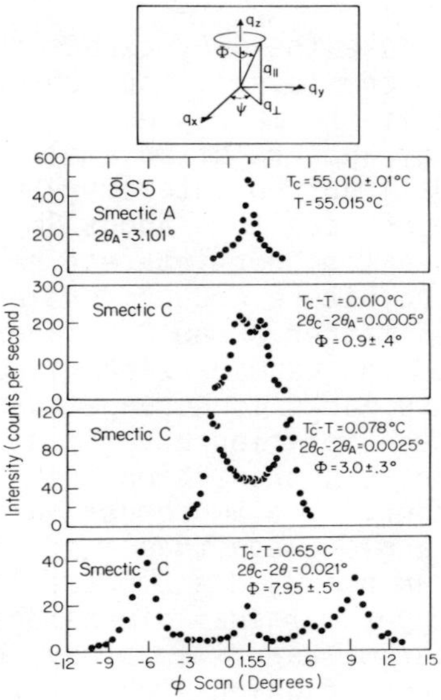

Fig. 12. Opening of the SmA peak into a ring in reciprocal space as the SmC phase of $\bar{8}S5$ is entered.

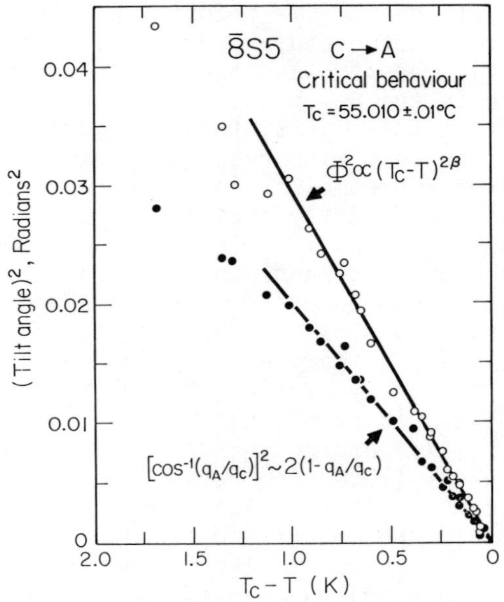

Fig. 13. Temperature dependence of the square of the tilt angle and the square of the change in layer spacing in $\bar{8}S5$; the solid lines represent mean field behavior.

where the correlation length diverges as $\xi_0 t^{-1/2}$ and ΔC is the heat capacity jump at the phase transition. From the experiments [21] of Schantz and Johnson on $\overline{8}S5$, $\Delta C \simeq 1$ Joule/cm^3. The critical fluctuations for the A-C transition are those in the molecular tilt. These can be estimated from Eq. (7) for the SmA phase free energy. If we calculate tilts Φ_1 (in the plane of $\vec{n}$ and $\vec{q}$) and Φ_2 (normal to $\vec{n}$ and $\vec{q}$) with respect to the SmA layers we find

$$\langle\Phi_1^2(q)\rangle = \frac{kT}{D(1 + \xi_{1\perp}^2 q_\perp^2 + \xi_{1\parallel}^2 q_\parallel^2)} \tag{18a}$$

$$\langle\Phi_2^2(q)\rangle = \frac{kT}{D(1 + \xi_{2\perp}^2 q_\perp^2 + \xi_{2\parallel}^2 q_\parallel^2)} \tag{18b}$$

with $\xi_{1\perp}^2 = K_1/D$, $\xi_{1\parallel}^2 = K_3/D = \xi_{2\perp}^2$, $\xi_{2\parallel}^2 = K_2/D$. At the SmA-SmC transition these diverge as $D \to 0$. If we use $\Delta C = 1$ J/cm^3 in Eq. (17), we find mean field behavior for $t < 10^{-3}$ if $\xi_0 \simeq 50$ Å. This is not inconsistent with values estimated from $K \sim 5 \times 10^{-7}$ dynes and $D \sim 10^{+6}$ erg/cm^3 that give $\xi \sim 70$ Å in the SmA phase well away from the C-A transition; it is also consistent with values of $\xi \sim 300$ Å determined by undulation instabilities in the SmC phase by Johnson and Saupe [28]. Thus it seems plausible that most SmA-SmC transitions show mean field behavior for the same reason as superconductors do - the intrinsic coherence lengths are long and reduce the fluctuations until very close to T_c. It is worth mentioning that Delaye [29] reports He like exponents in nonyloxybenzoate-butyloxyphenyl, and also values of $\xi_0(\parallel) \sim 8$ Å, $\xi_0(\perp) \sim 5$ Å which would still be consistent with the Ginzburg criterion; thus at least one SmC material apparently exhibits non-classical behavior due to fluctuations.

Finally we should mention that there are experiments [30,31] which report no change in the smectic layer spacing in the SmC phase. This implies either that the nematic order parameter increases in just such a way as to keep the layer thickness constant or else that the molecules are already tilted in the SmA phase, but randomly, so that the SmA-SmC phase would be described by a 3D x-y model. It is also possible that this result could be the caused by insufficient resolution normal to the scattering plane in the x-ray experiments. Thus smectic C phases may come in several fundamentally different varieties; we need more experiments to find out.

V. LIQUID CRYSTALS AND LOWER DIMENSIONAL PHYSICS

A) The present interest in lower dimensional ($d < 3$) physical systems arises because fluctuations play such an important role and one can hope to test many of our current ideas about the effects of fluctuations. Liquid crystal phases may be interesting to study not only because in the SmA and SmC phases do fluctuations play the same role for $d = 3$ as they do in other materials for $d = 2$, but also because it is possible to prepare smectic samples which should be two dimensional. The method has been pioneered by the former Harvard group [32] who showed that it is possible to prepare free smectic films from two molecules on up in thickness. The film is freely suspended like a soap bubble, and if one chooses a SmC phase whose molecules have a tilt angle θ_T with the normal to the layer (the z axis), then the component of the director in the plane of the film is orientationally degenerate and for thin films should be a good approximation to a two dimensional nematic and a physically realizable example of the 2D x-y model. If we consider such a film, let $\hat{n} = \cos\phi\,\vec{1}_x + \sin\phi\,\vec{1}_y$ be the two dimensional director, $\hat{\nabla} = \vec{1}_x\,\partial_x + \vec{1}_y\,\partial_y$, θ_T be constant, and $\partial_z \equiv 0$, then the elastic free energy density of Eq. (1) becomes

$$\Phi_N = \frac{1}{2}\left\{K_1\sin^2\theta_T(\hat{\nabla}\cdot\hat{n})^2 + (K_2\sin^2\theta_T\cos^2\theta_T + K_3\sin^4\theta_T)(\hat{\nabla}\times\hat{n})^2\right\} \quad (19)$$

One of the compounds studied at Harvard was 80.5* (octyloxybenylidene- -methylbutylaniline) which is a chiral molecule and thus the SmC phase is weakly ferroelectric. The free energy of the film must include the interactions of these electric dipoles. It can be written as

$$\begin{aligned} F = &\int dA\left\{\frac{1}{2}K_s(\hat{\nabla}\cdot\hat{n})^2 + \frac{1}{2}K_b(\hat{\nabla}\cdot\hat{n})^2 - \vec{P}_0\cdot\vec{E}\right\} \\ &+ \int dA\left\{\frac{\hat{\nabla}\cdot\vec{P}_0(r)\;\hat{\nabla}\cdot\vec{P}_0(r')}{2|r-r'|}\right\} \end{aligned} \quad (20)$$

where the film thickness is h, $\vec{P}_0$ is h times the bulk polarization $\vec{P}$, $K_s = hK_1\sin^2\theta_T$, and $K_b = hK_2\sin^2\theta_T\cos^2\theta_T + hK_3\sin^4\theta_T$. The Goldstone modes are two dimensional splay and bend distortions of $\hat{n}$. The splay occurs for $\vec{q} \perp \hat{n}$ and the bend for $\vec{q} \,||\, \hat{n}$. The analogue of Eq. (2) is readily calculated to be

$$\langle\hat{n}_s(\vec{q},0)\hat{n}_s(\vec{q},\tau)\rangle = kT[K_s q_\perp^2 + P_0E]^{-1}\,e^{-\Gamma_s\tau} \quad (21a)$$

$$\langle\hat{n}_b(\vec{q},0)\hat{n}(\vec{q},\tau)\rangle = kT[K_b q_{||}^2 + 2\pi P_0^2|q_{||}| + P_0E]^{-1}\,e^{-\Gamma_b\tau} \quad (21b)$$

with an applied electric field E and decay rates

$$\Gamma_s = (K_s q_\perp^2 + P_0 E)/\eta_s \tag{22a}$$

$$\Gamma_b = (K_b q_{\parallel}^2 = 2\pi P_0^2 \, |q_{\parallel}| + P_0 E)/\eta_b$$

The dispersion relation and relaxation times for these modes have been verified experimentally by the Harvard group. The results show that K_b, K_s vary with film thickness and are somewhat larger than one would estimate from bulk film values, presumably because of effects at the film surface which cause the tilt angle to vary. For simplification we may take $K_s = K_b = K$ to calculate the fluctuations in the angle ϕ (analogous to Eqs. (12) and (13) for the SmA) for E = 0

$$\begin{aligned} \langle\phi^2\rangle &\sim \frac{kT}{K(2\pi)^2} \int \frac{d^2q}{q^2 + (2\pi P_0^2/K)q_{\parallel}} \\ &\simeq \frac{kT}{8\pi K} \ln\left(\frac{4K}{P_0^2 h}\right) \end{aligned} \tag{23}$$

Thus the ferroelectric dipolar interactions stabilize the long range order and the ferroelectric SmC films have a lower marginal dimensionality $d^0 = 1$. However if the dipole moment P_0 were zero, this would be an exact example of the 2D x-y model and Eq. (23) would become

$$\langle\phi^2\rangle \sim \frac{kT}{8\pi K} \ln\left(\frac{L}{h}\right) \tag{24}$$

indicating $d^0 = 2$. All of the theoretically predicted effects for the 2D x-y model should then occur in thin non-ferroelectric SmC films. For the 80.5* films studied in the Harvard experiments $(P_0^2/4K)^{-1}$ ranged from 30 to 130 cm; thus for all practical purposes the 2D x-y model should be a good description of these films. The special effects that occur at d^0 are subtle and difficult to observe, as we know from our SmA experiments, but thin films of smectic phases should be very interesting materials in which to study them.

B) It may also be that concepts from two dimensional physics manifest themselves in some of the better ordered smectics, phases like SmB, SmH, SmD, SmE, etc. Birgeneau and Litster [33] have proposed a model for these phases based on calculations for two dimensional ordering carried out by Halperin and Nelson [34]. The latter authors found three types of phases could exist in two dimensions when the anisotropy of a crystal lattice was explicitly considered. To discuss these quantitatively one defines both translationan and orientational order parameters, the orientational one refers to the orientation of nearest neighbor bonds between molecules (specified by the angle $\theta(\vec{r})$ with respect to some fixed axis). If $\vec{G}$ is a reciprocal lattice vector and $\vec{u}(\vec{r})$ the displacement of an atom from its lattice site then the positional order parameter is defined in the usual way

$$P(\vec{G},\vec{r}) = \langle e^{i\vec{G}[\vec{u}(\vec{r})-\vec{u}(0)]}\rangle \quad .$$

For a hexagonal lattice one defines an orientational order parameter

$$O(\vec{r}) = \langle e^{i6[\theta(\vec{r})-\theta(0)]}\rangle \quad .$$

Halperin and Nelson found a low temperature phase (i) in which $O(\vec{r})$ has true long range order while $P \sim r^{-\eta(G)}$ has the algebraic decay of the Kosterlitz-Thouless [35] topological order. Then, on warming there is an intermediate phase (ii) with short range positional order $P \sim e^{-r/\xi}$ and algebraic decay of $O \sim r^{-\eta(6)}$. Finally the system becomes a 2D liquid with short range order of both O and P. Birgeneau and Litster proposed to explain the various smectic phases by stacking up layers of these two dimensional phases. To understand the result, we need to recall that whenever a correlation function decays algebraically, the associated susceptibility is infinite. Stacking phase (i) with even an infinitesimal interaction between layers would therefore result in a 3D solid with true long range positional order. However stacking phase (ii) with a sufficiently weak interaction between the layers would result in a system with true long range order for O and short range order for P. This phase, depending upon the basic 2D crystal structure, tilt of the molecules with respect to the layers, and so on, is proposed to account for all of the well ordered smectic phases (SmB, SmH, etc.). Within this picture a quite natural transition to SmA or SmC phases occurs when O also becomes short range; the SmA and SmC phases are thus regarded as stacked two dimensional liquids. This is an attractive model which was consistent with the known experimental data when proposed. It explained the absence of higher order Bragg peaks and the rather diffuse nature of those peaks observed in SmB phases, while at the same time the bond orientational long range order explained how the rotational symmetry of the hexagonal lattice could be seen. We don't yet know if it is correct, and experiments on films of SmB presently underway at Bell Labs (D. Moncton and R. Pindak) as well as bulk SmB samples at M.I.T. (P. Pershan, G. Aeppli, R. Birgeneau, and D. Litster) of the compound butoxybenzylidene-octylaniline (BBOA) indicate that in this material the SmB phase may possess long range positional correlations. The M.I.T. experiments, for example, show resolution limited Bragg peaks from which one may conclude that the positional order extends at least 2000 Å, depending on direction in the sample. It therefore appears that very high resolution x-ray studies will be needed to test this model.

ACKNOWLEDGEMENTS

The M.I.T. work described in this paper has been supported in part by the National Science Foundation under grants DMR-76-80895 and DMR-76-18035 as well as the Joint Services Electronics Program, contract No. DAAG-29-78-C-0020.

REFERENCES

1. P.G. de Gennes, The Physics of Liquid Crystals, Oxford University Press, 1977.
2. S. Chandrasekhar, Liquid Crystals, Cambridge University Press, 1977.
3. C.W. Oseen, Trans. Faraday Soc. 29, 883 (1933) and F.C. Frank, Disc. Faraday Soc. 25, 19 (1958).
4. Groupe d-Etudes des Cristaux Liquides, J. Chem. Phys. 51, 816 (1969).
5. W. McMillan, Phys. Rev. A4, 1238 (1971).
6. K.K. Kobayashi, Phys. Lett. 31A, 125 (1970); J. Phys. Soc. Japan 29, 101 (1970); Mol. Cryst. Liq. Cryst. 13, 137 (1971).
7. B.I. Halperin and T.C. Lubensky, Sol. St. Comm. 10, 753 (1972).
8. P.G. de Gennes, Sol. St. Comm. 10, 753 (1972).
9. A. Schmid, Phys. Rev. 180, 527 (1968).
10. K.G. Wilson and M.E. Fisher, Phys. Rev. Lett. 28, 289 (1972).
11. H.E. Stanley, Introduction to Phase Transitions and Critical Phenomena, Oxford University Press, New York, 1971.
12. T.C. Lubensky and Jing-Huei Chen, Phys. Rev. B17, 366 (1978); Jing-Huei Chen, T.C. Lubensky, and D.R. Nelson, Phys. Rev. B, in press.
13. F. Brochard, J. de Physique, 34, 411 (1970).
14. J. Als-Nielsen, R.J. Birgeneau, M. Kaplan, J.D. Litster, and C.R. Safinya, Phys. Rev. Lett. 39, 352 (1977).
15. J.D. Litster, J. Als-Nielsen, R.J. Birgeneau, S.S. Dana, D. Davidov, F. Garcia-Golding, M. Kaplan, C.R. Safinya, and R. Schaetzing, to be published in Colloque Bordeaux issue of J. de Physique, 1979.
16. D. Davidov, C.R. Safinya, M. Kaplan, S.S. Dana, R. Schaetzing, R.J. Birgeneau, and J.D. Litster, Phys. Rev. B, in press for April 1979.
17. A. Caillé, Comptes rendus Ac. Sc. Paris 274B, 891 (1972).
18. H. Birecki, R. Schaetzing, F. Rondelez, and J.D. Litster, Phys. Rev. Lett. 36, 1376 (1976).
19. D.L. Johnson, C.F. Hayes, R.J. deHoff, and C.A. Schantz, Phys. Rev. 18B, 4902 (1978).
20. C.W. Garland, G. Kasting, K. Lushington, private communication.
21. C.A. Schantz and D.L. Johnson, Phys. Rev. A17, 1504 (1978).
22. Y. Galerne, to appear in Colloque Bordeaux Issue of J. de Physique.

23. Z. Luz and S. Meiboom, J. Chem. Phys. 59, 275 (1973); S. Meiboom and R.C. Hewitt, Phys. Rev. 15A, 2444 (1977).
24. D. Guillon and A. Skoulios, J. de Physique 38, 79 (1977).
25. D. Bartolino, J. Doucet, G. Durand, Ann. de Phys. 3, 389 (1978).
26. M. Kaplan, C.R. Safinya, J. Als-Nielsen, D. Davidov, D.L. Johnson, J.D. Litster, and R.J. Birgeneau, Bull. Am. Phys. Soc. 24, 251 (1979).
27. V.L. Ginsburg, Sov. Phys. Sol. St. 2, 1824 (1960).
28. D.L. Johnson and A. Saupe, Phys. Rev. 15A, 2079 (1977).
29. M. Delaye, in Colloque Bordeaux issue of J. de Physique, in press.
30. W.H. de Jeu and J.A. de Poorter, Phys. Let. 61A, 114 (1977).
31. A. de Vries, Mol. Cryst. Liq. Cryst. 41, 27 (1977).
32. C.Y. Young, R. Pindak, N.A. Clark, R.B. Meyer, Phys. Rev. Lett. 40, 773 (1978).
33. R.J. Birgeneau and J.D. Litster, J. de Physique Lettres 39, L-399, (1978).
34. B.I. Halperin and D.R. Nelson, Phys. Rev. Lett 41, 121, 519 (1978).
35. J.M. Kosterlitz and D.J. Thouless, J. Phys. C6, 118 (1973).

DISLOCATIONS AND DISCLINATIONS IN SMECTIC SYSTEMS

S. T. Lagerwall and B. Stebler

Physics Department
Chalmers University of Technology
S-412 96 Göteborg, Sweden

INTRODUCTION

The dislocation theory of melting, revived and extended by R.M. Cotterill[1], has, combined with renormalization theory, been the basis for an intense recent development in the description of phase transitions, notably in the work of J.M. Kosterlitz and D.J. Thouless[2] on the xy-model, and its elaboration by B.I. Halperin and D.R. Nelson[3]. The variety of order in liquid crystals, especially in the topological sense, makes these systems interesting in several respects, regarding phase transitions and involved defects, as well as regarding the general physical behaviour of low-dimensional systems.

Smectics are periodic in one dimension and should therefore contain translational defects, dislocations. Single elementary dislocations can now be studied by a recently developed technique[4,5,6]. In the smectic C phase, in addition to dislocations there exist disclinations, singularities in the orientational field. Single pairs of disclinations - they are always created in pairs - can be generated under suitable conditions[7]. In the following we give the background and report on both kinds of observation.

TRANSLATIONAL DEFECTS

A highly schematic distinction of the two simplest and most important smectic phases, A and C, is made in Fig. 1. The unit vector $\hat{n}$, called the director, describes the local average direction of the molecular axis. Fig. 2a shows how dislocations are generated: by locking the boundary conditions (attained by surface treatment) and forming a wedge, one forces the creation of a quasi-regular array of defects, by which the layer thickness μ is kept constant.

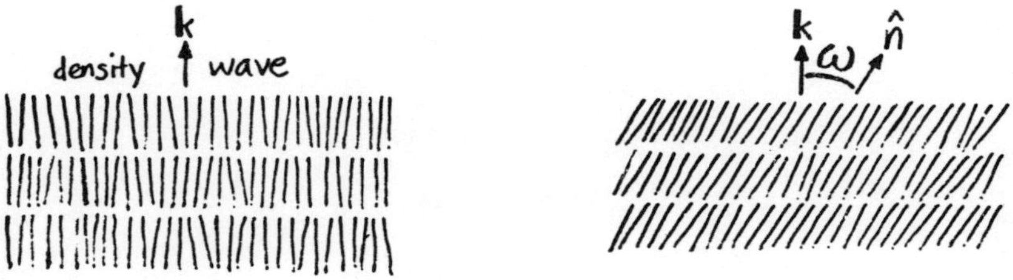

Fig. 1. Smectic A (left) and smectic C (right) order. The order parameter for the transition A to C is the angle ω between the wave vector $\vec{k}$ (expressing the periodicity) and the director $\hat{n}$. ω is zero in the A phase and increases with decreasing temperature in the C phase.

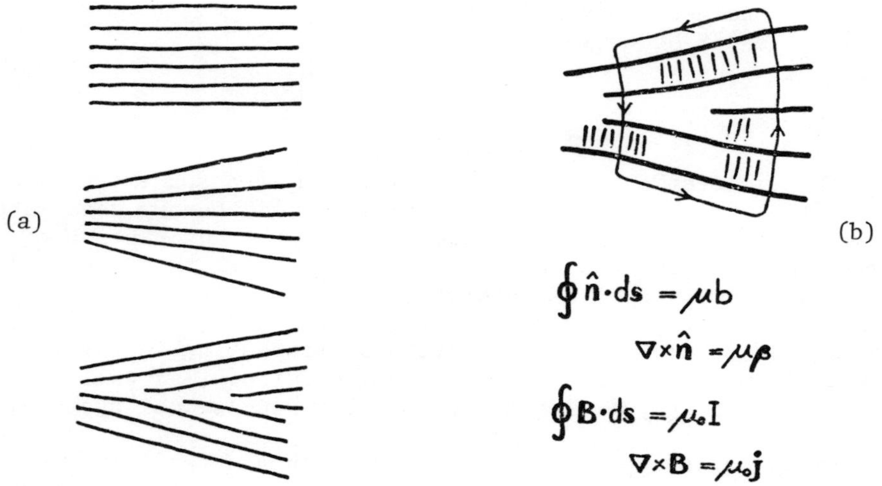

Fig. 2. Generation of dislocations in the described experiments (a); Burgers vector definition (b). The configuration of the core region may be different - cf the other version to the right in Fig. 5 - and has to change in glide motion (i.e. across layers).

(The layers can, to a first approximation, be treated as incompressible.) In Fig. 2b the Burgers vector of the dislocation is defined by taking the line integral of the director field $\hat{n}(\vec{r})$ around a closed loop containing the dislocation (in this case of one unit). The similarity with electromagnetism becomes clear from a comparison with Ampères law, giving the line integral of the magnetic field by going around a current I. The analogue of the corresponding Maxwell equation, curl $\vec{B} = \mu_0 \vec{J}$, means that curl $\hat{n}$ is equivalent to a dis-

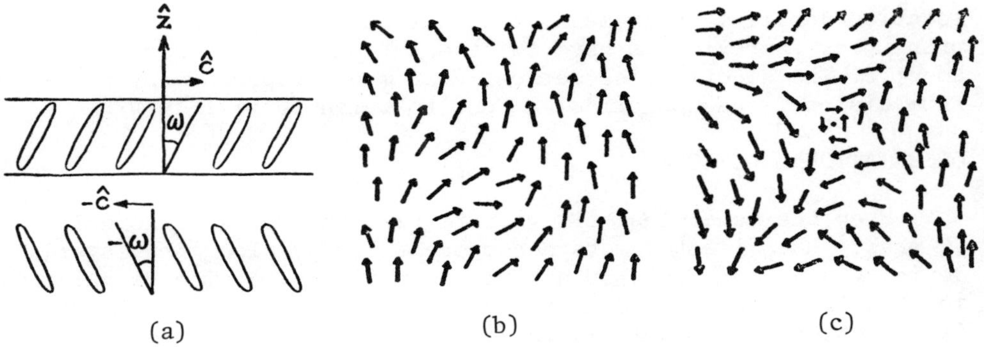

Fig. 3. The tilt generates a director component along the plane of the layer (a). This component defines a two-dimensional vector field, illustrated in two versions, (b) and (c), of which only the first one has perfect topological order.

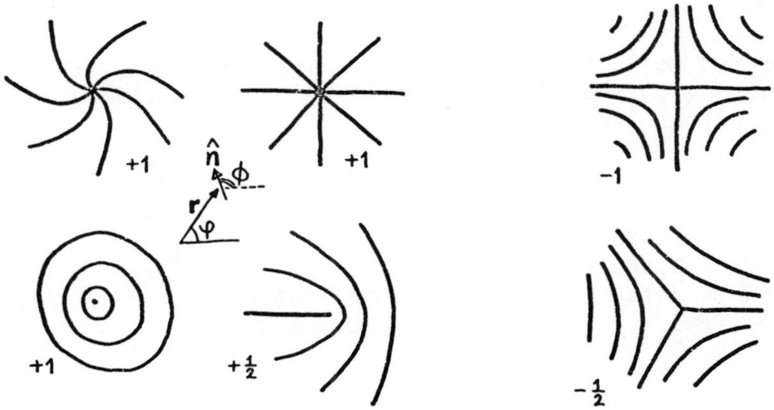

$$\oint \nabla\phi \cdot dl = 2\pi \cdot s$$

Fig. 4. Orientational field singularities of strength 1/2 and 1, occurring in liquid crystal and/or spin systems. Higher strengths are found occasionally but are rare. The strength S is a topological quantum number, defined by a line integral around a "Burgers" circuit. Disclinations in liquid crystals provided the first examples of topological quantum numbers to be studied in physics (G.Friedel around 1920).

location density, which means that such a density appears, in principle, wherever we have twist and bend deformations in the smectic. The smectic dislocation is in many respects very different from dislocations in solids, a fact to which we will return.

ORIENTATIONAL DEFECTS

When the molecules tilt with respect to the layer normal $\hat{z}$ (C phase) the director gets a component $\vec{c}(x,y)$ along the layer, defining a vector field of a kind shown in Fig. 3. The singularities in this field are called disclinations and some further examples of disclinations are displayed in Fig. 4 together with a definition of the disclination "strength" S corresponding to the definition of b in Fig. 2b. Only integer S values are possible in a C (or spin) phase, half integer values require omission of the arrow-heads in Fig. 4 and are thus possible in the nematic phase, which is invariant under $\hat{n} \rightarrow -\hat{n}$. The difference in order between Fig. 3b and 3c is a topological one: the +1 disclination cannot be "transformed away". On the other hand, all three +1 configurations of Fig. 4 are topologically equivalent: they can be obtained from each other by continuous transformations.

NON-ELEMENTARY DEFECTS

Layered media are also characterized by a number of composite defects, some of which can be geometrically very complicated[8]. Of great principal (and practical) interest is the translational defect, (also commonly called dislocation) with $b \geq 2$ according to Fig. 5. The Burgers vector as before corresponds to the number of inserted layers but, as can be seen, $\hat{n}(z,x)$ is a continuous function, even in the "core region" , except along two lines, which are actually two disclination lines of opposite sign, $\pm 1/2$, a distance $b/2$ apart. This kind of defect tends to be more stable the higher the b value, and in several studies very large deformed regions have been analyzed

Fig. 5. Composite dislocations compared to an elementary dislocation (far right). The elementary defect is analogous to a crystal dislocation. The composite defects are often called dislocations but are essentially a new kind of defect, characteristic for smectic and cholesteric liquid crystalline phases.

in terms of "effective" b values of the order of 100 to 1000. These large scale defects are especially important in the creation of what is called confocal domains, dominating the visual appearance of non-aligned smectics under a microscope.

OBSERVATION OF DISLOCATIONS

Dislocations are defects on a molecular scale, which means that their cross diameter is 30 - 50 Å in ordinary liquid crystals. Being two orders of magnitude smaller than the wavelength of light, single defects would seem exempt from any possibility of ever becoming observed in a microscope, which would be unfortunate, because light microscopy is for liquid crystals what electron microscopy is for metals. Nevertheless they have recently been observed and, surprisingly enough, by a light microscopy technique. How this might be possible was conceived by R. B. Meyer. The method takes advantage of a phase transition of second order, and it demonstrates that second order phase transitions can have quite "practical" applications. The basic idea is to combine two factors illustrated in Fig.6(b,c). The deformation strain changes sign on crossing the dislocation: it is positive ($s > 0$) on the dilational side (left in b) and negative ($s < 0$) on the compressional side. Second, there must be a coupling between strain and tilt (c), because tilting of the molecules releaves a compressional strain. The phase transition depicted in (c) can then be driven by two parameters, the temperature and the strain, and the

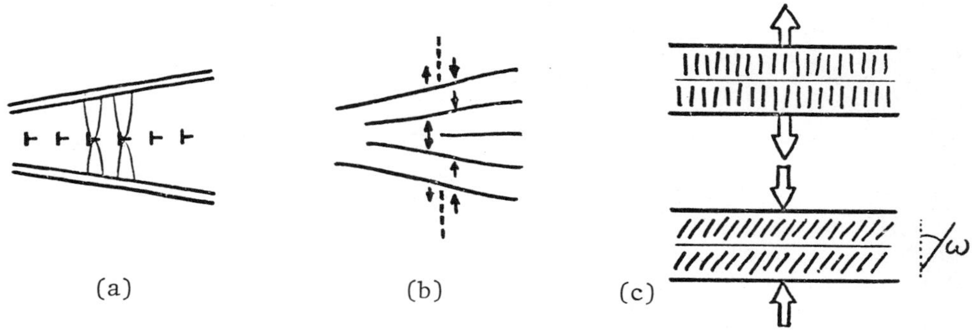

Fig. 6. Array of elementary dislocations generated in a wedge. A dislocation is generated roughly midway between points of fit (m, m+1, m+2... layers equals the local sample thickness). s is thus maximum in magnitude (and changes sign) at dislocations, whereas s=0 midways between. Image forces from the boundaries hold the defects in the middle plane of the sample. The highly anisotropic stress fields are also indicated (a). In (b) the change in local strain from dilatation (s>0, left) to compression (s<0, right) is shown, and in (c) the coupling strain-tilt.

transition temperature should be different on either side of the dislocation. Because the A-C transition is second order, the free energy can be expressed by a Landau development, containing only even powers, in the order parameter

$$F - F_o = \frac{1}{2}\frac{T - T_c}{T_c}c\omega^2 + \frac{1}{4}B\omega^4 + \frac{1}{2}C\left(s + \frac{1}{2}\omega^2 \right)^2 + \ldots.$$

The two first terms on the right hand side constitute the Landau expression for a strain free sample ($s = 0$) and, on minimizing F with respect to ω, give the ordinary thermodynamic transition temperature T_c. The last term is due to the coupling illustrated in Fig. 6(c): the relative decrease in layer thickness goes as ω^2 for small values of ω on assuming a simple cosine variation of the layer thickness μ itself. The (s, ω)-term then just expresses the fact that the free elastic energy must be a quadratic function of the mismatch between applied strain and tilt response. Again, minimizing F with respect to ω gives the transition temperature, now different from the case $s = 0$. The difference turns out to be proportional to the product $s \cdot T_c$, and can also be written

$$T \cdot \Delta z = \frac{1}{2} b T_c$$

Here, T_c is the transition temperature of the strain-free medium, b the Burgers vector of the dislocation and Δz the sample thickness.

A specimen about 5 μm thick having 50 Å layer spacing results in a strain amplitude of $\pm 1/2 \cdot 10^{-3}$ for a $b = 1$ dislocation. Thus on either side of the dislocation line the transition temperature shift is $T_c \approx \pm 1/2 \cdot 10^{-3}\ T_c$. $T_c = 350$ K then gives a total shift of about 0.35 K across a line. The consequences for the tilt as a function of temperature is seen from Fig. 7. The characteristic order parameter parabola is split up in two curves, the separation of which is inversely proportional to the sample thickness. At any specific temperature below T_2 (which lies slightly above the normal transition temperature) there is a difference in tilt (which means that the optical axis is oriented differently in space) between the parts of the medium bordering at the dislocation. The twist in the tilt (about 2 degrees according to the Landau expression) turns the polarization plane of suitably polarized light in a way that the line can be seen. There are in fact several contrast mechanisms involved[4,5,6], but we will leave out the discussion in this context. The maximum twist that can be achieved corresponds roughly to $T = T_1$ where the tilt is quite considerable on the compressional side whereas the medium is still in the A phase ($\omega = 0$) on the dilatational side (cf Fig. 7(a)). The peculiar variation in space of the director for this situation is very schematically indicated in Fig. 7(b) and roughly corresponds to the condition at which the following micrographs were taken. Decreasing the temperature below T_1 diminishes the contrast, but the dislocations can sometimes be seen through the whole range of the C phase, until the sample crystallizes. Some

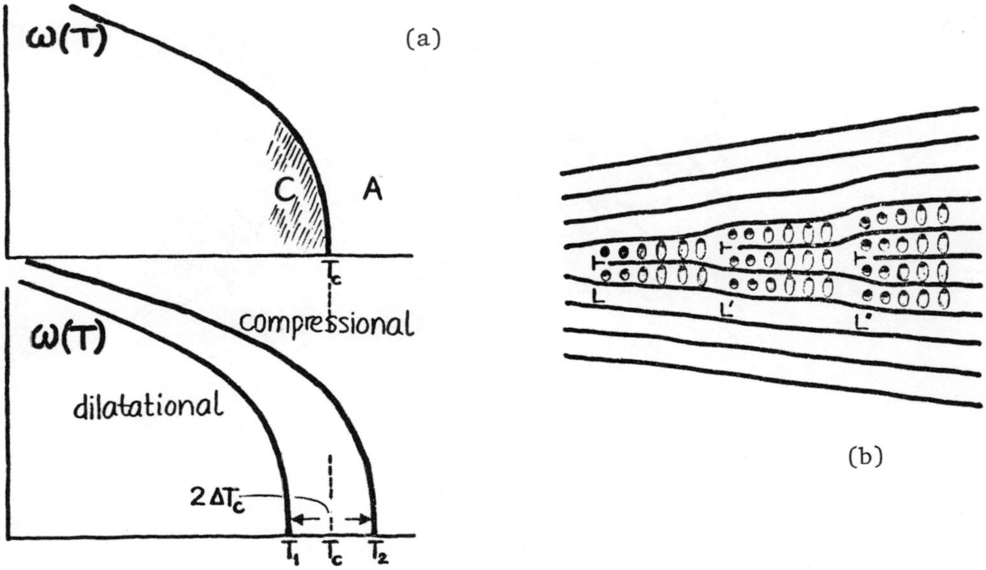

Fig. 7. The characteristic second order temperature dependence of the order parameter (a) is split up in two curves in a local strain field of amplitude s, (a), corresponding even to different phases, A/C for +s/−s, i.e. bordering the line, in the small temperature interval (T_1, T_2), illustrated schematically in (b).

representative observations of dislocation sequencies generated in thin wedges are shown in Fig. 8. Interference measurements revealed that the lines are elementary, i.e. corresponding to the Burgers vector b equal to one.

DISLOCATION MOTION

As for solids we distinguish between climb and glide in dislocation motion, referring to the line moving perpendicular or parallel to the Burgers vector, respectively. As expected (and in contrast to the solid case) climb is an easy motion and is readily observed in our samples. The speed of the process up to about 10μm-100μm per second (especially rapid prior to crystallization) rules out the possibility that the climb mechanism can involve layers sliding along layers (e.g. horizontally in Fig. 8(b). It confirms the reasonable idea of a permeation mechanism operating, in which molecules from neighbouring layers move without breaking of layers into the plane of the dislocation, reminding of particle motion in the propagation of a transverse wave. No quantitative measurements have been made so far.

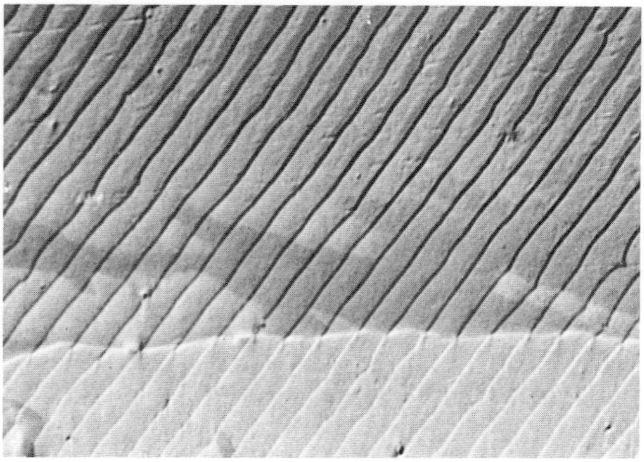

(a)

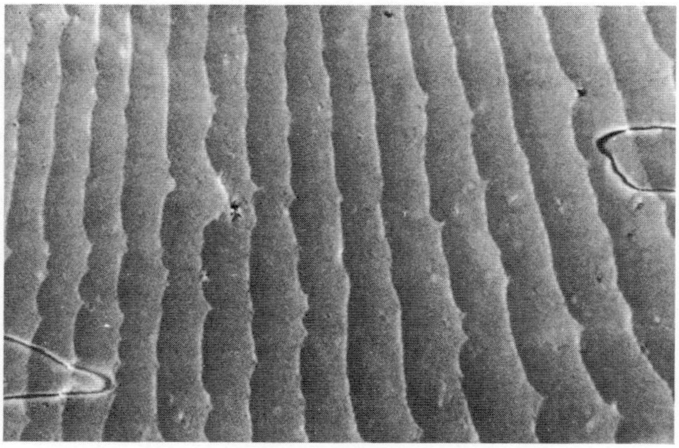

(b)

Fig. 8. Two examples of observations of dislocations near the A-C transition. In (a) case of tilt along dislocation lines. Wedge and tilt co-operate to give smoth dislocation lines. Two large domains are seen with tilt going north-east and south-west respectively. On passing the border the dislocations change contrast from black to white. In (b) case of main tilt perpendicular to dislocation lines. The wedge now opens along the tilt, which facilitates pinning and leads to irregular lines. Some small closed inversion walls are also seen. The Burgers vector is equal to one in both cases. Almost crossed polarizers(Decrossing by about 2 degrees).

"PAIR CREATION" OF DISCLINATIONS

In the smectic C phase, with careful precautions, disclinations can be generated in a controlled fashion "out of nothing", i.e. in geometries essentially corresponding to infinite defect-free regions. To see this we consider the alignment of the molecules produced by the wedge, by the presence of the dislocations and by special surface treatment. The complete order parameter has two components and can be written $\Psi = \omega \cdot e^{i\varphi}$, where $\omega(T)$ is the tilt angle according to Fig. 3(a), and φ is a phase variable (angular co-ordinate in the plane of Fig. 3(b) or (c)), whose giant fluctuations give the C phase director a nematic appearance. In absence of any aligning factors the director is thus free to rotate on a cone representing constant ω. From simple symmetry arguments it is immediately seen (cf. Fig. 9(a)) that the wedge removes part of the degeneracy and favours alignment parallel or perpendicular to the wedge axis. Half of the remaining degeneracy is again removed in the vicinity of the dislocations which favour tilt along their extension (cf. Fig. 9(b)). Additional surface treatment may strengthen or weaken these tendencies. In practice, tilt parallel to the dislocation lines is dominating, but also the perpendicular case is encountered (cf. Fig. 8). The tilt direction can experimentally be determined by conoscopy (Fig. 10), but a special spatial direction filtering technique has also been developed[6,9].

Control of the alignment parameters thus permits the preparation of homogeneous smectic C samples with only two-fold tilt degeneracy , corresponding to two equivalent domains. The system in many ways resembles a spin system with two opposite "easy" magnetization directions. A closed domain in such a system is topologically

(a)

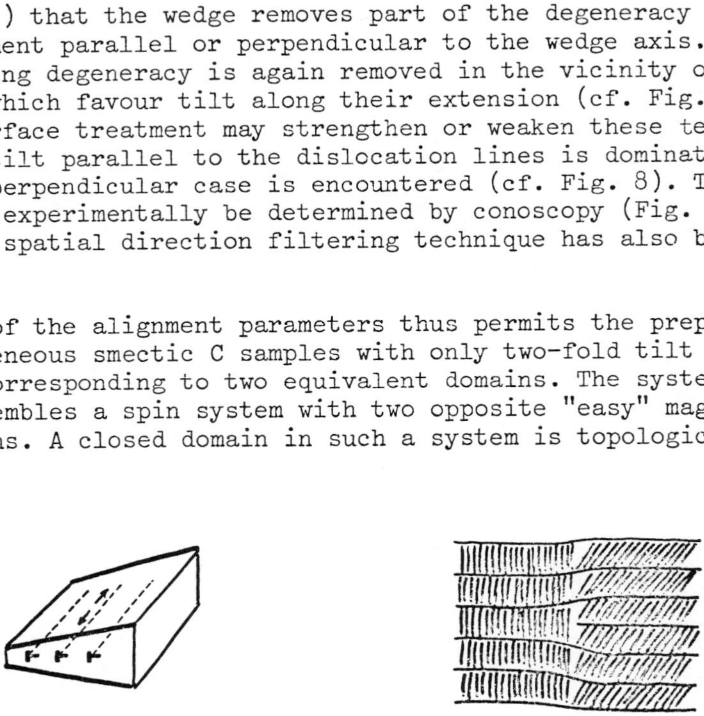

(b)

Fig. 9. Wedge geometry favours tilt alignment along dislocations, according to the arrows, or else in planes perpendicular to the arrows (a). The first case is furthermore favoured by the presence of the dislocations themselves, (b). The splay-character tilt shown in the figure violates the condition of constant density and is thus energetically unfavourable compared to the pure twist-tilt in Fig. 7(b).

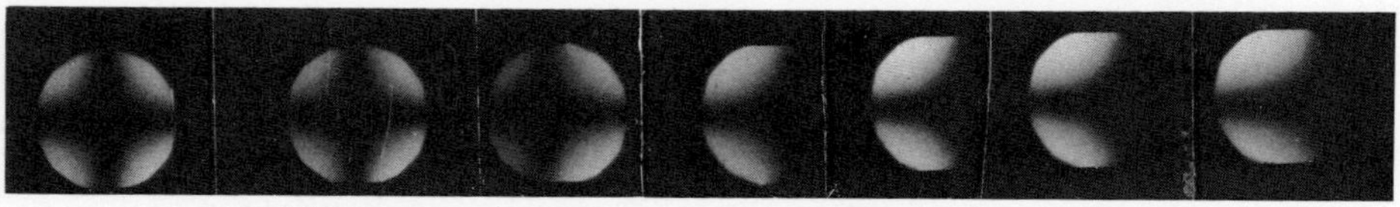

T_c

+0.3 +0.1 -0.1 -0.3 -0.5 -0.7 -0.9

Fig. 10. The temperature variation of the order parameter (molecular tilt angle) characteristic of a second order phase transition (cf. Fig. 7(a)) can be seen directly from this concscopy sequence. The optic axis points into the observers eye and tilts increasingly when T decreases below T_c. The pictures also revealed that the tilt in this case was strictly along the dislocation lines.

equivalent to a family of field singularities, the strengths of which add up to zero, as indicated in Fig. 11. Because only integer strengths S can appear and because singularities with S>1 will split up in elementary ones due to the S^2-dependence of their energy, the singularities in each family must consist of +1 and -1 disclinations in equal number, and disclinations must thus be created under the requirement of "charge conservation". With decreasing T and increasing tilt the inversion wall surrounding the domain gets more and more costly . When a small misaligned domain is then forced to anneal out (on gently decreasing the temperature) by turning its molecules and adopting the majority alignment of the surrounding medium, it must do so on preserving the topology. In principle, nothing will be changed if we carry away the border to infinity , although that situation will not permit any experimental observations. This means that each small isolated domain should break down into an even number N = 2, 4, 6... point singularities (the case N=0 is possible but will not be discussed here) equivalent to a pair-wise creation of plus and minus disclinations. That this is the case can be seen from Fig. 12, where three misaligned smectic C domains have transformed into three families of disclinations with N equals to 2, 4 and 8, respectively. Close inspection of these and similar pictures show disclination cores of alternating black and white contrast corresponding to plus and minus sign. As is seen from Fig. 13, in the core region of a -1 disclination there is always some effective tilt, whereas the tilt has to go to zero in the core of a +1 disclination, giving a black extinction. Although the core region is normally very small its radius diverges in the same way as the penetration length at $T=T_c$ (i.e. as $|T-T_c|^{-1/3}$), which may explain the possibility of determining the disclination sign (not too far from the transition).

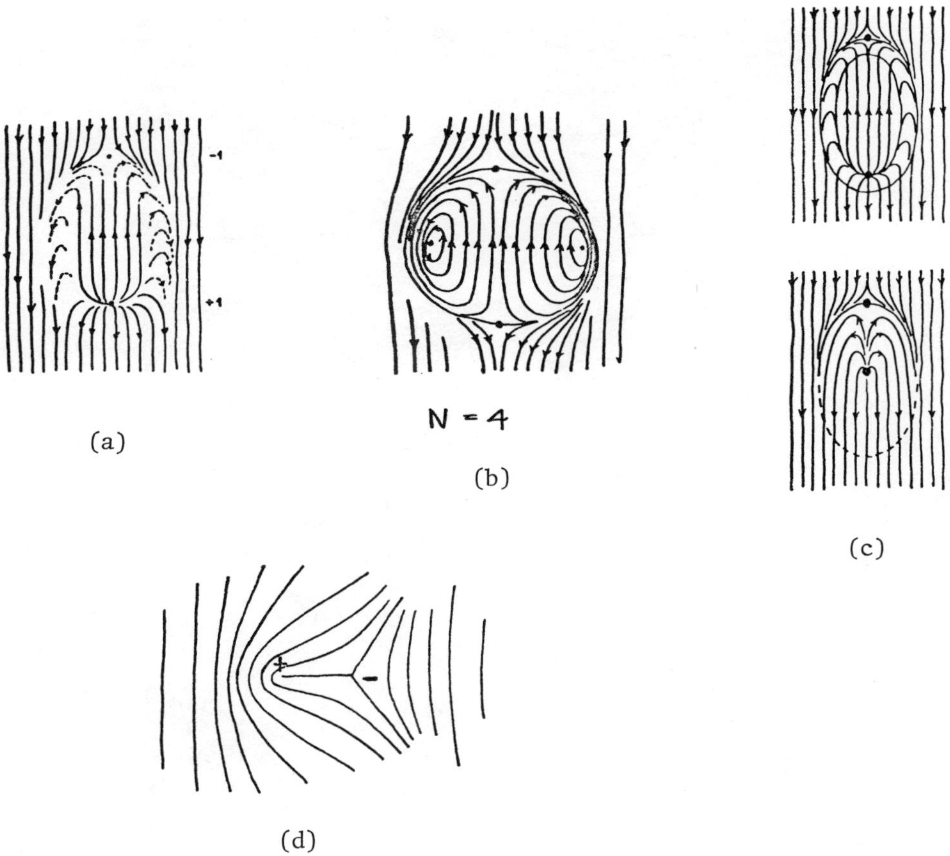

Fig. 11 An isolated domain with opposite tilt relative to its surrounding medium is bounded by a closed inversion wall, inherently equivalent to a point disclination structure. Examples are shown for (a) N=2 and (b) N=4 domains. In (c) is shown a N=2 domain in the process of annealing. Going in the opposite direction the process means creation of a vortex pair.
Finally, (d) gives another example of how rapidly the elastic distorsion falls off around a dipolar configuration, here a ±1/2 pair in a nematic.

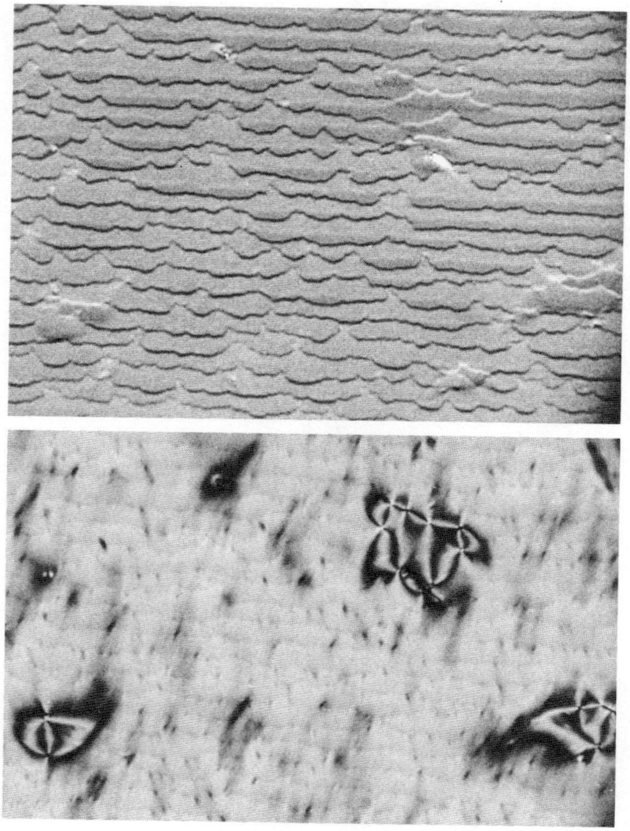

(a)

(b)

Fig. 12. Transformation of isolated small smectic C domains (a) into families of point singularities (b) (with N equal to 2, 4 and 8), on decreasing the temperature between (a) and (b) by 3 degrees.

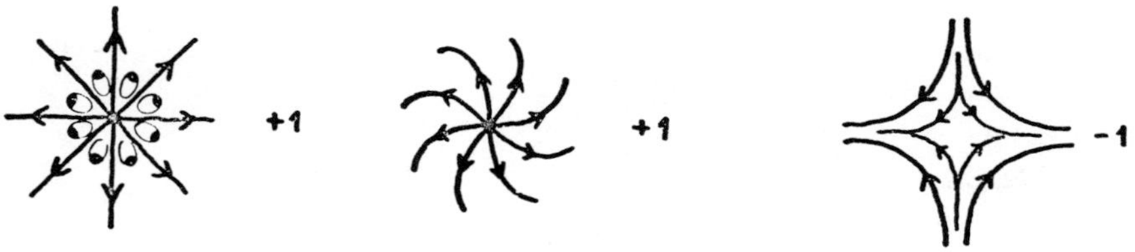

Fig. 13. The molecular tilt has to go to zero in the core region of a +1 disclination due to its symmetric splay. This is not the case for a -1 disclination due to its different saddle-splay (two-fold) symmetry.

DEFECTS AND PHASE TRANSITIONS

The idea of describing phase transitions as a breaking up of order by invading linear defects is intuitively appealing and considerable progress has been made recently on a number of such models. It is interesting to note certain common features of phase transition theory and the modern version of defect theory, especially in the very general classification of ordered media and their topologically stable defects worked out by M.Kléman, L.Michel and G.Toulouse[10] using homotopy group methods. In both descriptions the dimensionality of space and the nature of the order parameter play a decisive role.

In a three-dimensional crystal lattice disclinations would mean prohibitively large distorsions and could only appear immediately prior to the breaking up of the lattice. In two dimensions they ought to be more important and in the Kosterlitz-Thouless-Halperin-Nelson description they are assumed to exist at a late stage when the elastic screening by the already created dislocations makes their presence energetically possible. The systems for testing the 2d xy-model best ought to be thin films of media in which only disclinations can exist, e.g. helium, spin systems and liquid crystals. Of these, a thin smectic C film is especially interesting, since it can be prepared freely suspended, without a substrate, down to two smectic layers of thickness, probably the best achievable example today of a "true" two-dimensional physical system.

The various 2d systems can often be mapped on each other (identification of the disclinations of Fig. 11 with electric charges immediately shows the connection to the 2d Coulomb gas, for instance) and as W.Helfrich has shown, in his treatment of the smectic A - nematic transition[11], also three-dimensional phase changes may be mapped on the xy-model.

The Kosterlitz-Thouless transition in a spin system or in a smectic film means loss of topological order at a certain temperature. Far below this temperature we have thermally created, and tightly bound, vortex pairs but no free vortices. At the transition we then have spontaneous unpairing or, equivalently, spontaneous growth of free vortices (disclinations). The weak, local, distorsion of order caused by a vortex pair (dipole) as compared to the strong, long-range, distorsion caused by a free defect (charge) can be judged from a comparison of Fig. 11(d), or (a) and (c), with Fig.3(c).

The question if there is any disordering similar to a Kosterlitz-Thouless transition in a 3d smectic is still unclear. As already mentioned, a smectic dislocation is considerably different from a crystal dislocation. Due to the one-dimensional periodicity, its Burgers vector is essentially scalar. In the expression for the line integral of Fig. 2b, it also corresponds to the scalar current I. Within the smectic, the Burgers vector can have either of two directions (similar to the two flow directions of the current). A dislocation in a smectic thus has very much the same character as a disclination (with two possible rotational senses) or any vortex line in general. However, there is a striking and unique feature making the smectic dislocation different from other vortex lines as well as from other dislocations. The distorsion field energy of a single vortex line generally diverges (logarithmically) with the extension of the medium. This infinite self-energy can be cancelled by the simultaneous creation of similar vortex of opposite sign, giving a finite energy for the pair. Due to the very peculiar anisotropic elasticity, the self-energy of a single smectic dislocation is finite in an unbounded medium. This is in contradistinction to practically all other cases (the other exception from long-range, logarithmic, interaction are the vortex lines in a type II superconductor), and a clue to its understanding might be gained from Fig. 5. The composite-type dislocation is a pair of disclinations of opposite sign, whose infinities cancel. The elementary dislocation is a limiting case of the composite ones in a somewhat improper sense, but it certainly has preserved the dipolar character of the configuration (also cf the characteristic parabolic stress field indicated in Fig. 6(a)). Thus, whereas the disclinations in a smectic come in pairs, as even seen in the experiments, the dislocations do not, but can be thermally activated singly. Consequently there can be no de-pairing at any temperature and there seems to be no room for a disordering of a Kosterlitz-Thouless type taking place in the smectic A phase.

In "ordinary" media, capable of containing both dislocations and disclinations, de-pairing of each kind of defect could in principle produce disordering transitions at two different temperatures T_1 and T_2. There are here no topological defects beyond dislocations and disclinations that would give still more transitions at other temperatures, but in chiral media such possibilities may still have to be investigated.

REFERENCES

1. R.M.Cotterill and L.B.Pedersen, Sol.State Comm.10,439(1972); R.M.Cotterill, Phys.Blätter 31,571(1975), (with references to earlier work).
2. J.M.Kosterlitz and D.J.Thouless, J.Phys.C 6,1181(1973)
3. B.I.Halperin and D.R.Nelson, Phys.Rev.Lett.41,121 and 519(1978)
4. S.T.Lagerwall, R.B.Meyer and B.Stebler, Annales de Physique 3, 249(1978)
5. R.B.Meyer, B.Stebler and S.T.Lagerwall, Phys.Rev.Lett.41,1393 (1978)
6. S.T.Lagerwall, R.B.Meyer and B.Stebler, J.Phys.Paris, to be published.
7. S.T.Lagerwall and B.Stebler, J.Phys.Paris C3,40,53(1979)

8. A very general treatise of this field is M.Kléman, Points. Lignes. Parois.(Paris 1977)
9. S.T.Lagerwall and B.Stebler, Optics Comm., to be published.
10. M.Kléman, L.Michel and G.Toulouse, J.Phys.Lettr.Paris,38,L-195 (1977); cf also ref.8.
11. W.Helfrich, J.Phys.Paris,39,1199(1978)

FLUCTUATIONS AND FREEZING IN A ONE-DIMENSIONAL LIQUID: $Hg_{3-\delta}AsF_6$

J. D. Axe

Physics Department
Brookhaven National Laboratory
Upton, N.Y. 11973

INTRODUCTION

Many of the papers of this conference deal quite properly with systems at their critical dimensionality, d^*. (See, for example, the contributions of Young, Villain, Als-Nielsen, Litster, and Weeks.) In such systems the competing forces between organization and disorder are nearly equally balanced and the analysis of the resulting situation requires some subtlety. Not surprisingly, the situation is somewhat simplified when the dimensionality falls below d^*. For ordinary translational ordering of fluids (i.e. crystallization), $d^*=2$. In this paper we explore the properties of certain quasi-one-dimensional systems, which since they are effectively below d^*, resist the conventional crystalline order until abnormally low temperatures, and assume instead a state which we liken to a 1-dimensional liquid.

The circumstances which promote this unusual state arise in solids composed of two (or more) interpenetrating sublattices with spacings which are incommensurate one with another. The reason to suspect something out of the ordinary is shown by the following simple considerations. Imagine the two sublattices to be perfectly periodic and write their interaction energy as a product of the charge density, $\sigma_A(\vec{r})$, of one times the potential, $\Phi_B(\vec{r})$, of the other,

$$
\begin{aligned}
V_{AB} &= \int \sigma_A(\vec{r})\,\Phi_B(\vec{r})\,d\vec{r} \\
&= \sum_{GG'} \sigma_A(\vec{G})\,\Phi_B(\vec{G}') \int e^{i(\vec{G}-\vec{G}')\cdot\vec{r}}\,d\vec{r} \qquad (1) \\
&= \sum_{GG'} \sigma_A(\vec{G})\,\Phi_B(-\vec{G}')\,\delta_{G,G'}
\end{aligned}
$$

This shows that the two sublattices interact only by virtue of common reciprocal lattice vectors. Suppose that both sublattices can be thought of as two-dimensional arrays of chains arranged on a common rectangular lattice, but with different and incommensurate interatomic spacings along the common chain direction, z. It then follows trivially that the only common reciprocal lattice vectors have $G_z = G_z' = 0$ and the resulting forces, while constraining the chains in the x,y plane, do not fix the relative positions of the two sublattices along z. (In fact, the system can gain additional interaction energy by a mutual modulation of the natural period of one chain type with the period of the other. But this is a small effect and does not change the qualitative conclusion that the forces which act to localize the atoms on their chains are, at best, abnormally weak.)

Perhaps the best studied example to date of the type of structure we have in mind is the mercury chain compound $Hg_{3-\delta}AsF_6$. It consists of an ordered body-centered tetragonal (bct) lattice of AsF_6^- anions (the host lattice) through which pass linear chains of polymercury cations arranged in two identical perpendicular nonintersecting arrays, one parallel to $\vec{a}_L$, the other to $\vec{b}_L$. See Fig. 1. These will be referred to as the x- and y-arrays, respectively. Room temperature diffraction studies have shown in addition to the expected Bragg reflections, strong diffuse scattering arranged into series of thin sheets in reciprocal space [1-3]. Fig. 2 is a sketch of the (HK0) scattering plane. It is established that the diffuse sheets arise from the Hg-atoms and the narrow width of the sheets shows that the intrachain Hg-Hg distance, d, is well defined, and the nearly uniform distribution of intensity within a sheet shows that there is little or no interference between scattering from different chains [3,4]. Thus positions of the atoms along the chains are virtually uncorrelated from one chain to the next. Finally, from the spacing of the diffuse sheets, the interchain Hg distance, d = 2.67 Å, which is incommensurate with a_L = 7.53 Å. This results from a non-stoichiometric composition $Hg_{3-\delta}AsF_6$ with $3-\delta = (a_L/d) = 2.82$. (A puzzling fact is that chemical analyses consistently find $\delta = 0$. Whether this is due to "pools" of excess Hg, to random vacancies on the host lattice, or neither, is at present unresolved.)

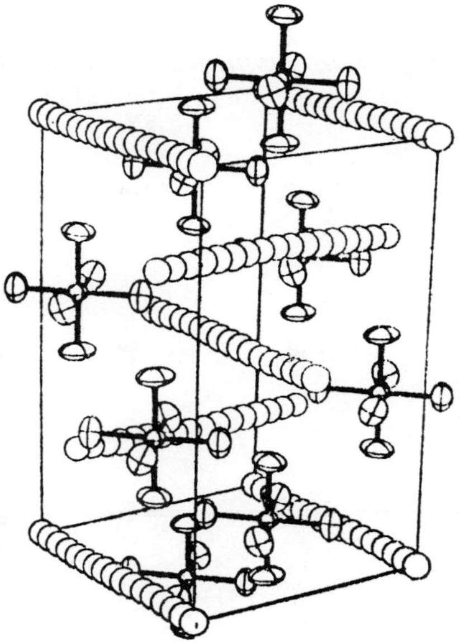

Fig. 1. Structure of $Hg_{3-\delta}AsF_6$. The octahedral AsF_6 groups carry one negative charge. The Hg-atoms on the chains are shown schematically. Above T_c = 120 K the average Hg density is uniform along the chains. After A. J. Schultz et al. (Ref. 3).

Further work by Hastings et al. [4] extended the diffraction study to low temperatures and concentrated on the behavior of the diffuse scattering in the m = 1 sheets. Fig. 3 shows that what is essentially a uniform distribution of intensity within a sheet at room temperature evolves into a pronounced modulation at 180 K. The modulation was interpreted as arising from short range correlations between the position of Hg atoms on nearby parallel chains. In the vicinity of T_c = 120 K sharp Bragg peaks grow out of the sheet of diffuse scattering with a temperature dependence typical of a continuous second order transformation (see inset, Fig. 3) and which must be associated with interchain ordering. Very peculiar, however, is the fact that the Bragg peaks do not develop at the positions on the sheets where the modulated diffuse intensity is strongest, but grow instead from regions of low intensity, i.e. the Bragg peaks are preceded by little or no "critical" scattering. The nature of the resulting ordering was deduced by Hastings et al. by noting that the positions of the Bragg peaks on the sheets were such that a reciprocal lattice vector from the x-array coincided with one from the y-array (at a point on the intersectionof the two m = 1 sheets). This fact, in conjunction with the theorem of the first paragraph, strongly implicates interactions between perpendicular chains as the dominant factor in the ordering. Unexplained, however, was the

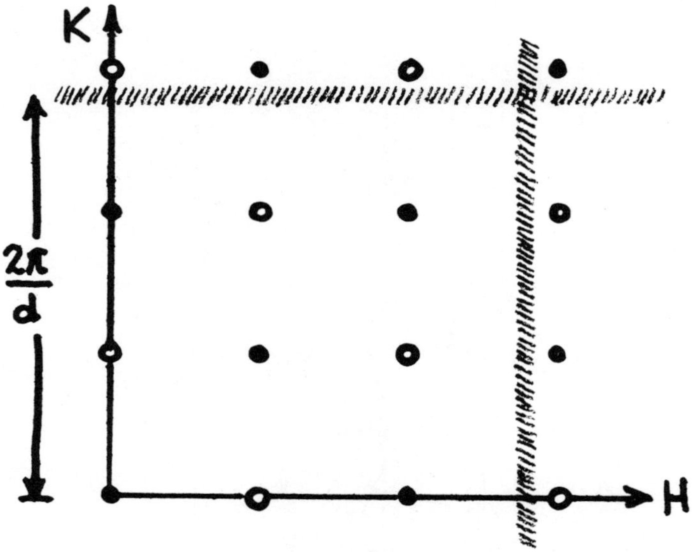

Fig. 2. A schematic representation of the diffraction pattern of $Hg_{3-\delta}AsF_6$ at room temperature. The straight lines represent the intersection of sheets of diffuse scattering lying perpendicular to the figure with the HKO scattering plane.

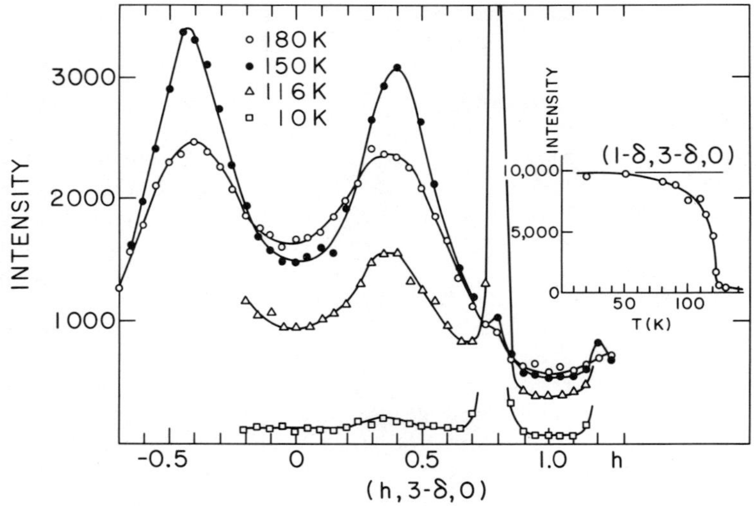

Fig. 3. Temperature dependent evolution of short and long range order as seen by the modulation of the m=1 diffuse sheet. Parallel chain interactions are responsible for the broad peaks at $h \approx \pm 0.4$. The long range order appears in the sharp Bragg peaks at $h = (1-\delta) \approx 0.82$. The inset shows the temperature dependent growth of the Bragg scattering below T_c. After J. M. Hastings et al. (Ref. 4).

apparent sudden reversal of the relative importance of the parallel chain interactions, responsible for the short range order, and perpendicular chain interactions responsible for the long range order.

At high temperatures Hastings et al. also found that emanating from all points of the diffuse scattering sheets are inelastic scattering surfaces with linear dispersion depending only upon the component, Q, of momentum along the chain direction. That is $\omega = \pm v\ |Q-Q_m|$ and $Q_m = 2\pi m/d$ specifies the position of the m'th diffuse sheet. They ascribed this scattering to 1-d longitudinal phonons propagating along the independent Hg chains and found that $v = 4.4 \times 10^5$ cm/sec.

The remainder of this talk is devoted to a discussion of a simple model developed and analyzed by Emery and Axe [5] for $Hg_{3-\delta}AsF_6$ (although with little modification it should be useful in thinking about other linear incommensurate phases as well). It incorporates competing parallel and perpendicular chain interactions, predicts correctly the long range order and clarifies the apparent failure of the system to anticipate this ordering in the fluctuations above T_c. In addition, it treats carefully the effects of one-dimensional fluctuations, and predicts that the Hg chains at high temperatures behave as a one-dimensional liquid. The subsequent phase transformation can be thought of as a freezing of the 1-d Hg liquid, and can be discussed in terms of self-consistent solutions of the sine-Gordon Hamiltonian.

THE MODEL HAMILTONIAN

The Hamiltonian is the sum of intra- and inter-chain contributions. $\mathcal{H}_{intra}$ assumes harmonic interactions, $\mathcal{H}_{intra} = \sum_{\ell_i} \mathcal{H}^0_{\ell i}$ where for example

$$\mathcal{H}^o_{\ell_x} = \frac{1}{2}\sum_{\alpha=1}^{N} \frac{\Pi^2(\ell_x,\alpha)}{m} + K(x(\ell_x,\alpha+1) - x(\ell_x,\alpha)-d)^2 \tag{2}$$

where $(\Pi(\ell_x,\alpha),\ x(\ell_x,\alpha))$ are the components of the momentum and position vectors of the α'th particle on the ℓ_x'th chain. (The subscript i = x,y is to be used to specify the x- or y-array of chains.) The effective near-neighbor stiffness constant $K = mv^2/d^2$ is chosen to give the measured 1-d phonon velocity. m is the bare Hg atom mass.

The configuration of the ℓ_x'th chain is specified by the particle density operators, for example

$$\rho_{\ell x}(x) = \sum_{\alpha} \delta(x-x(\ell_x,\alpha)). \tag{3}$$

In the disordered (high temperature) phase the Hg density is uniformly distributed along the chains, i.e. the thermodynamic averages $\langle\rho_{\ell x}(x)\rangle = \langle\rho_{\ell y}(y)\rangle$ = constant. In terms of the Fourier transformed variables, $\langle\rho_{\ell x}(Q)\rangle = \langle\rho_{\ell y}(P)\rangle = 0$ except for $P = Q = 0$. The quantities $(\langle\rho_{\ell x}(Q_m)\rangle, \langle\rho_{\ell y}(P_m)\rangle$ for $(P_m, Q_m) = 2\pi m/d$ can be taken as a complete set of order parameters specifying the chain ordering transformation. We will see that the instability is associated with the primary order parameters $(\langle\rho_{\ell x}(Q_1)\rangle, \langle\rho_{\ell y}(P_1)\rangle)$. Note that we have retained the notion of a local chain variable by Fourier transforming only the position along the chain direction. Although it is useful in what follows to introduce wave vector components perpendicular to the chain directions as well, it is still important to distinguish between parallel and perpendicular components, as the latter are conjugate to discrete chain positions and can be restricted to the first Brillouin zone, whereas the former is associated with a continuous distribution along the chains and are thus unrestricted.

We introduce coupling between chains of the form

$\mathcal{H}_{inter} = \mathcal{H}_{xx} + \mathcal{H}_{yy} + \mathcal{H}_{xy}$ where

$$\mathcal{H}_{xx} = \frac{1}{2} \sum_{\ell x, \ell' x} \int dx \int dx' v^{\parallel}_{\ell x, \ell' x}(x-x')\rho_{\ell x}(x)\rho_{\ell' x}(x') \tag{4a}$$

$$\mathcal{H}_{xy} = \frac{1}{2} \sum_{\ell x, \ell y} \int dx \int dy v^{\perp}_{\ell x, \ell y}(x-x^{o}_{\ell y}, y-y^{o}_{\ell x})\rho_{\ell x}(x)\rho_{\ell y}(y) \tag{4b}$$

These equations can be rewritten in terms of their Fourier transforms, e.g.

$$\rho_{\ell x}(Q) = N^{-1/2}\sum_{\alpha} e^{-iQx(\ell_x,\alpha)}$$

and for example

$$\mathcal{H}_{xx} = \frac{1}{2}\sum_{Q} v_{\ell x, \ell' x}(Q)\rho_{\ell x}(Q)\rho_{\ell' x}(-Q) \tag{4c}$$

where N is the number of atoms per chain.

1. Range of Interactions. We find that only rather near-neighbor coupling is necessary to explain the observed behavior of $Hg_{3-\delta}AsF_6$. The short range of the interchain coupling is understandable. If we associate a charge density, $\sigma_{\ell x}(x) = e^*\rho_{\ell x}(x)$ with

the atomic density and calculate the Coulomb coupling between two parallel chains ($\ell x, \ell x'$) separated by a distance R, we find

$$v^{\parallel}_{x,x'}(Q) = \frac{2e^{*2}}{d}\int_{-\infty}^{\infty} \frac{\cos Q(x-x')}{[(x-x')^2+R^2]^{1/2}}\, dx = \frac{2e^{*2}}{d} K_o(QR) \quad (5a)$$

$$\approx \frac{2e^{*2}}{d} \left(\frac{\pi}{2QR}\right)^{1/2} e^{-QR} \quad (QR >> 1) \quad (5b)$$

where $K_o(z)$ is a Bessel function. This shows that the coupling between charge modulations on parallel chains is exponentially small if the wave vector of the modulation is large compared to the inverse interchain spacing, R^{-1}. The coupling between perpendicular chains shows similar behavior. The important charge fluctuations are at multiples of $Q_1 = (2\pi/d)$ and for near-neighbor parallel chains $Q_1R = 2\pi(a_L/d) = 2\pi(3-\delta)$. Although neighbor perpendicular chains are closer, we are justified not only in neglecting interactions between widely separated chains, but also in neglecting interactions involving harmonics of the fundamental chain spacing even on nearby chains. That is, the secondary order parameters $<\rho_{\ell x}(Q_m)>$, etc. with $m > 1$ play a vanishingly small role in the interchain coupling.

HIGH TEMPERATURE PROPERTIES ($T > T_c$)

We discuss the thermodynamics using a generalized mean field theory in which the interchain coupling is approximated by a mean field but the resulting one-dimensional chain problem is solved exactly [6]. At low temperatures, where the full nonlinear response of the chains is important, this formulation leads to a sine-Gordon Hamiltonian, and thus is of most direct relevance for this conference. It is worthwhile, however, to sketch some results for $T > T_c$ since they display several unusual features of this system and establish much of the necessary justification for the model itself. For $T > T_c$ we need only the linear response, χ^o, of the harmonic chain, so that

$$<\rho_x(\vec{q})> = \chi^o(\vec{q})h^{eff}(\vec{q}) \quad (6a)$$

$$h^{eff}(\vec{q}) = h^o(\vec{q}) - v^{\parallel}(\vec{q})<\rho_x(\vec{q})> - \sum_{p} v^{\perp}(-\vec{q},\vec{p})\,\Delta(\vec{p}-\vec{q})<\rho_y(\vec{p})> \quad (6b)$$

where we have now introduced Fourier components perpendicular to the chain directions, so that for the x-array $\vec{q} \equiv (Q,q_y,q_z)$ and for the y-array $\vec{p} \equiv (p_x,P,p_z)$. The notation emphasizes the mixed nature of the momentum variables, with the components represented by lower case symbols being defined modulo a reciprocal lattice vector and thus reducible to the first Brillouin zone. This mixed momentum representation is also in evidence through the function

$$\Delta(\vec{p}-\vec{q}) \equiv 1 \text{ if } \vec{p}_x = \vec{Q}(\text{mod}\vec{G}); \vec{q}_y = \vec{P}(\text{mod}\vec{G}'); \vec{p}_z = \vec{q}_z;$$

$$\equiv 0 \text{ otherwise.}$$

where $\vec{G}(\vec{G}')$ is a reciprocal lattice vector of the x(y)-array.

Eq. (6), together with a similar set defining $\langle\rho_y(\vec{p})\rangle$ are to be solved for the coupled response $\chi(\vec{q}) \equiv \langle\rho_x(\vec{q})\rangle/h^o(\vec{q})$, or equivalently the pair correlation functions $\langle\rho_x(\vec{q})\rho_x(-\vec{q})\rangle = kT\chi(\vec{q})$. (We will justify shortly the use of the classical form of the fluctuation-dissipation theorem.) Because of the umklapp momentum terms, the solutions can only be developed perturbatively. Their character depends upon the relationship of the momenta components along the two chain directions.

1. Uncoupled Solutions. In regions of reciprocal space such that P and Q are not approximately equal the two chain arrays are effectively decoupled, and for the x-array

$$\langle\rho_x(\vec{q})\rho_x(-\vec{q})\rangle = \frac{S^o(Q)}{1+\beta v^{\parallel}(\vec{q})S^o(Q)} \tag{7}$$

where $\beta \equiv (kT)^{-1}$ and a similar expression holds for the y-array. $S^o(Q) = kT\chi^o(Q)$ is the pair correlation function for an independent one-dimensional harmonic chain. It is given by

$$S^o(Q) = \sum_\alpha e^{iQ(x^o_\alpha - x^o_o)} \langle e^{iQu_\alpha} e^{iQu_o}\rangle = \sum_\alpha e^{iQd\alpha} e^{-\frac{1}{2}Q^2\langle(u_\alpha-u_o)^2\rangle} \tag{8}$$

and $\langle(u_\alpha-u_o)^2\rangle$ may be evaluated as an ensemble average over the single chain Hamiltonian, $\mathcal{H}_o$

$$\langle(u_\alpha-u_o)^2\rangle = \frac{d^2kT}{4\pi mv^2}\int_{-\pi/a}^{\pi/a} dq \frac{(1-\cos qd)}{\sin^2(\frac{qd}{2})} \equiv |\alpha|\sigma^2 \tag{9}$$

where $\alpha^2 = \frac{kT}{mv^2}d^2$ is the mean square fluctuation in nearest neighbor distance. As is well known, even though there is a well-defined average spacing, αd, for α'th neighbors, the harmonic 1-d chain lacks long range order since the mean square fluctuation about αd increases linearly with $|\alpha|$. Substituting (9) into (8) yields a geometric series which can easily be summed to give

$$S^0(Q) = \frac{\sinh(\frac{1}{2}\sigma^2 Q^2)}{\cosh(\frac{1}{2}\sigma^2 Q^2) - \cos(Qd)} \tag{10}$$

This is a typical liquid-like scattering function (see Fig. 4). Using the measured phonon velocity, we find for $Hg_{3-\delta}AsF_6$, $(\sigma/\alpha)^2 = 6.4 \times 10^{-4}$ at room temperature, which justifies the use of the harmonic approximation within the chain.

In the high temperature limit (somewhat above room temperature for $Hg_{3-\delta}AsF_6$) we may set the denominator of (7) to unity and we recover the independent chain limit. For $Qd >> (\sigma/d)^2$ which is easily fulfilled in this case, $S^0(Q)$ consists of a series of nearly Lorentzian peaks (the sheets of scattering) centered at $Q_m = 2\pi m/d$ with a half width at half maximum, κ_m, given by $\kappa_m d = 2\pi^2(\sigma/d)^2 m^2$.

The above prediction, one of several made by this model, was put to the test in a second series of neutron scattering experiments by Heilmann, et al. [7]. Figure 5, taken from their paper, shows that the measured linewidth κ_m does increase as m^2. Furthermore, the absolute value of κ_m is quite close to that calculated in the preceding paragraph. (Much of the $\sim$ 20% discrepancy arises from a downward revision of v from 4.4 to 3.6 x 10^5 cm/sec, based upon more careful measurements and resolution corrections. But considering the residual uncertainty in v, it is by no means clear that the apparent discrepancy is to be taken seriously. Measurements of the temperature dependence of κ_m for m=3 also verify the predicted linear behavior for 120 K < T < 300 K. It thus appears that to a very good approximation, the high temperature thermal behavior of the Hg-chains is that of a 1-d harmonic liquid.

As the temperature is lowered, the form of (7) and (8) shows that the effect of parallel chain interaction is first evidenced near $Q = Q_1$ since successive maxima in $S^0(Q_m) = S^0(Q_1)/m^2$. (This explains the failure to observe modulation on the m=2 sheet at temperatures where such modulation was pronounced at m=1.) The modulation along the sheet is determined by $v''(\vec{q})$ and the existing data can be fit semiquantitatively with contributions from near neighbor and next near neighbor chains only, with $v_{nnn} \approx -2v_{nn} \approx 0.14$ K. (The interaction seems other than direct Coulomb as v_{nn} is the wrong sign and both are $\sim$ 50 too large.) Although the interactions are weak, they are sufficiently enchanced by the long coherence length within a chain as to tend toward an ordered state only a few degrees below T_c = 120 K.

2. Coupled Solutions. The character of the solutions of (6), together with the corresponding ones for $\rho_y(\vec{p})$ are of a different character if the momenta along the two chains are nearly equal,

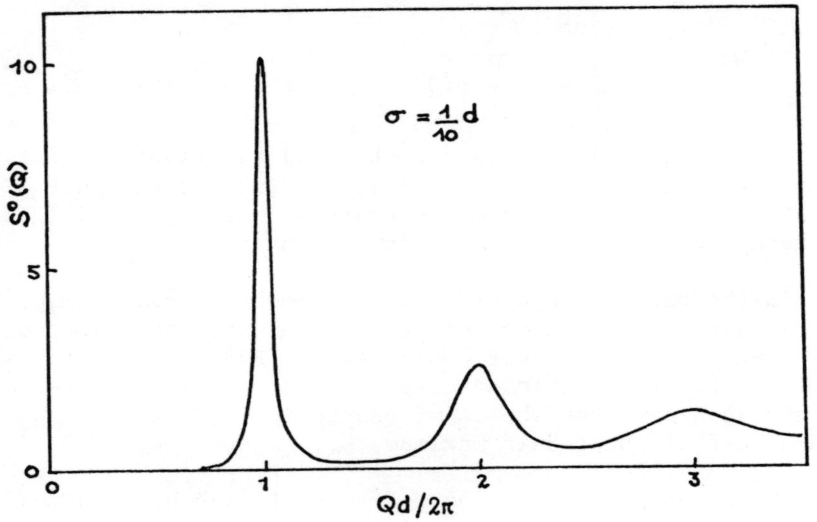

Fig. 4. The scattering function, $S^o(Q)$, for a 1-d harmonic model (see Eq. (10) shows a typical liquid-like pattern. For this case, $\sigma/d = 1/10$, the correlations are weak.

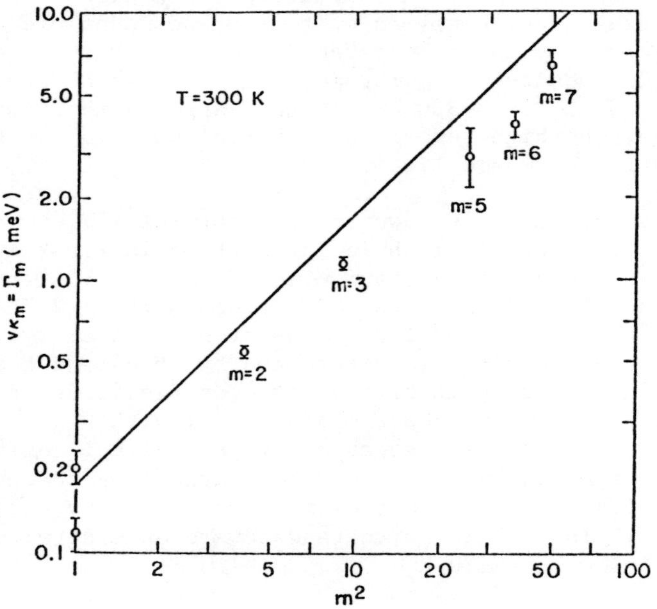

Fig. 5. Q width ($\kappa_m = \Gamma_m/v$) of successive planes of 1-d scattering in $Hg_{3-\delta}AsF_6$. The solid line represents the theoretical value based upon the independently measured phonon velocity, v.

$P = Q$. For these momenta, the x- and y-arrays are strongly coupled, giving rise to new fluctuation modes, $\rho^{\pm}(\vec{q}) = [\rho_x(\vec{q}) \pm \rho_y(\vec{q})]$ and the fluctuation scattering is proportional to

$$\sum_{i,j} \langle\rho_i(\vec{q})\rho_j(-\vec{q})\rangle = \frac{S^o(Q)}{1+\beta(v^{\parallel}(\vec{q})+v^{\perp}(\vec{q}))S^o(Q)} \qquad (11)$$

which for reasons discussed above is enhanced for $Q = Q_m$, that is along the line of intersection of the m'th sheets (a reflection of the simple physics of (1)) and most enhanced for m=1. Whether an instability first arises on the m=1 sheet at (Q_1,Q_1,q_z) due to perpendicular chain coupling or at a more general position (Q_1,q_y,q_z) due to parallel chain coupling depends upon whether the denominator is smaller in (11) or (7); the former is the case for $Hg_{3-\delta}AsF_6$. We believe that the apparent failure to observe critical scattering above T_c is the result of the fact that the region of enhanced scattering is restricted to a linear dimension of order $2\kappa_1$ in the (a_T,b_T) plane. Since this width is below the existing experimental resolution, the basal plane scans should have the appearance of weak Bragg scattering persisting above T_c. Just such scattering has been observed, and it should be possible to establish its true character by determining whether the scattering is broad or narrow in the z direction, perpendicular to both chain arrays. In both sense (repulsive) and magnitude $v^{\parallel}$ seems roughly consistent with Coulombic interactions.

The fact that the instability occurs exclusively on the m=1 sheet means that the long range order first appears as a weak purely sinusoidal modulation of the otherwise uniform average mass density of a Hg-chain, $\langle\rho_{\ell x}(x)\rangle$. A similar sinusoidal mass density wave also breaks the continuous translation symmetry of a liquid crystal in the nematic-smectic transformation, as discussed by Litster in this conference. Indeed much of the physics is the same, although there are also differences connected with the fact that the liquid crystal system is at its critical dimensionality, $d^*=3$.

LONG RANGE ORDER

As usual, we associate the order parameter with the mode giving rise to the divergent fluctuations (i.e. with the coupled mode solutions discussed above) and thus define a complex order parameter, $\eta_1 e^{i\psi} = \langle\rho_x(\vec{q}_c)\rangle = \pm\langle\rho_y(\vec{q}_c)\rangle$. The arbitrary phase factor $e^{i\psi}$ plays no role in determining the energetics of the system and is associated with a zero energy "sliding mode," familiar in incommensurate systems. For convenience, we set $\psi = 0$. η_1 specifies the amplitude of the sinusoidal modulation of the mean atomic density on a chain, e.g. $\langle\rho_{\ell x}(Q_1)\rangle = \eta_1 e^{i\phi^o_{\ell x}}$, where $\phi^o_{\ell x}$ is a phase associated with the perpendicular components of $\vec{q}_c$ and can be made to vanish by an appropriate choice of origin for each chain. Using this convention the mean field potential $\bar{v}$, obtained by replacing one of the density operators in (4) by its mean value, is identical for each chain. This allows us in what follows to suppress the chain index ℓ_i, and we are left with the problem of a 1-d harmonic chain in a (commensurate) staggered field,

$$\bar{v}(\eta_1) = \sum_\alpha \left\{ \frac{K}{2} (u_{\alpha+1} - u_\alpha)^2 + \eta_1 h \cos Q_1 u_\alpha \right\} \tag{12}$$

where $h = 2(v^{\parallel}(\vec{q}_c) + v^{\perp}(\vec{q}_c))$. To discuss the evolution of the low temperature phase we must calculate the growth of all of the Fourier components of the atomic density on a chain. This can be done classically using transfer matrix techniques [8] since except at very low temperatures the effect of zero point fluctuations are negligible. The long intrachain coherence length, $\kappa_1^{-1} \approx 200$ d, allows us to pass to the continuum limit $((u_{\alpha+1} - u_\alpha) \to d(\partial u(x)/\partial x))$ and (12) reduces to the classical sine-Gordon potential and we must calculate

$$\eta_m = \langle \sum_\alpha \cos Q_m u_\alpha \rangle = \sum_\alpha \int du_1 \ldots du_N \cos(Q_m u_\alpha) e^{-\beta \bar{v}} \tag{13a}$$

$$= \langle \Phi_o | \cos Q_m u | \Phi_o \rangle / \langle \Phi_o | \Phi_o \rangle \tag{13b}$$

where $\Phi_o \equiv ce_o(q,v)$ is the lowest eigen vector of the transfer matrix and satisfies the Mathieu equation.

$$\left[\frac{d^2}{dv^2} + (a_o - 2q\cos 2v) \right] \Phi_o(v) = 0 \tag{14a}$$

$$q = \frac{4K\beta^2 h\eta_1}{Q_1^2} = -2 \left(\frac{\beta}{\beta_c} \right)^2 \eta_1 \tag{14b}$$

where $2v = Q_1 u$. The transformation temperature $T_c = (k\beta_c)^{-1} = [-2Kh]^{1/2}/kQ_1$ is obtained by setting the denominator of (9) to zero for $\vec{q} = \vec{q}_c$. Note that T_c^2 is the geometric mean of the harmonic stiffness, K, and the ordering field, h.

Eq. (13) can be readily evaluated by developing $\Phi_o(v)$ in a Fourier series. When m=1 (13) must be solved self-consistently with (14b). The temperature dependence of the first three Fourier components of the atomic density are shown in Fig. 6. For small η_1 $(T \approx T_c)$

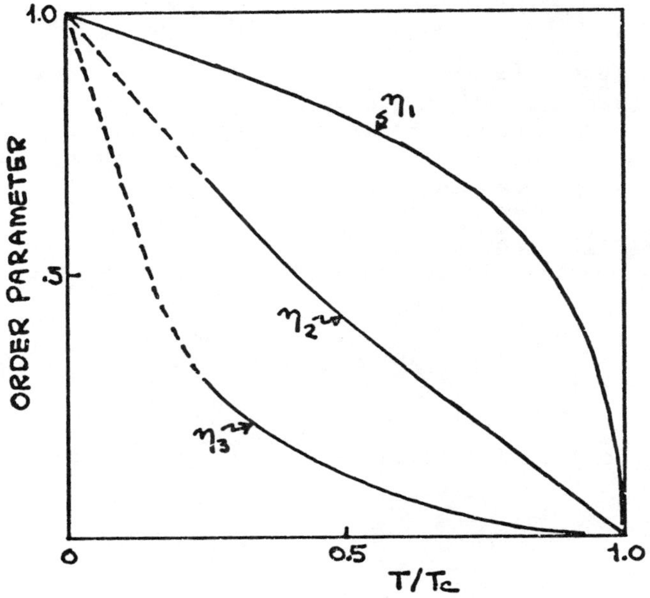

Fig. 6. Self-consistent solutions of the sine-Gordon Hamiltonian for the order parameter, η_m, associated with the first three density wave components.

$$\eta_1^2 \approx \frac{16}{7}\left[1 - \frac{T}{T_c}\right] \tag{15}$$

and η_m is proportional to $\eta_1{}^m$, while near T = 0

$$\eta_m = \eta_1{}^{m^2} \approx 1 - \frac{m^2}{\sqrt{8}}\left(\frac{T}{T_c}\right) .$$

As T → 0, the density distribution on the chain approaches that of a sum of Gaussians centered at x_α = nd with a mean square flucutation $<(x-x_\alpha)^2> = (T/\sqrt{2}\ Q_1^2 T_c)$.

The temperature dependence of the Bragg scattering associated with Q_1 as studied by Hastings et al. (see inset Fig. 3), rises much more quickly than predicted by [15]. We believe that the reason for this is that the fluctuations associated with v" may not be neglected for T $\approx$ T_c because they are divergent at T = T_1 a few degrees below T_c. The coupling of the two types of order parameters not only has the effect of promoting a more rapid growth in $\eta_1(T)$ but also in suppressing the v" fluctuations below T_c, a feature which is very noticeable in Fig. 3. As this aspect of the theory is specialized to $Hg_{3-\delta}AsF_6$ we will not pursue it further here.

DYNAMICS

We conclude with a brief discussion of the dynamical properties that are to be expected in a system of loosely coupled harmonic chains. The dynamics are readily susceptible to calculation and contain several novel features which one can compare with neutron scattering experiments in progress.

In the high temperature limit, it is possible to redo the calculations summarized in (8) and (9) for the time dependent pair correlations

$$S^o(Q,\omega) = (2\pi)^{-1}\int dt e^{i\omega t}\sum_\alpha \langle e^{iQx_\alpha(t)} e^{iQx_o(0)}\rangle \equiv kT\chi^o(Q,\omega)$$

The result, in the vicinity of the m'th diffuse sheet (i.e. $|\Delta Q_m| = |Q-Q_m| << d^{-1}$) is

$$S^o(\Delta Q_m,\omega) = \frac{4}{\pi d v}\left\{\frac{\kappa_m}{(\Delta Q_m - \frac{\omega}{v})^2 + \kappa_m^2}\right\}\left\{\frac{\kappa_m}{(\Delta Q_m + \frac{\omega}{v})^2 + \kappa_m^2}\right\} \qquad (16)$$

Eq. (16) deserves several comments.

1. The unusual product-of-Lorentzian form is characteristic of correlation functions of one-dimensional problems [9].

2. In deriving (16) one cannot proceed through the familiar separation into a product of a time dependent and time independent parts, as the latter (Debye-Waller) term vanishes while the time dependent fluctuations diverge. Similarly, there is no separation into one- and multiphonon terms. Eq. (16) represents the total density response.

3. When (16) is integrated over frequency one recovers (10), and, as with a 3-d liquid, there is no truly elastic scattering (i.e. no term proportional to $\delta(\omega)$).

It is possible to extend the above results to include interchain coupling in the random phase approximation. The dynamical analogs of (7) and (11) are obtained by replacing $\langle\rho_i(\vec{q})\rho_j(\vec{q})\rangle$ by

$$S_{ij}(\vec{q},\omega) \equiv (2\pi)^{-1}\int e^{i\omega t}\langle\rho_i(\vec{q},t)\rho_j(-\vec{q}0)\rangle dt$$

and $S^o(Q)$ by $S^o(Q,\omega)$ in those expressions.

Below T_c the dynamics can be discussed in terms of weakly coupled sine-Gordon systems. For an individual chain there are two sorts of excitations to consider [10]. The first are free solitons for which

$$\omega_s^2 = \Delta_s^2 + v^2Q^2; \; \Delta_s^2 = \frac{16\, mv^2 h\eta_1}{(\pi\hbar)^2}$$

At low temperatures, the minimum energy necessary to create a soliton is so large ($\hbar\Delta_s/k \sim 700$ K) that these are not important thermal excitations, but the gap vanishes as $\eta_1^{1/2}$ near T_c. It may be possible to directly excite these soliton defect pairs with neutrons, or at least to observe the scattering from the thermally excited pairs near T_c. This latter experiment would be directly analogous to the experiments described by Steiner in this volume on the 1-d ferromagnetic system $CsNiF_3$. The second kind of excitations can be described as bound soliton-antisoliton pairs or doublets for which

$$\omega_\nu^2 = \Delta_\nu^2 + v^2Q^2; \; \Delta_\nu^2 = 4\Delta_s^2 \sin^2(\frac{\pi\theta\nu}{2})$$

where $\nu = 1,2,\ldots\theta^{-1}$ and $\theta = (\pi\hbar/2mvd)$. The maximum value of ν is the boundary of stability for breakup into a free soliton-antisoliton pair, whereas for small ν, $\omega_\nu \approx \pi(\frac{4h\eta_1}{md^2})^{1/2}\,\nu$ and the excitations can be thought of as ordinary phonons near the bottom of the sinusoidal potential. These single-chain excitations form the basis for coupled collective modes which satisfy the lattice translational symmetry. (In particular there will still be a collective gapless Goldstone mode representing motion of the chains without change of the relative phase relation between them.) The appearance of a gap in the 1-d phonon spectrum below T_c has been recently observed [11]. As shown in Figure 7, the gap drops rapidly as $T \to T_c$, in at least qualitative agreement with the prediction, $\Delta \sim \eta_1^{1/2}$. Further inelastic neutron scattering experiments are planned.

ACKNOWLEDGEMENTS

It is a pleasure to acknowledge the collaboration of V. J. Emery, who brought to this endeavor not only many of the ideas presented here, but a considerable sophistication which is missing in this account of it. We both profited from discussions of the experiments with J. M. Hastings, I. U. Heilmann, J. P. Pouget, and G. Shirane. Research at Brookhaven was supported by the Division of Basic Energy Sciences, U. S. Department of Energy, under Contract No. EY-76-C-02-0016.

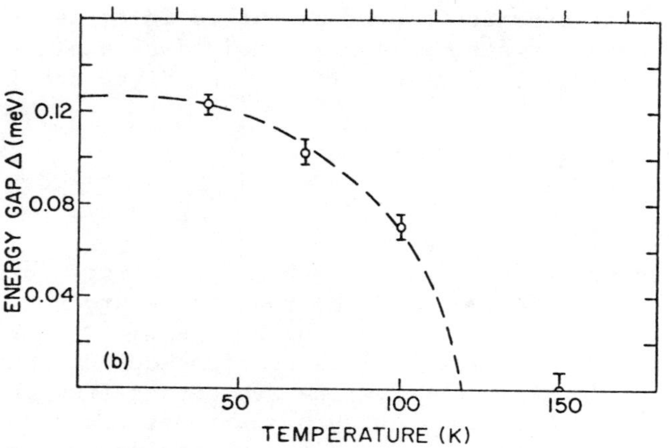

Fig. 7. Temperature dependence of energy gap in Hg-chain spectrum induced by long range order. The dashed curve is merely a guide to the eye.

REFERENCES

1. I. D. Brown, B. D. Cutforth, C. G. Davies, R. J. Gillespie, P. R. Ireland, and J. E. Verkris, Can. J. Chem. 52, 791 (1974).
2. C. K. Chiang, R. Spal, A. Denenstein, A. J. Heeger, N. D. Miro, and A. G. MacDiarmid, Solid State Commun. 22, 293 (1977).
3. A. J. Schultz, J. M. Williams, N. D. Miro, A. G. MacDiarmid, and A. J. Heeger, Inorg. Chem., March, 1978.
4. J. M. Hastings, J. P. Pouget, G. Shirane, A. J. Heeger, N. D. Miro, and A. G. MacDiarmid, Phys. Rev. Lett. 39, 1484 (1977).
5. V. J. Emery and J. D. Axe, Phys. Rev. Lett. 40, 1507 (1978).
6. This approximation was first described in a systematic way by D. J. Scalapino, Y. Imry, and P. Pincus, Phys. Rev. B 11, 2042 (1975).
7. Neutron scattering measurements have verified these predictions. (Private communication, I. U. Heilmann, G. Shirane, and J. D. Axe.)
8. S. F. Edward and A. Lenard, J. Math. Phys. 3, 778 (1962); N. Gupta and B. Sutherland, Phys. Rev. A 14, 790 (1976).
9. See, for example, S. A. Brazovskii and I. E. Dzyaloshinskii, Zh. Eksp. Teor. Fiz. 71, 2338 (1976). (English transl., Sov. Phys. JETP 44, 1233 (1977)); A. Luther and I. Peschel, Phys. Rev. B 9, 2911 (1974).
10. R. F. Dashen, B. Hasslacher, and A. Neveu, Phys. Rev. D 11, 3424 (1975); A. Luther, Phys. Rev. B 14, 2153 (1976).
11. I.U. Heilmann, J.M. Hastings, G. Shirane, A.J. Heeger and A.G. MacDiarmid, Solid State Comm. 29, 469 (1979).

THE EFFECT OF PRESSURE ON THE MODULATED PHASES OF TTF-TCNQ

S. Megtert[x], R. Pynn[+], C. Vettier[+], R. Comès[x] and A.F. Garito[†]
[+]Institut Laue-Langevin, 156X Centre de Tri, 38042 Grenoble, France
[x]Physique de Solides, Faculté des Sciences, Bâtiment 510, 91400 Orsay, France
[†]University of Pennsylvania, Philadelphia, Penn. 19104, USA

ABSTRACT

A satellite reflection in the {hk0} zone of TTF-TCNQ has been measured by neutron scattering at pressures up to 0.46 G.Pa. At 10K the a^* component of the satellite wavevector changes from $0.25a^*$ to $0.5\ a^*$ at a pressure between 0.3 GPa & 0.46 GPa. The satellite is found to disappear continuously at $\sim$56K at the latter pressure and to show no signature of the transition observed at 32K by transport measurements. As pressure is applied the charge transfer increases from 0.59 electrons at ambient pressure to 0.616 electrons at 0.46 GPa. Extrapolating linearly, 2/3 electron transfer is expected around 1.7 GPa. This confirms that the first-order transition observed between 1.7 GPa and 2.3 GPa by transport measurements is indeed to a commensurable state.

\+ + + + + + + + +

X-ray and neutron diffraction measurements have established the existence of three, distinct modulated phases in TTF-TCNQ at low temperatures and ambient pressure[1]. Phase transitions between these phases occur at 54K (T_H), 49K (T_M) and 38K (T_L). Above T_H, TTF-TCNQ is an organic conductor while the low-temperature phases are all Peierls insulators[2,3]. Recent conductivity measurements[4] have mapped the evolutions of the various phase transitions as a function of pressure. Between 0 and 0.4 GPa (1GPa ≡ 10k bar) the T_M transition was found to "disappear" and was presumed to have coalesced with the transition at T_L. The remaining T_L and T_H transitions persisted individually below 1.5 GPa at which pressure they merged to a single transition. Our data shows clearly that this simple picture is incomplete and that several new phase boundaries need to be added. An important feature of the conductivity measurements[4] was the finding that at pressures between 1.7 GPa and 2.3 GPa the (single) metal-insulator transition was of first order. This indicates that the low-temperature distortion, caused by the formation of charge density waves (CDW) on the conducting chains, might be commensurable with the lattice. The wavevector of the CDW in the direction, b^*, parallel to the chains is $2k_F$ for a Peierls transition. At ambient pressure the value $2k_F = 0.295\ b^*$ is found. Commensurability could thus be achieved if $2k_F$ moved either towards $b^*/4$ or $b^*/3$ under pressure. Our results show, as conjectured in ref. 4, that $2k_F$ increases with pressure and tends towards $2k_F = b^*/3$ at p ∿ 2G.Pa.

All measurements were performed with the three-axis spectrometer IN8 at the Institut Laue-Langevin. Incident neutrons of 13.8 meV were used and higher-order contamination was reduced by a pyrolytic graphite filter placed after the sample. Pyrolytic graphite (002) was used as monochromator and analyser. The monochromator crystal was vertically curved while the analyser was flat. Beam collimations were 50' (FWHM) in-pile, 20' between monochromator and sample, 20' between sample and analyser and 60' before the detector.

The sample was an untwinned, fully-deuterated, TTF-TCNQ single crystal[5] of volume 4 x 3 x 0.1 mm^3 contained in an Al-foil envelope.[6] This method of mounting avoids complications which might result from shear strains imposed by gluing the crystal to a support. The sample envelope was placed within an Al high-pressure cell[7] which used He as a pressure medium and which could achieve pressures up to 0.5 G.Pa. As long as the pressure medium was fluid, pressure was measured using a manganin gauge; at lower temperatures pressure was determined from the output of a strain-gauge bridge which measured the deformation of the pressure-cell walls. All measurements were performed in the {hk0} zone of the reciprocal lattice. With this choice, $2k_F$ lattice modulations polarised parallel to c^* (which condense at T_H at ambient pressure) could not be observed. However, $2k_F$ longitudinal modulations, polarised along b^* (which condense at T_M at ambient pressure) were accessible. More importantly, effects of interchain coupling, which ought to depend sensitively on pressure , could be observed by their effect on the a^* component of the lattice modulation. For reasons of convenience we chose to follow the development of the satellite reflection which occurs at (0.25 a^*, $1-2k_F$,0) below T_L at ambient pressure.

For purposes of crystal alignment we have measured the (020) and (300) Bragg reflections at each temperature and pressure. These measurements give lattice compressibilities of -1.4×10^{-2}/G.Pa and -1.6×10^{-2}/GPa in the a^* direction at 10K and 300K respectively. Corresponding results in the b^* direction are -2.6×10^{-2}/GPa and -6×10^{-2}/GPa. Although the compressibility along b^* agrees with earlier measurements[8,9], that in the a^* direction disagrees both with the elastic-constant measurements of Pouget[8] and the powder diffraction data of Debray[9].

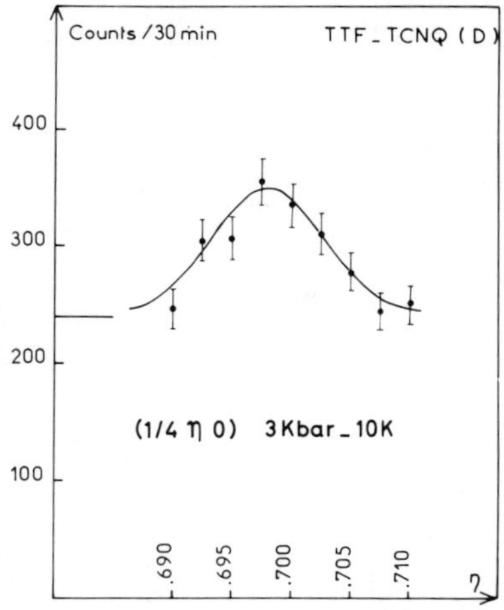

Fig. 1(a)

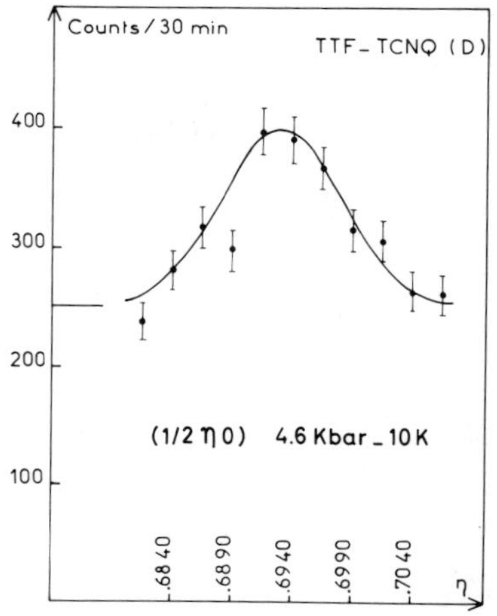

Fig. 1(b)

Typical longitudinal scans through the satellite reflection

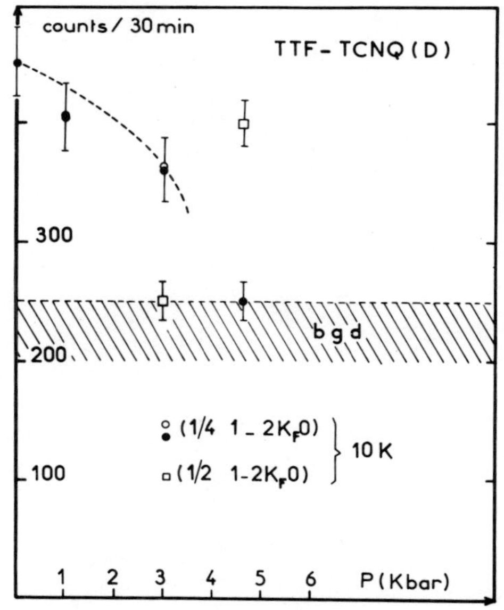

Fig. 2 Variation of satellite intensity with pressure

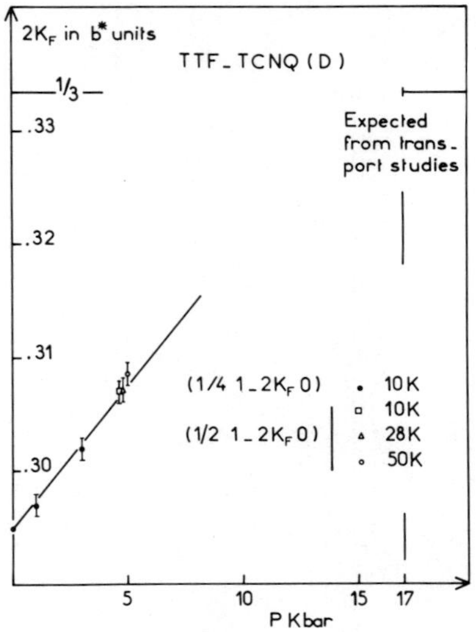

Fig. 3

Pressure dependence of b^{*} component of satellite wavevector

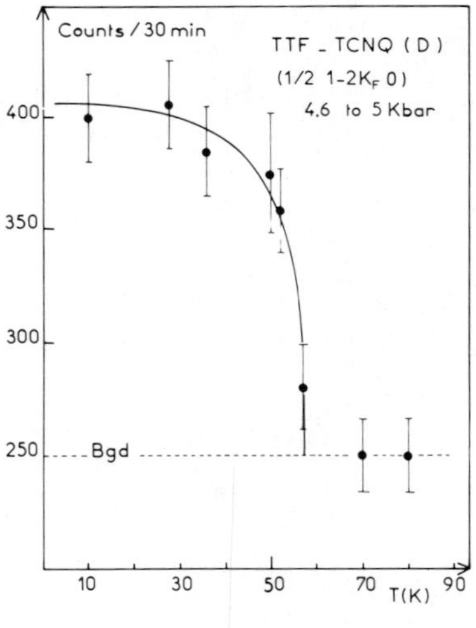

Fig. 4

Temperature dependence of the intensity of the $(0.5a^{*}, 1-2k_F, 0)$ satellite

A typical scan of the $(0.25a^{*}, 1-2k_F, 0)$ satellite at 10K and 0.3 G.Pa is shown in Figure 1a. Measurements of this satellite at 10K and at pressures below 0.3 G.Pa show a monotonic decrease of satellite intensity with pressure, as shown in Figure 2. The measurements suggest that the satellite should disappear at a pressure $\sim$ 0.5 G.Pa. Although the a^{*} component of the satellite wavevector remains unchanged for pressures below 0.3 G.Pa, the b^{*} component decreases (i.e. $2k_F$ increases) as shown by the filled symbols in Figure 3. Linear extrapolation of the data of Figure 3 indicates that $2k_F$ tends towards the value $b^{*}/3$ at 1.7 G.Pa. This change, which represents an increase in charge transfer from the TTF to the TCNQ chains, confirms the suggestion made on the basis of transport measurements[4].

Scans of the $(0.25a^{*}, 1-2k_F, 0)$ satellite at 0.46 G.Pa and 10K revealed no intensity above background, confirming the trend observed in figure 2. However, a scan around $(0.5a^{*}, 1-2k_F, 0)$ (Figure 1b) revealed a satellite which had not been present at pressures up to 0.3 G.Pa. The value of $2k_F$ deduced from the position of this satellite is represented by an open circle in Figure 3. Evidently, the change of $2k_F$ with pressure observed below 0.3 G.Pa is continued at higher pressure in spite of the pressure-induced phase transition which has occurred at 10K between 0.3 G.Pa and 0.46 G.Pa. The observation of this new phase transition is probably the most unexpected result of this study.

We have followed the development of the $(0.5a^{*}, 1-2k_F, 0)$ satellite as a function of temperature at the highest pressure available with our cell (0.46 G.Pa below 38 K and 0.5 G.Pa above 38 K). Our results are shown in Figure 4. The satellite disappears in an apparently continuous manner at the <u>upper</u> phase transition T_H. Although k_F changes slightly between 10K and 50K (c.f. Figure 3) there is no abrupt change of either position or intensity of the satellite to announce the T_L phase transition. Nevertheless, this transition, which is of the lock-in type at ambient pressure, has been observed in conductivity measurements[4].

The geometry of our experiment is such that modulations polarised perpendicular to the a^{*}-b^{*} plane are not detectable. In practice this implies that modulations polarised along c^{*} are invisible and that, at ambient pressure, we are unable to observe the satellite reflections which develop immediately below T_H. Only at T_M, when a modulation polarised along b^{*} condenses[1], do we observe satellites in the {hko} zone at ambient pressure. At 0.46 GPa, the satellite at $(0.5a^{*}, 1-2k_F, 0)$ is observed throughout the temperature range $10K < T < T_H$. This implies that, at this pressure, the T_H transition ivolves a modulation whose polarisition has a component in the a^{*}-b^{*} plane. Since at ambient pressure the T_H transition corresponds to a c^{*}-polarised mode there is apparently a change in the character of this transition with pressure. Similarly, pressure has a qualitative effect on the T_L transition. At ambient pressure, the signature of this transition is the lock-in of the a^* component of the modulation wavevector to the value $a^*/4$. At 0.46 GPa the a^* component of the modulation wavevector is temperature-independent. The fact that the satellite intensity does not change at T_L suggests that, at 0.46 GPa, the lowest transition may involve the condensation of a c^*-polarised modulation.

It is worth pointing out that the lock-in to a commensurable value of $2k_F = b^*/3$ at $p \sim 1.7$ G.Pa implies that the low-temperature value of the a^* component of the modulation cannot be either $a^*/4$ or $a^*/2$ at this pressure. The lock-in to $b^*/3$ is driven by a third-order Umklapp term in the free energy. Such a term can only be translationally invariant if each component of the modulation wavevector is n/3 (where n is an integer, or zero) times a reciprocal lattice vector of the undistorted structure. Thus, at temperatures below T_L, the satellite wavevector must change from $(a^*/4, 2k_F, 0)$ at ambient pressure to $(na^*/3, b^*/3, 0)$ at $p \sim 1.7$ G.Pa. Since the transition observed between 0.3 GPa and 0.46 GPa does not accomplish the appropriate change of modulation wavevector, there must be at least one more pressure-induced transition between 0.5 GPa and 1.7 GPa. Recent measurements of high-pressure thermopower [10] indicate that the true story may be yet more complicated. Whereas at ambient temperature the TCNQ chains are the first to order (at T_H), the thermo-power data indicate that, as pressure is increased, the honour of being first chain may alternate several times between the TTF and the TCNQ.

Whatever the explanation, the behaviour we have observed adds a new dimension to the already complicated sequence of phase transitions which occurs in TTF-TCNQ. The application of hydrostatic pressure which, not unreasonably, has a strong effect on the interchain coupling, is a powerful tool for the investigation of such coupled, low-dimensional systems.

Acknowledgements

We would like to thank S. Barisic, D. Jerome and P. Chaikin for several useful discussions and A. Brochier for his assistance with the experiment.

References

1. R. Comès and G. Shirane in "Highly conducting one dimensional solids" edited by J.T. Devreese, Plenum,(1979)(and references cited).

2. P. Bak and V.J. Emery, Phys. Rev. Lett. 36, 978 (1976)

3.- E. Rubaczewski, S. Smith, A.F. Garito, A.J. Heeger and B. Silbernagel, Phys. Rev. B14, 2746, (1976)

- Y. Tomkiewicz, A.R. Taranko, J.B. Torrance, Phys. Rev. Lett. 36, 751 (1976)

- P.M. Chaikin, J.J. Kwak, R.L. Greene, S. Etemad, E.M. Engler, Sol. St. Commun. 19, 954 (1976)

4. R.H. Friend, M. Miljak and D. Jérome, Phys. Rev. Lett. 40, 1048 (1978)

5. P.J. Nigrey, J. Cryst. Growth 40, 253 (1977)

6. P. Bak, Phys. Rev. Lett. 37, 1071 (1976)

7. J. Paureau and C. Vettier, Rev. Sci. Inst. 46, 963 (1975)

8. J.P. Pouget, S.M. Shaprio, G. Shirane and A.F. Garito to be published.

9. D. Debray, R. Millet, D. Jerome, S. Barisic, J.M. Fabre, L. Giral, J. Phys. Lett. 38, L-277 (1977)

10. P. Chaikin, C. Weyl and D. Jérome (private communication)

SPIN GLASSES:

A BRIEF REVIEW OF EXPERIMENTS, THEORIES, AND COMPUTER SIMULATIONS

K. Binder

IFF, KFA Jülich
Postfach 1913
D-5170 Jülich, W-Germany

1. SPIN GLASS MATERIALS AND EXPERIMENTS

Spin glasses are magnets where interactions between the spins are "in conflict" with each other, due to some disorder in the system. No conventional long range order {(anti-)ferromagnetism etc.} can occur but only a "transition" to a "state" where the spins are "frozen in" in random directions. The nature of both this transition and this state are currently debated /1,2/. The present lectures give a (subjective!) selection of results and problems.

Diluting a nonmagnetic metal with a few % magnetic ions one gets a "classical" spin glass. The spins have Ruderman-Kittel (RKKY) interaction J, which strongly oscillates with distance R. Due to the randomness of R sometimes $J>0$, sometimes $J<0$: hence a rather random distribution P(J) results. First the long range of **J** $[|J| \propto R^{-3}$ as $R \to \infty]$ was thought to be essential (e.g. /3/). Then it became clear that it is only the randomness of J which matters (e.g. /4/). In fact, spin glasses were found by dilution of non-metallic hosts, e.g. Eu in SrS /5-7/. There superexchange makes $J>0$ for nearest neighbors and $J<0$ for somewhat larger distances. Also concentrated amorphous systems (e.g. $Gd_{0.37}Al_{0.63}$) and glasses (e.g. $MnO.Al_2O_3.SiO_2$) show spin-glass behavior /1/: there J is random due to structural disorder rather than due to dilution. Even diluted nonmagnetic systems with "pseudospins" (ferroelectrics /8/, molecular crystals: NH_4Cl /9/, oH_2-pH_2 mixtures /10/, doped semiconductors /11/ may have similar properties.

In magnetic spin glasses neutron diffraction proves that conventional long range order is absent, while short range order leads to a wave-vector dependence of the small-angle scattering /1/. Since the separation of strictly elastic and quasielastic parts of the scattering is delicate /12/, and theories mostly disregard the correlations between local geometries and the magnitude of J, we shall not consider neutron scattering further. Interest is rather focused on the susceptibility χ, Fig. 1A, and its fairly

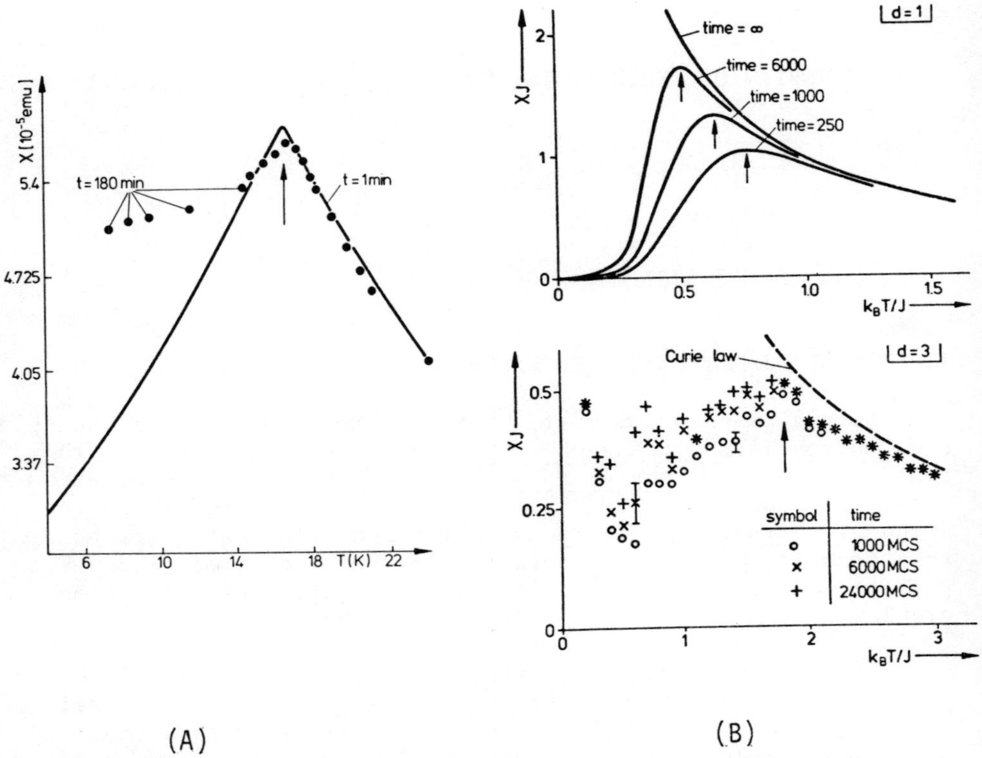

Fig. 1 A) Magnetization of AuFe plotted vs. T at two times during which a field H=10 gauss was applied /13/.
B) Susceptibility of $\pm$J Ising spin glasses plotted vs. T at various times for two dimensionalities /18/. Arrows mark T_f.

sharp "cusp" at a "freezing temperature" T_f. The shape of this peak depends on observation time (or frequency of ac-measurements) /13/, sometimes also its location /5,14/ - but not always /15/. A similar ambiguity concerns computer simulations, Fig. 1B (e.g. /4, 16-20/), as will be discussed below. This peak is strongly rounded by weak magnetic fields /1/. For $T<T_f$ remanent magnetization occurs /3,5/. Its value M_r not only depends on the field H but also on the sample's history, Fig. 2: Cooling at H=0 and then applying a field yields M_r(="IRM") different from that (="TRM") obtained by

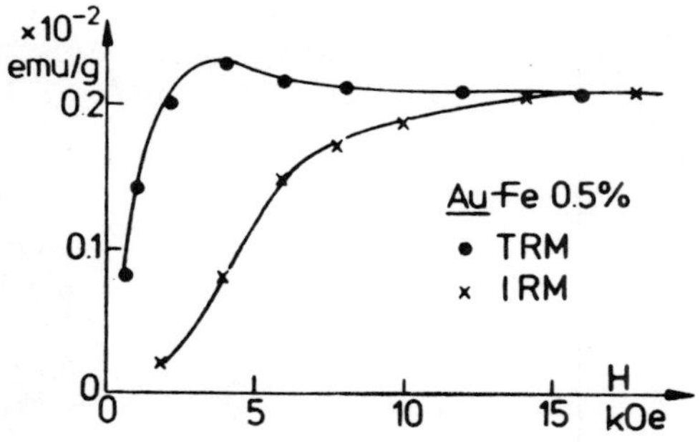

(A)

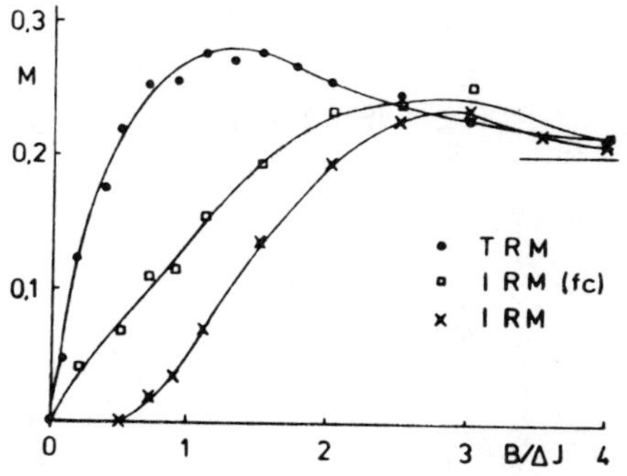

(B)

Fig. 2. A) M_r plotted vs. field for Au-Fe 0.5% /3/ B) Monte Carlo results for M_r for a two-dimensional Ising spin glass with nearest-neighbor random gaussian exchange /21/.

cooling at $H \neq 0$. Similar behavior is seen in simulations, Fig. 2B. M_r amounts to a small fraction of the saturation moment only. It decays with time t typically on a scale of $1-10^5$ sec /1,3,5/; this decay is nonexponential (decrease $\propto \ln t$ /3/). The frequency spectrum of fluctuations extends also to the short frequencies seen in ESR and inelastic neutron scattering /1/. While a rather sudden freezing of the spins upon cooling is seen by Mössbauer studies, μ^+-precession, NMR, anomalous Hall effect /1/, the T_f's such determined exceed the T_f's where χ is peaked distinctly, at least sometimes /5/.

While the peaks of χ have been linked to phase transition at T_f, no pronounced anomalies could be seen there in the specific heat C, electrical resistivity and ultrasonic attenuation /1/. All these quantities are consistent with the picture that fairly strong magnetic correlations, (dynamical) "clusters" develop already at $T \gg T_f$, where χ still follows a Curie (-Weiss) law or nearly so. But from experiment no general agreement is reached on what precisely happens at $T \approx T_f$ and $T < T_f$ /1/. An interesting analogy to "window glasses" is the (nearly) linear variation of C with T at low temperatures /22/, also confirmed by simulations /23/ (see Fig. 10).

An interesting parameter here is the concentration x of magnetic ions /1/. We shall not discuss very dilute RKKY-systems where simple "scaling laws" describe the x-dependence /1,2/, but rather focuse attention to high x where a transition to ferromagnetism occurs (e.g. /6/, Fig. 3). Note that for some x upon cooling one first reaches a ferromagnetic phase and then the spin glass.

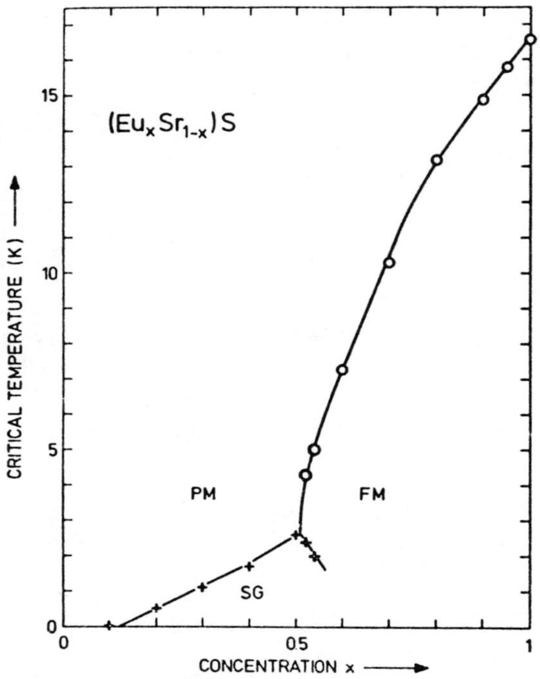

Fig. 3.

Phase diagram of EuS diluted with Sr /6/

Inspite of a wealth of experimental data /1/ only rather indirect evidence on the nature of magnetic correlations in the "frozen" state exists; therefore the discussion on the appropriate model description of these systems continues /2/.

2. THEORETICAL MODELS AND CONCEPTS

As a simple model of diluted magnets with competing interactions, we consider the Ising square lattice with nearest and next nearest neighbor exchange $J_1>0$, $J_2<0$ /24/ {a diluted fcc Heisenberg magnet with $J_1>0$, $J_2<0$ would model $(Eu_x Sr_{1-x})S$}. Fig. 4 shows the phase diagram at T=0. The pure system is ferromagnetic for $J_2/J_1>-1/2$, while for $J_2<-J_1/2$ it is "superantiferromagnetic". For $J_2\geq 0$ ferromagnetism would extend up to the percolation threshold x_c ($x_c^{nnn} = 0.41$ or $x_c^{nn} = 0.59$ if $J_2=0$ /25/). For $x<x_c$ we have finite clusters of ferromagnetically coupled spins ("superparamagnetism" /26/) while for $x>x_c$ an infinite cluster carries the ferromagnetic phase. For $J_2<0$ an infinite cluster appears at x_c^{nnn} but stays paramagnetic up to x_{cl} (J_2/J_1): "frustration" /27/

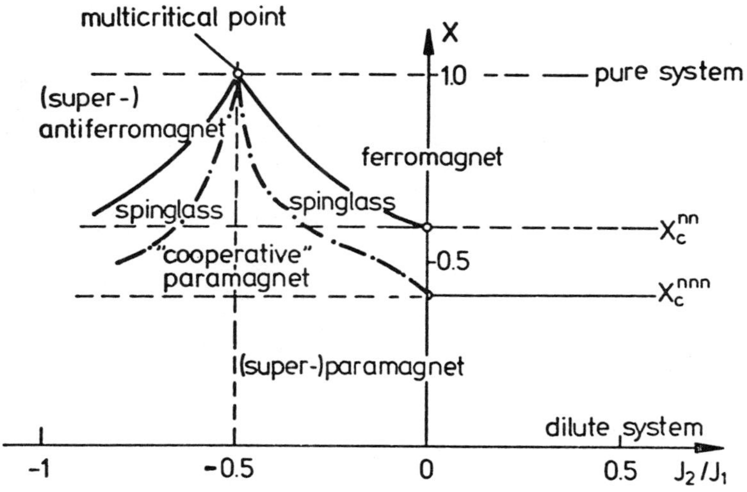

Fig. 4. Qualitative ground state phase diagram of the diluted Ising square lattice /24/. Dash-dotted curve denotes $x_{c1}(J_2/J_1)$

prevents the build-up of long-range correlation in this "cooperative paramagnet" /28/. This behavior is well-known from the triangular Ising system with $J_1<0$, $J_2=0$: $T_c=0$, the ground-state-energy is enhanced and -entropy finite (due to high degeneracy), and correlations behave as $\langle S_{\vec{R}_o} S_{\vec{R}_o+\vec{R}}\rangle \propto R^{-1/2}$ /29/. In our case we suspect exponential decay of correlations in this phase and a power-law decay only at the boundary of the spin-glass phase, where the "susceptibility" $k_BT\,\chi_{EA} = \sum_{jk} \langle S_j S_k\rangle_T^2/N$ diverges. We may define a spin glass by

$$M^2(\vec{Q}) \equiv \sum_{jk} e^{i\vec{Q}\cdot(\vec{r}_j-\vec{r}_k)} \langle S_j\rangle_T \langle S_k\rangle_T/N^2 = 0 (\text{all } \vec{Q}), \quad q=\sum_j \langle S_j\rangle_T^2 \neq 0: \quad (1)$$

i.e., no <u>periodic</u> long-range order, but nevertheless $\langle S_j\rangle_T \neq 0$ (for a finite fraction of sites), and thus $\langle S_j\rangle_T\langle S_k\rangle_T \neq 0$ for a finite fraction of pairs even for $|\vec{r}_j-\vec{r}_k|\to\infty$. q is the Edwards-Anderson order parameter /30/, $M(\vec{Q})$ a (staggered) magnetization. Existence of a spin glass is not rigorously established for our model even at T=0 {the paramagnetic phase could extend up to the boundary of the (anti)ferromagnetic phase} but is assumed in the following.

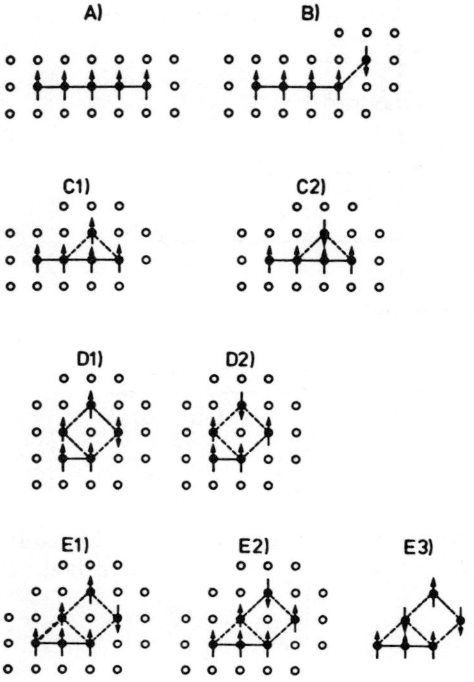

Fig. 5

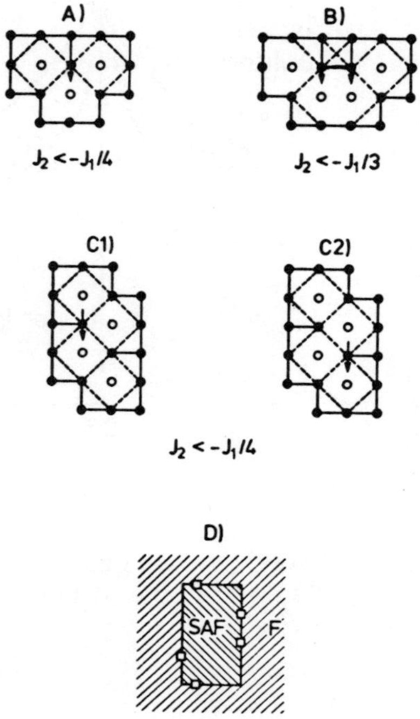

Fig. 6

Fig. 5. Various clusters on the square lattice with nearest neighbor exchange J_1 (full) and next nearest neighbor exchange J_2 (broken). Full circles denote occupied sites, open circles empty ones. Arrows indicate ground state arrangement /24/.

Fig. 6. Antiparallel spins near empty sites in a dilute ferromagnet with competing interactions /24/.

For $J_2<0$ in the superparamagnetic regime several types of clusters occur (Fig. 5): all spins parallel (A) or some antiparallel (B); or all spins parallel for $J_2>-J_1/2$ (C1), some antiparallel for $J_2<- J_1/2$ (C2); then for $J_2=-J_1/2$ both cases are degenerate (the cluster effectively splits in two decoupled clusters with 1 and 4 spins); or several ground states are degenerate for a whole range of interactions (D_1, D_2). In both states some interactions are unfavorable ("frustrated"). High ground state degeneracy due to frustration is a basic feature of spin glasses /27/. Some clusters (Fig. 7E) are degenerate for $J_2>-J_1/3$ (E1,E2) but nondegenerate for $J_2<-J_1/3$ (E3), etc. Similar frustrated loops occur also in the infinite net for $x>x_c^{nnn}$; since $<S_iS_j>_{T=0}$ is averaged over all ground states the correlation is smaller than in the unfrustrated case. In fact, under suitable conditions the infinite net breaks up into finite clusters of aligned spins, and magnetic correlations are only short range. Clusters (B,C) contribute more to $k_BT\,\chi_{EA}$ than to $k_BT\,\chi = \sum_{j\ell} <S_jS_\ell>_T/N$ at T=0; therefore χ_{EA} (sensitive to <u>any</u> long-range alignment) may diverge at smaller x than χ (sensitive to ferromagnetic alignment only).

For weak dilution and $J_2>-J_1/2$ the state is ferromagnetic, but near dilution sites "parasitic" antiparallel spins (Fig. 6A) and clusters (B) occur, as well as degenerate configurations, (C1, C2). Spins or clusters magnetically decoupled from the otherwise "frozen" net occur also in the spin glass phase. Ferromagnetism becomes more and more unstable against dilution as $J_2\to -J_1/2$, where it is degenerate with the SAF phase and also the interface energy between these phases vanishes /31/. The main energy cost for a large SAF domain is due to corners for J_2 close to $-J_1/2$: if the system contains defects, energy may be won at the defect sites (x) if the domain boundary crosses these sites (E); ferromagnetism is destroyed forming many such domains (due to their mismatch no overall SAF phase will be stabilized). Near a multicritical point small dilution suffices to make the system a spin glass (or cooperative paramagnet?), Fig. 4. Thus we may understand that in $(Eu_xSr_{1-x})S$ (Fig. 3) the spin glass at T=0 exists over a broad range of x ($x_c^{nnn} \cong 0.136$ /25/ there). Our picture of the spin glass at T=0 in such diluted systems with competing interactions is: "clusters" of strongly (anti-)parallel aligned spins are coupled together by partially frustrated (and hence weaker) links, giving rise to many equivalent ground states. The size of ferromagnetic clusters becomes infinite at the transition to the ferromagnetic state.

Neglecting the cluster-cluster interactions the whole regime outside the (anti-)ferromagnetic phases were superparamagnetic as suggested by experimentalists /1,32/. But fitting M(H) to Brillouin functions mean cluster size and concentration are strongly temperature dependent /32/, reflecting the growth of magnetic correlations without ordinary order. Of course, this approach is no <u>theoretical</u> explanation relating T-dependence to T-independent interactions. Further assumptions on energy barriers preventing free cluster rotation (mostly one dilutes Heisenberg rather than Ising systems) make this phenomenological approach an obscure fitting procedure.

For $(Eu_xSr_{1-x})S$ this approach could be pursued on a more fundamental level, however /7/: Counting the numbers of all small clusters and calculating energy barriers due to <u>intra-cluster</u> dipolar energy one gets the time-dependent susceptibility <u>without adjustable parameters</u> in good agreement with experiment (Fig. 7).

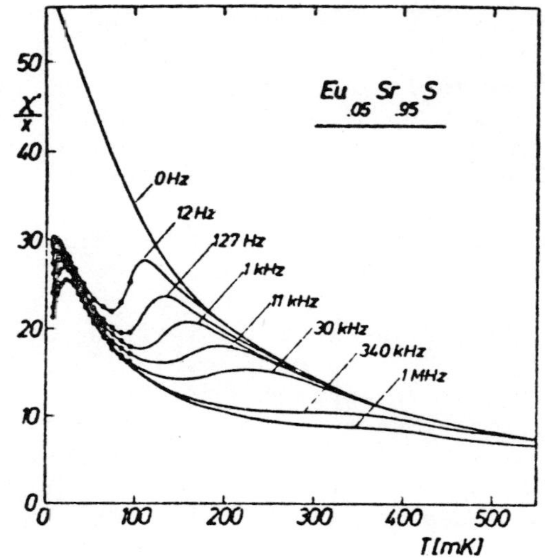

(A)

Fig. 7

Dynamic susceptibilities of $Eu_{0.05}Sr_{0.95}S$ plotted vs. T as measured (A) and calculated (B) /7/. Parameters of the curves are measurement frequency or time (in seconds) up to which the field is applied.

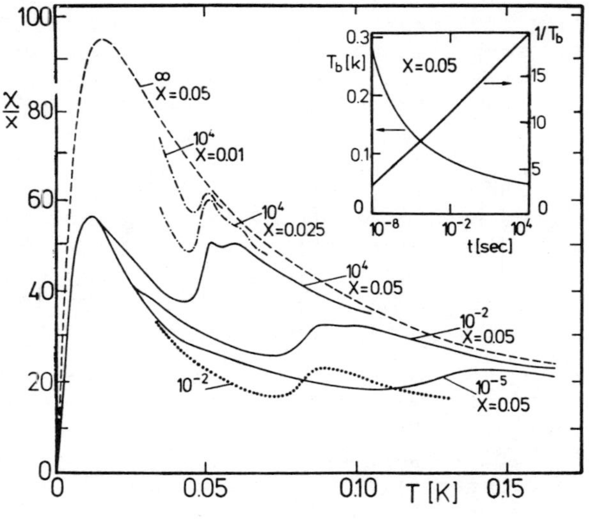

(B)

Isolated Eu-pairs, triplets etc. produce a peak of χ at higher T. But at lower T a second peak is seen which is shown to be due to inter-cluster dipolar interactions /7/: there energy barriers depend on relative cluster orientations, one has thus a cooperative phenomenon (dipolar spin glass) and no single cluster-superparamagnetism. This analysis along the lines of earlier qualitative suggestions /17/ shows that one not only must do the statistical mechanics of small clusters exactly /33/ but also take their configurations realistically into account. Comments: (i) usual superparamagnetism in principle is always inconsistent (long-range weak dipolar forces always couple clusters together) but in practice may yield excellent results at high T (ii) Properties due to single clusters and due to cluster-interactions need not always be as cleanly separated as here: superparamagnetic effects need not always be of equal importance. Perhaps this is the reason why χ sometimes strongly depends on t /5,14/, sometimes not /15/. (iii) In the XY model (where spins can be rotated in a plane) frustrated clusters may have equivalent ground states separated by an energy barrier ("two-level-system") /34/. In a "plaquette" with one J<0 (Fig. 8) the spin directions at T=0 are not colinear but canted

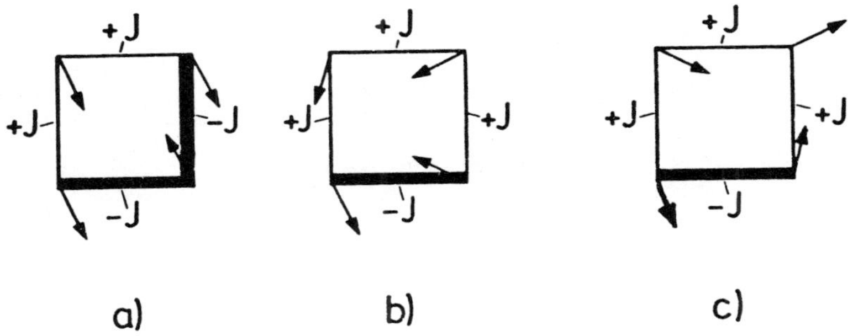

Fig. 8. Ground states of a nonfrustrated (a) and a frustrated (b, c) plaquette in the XY model /34/

by an angle of $\pm\pi/4$ from one spin to the next {+ in case b), - in case c)}. Representing the signs of this "chirality" by an Ising spin $\tau_j=\pm 1$ the XY system with random nearest-neighbor exchange transforms to an Ising spin glass (with nonrandom long-range forces but random positions of the τ_j's) /34/.

In the following we disregard isolated clusters but concentrate on many-body aspects. Since the model of Figs. 4-6 is complicated we try to find a simpler one. Consider a random binary alloy (AB) the exchange between AA-, AB- and BB-pairs being J_{AA}, J_{AB} and J_{BB}. Thus {c_j=+1 if j is an A-atom and zero else} $J_{jk}^{AB}=J_{AA}c_jc_k+J_{AB}[c_j(1-c_k)+c_k(1-c_j)]+J_{BB}(1-c_j)(1-c_k)$. For $J_{AA}=J_{BB}=-J_{AB}=J$ the hamiltonian becomes /35/

$$\mathcal{H}=-J\sum_{jk}\varepsilon_j\varepsilon_k S_j S_k\ , \qquad \varepsilon_j=2c_j-1=\pm 1 \text{ (random)}. \tag{2}$$

Taking $S_j'\equiv\varepsilon_j S_j$ as a new variable, Eq. (2) becomes an Ising ferromagnet, and thus $\psi=\langle\langle S_j'\rangle\rangle\neq 0$ for $T<T_c$, $\langle\langle S_j\rangle\rangle=\langle\varepsilon_j\rangle\psi=0$ {the inner

bracket denotes thermal averaging, the outer averaging over the ε_j's; 50%A:50%B is assumed and thus $\langle\varepsilon_j\rangle=0$}. Although no (sublattice) magnetization exists, we have $q=\langle\langle S_i\rangle^2\rangle=\langle\varepsilon_i^2\rangle\psi^2=\psi^2$ and hence Eq. (1) is satisfied: but this Mattis spin glass has neither frustration nor enhanced ground-state degeneracy, because it is a ferromagnet in disguise! Hence the <u>disorder</u> in Eq. (2) is in a sense <u>"irrelevant"</u>. We express this idea more clearly by defining <u>gauge transformations</u> as /27,36,37/ $\{S_j,\text{ all } J_{jk}\}\to\{-S_j,-J_{jk}\}$. Clearly this leaves the hamiltonian invariant (for H=0). Only such disorder is relevant which cannot be gauged away upon making the ground state ferromagnetic with such transformations. Studying consequences of relevant (=gauge-invariant) disorder, we should consider gauge-invariant correlations, such as

$$g_{jk}^{SG} = \langle S_j S_k\rangle_o \ \langle S_j S_k\rangle_T, \qquad g_{jk}^{EA} = \langle S_j S_k\rangle_T^2 \tag{3}$$

We introduce a Mattis-like order parameter ψ by /4,17/ projecting a state $\vec{X}\equiv(S_1,\ldots,S_N)/\sqrt{N}$ on particular ground state configurations $\vec{\phi}^{(\ell)}\equiv(S_1^{(\ell)},\ldots,S_N^{(\ell)})/\sqrt{N}$ (ℓ-th ground state)

$$\psi^{(\ell)} \equiv \langle\vec{X}\cdot\vec{\phi}^{(\ell)}\rangle = \sum_j S_j^{(\ell)} \ \langle S_j\rangle_T/N \tag{4}$$

The associated susceptibility is /17/ $k_BT\,\chi^{(\ell\ell)}=\sum_{jk} S_j^{(\ell)}S_k^{(\ell)}$ $\langle S_jS_k\rangle_T/N$. In the disordered phase where all L ground states are equivalent an average over ℓ yields a "spin-glass susceptibility" /38/

$$k_BT\,\chi^{SG} = \sum_{jk} \langle S_j S_k\rangle_o \ \langle S_j S_k\rangle_T/N\ , \tag{5}$$

thus involving the first correlation of Eq. (3) while χ^{EA} involves the second {χ^{EA} is easier calculated since no ground state information is needed}.

For XY and Heisenberg spin glasses progress was made by treating the gauge-invariant disorder as a "field" in addition to the "spin field" /39,40/. By this approach the thermodynamics of spin glass freezing could not yet been treated. But it sheds light on the nature of elementary excitations for T→0 /39/: Apart from propagating modes (wavevector $\vec{q}$) with energy $\omega(q)\propto q$, which are found also from other approaches /41,42/, modes occur with $\omega(q)\propto q^3$, implying a density of states $f(\omega\to 0)\to$const, and hence a specific heat $C\propto T$ for $T\to 0$. This picture qualitatively agrees with simulations /23/ where both "localized" modes with rather large ω and "delocalized" ones with small ω were found (since N=96 spins were used these results are rather qualitative, however).

This promising approach of gauge theory will not be pursued however, but rather we discuss now the more conventional approach /2,30,43/. Instead of site disorder in systems with competing exchange one assumes bond disorder, usually J is assumed to be distributed as $P(J_{jk}) \propto \exp[-(J_{jk}-\bar{J})^2/2(\Delta J)^2]$, and one first tries to develop a mean-field approximation (MFA). Hence one should calculate $F_{\{J_{jk}\}} = -k_BT \,\ell n Z_{\{J_{jk}\}} = -k_BT \,\ell n \mathrm{Tr} \exp(-\mathcal{H}/k_BT)$ for a fixed set $\{J_{jk}\}$, and then use $P(J_{jk})$ to average, $F = \langle F_{\{J_{jk}\}}\rangle$. This procedure is very difficult even in MFA, and thus one uses the "replica trick" /2/: One calculates $\langle Z^n{}_{\{J_{jk}\}}\rangle$ for positive integer n and then gets F by analytic continuation to $n \to 0$:

$$F = -k_BT \langle \ell n Z_{\{J_{jk}\}}\rangle = -k_BT \lim_{n\to 0} \frac{1}{n} (\langle Z^n{}_{\{J_{jk}\}}\rangle - 1) \tag{6}$$

Since $Z^n{}_{\{J_{jk}\}} = \mathrm{Tr} \prod_{j,k} \prod_{\alpha=1}^{n} \exp(J_{jk} S_j^\alpha S_k^\alpha / k_BT)$, the gaussian average is easily performed, yielding $\langle Z^n\rangle = \mathrm{Tr} \exp(-\mathcal{H}_n^{eff}/k_BT)$ with

$$\mathcal{H}_n^{eff}/k_BT = -(\bar{J}/2k_BT) \sum_{j\neq k} \sum_{\alpha=1}^{n} S_j^\alpha S_k^\alpha - (\Delta J/2k_BT)^2 \sum_{j\neq k} \sum_{\alpha,\beta} S_j^\alpha S_k^\alpha S_j^\beta S_k^\beta \, . \tag{7}$$

Rather than treating the statistical mechanics of inhomogeneous systems one thus considers an equivalent translationally invariant hamiltonian, Eq. (7). Due to the disorder the n replicas of each spin "interact" now with each other. One treats Eq. (7) in MFA with the order parameters $m = \langle S_i^\alpha \rangle$, $q = \langle S_i^\alpha S_i^\beta \rangle$ $(\alpha \neq \beta)$; for $\bar{J} > \sqrt{z}\,\Delta J$ (z=coordination number) the system becomes ferromagnetic at $k_BT_c = z\bar{J}$, while for $\bar{J} < \sqrt{z}\,\Delta J$ spin glass order sets in at $k_BT_f = \sqrt{z}\,\Delta J$. There $\chi = \chi_{sym}/(1 - z\bar{J}\chi_{sym})$ with $\chi_{sym} = \mu_B^2(1-q)/k_BT$ the susceptibility for $\bar{J}=0$. Thus χ has a cusp, since $q \propto (1-T/T_f)$ near T_f. C also has a cusp as pronounced as that of χ, in contrast to experiment. For nonzero field both cusps get rounded, but the fields required to produce the rounding are 20 times too strong as compared to experimental ones. Only this latter discrepancy is removed by interpreting $\mathcal{H}$ as a model of randomly interacting clusters rather than individual spins /17/. With respect to the cusp of C, we note that renormalization group expansions /44/ in terms of $\varepsilon = 6-d$ [d = dimensionality] yield the exponent α of the singularity $[C_{sing} \propto |T-T_f|^{-\alpha}]$ as $\alpha = -1-2\,\varepsilon$: i.e. the slope of C is continuous at T_f, only higher derivatives of C at T_f would show a divergence. But we shall see that the relevance of this result for d=3 is doubtful.

The most serious difficulty of the above MFA, however, is that the entropy becomes negative for $T \to 0$ /43/, an impossible result for Ising spins. Also some correlation functions become negative below T_f /45,46/. This is surprising, as usually MFA become exact as $z \to \infty$.

Sometimes the difficulty is attributed to the replica trick, particularly its interchange of limits $n \to 0$, $N \to \infty$ /47/, but more or less equivalent results follow also from diagrammatic methods /48/ and the Lagrangian method of critical dynamics /49/.

Among various attempts to solve this puzzle, Thouless et al. /50/ (TAP) consider the MFA for $z \to \infty$ without "replica trick". For $T > T_f$ they get identical results as before, as expected since expansions in (1/T) obtained with and without replica tricks agree term by term /51/. But for $T \lesssim T_f$ the results differ: while the replica solution amounts to take a maximum of F rather than a minimum as usual, TAP find a saddle point of F, implying χ_{EA} to diverge also below T_f. For $T \to 0$ TAP partially relied upon numerical techniques to find zero ground state entropy. Their results are consistent with computer simulations Fig. 9 /52/. Attempts to solve

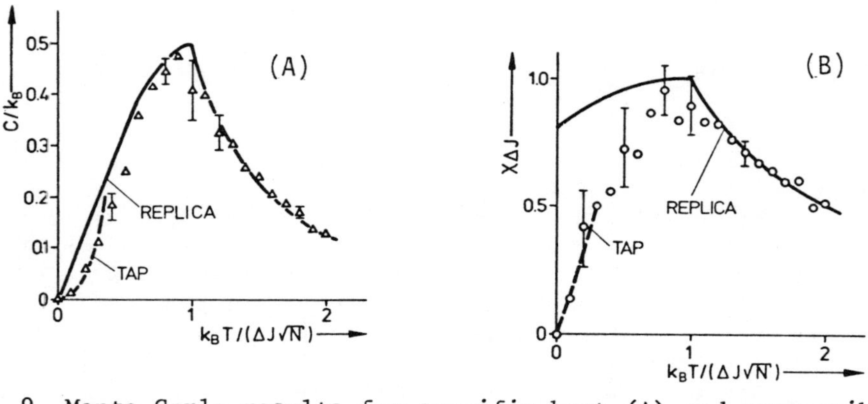

Fig. 9. Monte Carlo results for specific heat (A) and susceptibility (B) of the infinite range spin glass, as compared to replica solution and TAP /52/.

TAP's equations below T_f analytically yielded different results /53/ but fail to satisfy the symmetry relation $\chi = (\mu_B^2/k_B T)(1-q)$ which should hold for $\bar{J}=0$ /1/. Other attempts rely on the concept of breaking the symmetry among the replicas /54,55/: Suppose two groups containing m and n-m replicas and regard two replicas belonging to the same group as equivalent /55/. Instead of an unique q one now has $q_1 = \langle S_i^\alpha S_i^\beta \rangle$ $(\alpha < \beta \le m)$, $q_2 = \langle S_i^\alpha S_i^\beta \rangle$ $(\alpha \le m < \beta)$ and $q_3 = \langle S_i^\alpha S_i^\beta \rangle$ $(m < \alpha < \beta)$. Taking first $m \to \infty$ and then $n \to 0$ "restores" the symmetry $(q_1 = q_2 = q_3 = q)$ and removes instabilities occurring in the symmetric solution in the order $(1-T/T_f)^2$. But it is unclear whether this trick can be extended to low T, and also its physical significance is obscure. I consider it as a weakness of all these approaches relying on the use of the order parameter q that physical concepts important for spin glasses (frustration, loose clusters in the rigid network, ground state degeneracy, gauge invariance) do not enter explicitly enough in all these treatments.

III. SPIN-GLASS FREEZING: PHASE TRANSITION OR NONEQUILIBRIUM EFFECT?

For $z \to \infty$ there is no doubt that spin glasses exist, only their properties are debated. For short-range exchange and d=3 even the very existence of spin glass phases is questionable. While real-space renormalization suggests for Ising spin glasses a transition with $T_f \neq 0$ for d=3, the results for d=2 are controversial /56/. For the Heisenberg case similar arguments yield as "lower critical dimensionality" $d_c=3$ $\{T_f=0$ for $d \leq d_c\}$ /57/, consistent with implications of the q^3-modes of gauge theory /39/. But high temperature series extrapolations of χ_{EA} can locate a singularity for d=5, not for d<4 /51,58/. Replica symmetry breaking also yields $d_c=4$ /55/. As further evidence for $d_c=4$ were noted numerical studies of the boundary energy/boundary site E_b, indicating $E_b(T=0)=0$ /59/. However, an exactly soluble counterexample with $T_c \neq 0$ but $E_b(T=0)=0$ /60/ shows that the boundary free energy $F_b(T=0) \neq E_B(T=0)$ due to the nonzero ground-state entropy. For other arguments concerning $E_b(T=0)=0$ see /19/.

We are not going in the details of all these discussions, but rather describe now the results of computer simulations although their interpretation also is somewhat ambiguous /4,16-20,61/. Fig. 10A shows χ and C for classical Heisenberg spin glasses with nearest neighbor random gaussian exchange. The smallness of "T_f" where χ is peaked, the nonequilibrium effects $\{C=(\langle\mathcal{H}^2\rangle-\langle\mathcal{H}\rangle^2/(Nk_B^2T^2)$ points and $C=\partial\langle\mathcal{H}\rangle/\partial T$, broken curves disagree strongly$\}$ and the very gradual freeze-in of q(t) {defined in simulations via $\sum_i (\int^t \vec{S}_i(t')dt'/t)^2/N\}$ are better consistent with $T_f=0$ rather than $T_f \neq 0$ as suggested by some authors /62/. The RKKY results (Fig. 10B) /63/ are even less conclusive since N=96. But the peak of χ roughly occurs at the experimental T_f (the magnitude of χ for $T \to \infty$ was adjusted but the strength A of RKKY exchange had its known value). While C agrees with the coefficient of the 1/T-law for $T \to \infty$ /64/, $C \to 1$ for $T \to 0$ (as expected for classical systems). One gets a more realistic C in agreement with experiments /22/ from a numerical spin-wave analysis based on ground state spin configurations /23/.

Fig. 11 shows q(t),ψ [observation "time" t=2000 Monte Carlo steps/spin] for nearest neighbor gaussian Ising spin glasses /20/. The data suggest freezing at $k_BT_f/\Delta J \approx d/2.1$; note $\psi<q(t)$ in contrast to Mattis spin glasses where $\psi=q^{1/2}$. Fig. 12 shows ψ,M plotted

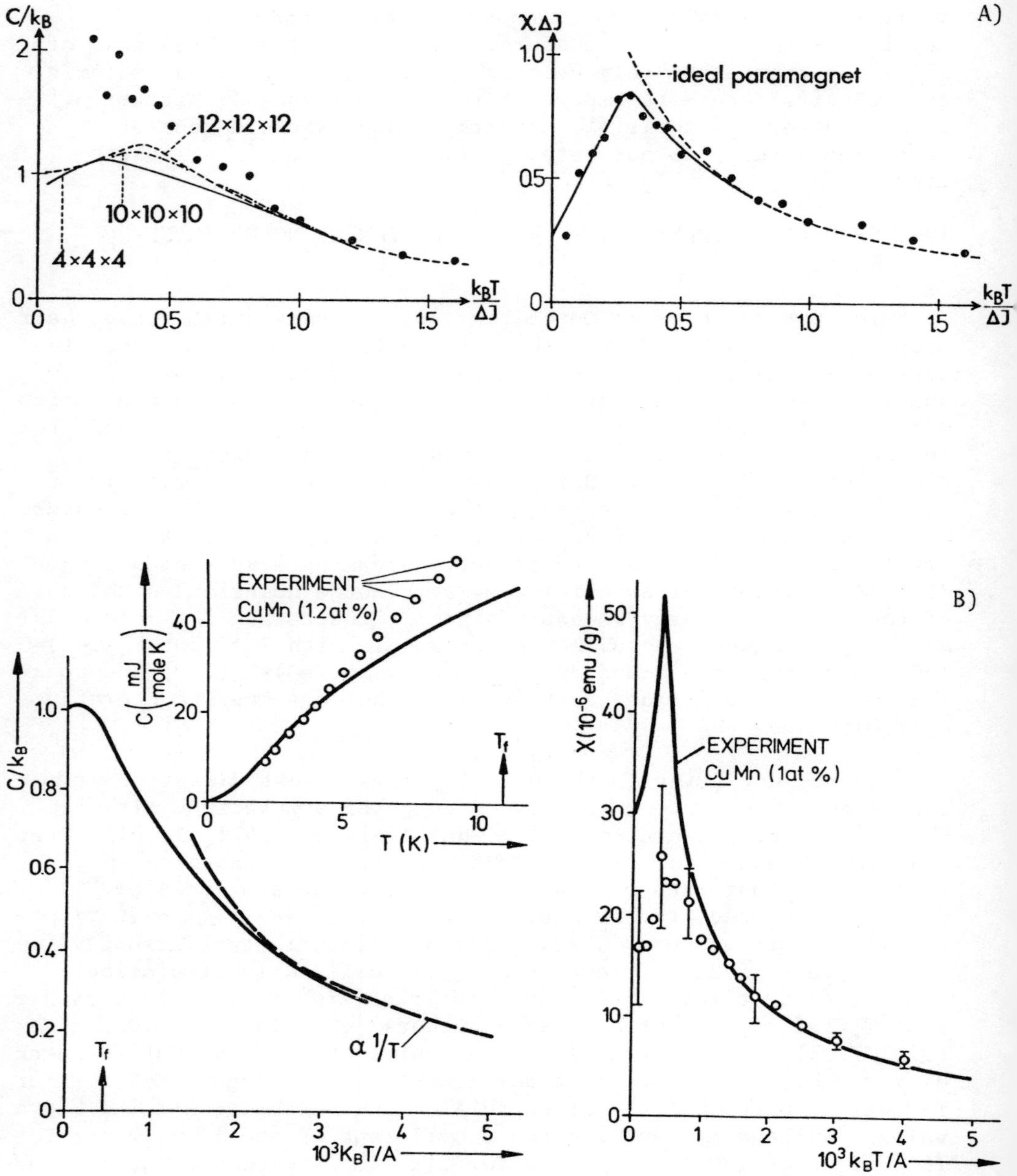

Fig. 10. Monte Carlo results for specific heat C and susceptibility of χ of Heisenberg spin glasses A) nearest neighbor gaussian exchange /17/ B) RKKY /63/.

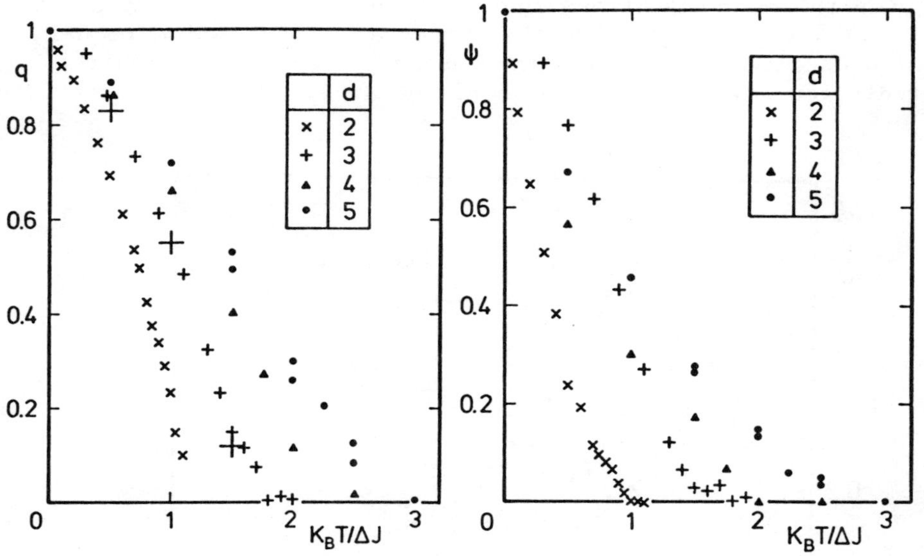

Fig. 11. Monte Carlo results for order parameters q(A) and ψ(B) plotted vs. T for Ising spin glasses /20/.

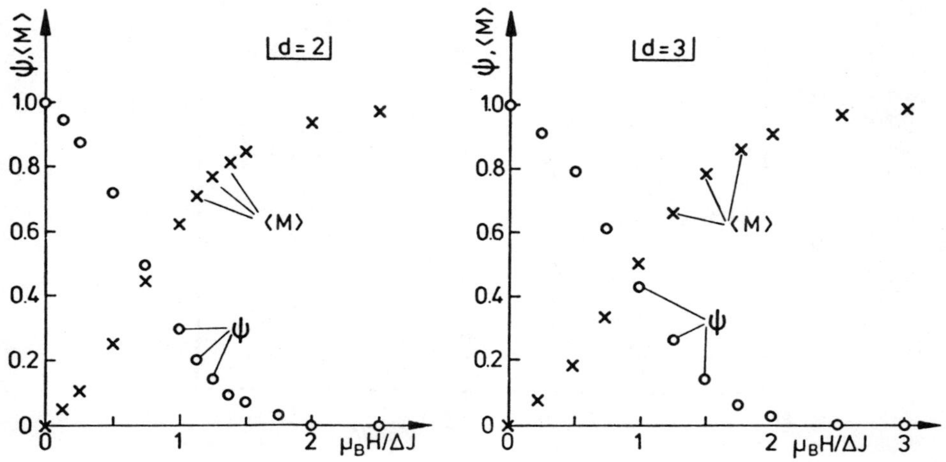

Fig. 12. ψ,M plotted vs. field at T=0 /19/.

vs. field /19/. Since $\chi(H=0)=\partial M/\partial H|_0$ is finite and $\chi(H=0)=\mu^2(1-q)/k_BT$, $q(T=0)=1$ is implied: at least for T=0 a spin glass phase exists, in contrast to suggestions /65/ that spin glasses are metastable even at T=0 Fig. 13A shows that C has a broad maximum for d=2, which becomes sharper with increasing d, consistent with the prediction /44/ that a cusp should occur for $d\geq 6$. For large T the

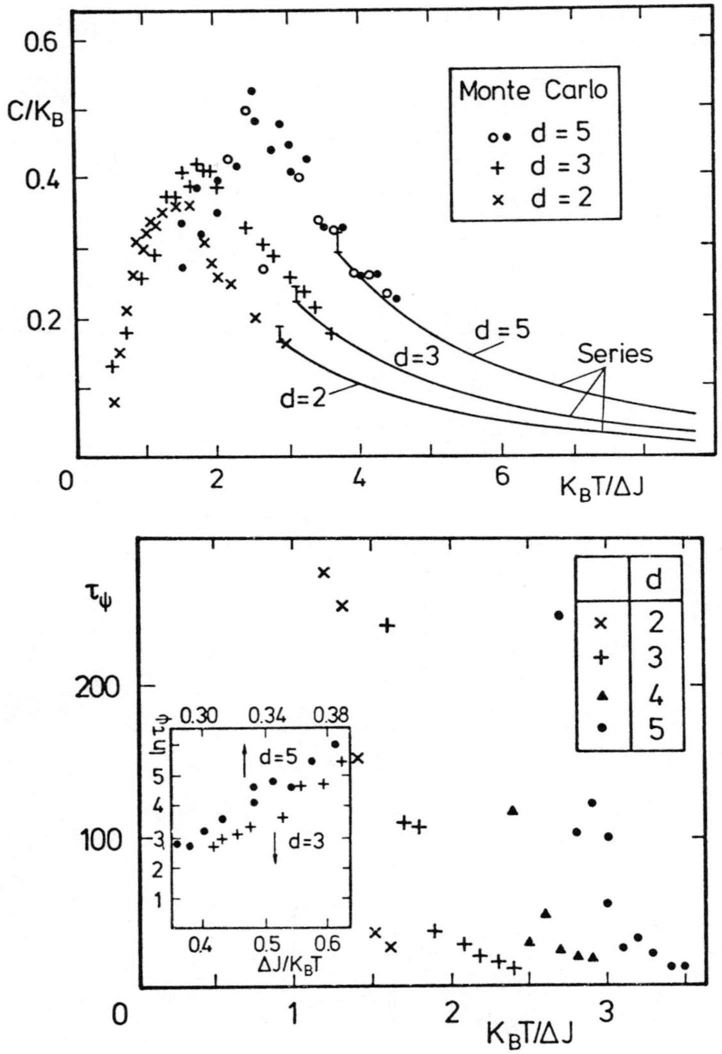

Fig. 13. A) C plotted vs. T for Ising spin glasses (/20/, Series from /51/) B) τ_ψ plotted vs. T /20/.

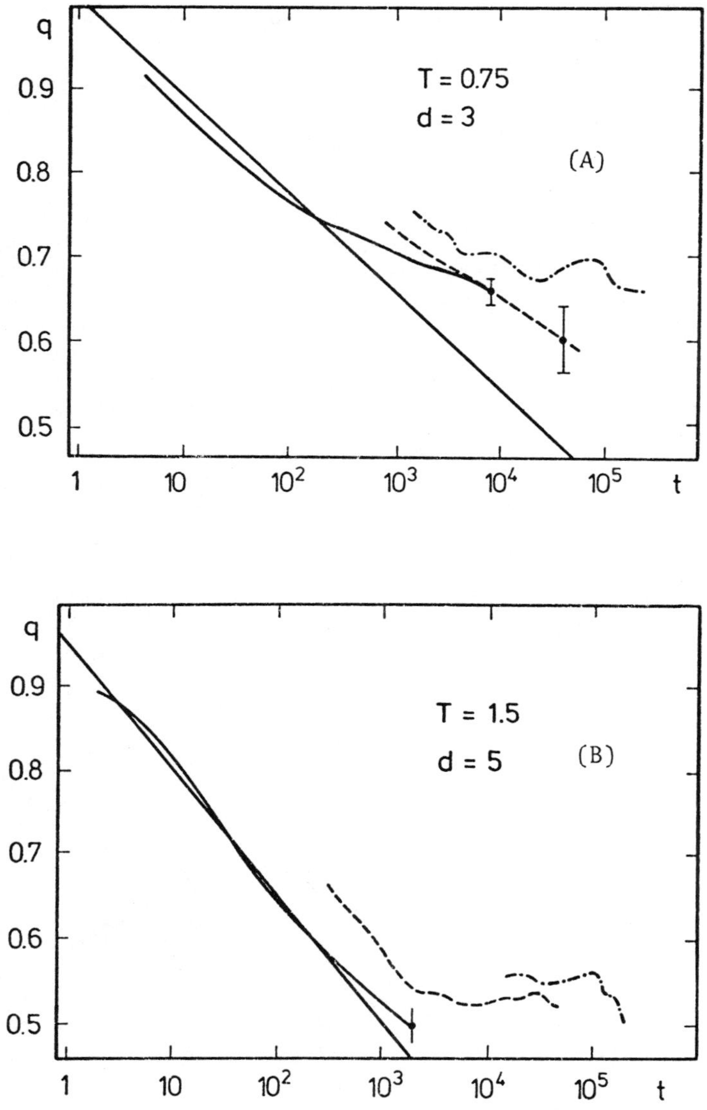

Fig. 14. Semilog plot of q vs. t for d=3(A) and d=5(B). Straight-line is Eq.(8), other curves represent various Monte Carlo runs /20/.

data nicely agree with truncated series /51/. For $T>T_f$ a relaxation time τ_ψ was found fitting the decay of $\psi(t)$ {after heating up from a ground state to the considered T} to $\exp(-t/\tau_\psi)$ at large times (Fig. 13B). τ_ψ strongly increases near T_f (perhaps diverging there). Similarly χ^{SG}_ψ {Eq. (5)} indicates the onset of strong spatial correlations as $T \to T_f$ /17/. Of course, Monte Carlo methods /66/ never clearly distinguish a true 2nd order phase transition from a"transition" where correlations and relaxation times increase several orders of magnitude but stay finite.

Doubts in the existence of a transition arose from the strange time-dependence of χ for $T<T_f$ (Fig. 1B) /18/, leading to the suggestion that $q(t) \underset{t\to\infty}{\to} 0$ for $d<4$. Fig. 14 shows that no qualitative difference between d=3 and d=5 occurs, however /20/: q(t) decreases $\propto \ln t$ for several decades of t, then deviations occur but fluctuations prevent us from unambiguously estimating the final behavior. Bray and Moore attribute this decay to single-spin relaxation, deriving a law /61/

$$dq(t)/d(\ln t) = -k_B T\, P(0,0) \; . \tag{8}$$

where $P(H_{eff},T)$ is the distribution of effective fields, Fig. 15. For $z\to\infty$ $P(H_{eff},0) \propto H_{eff}$ as $H_{eff} \to 0$ /52/, {in contrast to /43/ and equivalent work without replica trick /65/ and the "mean random field" method /67/ which all predict $P(H_{eff},T)$ to be gaussian}. But for z finite $P(0,0) \neq 0$ /20/ and thus Eq. (8) is meaningful, and implies /20/ $dq(t)/d(\ln t) \approx -T/(10T_f)$. Eq. (8) follows from the

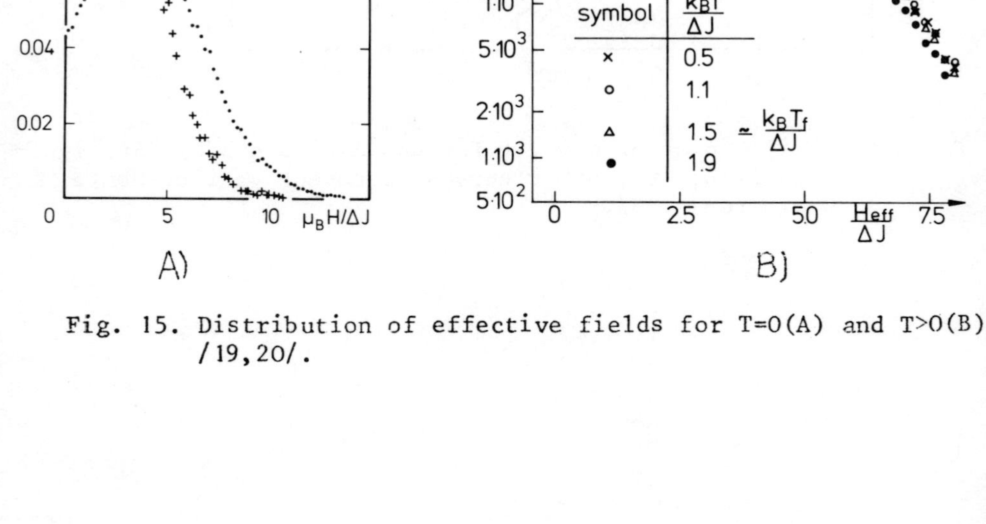

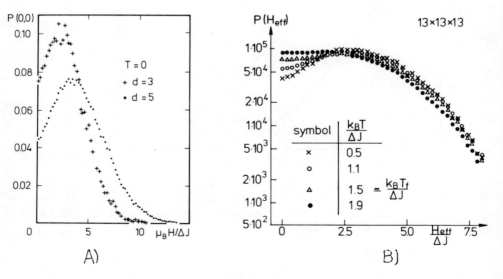

Fig. 15. Distribution of effective fields for T=0(A) and T>0(B) /19,20/.

equation exact for a kinetic Ising model /61/

$$(d/dt)\langle S_i(t)S_i(0)\rangle = -\langle S_i(t)S_i(0)\rangle + \langle S_i(t)S_i(0)\tanh\{\mu_B H_i^{eff}(0)S_i(0)/k_BT\}\rangle \quad (9)$$

by assuming $H_i^{eff}(0)S_i^{eff}(0) \approx H_i^{eff}(0)$ for $T \ll T_f$, and factorizing the second term on the r.h.s. of Eq. (9) as $\langle S_i(t)S_i(0)\rangle \tanh\{\mu_B H_i^{eff}(0)/k_BT\}$, which is accurate for $t \to 0$. Integrating Eq. (9) and averging with $P(H_{eff},T)$ one gets Eq. (8) /61,20/. Inspite of the crude approximations Eq. (8) works quite well, Fig. 14. But there is nothing specific to $d<4$ in these approximations. At late t Eq. (8) cannot hold. In order to discuss the question of a transition, one needs a theory including collective degrees of freedom. MFA yields /68/ $\langle S_i(t)S_i(0)\rangle - q \propto \exp(-t/\tau)$, the relaxation time τ diverging at T_f. This decay law is not in good agreement with the data /4/. More advanced treatments /69/ imply $q(t)-q(\infty) = (t_o/t)^{1/2}$, consistent with our data for even longer times (Fig. 16) than Eq. (8). But these theories do not yield t_o explicitly. Thus it is not yet clear to what extent single-spin and collective excitations contribute to the dynamics. Other sources of relaxa-

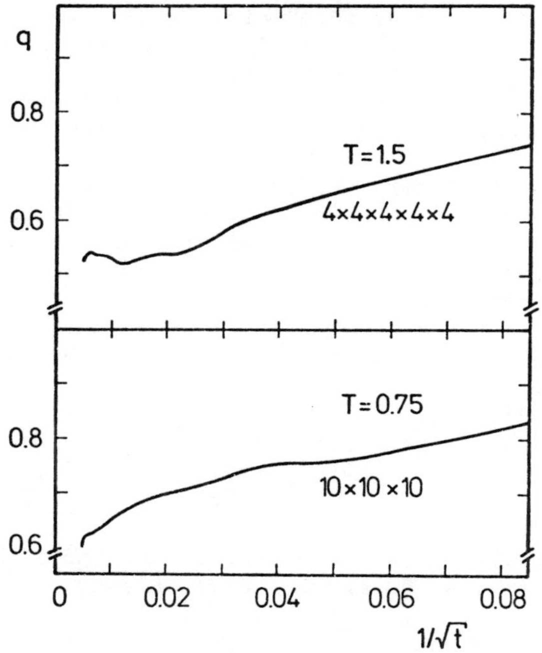

Fig. 16. Same data as Fig. 14 plotted vs. $t^{-1/2}$ /20/

tion phenomena (finite size, domain formation kinetics /19/) are also conceivable. In view of this situation, we do not discuss the work on the dynamics of Heisenberg spin glasses /70/. In conclusion, we doubt that the anomalous relaxation of spin glasses below T_f (seen in experiments and simulations) is an argument for $d_c=4$, since it occurs for d=5 as well as for d=2,3. More analytic work on the theory of spin glasses is necessary to clarify the situation.

IV. CONCLUSIONS AND OUTLOOK

Large classes of materials were shown to exhibit spin-glass behavior. Both experiment and computer simulations show anomalous slow relaxation phenomena, which are not yet fully understood. Neither experiment nor theory can convincingly answer if spin-glass freezing is associated with a phase transition at $T_f \neq 0$, or only a nonequilibrium phenomenon (true transition at $T_f=0^+$). But the statistical mechanics of these inhomogeneous systems involves many challenging questions: how to deal with the nonuniformity of states where some spins are rigidly aligned in clusters but others are loose or coupled only in frustrated loops? what effects has the high ground state degeneracy? may one "forget" about this nonuniformity by use of the replica trick? how shall one exploit concepts like gauge invariance? Etc. Finally we mention possible relations to other exciting problems like systems with random anisotropies /71/, the roughening transition /72/, Anderson localization /73/ etc. Clarification of the theory of spin glasses may help our understanding of these other problems and vice versa.

Acknowledgements: Sincere thanks are due to D. Stauffer and W. Kinzel for their collaboration on work, on which part of these lectures are based, and use of unpublished material; to them and numerous colleagues in this field I am grateful for useful discussions and preprints.

1. Recent reviews on experiments are: J.A. Mydosh, J. Mag. Magn. Mat. 7, 237 (1978); A.P. Murani, J. phys. (Paris) 39, C6-1517 (1978); J. Souletie, J. phys. (Paris) 39, C2-3 (1978); P.A. Beck, to be published.
2. Recent theoretical reviews are: A. Blandin, J. phys. (Paris) 39, C6-1499 (1978); K. Binder, ibid. C6-1527, and in Festkörperprobleme XVII, 55 (J. Treusch, ed.) Vieweg, Braunschweig 1977; P.W. Anderson, in Amorphous Magnetism II, 1 (R.A. Levy, R. Hasegawa, eds.), Plenum, New York 1977, and J. Appl. Phys. 49, 1599 (1978).
3. J.L. Tholence and R. Tournier, J. Phys. (Paris) 35, C4-229 (1974).
4. K. Binder and K. Schröder, Solid State Comm. 18, 1361 (1976); Phys. Rev. B14, 2142 (1976).

5. H. Maletta and W. Felsch, J. Magn. Mag. Mat., in press.
6. H. Maletta and P. Convert, Phys. Rev. Lett. 42, 108 (1979).
7. G. Eiselt et al., Phys. Rev. B , (1979).
8. K.L. Ngai and T. Reinecke, Phys. Rev. Lett. 38, 74 (1977).
9. W. Press, private communication.
10. N.S. Sullivan and M. Devoset, J. phys. (Paris) 39, C6-72 (1978)
11. R.B. Kummer et al., Phys. Rev. Lett. 40, (1978).
12. C.M. Soukoulis et al., Phys. Rev. Lett. 41, 568 (1978).
13. C.N. Guy, J. Phys. F5, L242 (1975).
14. H. v. Löhneysen et al., J. Phys. (Paris) 39, C6-992 (1978).
15. E.D. Dahlberg et al., Phys. Rev. Lett. 42, 401 (1979).
16. K. Binder and D. Stauffer, Phys. Lett. 57A, 1977 (1976).
17. K. Binder, Z. Phys. B26, 339 (1977).
18. A.J. Bray et al., J. Phys. C11, 1187 (1978).
19. D. Stauffer and K. Binder, Z. Phys. B30, 313 (1978).
20. D. Stauffer and K. Binder, Z. Phys. B , (1979).
21. W. Kinzel, J. Phys. 39, C6-905 (1978); Phys. Rev. B (1979).
22. L.E. Wenger and P.H. Keesom, Phys. Rev. B13, 4053 (1976).
23. L.R. Walker and R.E. Walstedt, Phys. Rev. Lett. 38, 514 (1977).
24. K. Binder, W. Kinzel and D. Stauffer, to be published.
25. V.K. Shante and S. Kirkpatrick, Adv. Phys. 20, 325 (1971).
26. L. Néel, Ann. Geophys. 5, 99 (1949).
27. "Frustration" is defined by G. Toulouse, Commun. Phys. 2, 115 (1977).
28. This term is defined in J. Villain, Z. Phys. B32, (1979).
29. J. Stephenson, J. Math. Phys. 11, 420 (1970).
30. S.F. Edwards and P.W. Anderson, J. Phys. F5, 965 (1975).
31. K. Binder and D.P. Landau, to be published.
32. A.K. Mukopadhyay et al., J. Less-Common Metals 43, 69 (1975).
33. C.M. Soukoulis and K. Levin, Phys. Rev. Lett. 39, 581 (1977).
34. J. Villain, J. Phys. C10, 4793 (1977); C11, 745 (1978).
35. D.C. Mattis, Phys. Lett. 56A, 421 (1976).
36. S. Kirkpatrick, Phys. Rev. B16, 4630 (1977).
37. E. Fradkin, B.A. Huberman and S.H. Shenker, Phys. Rev. B18, 4789 (1978).
38. A. Aharony and K. Binder, to be published.
39. I.E. Dzyaloshinskii and G.E. Volovik, J. phys. 39, 693 (1978), and preprint.
40. J.A. Hertz, Phys. Rev. B18, 4875 (1978).
41. S.F. Edwards and P.W. Anderson, J. Phys. F6, 1927 (1976).
42. B.I. Halperin and W.M. Saslow, Phys. Rev. B16, 2154 (1977).
43. D. Sherrington and S. Kirkpatrick, Phys. Rev. Lett. 35, 1972 (1975).
44. A.B. Harris et al., Phys. Rev. Lett. 36, 415 (1976).
45. J.R.L. de Almeida and D.J. Thouless, J. Phys. A11, 983 (1978).
46. E. Pytte and J. Rudnick, to be published.
47. J.L. Van Hemmen and R.G. Palmer, to be published.
48. S.-K. Ma and J. Rudnick, to be published.

49. C. de Dominicis, Phys. Rev. B18, 4913 (1978).
50. D.J. Thouless et al., Phil. Mag. 35, 593 (1977).
51. R.V. Ditzian and L.P. Kadanoff, to be published.
52. S. Kirkpatrick and D. Sherrington, Phys. Rev. B17, 4384 (1978).
53. H.-J. Sommers, Z. Phys. B31, 301 (1978), and preprint.
54. A. Blandin, Ref. 2.
55. A.J. Bray and M.A. Moore, Phys. Rev. Lett. 41, 1068 (1978); J. Phys. C12, 79 (1979).
56. W. Kinzel and K.H. Fischer, J. Phys. C11, 2115 (1978), and references therein.
57. P.W. Anderson and C.M. Pond, Phys. Rev. Lett. 40, 903 (1978).
58. R. Fisch and A.B. Harris, Phys. Rev. Lett. 38, 785 (1977).
59. P. Reed et al., J. Phys. C11, L139 (1978); J. Vannimenus and G. Toulouse, J. Phys. C10, L537 (1977).
60. Derrida et al., preprint.
61. A.J. Bray and M.A. Moore, J. Phys. C (1979).
62. W.Y. Ching and D.L. Huber, Phys. Lett. 59A, 383 (1977).
63. W.Y. Ching and D.L. Huber, J. Phys. F8, L63 (1978).
64. A.I. Larkin and D.E. Khmelnitskii, Zh. Eksp. Teor. Fiz. 58, 1789 (1970).
65. S. Katsura, priv. communication.
66. K. Binder (ed.) Monte Carlo Methods in Statistical Physics Springer, Berlin 1979.
67. M.W. Klein, Phys. Rev. B14, 5008 (1976), and references therein.
68. W. Kinzel and K.H. Fischer, Solid State Comm. 23, 687 (1977).
69. S.-K. Ma and J. Rudnick, Phys. Rev. Lett. 40, 589 (1978).
70. For references see K.H. Fischer, Z. Physik B (1979).
71. R.A. Pelcovits et al., Phys. Rev. Lett. 40, 476 (1978).
72. G. Toulouse et al., to be published.
73. A. Aharony and Y. Imry, J. Phys. C L487 (1977); H.G. Schuster, preprint.

NOTE ADDED: Very recently it was noted that "better" replica results are obtained replacing factors n by n-1 suitably before letting $n \to 0$ /74/. However, a more rigorous justification is lacking. As a source of the irreversible behavior (Figs. 2, 14) it was shown /75/ that in spin-glass "ground-states" many small clusters exist, the overturning of which costs an energy δE ("two-level systems" as in window glasses). Such clusters indeed are seen in simulations, see e.g. /17/, Fig. 12. A theory on this basis seems to account for many observations /75/.

74. S. Fishman and A. Aharony, preprint.
75. C. Dasgupta et al., preprint.

RANDOM ANISOTROPY SPIN-GLASS

Erling Pytte
IBM Research
Yorktown Heights, N.Y. 10598, U.S.A.

In this seminar some recent work on a new type of spin-glass will be discussed.[1-4] The model consists of a conventional Heisenberg model together with a random uniaxial anisotropy term,

$$\mathcal{H} = -\sum J_{ij}\, S_i \cdot S_j - D \sum (\hat{n}_i \cdot S_i)^2 \tag{1}$$

Here $\hat{n}_i$ is a unit vector denoting the orientation of the uniaxial anisotropy axis on the i th site. The orientation of the axes is assumed to be random with no correlation between the axes on different sites. In general, a model with m spin components in d dimensions will be considered.

This model has played an important role in the field of amorphous magnetism. It was introduced by Harris, Plischke and Zuckerman[5] (HZP) to account for experimental data on amorphous $TbFe_2$.[6] The most striking observation was a large reduction in the magnetization and the transition temperature as compared to the crystalline material. For $TbFe_2$ the transition temperature was reduced from about 700° K to about 400° K. The subsequent great popularity of this model was to a large extent due to computer simulations on Fe-RE alloys using a Dense Random Packing of Hard Spheres algorithm together with point charge crystal field.[7] These simulations showed that there was indeed uniaxial anisotropy in these amorphous materials, and that its orientation was random with no apparent correlation between the axes on different sites. The HZP model has since been used extensively to describe the properties of amorphous rare earth alloys. A review of the experimental and theoretical work relevant to this model is contained in a recent issue of Physics Reports.[8]

The early experiments were all performed on Fe-RE alloys which are magnetically much more complicated than the simple model described by Eq. (1). Both the Fe-Fe exchange and the Fe-RE exchange need be included as well as the uniaxial anisotropy of the RE. Thus an important development in this field has been the recently made alloys in which the

RE ion is the only magnetic ion, such as RE-Ag,[9] RE-Al[10] and RE-Cu.[11] For these materials the simple model of Eq. (1) should apply much better.

The theoretical work on this model has been of two types 1) mean field theory and 2) computer simulations.

One of the points emphasized at this study institute is the existence of a lower critical dimension, d_ℓ, below which long range magnetic order does not exist,[12] whereas mean field theory yields long range magnetic order in all dimensions. For the Heisenberg model, for example, $d_\ell = 2$. Thus in three dimensions the mean field theory gives reasonable results and can be used, at least as a first approximation, whereas in one or two dimensions the mean field theory makes no sense at all. Below it will be shown that the lower critical dimension for the HZP model is four. Thus using the mean field approximation in three dimensions for this model is comparable to using mean field theory in one dimension for the Heisenberg model.

Three separate computer simulations have been reported for this model.[13-15] The first two concludes that the ground state has long range magnetic order, whereas the most recent conclude that the ground state is probably a spin-glass.[15]

To show that $d_\ell = 4$ for this model we present first a simple perturbation theory argument. We assume that the magnetization

$$M \equiv \langle\langle S_i^z \rangle\rangle_A \neq 0$$

where the inner brackets denote the thermal average and the outer brackets the average over the orientation of the anisotropy axes. Then we calculate $\langle\langle S_i^x \rangle^2 \rangle_A$ to lowest order in D,

$$\langle\langle S_i^x \rangle^2 \rangle_A \sim D^2 M^2 \int d^d k \, \chi_\perp^2(k) = D^2 M^2 \int \frac{d^d k}{k^4} \tag{2}$$

The integral on the right side diverges for $d \leq 4$. As the left side is clearly finite we have a contradiction unless $M \equiv 0$. This result depends on the fact that the transverse susceptibility for the Heisenberg model is gapless,

$$\chi_\perp(k) = \frac{1}{k^2} \quad ; \quad M \equiv 0 \tag{3}$$

This can be shown to be true also for the HZP model to all orders in D in the large m limit[1,4] as well as by a renormalization group calculation.[16]

A more rigorous derivation of $d_\ell = 4$ is obtained by use of the Bogoliobov inequality.[1,2] The calculation is closely analogous to that used by Mermin and Wagner[17] to show that there is no long range magnetic order for the Heisenberg model for $d \le 2$ and $m \ge 2$. Because we are dealing with a quenched random system, the replica method is used to derive an effective translationally invariant Hamiltonian. Then, the Bogoliobov inequality can be applied in the usual manner. For $m = 2$ the result is[1,2]

$$1 \ge k_B T M^2 \frac{1}{N} \sum_k \left\{ \frac{1}{\alpha k^2} + \frac{1}{(\alpha k^2)^2} \frac{D^2}{8 k_B T} \right\} \tag{4}$$

where α is defined by, $J(k) = J(0) + \alpha k^2$.

For $D = 0$ this is precisely the result obtained by Mermin and Wagner. Because the sum diverges for $d \le 2$, the inequality requires $M \equiv 0$ except when $T \equiv 0$. As is well known long range order is possible at $T \equiv 0$ for both the one and two dimensional Heisenberg model. When $D \neq 0$ the integral diverges for $d \le 4$. Note that in this case the inequality requires $M = 0$ also for $T = 0$. For $D \neq 0$ the long range order is destroyed not by thermal fluctuations, which vanish at $T = 0$, but by the random orientations of the uniaxial anisotropy axes.

Because the replica method gave rise to unphysical and/or unusual results when used in the Edwards-Anderson theory[18] of the conventional random exchange spin glass, questions concerning the reliability of this method have been raised. However, all the questionable results have subsequently been rederived, performing the averaging over the random variables, without use of the replica method. This suggests that the problems associated with current spin glass theories are of a more fundamental nature than the use of the replica method.[19] As used above, in order to derive Eq. (4), there is no reason to believe that the replica method does not give the correct result.

If there is no long range order for $d \le 4$ the question naturally arises as to the nature of the ground state. We suggest that the HZP model is a spin glass for all values of D/J. Explicit calculations have been carried out only in the limit of large m. In this limit it is found that for $d \le 4$

$$M \equiv 0 \quad ; \quad Q \neq 0 \quad \text{for any} \quad D \neq 0$$

consistent with $d_\ell = 4$ derived above. Here Q is the Edwards-Anderson order parameter defined by,[18]

$$Q = \langle \langle S_i \rangle \cdot \langle S_i \rangle \rangle_A$$

For $d > 4$ a spin glass phase or a magnetic phase is obtained depending on the value of $\Delta \equiv (D/J)^2$ giving rise to a phase diagram shown schematically in Fig. 1.

In the limit $\Delta >> 1$ Chen and Lubensky[20] previously derived a Landau-Ginzburg-Wilson free energy functional for the HZP model, based on the Edwards-Anderson order parameter, identical to that for

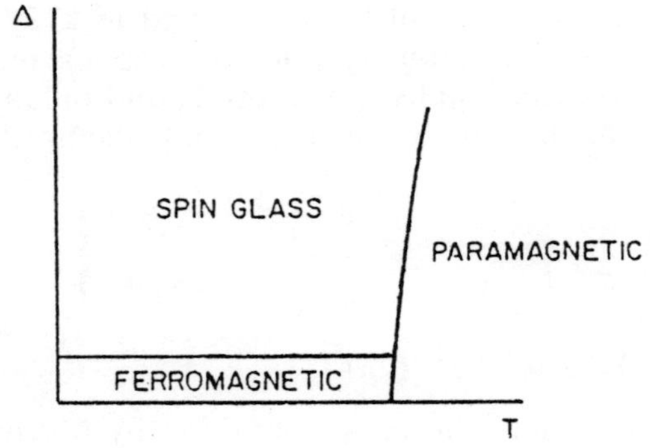

Fig. 1 – Phase diagram for a magnet with isotropically distributed random uniaxial anisotropy in more than four dimensions.

the Ising random exchange spin-glass. The role of frustration in the HZP model has recently been discussed by Alexander and Lubensky.[21]

Next the results of two renormalization group calculations will be discussed very briefly. Both calculations refer to the magnetic transition for $d > 4$. For the Heisenberg model $d_\ell = 2$ such that $T_c = 0$ for $d = 2$ and $m > 2$. This has given rise to a renormalization group expansion[22] for $d = 2+\varepsilon$ with $T_c \sim \varepsilon$. Because $d_\ell = 4$ for the HZP model an analogous expansion can be made for $d = 4+\varepsilon$ with in this case a critical value of Δ, $\Delta_c \sim \varepsilon$ as follows from the phase diagram in Fig. 1. The exponents obtained[1,2] for the transition in $(\Delta - \Delta_c)$,

$$\nu = 1/\varepsilon \ , \ \eta = \varepsilon/(m-2) \quad . \tag{5}$$

are identical in form to those for the Heisenberg model with $d = 2+\varepsilon$ in $(T - T_c)$.

The divergence in η for $m = 2$ results from a breakdown of the assumption that $\Delta_c \sim \varepsilon$. For $d = 4$, $m = 2$ there is a finite transition value $\Delta_c = \Delta_o$ for the HZP model. This is analogous to the finite transition temperature for the two dimensional Heisenberg model for $m = 2$ even though the Mermin-Wagner theorem shows that $M \equiv 0$ for this case. The resulting phase is characterized by a line of critical points.[12] For the HZP model the analogous phase for $d = 4$, $m = 2$ is characterized by a plane of critical points, as Δ and T both have finite critical values even though according to Eq. (4) $M \equiv 0$.

For $m > 2$ it follows from Eq. (5) that $\nu \to \infty$ as $d \to 4$. This gives rise to exponential dependence of the correlation length, $\xi = e^{1/\Delta}$, $\Delta_c = 0$. The analogous result for the Heisenberg model is $\xi = e^{1/T}$, $T_c = 0$ for $d = 2$, $m > 2$. Further because for Heisenberg model $\xi = 1/T$ for $d = 1$, $m > 2$ we expect,

$$\xi = \frac{1}{\Delta} \quad \text{for} \quad d = 3 \ , \ m > 2 \ .$$

for the HZP model. The result $\xi = 1/\Delta$ has also been obtained for $d = 3$, $m = 3$ by domain energy arguments.[13] Finally, we note that the usual scaling relation $d\nu = 2 - \alpha$ is for this transition replaced by, $(d - 2)\ \nu = 2 - \alpha$, as in the random field model.[23]

So far we have considered the transition as a function of Δ. The transition along the temperature axis for $T \gtrsim T_c$ has previously been studied by Aharony using the conventional $d = 4 - \varepsilon$ expansion.[24] No accessible stable fixed point was found for $d < 4$. For $d > 4$ the Gaussian fixed point is stable giving rise to mean field exponents. The absence of a stable fixed point can have many different interpretations but is certainly consistent with our result that there is no magnetic phase transition for $d \leq 4$. The fact that mean field exponents are obtained for $d > 4$ suggests the unusual case of a transition for which the upper and lower critical dimensions are the same, $d_u = d_\ell = 4$. Aharony considered $T > T_c$ only. It turns out that for $T < T_c$, the longitudinal susceptibility, χ_L, diverges on the coexistence curve, $H = 0$, for $4 < d \leq 6$. Thus the upper critical dimension is, in fact, six for this model, although for all other properties mean field exponents are obtained. Explicitly we find,

$$\chi_L \sim H^{-\varepsilon/2} \qquad \text{for} \quad d = 6 - \varepsilon$$

$$\sim \ \ln H \qquad \text{for} \quad d = 6$$

These results were derived[1,3] by recognizing that the series of most divergent diagrams for χ_L for the m-component random anisotropy model in the ordered phase is the same as that for the specific heat of an (m – 1) component random field model in the disordered phase. The latter was obtained by integrating up the recursion relations for that model. For the random field model $d_\ell = 4$ and $d_u = 6$, and we can expand about six dimensions in the usual way.

Spin-glass behaviour has been observed experimentally in various TbAg, DyAl and DyCu alloys.[9-11] These alloys are expected to be well described by the HZP model, particularly at lower RE concentrations.[25] Of these the best studied is Dy_xCu_{1-x}. These alloys show spin-glass behaviour for $x \lesssim 0.4$. They exhibit a cusp in the susceptibility and a broad smooth specific heat curve characteristic of other spin-glasses.

Thus, both experimentally and theoretically the HZP model appears to yield as good a spin-glass as that described by the more conventional random exchange model.

Unfortunately, there is at present no reliable analytic theory of spin-glasses for either model. Most recent theoretical work has been based on the theory of Edwards and Anderson,[18] and the state described by the Edwards-Anderson order parameter is now believed to be unstable. For the infinite range random exchange model,[26] Alameida and Thouless[27] found a mode with a negative gap when diagonalizing the quadratic form describing the fluctuations about the state characterized by the Edwards-Anderson order parameter, such that the free energy was neither positive nor negative definite. For the short range model a renormalization group calculation for the ordered phase has recently been carried out for $d = 6-\varepsilon$[28] extending previous calculations for $T > T_c$.[29,20] Even though the term giving rise to the instability in the infinite range model is irrelevant for $d \sim 6$ an instability is generated in higher order for $d < 6$. Although there has been a great deal of discussion as to what the lower critical dimension is for this model, it was generally believed that at least for $d \lesssim 6$ the spin glass phase could be described by the Edwards-Anderson order parameter.

The instability for $d < 6$ was found to be directly related to the completely unacceptable result of a negative value for a correlation function (spin-glass susceptibility) which by definition is positive.[28] For $T < T_c$ two spin-glass susceptibilities are obtained,[30]

$$\begin{aligned} \chi_1 = \frac{1}{N}\sum_{ij} \; & \{<<S_iS_j>^2>_J \\ & - 4 <<S_iS_j> <S_i> <S_j>>_J \\ & + 3 <<S_i>^2 <S_j>^2 >_J \} \\ \chi_2 = \frac{1}{N}\sum_{ij} \; & < \{ <S_iS_j> - <S_i> <S_j>\}^2>_J \end{aligned} \qquad (6)$$

Whereas χ_1 is well behaved, χ_2 is found to be negative for $d < 6$. Subsequent calculation of χ_2 for the infinite range model[31] showed similarly that $\chi_2 < 0$, and that this result is directly related to the negative mode discussed above. Both these calculations of χ_2 made use of the replica formalism. Khurana and Hertz have obtained the same negative result for χ_2 for several kinds of mean field theories, without use of replicas.[32]

It is interesting to note that this instability vanishes in the limit $m \to \infty$.[33] Finally the symmetry breaking approximation of Bray and Moore[34] should be mentioned who obtain $\chi_2^{-1} = 0$ by their limiting procedure. This result is possibly related to the fact that $\chi_2^{-1} = 0$ also in the limit $m \to \infty$.

Although $M \equiv 0$ for $d \leq 4$ for arbitrarily small values of (D/J), the correlation length diverges as $\xi/a = 1/\Delta = (J/D)^2$, as $D \to 0$, in the spin glass phase, where a is the lattice constant. In soft magnetic materials $D/J \sim 10^{-4}$ is not uncommon, in which case ξ may be of the order or larger than the sample size. Such materials will appear to have long range magnetic order, although in principle, for an infinite sample, $M \equiv 0$. This suggests an explanation of a long standing problem. Magnetic transitions in random systems are predicted to be sharp, whereas experimentally they are found to be rounded. This discrepancy may be due to weak interactions, present in the experimental system but not included in the theoretical model, which destroys magnetic long range order for the infinite sample and gives rise to a rounded transition characteristic of a spin-glass, sufficiently close to the transition point. The HZP model is a fairly restricted model. However, Aharony[35] has shown that any model of the form, $\Sigma J_{ij}^{\alpha\beta} S_i^{\alpha} S_j^{\beta}$, with random off-diagonal elements, $J^{\alpha\beta}$, will have $M = 0$ for $d \leq 4$. This includes random dipolar interactions present in all amorphous materials. Random fields also destroy long range magnetic order for $d \leq 4$.[36] From this point of view, amorphous magnets, with rounded transitions, do not possess true long range order.

Finally two generalizations of the HZP model will be mentioned. It is easy to show that a model for RE-Fe alloys which includes Fe-Fe exchange and Fe-RE exchange together with the random RE uniaxial anisotropy still yields $M = 0$ for $d \leq 4$.[16] Further this remains true if short range correlations between anisotropy axes on different sites are included.

List of References

1. R. A. Pelcovits, E. Pytte and J. Rudnick, Phys. Rev. Lett., **40**, 476 (1978).
2. R. A. Pelcovits, Phys. Rev. B**19**, 465 (1979).
3. E. Pytte, Phys. Rev. B**18**, 5046 (1978).
4. J. Rudnick, to be published.
5 R. Harris, M. Plischke and M. J. Zuckermann, Phys. Rev. Lett. **31**, 160 (1973).
6. J. J. Rhyne, S. J. Pickart and H. A. Alperin, Phys. Rev. Lett. **29**, 1562 (1972).
7. R. W. Cochrane, R. Harris and M. Plischke, J. Non-C Solids **15**, 239 (1974); R. W. Cochrane, R. Harris, M. Plischke, D. Zobin and M. J. Zuckerman, J. Phys. F **5**, 763 (1975).
8. R. W. Cochrane, R. Harris and M. J. Zuckermann, Physics Reports **48**, 1 (1978).
9. B. Y. Boucher, J. Physique **37**, L345 (1976); B. Y. Boucher, Phys. Stat. Sol. A**10**, 179 (1977).
10. T. R. McGuire, to be published.
11. J. M. D. Coey and S. von Molnar, J. Phys. (Paris) **39**, L327 (1978); T. R. McGuire and J. M. D. Coey, to be published; S. Von Molnar, C. N. Guy, R. Gambino and T. R. McGuire, to be published. See also J. M. D. Coey, J. Appl. Phys. **49**, 1646 (1978).
12. See for example, A. P. Young, Phase Transitions in Low-Dimensional Systems at this Study Institute and references listed therein.

13. M. C. Chi and R. Alben, J. Appl. Phys. **48**, 2987 (1977).
14. R. Harris, S. H. Sung and M. J. Zuckermann, IEEE Trans. Mag-**14**, 725 (1978).; R. Harris and S. H. Sung, J. Phys. F**8**, L299 (1978).
15. M. C. Chi and T. Egami, J. Appl. Phys., March (1979) and references listed therein.
16. E. Pytte, unpublished.
17. N. D. Mermin and H. Wagner, Phys. Rev. Lett. **17**, 1133 (1966).
18. S. F. Edwards and P. W. Anderson, J. Phys. F**5**, 965 (1975).
19. For a more detailed discussion see for example, E. Pytte and J. Rudnick, Phys. Rev. B**19**, April 1 (1979).
20. J. H. Chen and T. C. Lubensky, Phys. Rev. B**16**, 2106 (1977).
21. S. Alexander and T. C. Lubensky, Phys. Rev. Lett. **42** 125 (1979).
22. A. A. Migdal, Sov. Phys. JETP **42**, 413 and 743 (1976); A. M. Polyakov Phys. Lett. B**59**, 79 (1975); D. R. Nelson and R. A. Pelcovits, Phys. Rev. B**16**, 2191 (1977).
23. G. Grinstein, Phys. Rev. Lett., **37**, 944 (1976); A. Aharony, Y. Imry and S. Ma, Phys. Rev. Lett. **37**, 1364 (1977).
24. A. Aharony, Phys. Rev. B**12**, 1038 (1975).
25. Based on recent computer simulations, M. J. Zuckermann, private communication.
26. D. Sherrington and S. Kirkpatrick, Phys. Rev. Lett. **35**, 1792 (1975); S. Kirkpatrick and D. Sherrington, Phys. Rev. B**17**, 4384 (1978).
27. J. R. L. de Alameida and D. J. Thouless, J. Phys. A**11**, 983 (1978).
28. E. Pytte and J. Rudnick, Phys. Rev. B**19**, April 1 (1979).
29. A. B. Harris, T. C. Lubensky and J. -H Chen, Phys. Rev. Lett. **36**, 415 (1976).
30. For $T > T_c$ χ_1 and χ_2 become degenerate. It is interesting to note that this susceptibility differs from that considered by R. Fisch and A. B. Harris, Phys. Rev. Lett. **38**, 785 (1977) but is equivalent to χ^{EA} discussed by K. Binder at this Study Institute.
31. D. J. Thouless and J. R. L. Alameida, to be published.
32. A. Khurana and J. A. Hertz, Bull. Am. Phys. Soc. **24**, 304 (1979).
33. J. R. L. de Alameida, R. C. Jones, J. M. Kosterlitz and D. J. Thouless, J. Phys. C**11**, L 871 (1978).
34. A. J. Bray and M. A. Moore, Phys. Rev. Lett. **41**, 1068 (1978).
35. A. Aharony, to be published.
36. Y. Imry and S. Ma, Phys. Rev. Lett. **35**, 1399 (1975).

EXACT RESULTS FOR A ONE-DIMENSIONAL RANDOM-ANISOTROPY SPIN GLASS

H. Thomas

Institut für Physik, Universität Basel

Spin-glass behaviour induced by random uniaxial anisotropy has recently been discussed by Pelcovits, Pytte and Rudnik /1/. It was found that a random distribution of anisotropy axes destroys the ferromagnetic state and generates a spin-glass state in fewer than four dimensions. This note gives an exact solution for the one-dimensional case. The ground-state spin configuration is represented as a random walk in spin space, which demonstrates clearly the absence of ferromagnetic order. There exists a spectrum of low-lying excitation extending down to zero energy which gives rise to specific properties characteristic of a one-dimensional spin-glass ground state.

Model

We consider an open chain of classical ($n \geq 2$) -component spins $\hat{s}_\ell$, ($|\hat{s}_\ell|=1$), with random-direction single-site uniaxial anisotropy energy of strength D, coupled by isotropic nearest-neighbour exchange interactions J. The system is described by the Hamiltonian

$$\mathcal{H} = -J\sum_\ell \hat{s}_\ell \cdot \hat{s}_{\ell+1} - \tfrac{1}{2}D\sum_\ell\left((\hat{a}_\ell \cdot \hat{s}_\ell)^2 - 1\right) - \vec{H}\cdot\sum_\ell \hat{s}_\ell . \quad (1)$$

The directors $\hat{a}_\ell$ of the easy axes in spin space are fixed by the convention that their component along a specified direction is positive. They are assumed to be statistically independent and uniformly distributed over the unit half-sphere. Configurational averages over functions of the $\hat{a}_\ell$ may be extended over the full sphere by even continuation.

We restrict the discussion to the case of large anisotropy D>>J such that

$$\hat{s}_\ell = \hat{a}_\ell \sigma_\ell, \quad \sigma_\ell = \pm 1. \tag{2}$$

In this limit, the model becomes equivalent to an Ising model

$$\mathcal{H} = - \sum_\ell J_{\ell,\ell+1} \sigma_\ell \sigma_{\ell+1} - \sum_\ell H_\ell \sigma_\ell \tag{3}$$

with random interactions

$$J_{\ell,\ell+1} = J(\hat{a}_\ell \cdot \hat{a}_{\ell+1}) \tag{4}$$

and random external fields

$$H_\ell = H \cdot \hat{a}_\ell . \tag{5}$$

In order to find the ground state, the energy has to be minimized for given configuration $\{\hat{a}_\ell\}$. The sequence of operations (first minimization, then configurational average) is crucially important: If (3) is configurationally averaged for fixed σ_ℓ, one obtains a ferromagnetic ground state.

Zero-Field Ground State

For H=0, the ground state is found by minimizing each bond separately. The minimum occurs for

$$(\hat{s}_\ell \cdot \hat{s}_{\ell+1})_o = |\hat{a}_\ell \cdot \hat{a}_{\ell+1}| = (\hat{a}_\ell \cdot \hat{a}_{\ell+1}) \operatorname{sign}(\hat{a}_\ell \cdot \hat{a}_{\ell+1}), \tag{6}$$

i.e. for $\hat{s}_\ell$ and $\hat{s}_{\ell+1}$ forming an acute angle. For a succession of acute angles one has

$$(\hat{s}_\ell \cdot \hat{s}_{\ell+\lambda})_o = (\hat{a}_\ell \cdot \hat{a}_{\ell+\lambda}) \prod_{\ell'=\ell}^{\ell+\lambda-1} \operatorname{sign}(\hat{a}_{\ell'} \cdot \hat{a}_{\ell'+1}). \tag{7}$$

The ground state may thus be visualised as a random walk on the n-dimensional unit sphere with random step width uniformly distributed between 0 and $\pi/2$, covering gradually the whole sphere. This demonstrates clearly the absence of ferromagnetic order: The magnetization vanishes in the thermodynamic limit as

$$m_o = N^{-1}\sum_\ell (\hat{s}_\ell)_o \propto N^{-1/2}, \tag{8}$$

whereas the spin-glass order parameter is trivially non-zero,

$$q_o = N^{-1}\sum_\ell (\hat{s}_\ell)_o^2 = 1. \tag{9}$$

Since (6) and (7) are even functions of the $\hat{a}_\ell$, the configurational average $\langle\rangle_c$ may be extended over the full spheres. One finds

$$C_{o,1} = \langle(\hat{s}_\ell \cdot \hat{s}_{\ell+1})_o\rangle_c = \int_{-1}^{+1} |\zeta| p(\zeta) d\zeta \tag{10}$$

where

$$p(\zeta) = (O_{n-1}/O_n)(1-\zeta^2)^{(2-3)/2} \tag{11}$$

is the distribution function of $\zeta = \hat{a}_\ell \cdot \hat{a}_{\ell+1}$, and $O_n = 2\pi^{n/2}/\Gamma(n/2)$ is the surface of the n-dimensional sphere. Integration yields

$$\begin{aligned} C_{o,1} &= \frac{1}{\sqrt{\pi}} \Gamma(\tfrac{n}{2})/\Gamma(\tfrac{n+1}{2}) \\ &= 2/\pi \qquad (n=2) \\ &= 1/2 \qquad (n=3) \\ &\simeq (2/(n\pi))^{1/2} (n>>1), \end{aligned} \tag{12}$$

and the ground-state energy is

$$E_o = -NJC_{o,1}. \tag{13}$$

Further, by observing

$$\langle\hat{a}_\ell \operatorname{sign}(\hat{a}_\ell \cdot \hat{a}_{\ell+1})\rangle_\ell = \langle|\hat{a}_\ell \cdot \hat{a}_{\ell+1}|\rangle_\ell \hat{a}_{\ell+1} \tag{14}$$

one obtains for the spin correlation function

$$C_{o,\lambda} \equiv \langle(\hat{s}_\ell \cdot \hat{s}_{\ell+\lambda})_o\rangle_c = C_{o,1}{}^\lambda = e^{-\lambda/\xi} \tag{15}$$

which yields a correlation length

$$\xi = 1/|\ell n C_{o,1}| \tag{16}$$

of the order of a lattice constant for small n and $\propto 1/\ell n\, n$ for n>>1.

There is no frustration in the ground state (for open boundaries), and only trivial twofold degeneracy. Higher degeneracy would arise from bonds with exactly orthogonal anisotropy directions, $\hat{a}_\ell \cdot \hat{a}_{\ell+1} = 0$, which occur with probability zero only. The specific properties distinguishing the spin-glass ground state from ordered phases arise from the presence of "weak links", i.e. of bonds with nearly orthogonal anisotropy directions, which give rise to a spectrum of low-energy excitations with constant density of states extending down to zero energy, whereas the excitation spectrum of an anisotropic ferromagnet has a gap.

Statistical Mechanics

The excited states of the system may be represented by the set $\{\tau\}$ of bond spins τ_ℓ for the bonds $(\ell,\ell+1)$, taking the value $\tau_\ell = +1$ for an intact (i.e. ground-state) bond and $\tau_\ell = -1$ for a broken (i.e. excited) bond. The energy of the state $\{\tau\}$ takes the form

$$E(\{\tau\}) = -J\sum_\ell |\hat{a}_\ell \cdot \hat{a}_{\ell+1}|\tau_\ell, \tag{17}$$

and the statistical distribution for thermodynamic equilibrium at temperature $T=1/k\beta$ is given by

$$\begin{aligned}\rho(\{\tau\}) &= Z^{-1}\exp\left(-\beta E(\{\tau\})\right) \\ &= \prod_\ell \left\{\frac{1}{2}\left(1+\tau_\ell \tanh(\beta J|\hat{a}_\ell \cdot \hat{a}_{\ell+1}|)\right)\right\}\end{aligned} \tag{18}$$

whence

$$\langle\tau_\ell\rangle = \tanh\left(\beta J|\hat{a}_\ell \cdot \hat{a}_{\ell+1}|\right). \tag{19}$$

The value of $\hat{s}_\ell \cdot \hat{s}_{\ell+\lambda}$ in state $\{\tau\}$ may be represented in terms of the bond spins as

$$\hat{s}_\ell \cdot \hat{s}_{\ell+\lambda} = (\hat{s}_\ell \cdot \hat{s}_{\ell+\lambda})_0 \prod_{\ell'=\ell}^{\ell+\lambda-1} \tau_{\ell'}. \tag{20}$$

Using (7) and (19), one obtains for the thermodynamic average

$$<\hat{s}_\ell \cdot \hat{s}_{\ell+\lambda}> = (\hat{a}_\ell \cdot \hat{a}_{\ell+\lambda}) \prod_{\ell'=1}^{\ell+\lambda-1} \tanh\left(\beta J(\hat{a}_{\ell'} \cdot \hat{a}_{\ell'+1})\right). \quad (21)$$

The same results may be obtained by a transfer matrix method.

By taking the configurational average and using a similar reasoning as in (14), one obtains the spin correlation function

$$C_\lambda(T) \equiv <<\hat{s}_\ell \cdot \hat{s}_{\ell+\lambda}>>_c = \left(C_1(T)\right)^\lambda \quad (22)$$

where

$$C_1(T) = <(\hat{a}_\ell \cdot \hat{a}_{\ell+1}) \tanh\left(\beta J(\hat{a}_\ell \cdot \hat{a}_{\ell+1})\right)>_c$$
$$= \int_{-1}^{+1} \zeta \tanh(\beta J \zeta) p(\zeta) d\zeta. \quad (23)$$

For the susceptibility, one finds for $N \to \infty$

$$kT<\chi(T)>_c \equiv \sum_{\lambda=0}^{\infty} C_\lambda = 1/\left(1-C_1(T)\right). \quad (24)$$

The internal energy is given by

$$<E(T)>_c = -NJC_1(T), \quad (25)$$

which yields for the specific heat per spin

$$<c(T)>_c = -JdC_1(T)/dT. \quad (26)$$

For low temperatures, one has

$$C_1(T) = C_{o,1} - (\beta J)^2 \int_{-\beta J}^{+\beta J} (|x| - x\tanh x) p(x/\beta J) dx$$
$$\simeq C_{o,1} - (\pi^2/12)(kT/J)^2 p(0). \quad (27)$$

Thus, one finds a susceptibility

$$<\chi(T)>_c \simeq (1-C_{o,1})^{-1}(kT)^{-1} \quad (28)$$

diverging like T^{-1} as in a paramagnet, and a linear specific heat

$$<C(T)>_c = (\pi^2/6)(k^2T/J)p(0). \quad (29)$$

For high temperatures, one has

$$C_1(T) \simeq \beta J<(\hat{a}_\ell\cdot\hat{a}_{\ell+1})^2> = \beta J/n \tag{30}$$

leading to a Curie-Weiss law

$$<\chi(T)>_c^{-1} = kT-J/n. \tag{31}$$

Magnetization at T=O

We calculate the field dependence of the magnetization at T=O by a qualitative argument: Weak links with excitation energies $\leq\varepsilon$ divide the chain into sections of length $L \propto \varepsilon^{-1}$. A magnetic field may break the bonds at these weak links and turn the magnetic moment $M_L \propto L^{1/2}$ of each section into a favourable direction, if the Zeeman energy gain M_LH compensates the expense in bond energy $\varepsilon \propto L^{-1}$. This yields

$$L \propto H^{-2/3} \tag{32}$$

$$M_L \propto H^{-1/3} \tag{33}$$

from which follows for the magnetization

$$m \propto M_L/L \propto H^{1/3}. \tag{34}$$

It is interesting to note that the critical exponents $\gamma = 1$, $\alpha = -1$ and $\delta = 3$ following from (28), (29) and (34) satisfy the zero-temperature scaling relation /2,3/

$$\gamma(\delta+1) = (1-\alpha)(\delta-1). \tag{35}$$

References

1. R.A. Pelcovits, E. Pytte, J. Rudnick, Phys.Rev. Letters 40, 476 (1978)
2. R. Oppermann, H. Thomas, Z.Physik B22, 387 (1975)
3. R. Morf, H. Thomas, Z.Physik B23, 81 (1976)

On Critical Slowing-Down in Spin Glasses

Scott Kirkpatrick
IBM Research
Yorktown Heights, N.Y. 10598, U.S.A.

It is rare to find an experimental feature in spin glasses which is both clear-cut and unchanged over a wide variety of materials. The rather simple critical slowing-down of the spin relaxation rates which Salamon and coworkers[1,2] have recently identified in the electron spin resonance data from a variety of spin glasses seems to be such a feature. I shall review Salamon's arguments, and then report new results from two theoretical models which support the view that the transition from paramagnet to spin glass should be marked by slow spin fluctuations, and that experimentally this critical slowing-down should be dominated by mean field behavior.

The typical temperature-dependence of the spin resonance linewidth (Γ) and static magnetic susceptibility is shown in Fig. 1. The resonance line broadens as one approaches the spin glass freezing temperature, T_f, defined by the susceptibility maximum. In some cases the esr line becomes too broad to measure before T reaches T_f. Large shifts in the resonant frequency are also seen as T approaches T_f. Two contributions to Γ are evident in Fig. 1. At high temperatures, Korringa relaxation, which is proportional to T, dominates, while a second mechanism takes over at low temperatures. The obvious candidate for the low temperature mechanism is exchange-narrowed inhomogeneous dipolar broadening.[1]

When spin relaxation, described by a correlation function $\varphi(t) \sim \exp(-t/\tau)$, is rapid compared to the resonance frequency, ω_0, we can estimate the damping of the resonance line by using the venerable "frequency modulation" approximation[3]:

$$i\Gamma(\omega_0) \propto \chi^{-1} \int_0^\infty dt \exp(-i\omega_0 t)\, \varphi(t)$$

$$\propto \chi^{-1}\, \tau/(1 - i\omega_0\tau) \quad , \tag{1}$$

and find a width and g-shift given by

$$\Gamma_2 \equiv T_2^{-1} \propto \tau/\chi \tag{2}$$

$$\Gamma_1/\omega_0 \equiv \Delta g/g \propto \tau^2/\chi \quad . \tag{3}$$

Thus, for this mechanism the natural way to extract τ is to study $(T_2\Delta g/g)$, since all other factors then drop out. When this is done,[2] one finds that $\tau^{-1} \rightarrow 0$ as $T \rightarrow T_f$ for those spin glasses for which esr data are available. These include "classic" spin glasses like $Cu_{.99}Mn_{.01}$, "mictomagnetic" alloys like $Cu_{.75}Mn_{.25}$, and amorphous intermetallic compounds like $Gd_{.3}Al_{.7}$. Furthermore, the decrease appears to be roughly linear in $(T - T_f)$. (See Fig. 2, which is borrowed from Ref. 2.) Salamon has interpreted this linear dependence as mean field behavior.

To see what the mean field critical dynamics of a spin glass should be, we consider a model of an Ising spin glass with infinite-ranged interactions, i.e. there are N spins, each coupled to all the remaining N−1 to minimize the importance of spatial fluctuations in the magnetization.[4,5] The Hamiltonian is

$$\mathscr{H} = \frac{1}{2} \sum_{i,j} J_{ij} S_i S_j \tag{4}$$

with the summation running over all i and j. Each J_{ij} is an independent random variable with mean value

$$\langle J_{ij} \rangle = 0 \tag{5}$$

and variance

$$\langle J_{ij}^2 \rangle \equiv \tilde{J}^2/N \quad . \tag{6}$$

If there is a mean field limit for spin glasses, this model should be it. It was introduced[4] as the natural analogue of the infinite-ranged uniform ferromagnet, for which the usual mean field theory is exact. A satisfac-

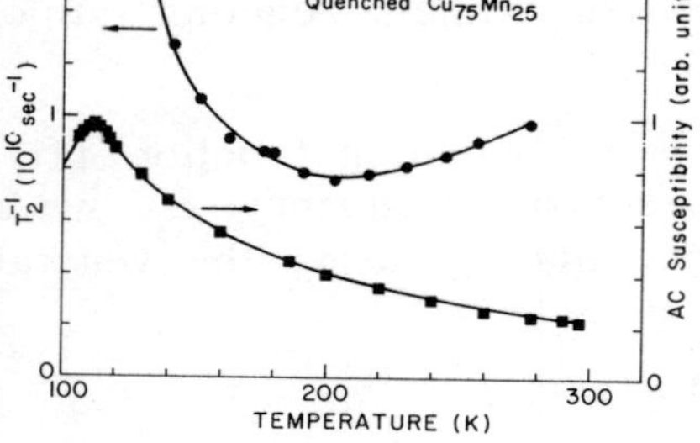

Fig. 1 (after Ref. 1)

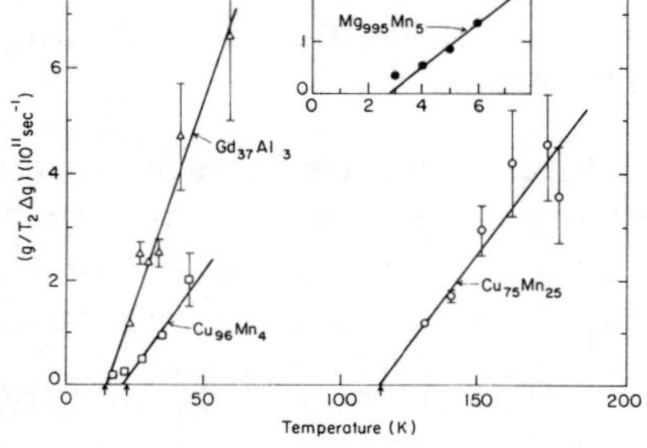

Fig. 2 (from Ref. 2)

tory solution to the model defined in (4)-(6) does not yet exist, but it is generally accepted that the system is paramagnetic down to $k_B T_f = \tilde{J}$, and has a novel sort of low temperature phase below T_f. I shall consider only the slowing down of spin autocorrelations in the paramagnetic phase, where all expectation values of magnetization are small, and a linearized transport theory[6,7] should be accurate. Such a theory is developed in Ref. 5. The essential features are now summarized:

We adopt Glauber[8] dynamics, in which each spin is assumed coupled independently to a thermal bath. The rate at which spin flips are produced by the thermal fluctuations must depend upon the associated energy cost in such a way that detailed balance is preserved and the system will tend towards thermodynamic equilibrium. We take this rate to be

$$W(S_i \to -S_i) = \frac{1}{2}(1 - \tanh\beta E_i) \quad , \tag{7}$$

where $\beta \equiv 1/k_B T$ and the unit of time is taken to be the frequency at which spin flips are attempted. This leads to a master equation for the decay of the autocorrelation function for a given spin:

$$\partial/\partial t <S_i S_i(t)> = - <S_i S_i(t)> + <S_i \tanh\beta h_i(t)> \quad , \tag{8}$$

with $h_i(t)$ the molecular field. For the infinite ranged model, defining h_i proves to be a bit tricky. To leading order, there are two terms

$$h_i(t) = \sum_j J_{ij} [S_j(t) - \chi_{ij} S_i(t)] \quad , \tag{9}$$

and, above T_f, $\chi_{ij} = \beta J_{ij}$. The second term is included to eliminate from the average restoring field, $h_i(t)$, the part which will change if S_i flips to $-S_i$, since that part doesn't provide any restoring tendency. Brout and Thomas[9] first showed that such a correction term is necessary to ensure that a susceptibility calculated in mean field theory would satisfy the fluctuation-dissipation theorem. Thouless, Anderson, and Palmer[10] have also derived the form (9) for spin glasses, using slightly different arguments. The correction is especially important in a spin glass model, since the first term in h_i fluctuates about a vanishing mean value while the second, which is $\propto \sum_j J_{ij}^2 \sim \tilde{J}^2$, is nonzero on the average.

The master equation is now

$$(\partial/\partial t + 1 + \beta^2\tilde{J}^2) <S_i S_i(t)> = \sum_j \beta J_{ij} <S_i S_j(t)> \quad , \tag{10}$$

and may be solved by expanding in terms of eigenvectors of the exchange matrix J_{ij}. Some useful properties of this random matrix are known -- for example, its spectrum extends from $-2\beta\tilde{J}$ to $2\beta\tilde{J}$, and its density of states is $\propto (4\beta^2\tilde{J}^2 - \lambda^2)^{1/2}$. Thus the slowest modes given by (10) have a relaxation rate

$$\tau_\alpha^{-1} = (1 - \beta\tilde{J})^2 = (1 - T/T_f)^2 \quad , \tag{11}$$

which vanishes at T_f. More rapidly relaxing modes also contribute to $<S_iS_i(t)>$, which can be expanded as

$$<S_iS_i(t)> = \int_{-2\beta J}^{2\beta J} d\lambda \exp[-(1 + \beta^2\tilde{J}^2 - 2\beta\tilde{J}\lambda)t] \times \frac{(4\beta^2 J^2 - \lambda^2)^{1/2}}{(2\pi\beta^2 J^2)(1 + \beta^2 J^2 - 2\beta J\lambda)} \tag{12}$$

in this linear approximation.[5] Calculating the mean relaxation as was done in (2) we obtain

$$\bar{\tau} = \int_0^\infty dt <S_iS_i(t)> = (1 - \beta^2\tilde{J}^2) \tag{13}$$

or

$$(\bar{\tau})^{-1} \sim (1 - T_f/T) \quad . \tag{14}$$

Therefore the linear decrease in relaxation time extracted by Salamon[1,2] is indeed the mean field result expected for a spin glass. The fact that the mean field model has a distribution of relaxation rates contributing may also prove to be typical of dynamics in more realistic models. We consider next the behavior of a 3D Ising system with nearest-neighbor interactions of constant magnitude, J, but random sign. The spin autocorrelation function $<S_iS_i(t)>$ was calculated by computer simulation on rather large (30x30x30) samples at temperatures above T_f, which for this model appears to be at about $k_BT_f = 2.0$ J. As in (13), the relaxation time was determined by integrating the autocorrelation function.

Several natural methods of graphical analysis of this data suggest themselves, and have been used in Fig. 3. To test for a simple form of slowing-down, like that seen in Fig. 2,

$$\tau \propto A/(T - T_f)^\Delta \quad , \tag{15}$$

one simply plots τ^{-1} against T. If the freezing of a spin glass is like that of a glass, rather than a conventional phase transition, one might expect to see in the relaxation time an exponentional dependence like that of Fulcher's law for the viscosity of a glass:

$$\tau \propto \exp[A/(T - T_f)^\Delta] \quad . \tag{16}$$

To test for this we plot $(\ln \tau)^{-1}$ against T. Finally, the time-dependent phenomena on rather long time scales (kHz to seconds) seen in some of the classic low-concentration spin glasses have been interpreted in terms of the Neel picture of superparamagnetism,[11] in which

$$\tau \propto \exp(A/T) \quad . \tag{17}$$

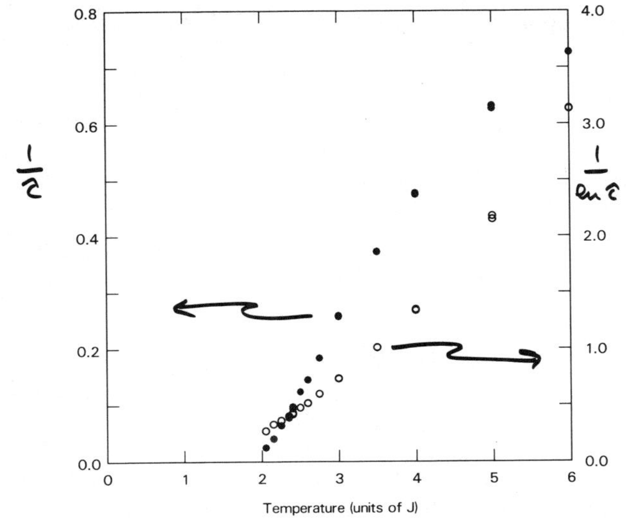

Fig. 3 Relaxation times for 3D Ising spin glass.

Figure 3 shows (15) to be the most plausible interpretation of the slowing-down, with the exponent Δ taking the mean field value of 1. The "glass-like" interpretation (16) cannot be ruled out by the data, but would require that the formula (16) involve a different T_f than the one seen in the static susceptibility. This model does not show the simple activated behavior envisioned in the Neel "paramagnetic clouds" picture, at least not on the short time scales considered.

REFERENCES

1. M. B. Salamon and R. M. Herman, Phys. Rev. Letts. **41**, 1506 (1978).
2. M. B. Salamon, preprint.
3. P. W. Anderson and P. R. Weiss, Revs. Mod. Phys. **25**, 169 (1953).
4. D. Sherrington and S. Kirkpatrick, Phys. Rev. Letts. **35**, 1792 (1975).
5. S. Kirkpatrick and D. Sherrington, Phys. Rev. **B17**, 4384 (1978).
6. M. Suzuki and R. Kubo, J. Phys. Soc. Jpn. **24**, 51 (1968).
7. W. Kinzel and K. H. Fischer, Solid State Commun. **23**, 687 (1977).
8. R. J. Glauber, J. Math. Phys. **4**, 294 (1963).
9. R. Brout and H. Thomas, Physics **3**, 317 (1967).
10. D. J. Thouless, P. W. Anderson, and R. G. Palmer, Phil. Mag. **35**, 593 (1977).
11. L. Neel, Ann. Geophysique **5**, 99 (1949), J. Phys. Soc. Jpn. **17** B1, 676 (1962).

PARTICIPANTS

AÏN, M.	D.Ph-S.R.M. C.E.N-S. B.P. No. 2, 91190 Gif-sur-Yvette, France
ALS-NIELSEN, J.	Risø National Laboratory, DK-4000 Roskilde, Denmark
ANDRESEN, A.F.	Institutt for atomenergi, 2007 Kjeller, Norway
AXE, J.D.	Brookhaven National Laboratory, Upton, L.I., N.Y. 11973, USA
BERRE, B.	Dept. of Physics, Agricultural University of Norway, 1432 Ås-NLH
BEYSENS, D.	D.Ph.-S.R.M. C.E.N.S, B.P. No. 2 91190 Gif-sur-Yvette, France
BINDER, K.	Inst. für Festkörperforschung der KFA, Postfach 1913, D-5170 Jülich 1, West-Germany
BRUCE, D.A.	Dept. of Physics, University of Edinburgh, James Clerk Maxwell Bldg., Mayfield Road, Edinburgh EH9 3JZ, UK
CARLTON, J.P.	D.Ph.-S.R.M. C.E.N.S, B.P. No. 2 91190 Gif-sur-Yvette, France
CHRISTENSEN, F.E.	Risø National Laboratory, DK-4000 Roskilde, Denmark
CLAUSEN, K.N.	Risø National Laboratory, DK-4000 Roskilde, Denmark

COMPANGNER, A. | Dept. or Appl, Physics, Delft University of Technology, Lorenzweg 1, Delft, The Netherlands

COTTERILL, R.M.J. | The Technical University of Denmark, Bldg. 307, 2800 Lyngby, Denmark

COWLEY, R.A. | Dept. of Physics, University of Edinburgh, James Clerk Maxwell Bldg. Mayfield Road, Edinburgh, EH9 3JZ,UK

DIEHL, H.W. | Inst. für Theor. Physik der Universität München, Theresienstr. 37 8000 München 2, West-Germany

DOLLING, G. | Chalk River Nuclear Laboratories, Chalk River, Ontatrio, Canada

DULTZ, W. | Fachbereich Physik, Universität Regensburg, Postfach 8400 Regensburg, West-Germany

EDWARDS, Sir Sam | Cavendish Laboratory, Madingley Rd. Cambridge, CB3 OHE, UK

FEINER, L.F. | Philips Research Laboratories, Eindhoven, The Netherlands

FRENKEL , D. | Dept. of Chemistry, University of California, Los Angeles, Cal. 90024 USA

GOOVAERTS, E.E.J. | Dept. of Physics, University of Antwerp, B-2610 Wilrijk, Belgium

HARLEY, R. | Clarendon Laboratory, University of Oxford, Parks Rd., Oxford OX1 3PU, UK.

HEMMER, P. Chr. | Dept. of Theoret. Physics, Technical University of Norway, 7034 Trondheim-NTH

HIIS HAUGE, E. | Dept. of Theoret. Physics, Technical University of Morway, 7034 Trondheim-NTH

HILHORST, H.J. | Dept. of Appl. Physics, Delft University of Technology, Lorentzweg 1, Delft The Netherlands

HÖCK, K.H. — Ruhr-Universität, Theor. Physik III
463 Bochum, West-Germany

JULLIEN, R. — Lab. Phys. Solides, Université
Paris XI, Bat. 510, 91405 Orsay,
France

JØSSANG, T. — Dept. of Physics University of
Oslo, Blindern, Oslo 3, Norway

KAKURAI, K. — Hahn-Meitner-Institut, Glienicker-
Str. 100, Berlin 39, Germany

KIRKPATRICK, S. — IBM Research Center, P.O. Box 218
Yorktown Heights, N.Y. 10598, USA

KJEMS, J. — Risø National Laboratory, DK-4000
Roskilde, Denmark

KNOPS, H.J.F. — Inst. für Theor. Physics, Catholic
University, Toernooiveld,
Nijmegen, The Netherlands

KRAGLER, R. — Laboratories RCA Ltd., Badenstr. 569
CH-8048 Zürich, Switzerland

KROLL, D.M. — Inst. fürTheor. Physik der Universität
München, Theresienstr. 37,
8000 München 2, West-Germany

LAGERWALL, S.T. — Chalmers University of Technology,
Fack, S-41296 Gothenburgh 5, Sweden

LE FEVER, H.T. — University of Leiden, Dept. of
Physics, Postbus 75, Leiden,
The Netherlands

LIEBMANN, R. — Inst. für Theor. Phys., University
of Frankfurt, Robert-Mayer-Str. 8-10
D-6000 Frankfurt 1, West-Germany

LINDGÅRD, P.A. — Risø National Laboratory, DK-4000
Roskilde, Denmark

LITSTER, J.D. — Dept. of Physics, M.I.T., Cambridge,
Mass. 02139, USA

Mc TAGUE, J.P. — Dept. of Physics, University of
California, Los Angeles, Cal.
90024, USA

MINNHAGEN, P.	NORDITA, Blegdamsvej 17, DK-2100 Copenhagen, Denmark
MÜLLER, G.	Dept. of Physics, University of Basel, Klingelbergstrasse 82, CH-4056 Basel, Switzerland
NAPIÓRKOWSKI, M.	Dept. of Theor. Physics. Technical University of Norway, 7034 Trondheim-NTH
OLAUSSEN, K.	Dept. of Theor. Physics, Technical University of Norway, 7034 Trondheim-NTH
PICHARD, J.L.	D.Ph.SRM. C.E.N.S, B.P. No. 2, 91190 Gif-sur-Yvette, France
PYNN, R.	Institut Laue-Langevin, B.P. 156, Centre de Tri, 38042 Grenoble-Cedex, France
PYTTE, E.	IBM Research Center, P.O. Box 218, Yorktown Heights, N.Y. 10598, USA
RAVNDAL, F.	Dept. of Physics, University of Oslo, Blindern, Oslo 3, Norway
RENNIE, R.	Cavendish Laboratory, Madingley Rd., Cambridge CB3 OHE, UK
RICCI, F.P.	Instituto de Fisica, Piazzale delle Science 5, Roma, Italy
RIKVOLD, T.A.	Dept. of Physics, University of Oslo Blindern, Oslo 3, Norway
SAITO, Y.	Inst. für Festkörperforschung der KFA, Postfach 1913, D-5170 Jülich 1, West-Germany
SAMUELSEN, E.J.	Dept. of Physics, Technical University of Norway, 7034 Trondheim-NTH
SCHOEMAKER, D.	Dept. of Physics, University of Antwerp, B-2610 Wilrijk, Belgium

SEABRA LAGE, E.J.	Dept. of Physics, University of Porto, Portugal
SEMMINGSEN, D.	Institutt for atomenergi, 2007 Kjeller, Norway
SHERRINGTON, D.	Institut Laue-Langevin, B.P. 156 Centre de Tri, 38042 Grenoble-Cedex France
SKJELTORP, A.	Dept. of Physics, University of Oslo, Blindern, Oslo 3, Norway
SOUTHERN, B.W.	Institut Laue-Langvin, B.P. 156 Centre de Tri, 38042 Grenoble-Cedex France
STEIJGER, J.J.M.	Netherlands Energy Research Foundation (ECN), Petten (NH), The Netherlands
STEINER, M.	Hahn-Meitner-Institut, Berlin, p.t. Yale University, Becton Center 15 Prospect Street, New Haven, Conn. 06520, USA
STØLAN, B.	Dept. of Physics, Technical University of Norway, 7034 Trondheim-NTH
THEODORAKOPOULOS, N.	Dept. of Physics, University of Konstanz, P.O.B. 7733, D-7750 Konstanz, West-Germany
THOMAS, H.	Dept. of Physics, University of Basel, Klingelbergstrasse 82, CH-4056 Basel, Switzerland
VILLAIN, J.	Centre d'Etudes Nucléarires, Rue des Martyrs, 38041 Grenoble-Cedex, France
WEEKS, J.D.	Bell Laboratories, 600 Mountain Ave., Murray Hill, N.J. 07974, USA
WRIGHT, G.	Office of Naval Research, Code 427, Arlington, VA 22217, USA

YEOMANS, J.M. Dept. of Theoret. Physics,
1 Keble Rd., Oxford OX1 3NP, UK

YOUNG, A.P. Dept. of Mathematics, Imperial
College, London SW7, UK

YOUNGBLOOD, R. Physics Dept., Brookhaven Nat. Lab.,
Upton, L.I., N.Y. 11973, USA

ZIMAN, T.A.L. Dept. of Theoret. Physics,
1 Keble Rd., Oxford OX1 3NP, UK

ORGANIZING COMMITTEE:

ANDERSEN, E.
JARRETT, G.
RISTE, T.
Institutt for Atomenergi, P.O.B. 40,
2007 Kjeller, Norway

INDEX